T0297608

Research in
MEDICAL AND BIOLOGICAL SCIENCES

Research in MEDICAL AND BIOLOGICAL SCIENCES

From Planning and Preparation to Grant Application and Publication

Edited by

Petter Laake
University of Oslo, Oslo, Norway

Haakon Breien Benestad
University of Oslo, Oslo, Norway

Bjorn Reino Olsen
Harvard University, Cambridge, USA

AMSTERDAM • BOSTON • HEIDELBERG • LONDON
NEW YORK • OXFORD • PARIS • SAN DIEGO
SAN FRANCISCO • SINGAPORE • SYDNEY • TOKYO
Academic Press is an imprint of Elsevier

Academic Press is an imprint of Elsevier
125, London Wall, EC2Y 5AS
525 B Street, Suite 1800, San Diego, CA 92101-4495, USA
225 Wyman Street, Waltham, MA 02451, USA
The Boulevard, Langford Lane, Kidlington, Oxford OX5 1GB, UK

ISBN: 978-0-12-799943-2

British Library Cataloguing-in-Publication Data
A catalogue record for this book is available from the British Library

Library of Congress Cataloging-in-Publication Data
A catalog record for this book is available from the Library of Congress

For information on all Academic Press publications visit our website at http://store.elsevier.com/

Typeset by MPS Limited, Chennai, India
www.adi-mps.com

Printed and bound in Great Britain

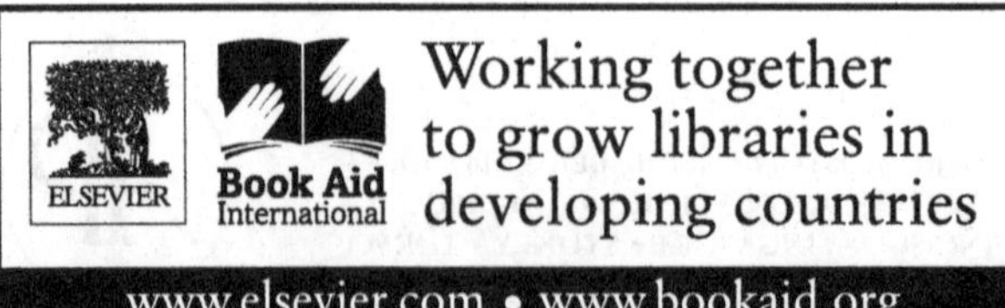

CONTENTS

LIST OF CONTRIBUTORS

Mahmood Amiry-Moghaddam
Department of Anatomy, Institute of Basic Medical Sciences, University of Oslo, Blindern, Oslo, Norway

Haakon Breien Benestad
Department of Physiology, Institute of Basic Medical Sciences, University of Oslo, Blindern, Oslo, Norway

Heidi Kiil Blomhoff
Department of Biochemistry, Institute of Basic Medical Sciences, University of Oslo, Blindern, Oslo, Norway

Ellen Christophersen
University of Oslo Library, Medical Library, Blindern, Oslo, Norway

Morten Wang Fagerland
Oslo Centre for Biostatistics and Epidemiology, Research Support Services, Oslo University Hospital, Oslo, Norway

Anne-Marie B. Haraldstad
University of Oslo Library, Medical Library, Blindern, Oslo, Norway

Bjørn Hofmann
Section for Health, Technology and Society, University College of Gjøvik, Gjøvik, Norway and Centre for Medical Ethics, Institute of Health and Society, University of Oslo, Blindern, Oslo, Norway

Søren Holm
Centre for Social Ethics and Policy, School of Law, The University of Manchester, England, Centre for Medical Ethics, Institute of Health and Society, University of Oslo, Blindern, Oslo, Norway and Department of Health Science and Technology, Aalborg University, Denmark

Petter Laake
Oslo Centre for Biostatistics and Epidemiology, Department of Biostatistics, Institute of Basic Medical Sciences, University of Oslo, Blindern, Oslo, Norway

Anne-Lise Middelthon
Department of Community Medicine, Institute of Health and Society, University of Oslo, Blindern, Oslo, Norway

Kåre Moen
Department of Community Medicine, Institute of Health and Society, University of Oslo, Blindern, Oslo, Norway

Bjorn Reino Olsen
Harvard School of Dental Medicine, Harvard Medical School, Boston, MA, USA

Ole Petter Ottersen
Department of Anatomy, Institute of Basic Medical Sciences, University of Oslo, Blindern, Oslo, Norway

Eva Skovlund
School of Pharmacy, University of Oslo, Blindern, Oslo, Norway

Dag S. Thelle
Oslo Centre for Biostatistics and Epidemiology, Department of Biostatistics, Institute of Basic Medical Sciences, University of Oslo, Blindern, Oslo, Norway

Kjell Magne Tveit
Department of Oncology, Oslo University Hospital, Nydalen, Oslo, Norway

Preface

The famous physician William Osler once said, while instructing a student during an autopsy, "it is always best to do a thing wrongly the first time." Indeed, in the fledging stages of our scientific careers we are likely to make not just one but many mistakes. Although one can learn from one's mistakes, we believe that some can be prevented by appropriate instruction and reading. Nobel Prize winner Max Perutz allegedly complained that in his opinion "many young scientists work too much, and read and think too little." If this was the case 50 years ago, it is probably a larger problem today. The pace of graduate studies has accelerated, and it is often difficult for students to find the time to read even the most relevant scientific articles, not to mention entire books that have been written on the various themes that we present in this book. We are not aware of any other textbook that includes most of the subjects dealt with in this volume. Our wish in creating this book is to make academic life easier for graduate students.

The history of this book began several decades ago, when the three of us started to arrange courses to help PhD students avoid some of the pitfalls that we ourselves experienced, and to teach them some basic rules, *lege artis*, for doing research. At the turn of this century, a national program was adopted in Norway that applied to all students enrolled in doctoral programs at Medical Schools. It included a compulsory course that covered a broad range of topics and that came to comprise the basis of a textbook, which was first published in Norwegian by Gyldendal Akademisk, and later in English by Elsevier, as a revised and expanded version of the Norwegian edition (*Research Methodology in the Medical and Biological Sciences*, 2007). Some chapters of the present book resemble those of the former book, but all chapters have been thoroughly revised. The following chapters are either completely new or have been written by new authors: "Ethics and scientific conduct," "Basic medical science," "Translational medical research," "Qualitative research," and "Evidence-based practice."

There are several reasons for the broad range of topics covered in this book. Master's and PhD programs should train students for competence in research as well as other professional pursuits that require scientific insight, including the basics of the research process. Students entering these graduate programs typically come from a variety of undergraduate programs, including studies in medicine, natural sciences, social sciences, or nursing. During their undergraduate studies they gain only narrow scientific capabilities, which can lead to difficulties communicating and working with colleagues who belong to other fields even after they have finished their graduate work. This is contrary to the goals established by the European Qualifications Framework (EQF) as to what a student at these levels should know, understand, and be able to do. It is therefore important that graduate studies include schooling in general scientific research, regardless of the specific discipline pursued.

This book is intended for students with varied professional backgrounds. Multidisciplinary communication and research cooperation is increasingly important, as the "learning outcomes" of the EQF imply. Scientists in any one scientific community should be familiar with and should respect the traditions of other scientific communities. This book outlines a possible curriculum for Master's and PhD students in the medical and biological fields. We believe that everyone in these fields should know a little about all the subjects covered in this book, in addition to the deeper knowledge of topics related to their own disciplines.

Petter Laake
Bjorn Reino Olsen
Haakon Breien Benestad

CHAPTER 1

Philosophy of Science

Bjørn Hofmann[1] and Søren Holm[2]
[1]Section for Health, Technology and Society, University College of Gjøvik, Gjøvik, Norway and Centre for Medical Ethics, Institute of Health and Society, University of Oslo, Blindern, Oslo, Norway
[2]Centre for Social Ethics and Policy, School of Law, The University of Manchester, England, Centre for Medical Ethics, Institute of Health and Society, University of Oslo, Blindern, Oslo, Norway and Department of Health Science and Technology, Aalborg University, Denmark

1.1 INTRODUCTION

The sciences provide different approaches to the study of man. Man can be scrutinized in terms of molecules, tissues and organs, as a living creature, and as a social agent and a cultural person. Correspondingly, the philosophy of science investigates the philosophical assumptions, foundations, and implications of the sciences. It is an enormous field that covers the formal sciences, such as mathematics, computer sciences, and logic; the natural sciences; the social sciences; and the methodologies of some of the humanities, such as history. Discussion in the current chapter is limited to the natural sciences (Section 1.2) and the social sciences (Section 1.13), and comprises a brief overview of the philosophical aspects salient to research in the medical and biological sciences.

1.2 PHILOSOPHY OF THE NATURAL SCIENCES

What does it mean when one says that "smoking is the cause of lung cancer"? What counts as a scientific explanation? What is science about, e.g., what is a cell? How does one obtain scientific evidence? How can one reduce uncertainty? What are the limits of science? These are but a few of the issues discussed in the philosophy of the natural sciences that will be presented in this chapter.

The traditional philosophy of science has aimed to put forth logical analyses of the premises of science; in particular the logical analysis of the syntax of basic scientific concepts. In the following sections, the principal traditional issues regarding the rationality, method, evidence, and the object of science (the world) are discussed. But first the core concepts of science, knowledge, and truth will be addressed: What is science? What is scientific knowledge, and what constitutes a scientific fact?

Research in Medical and Biological Sciences
DOI: http://dx.doi.org/10.1016/B978-0-12-799943-2.00001-X

1.3 WHAT IS SCIENCE? DIFFERENTIATING SCIENCE FROM NONSCIENCE

Science is traditionally defined as "the systematic search for knowledge," meaning that science has an aim (knowledge) and it is an activity (search) with certain qualifications (systematic). However, not everyone who carries out a systematic search for knowledge can obtain research funding. For example, though a religious man may search for knowledge through highly systematic meditation techniques, it is unlikely that this will be considered a project in need of research funding. Indeed, much of the time even well-funded scientists do not perform systematic searches for knowledge. They struggle with experimental designs, analyze data, present research results, argue with other researchers over conflicting results, and write funding proposals.

Therefore, a more complete definition of science may be: "Scientific research is the systematic and socially organized (a) search for, (b) acquisition of, and (c) use or application of knowledge and insight brought forth by acts and activities involved in (a) and (b)."[1] This definition better reflects what scientists actually do. They search for new knowledge, e.g., by investigating the possibility of using biomolecular tests of cell-free fetal DNA/RNA in the blood of pregnant women to find defects in the fetus (noninvasive prenatal testing, NIPT). They acquire knowledge by testing, accepting, or rejecting a hypothesis, e.g., that NIPT is better than combined ultrasound and serum tests for detecting fetuses with trisomy 13, 18, and 21. Finally, scientists apply knowledge when they argue that a certain study is either appropriate or flawed, and therefore its results either valid or invalid. Although this definition of science is closer to what scientists do, it may still be difficult to differentiate those doing science from those doing nonscience.

Throughout history a series of criteria has been used to demarcate science from nonscience (sometimes also referred to as pseudoscience), and thereby also to define science. Francis Bacon (1561–1626) defined science as a specific method, i.e., performing a systematic analysis of data without preconception. Data analysis framed by preconception was considered nonscience. However, in reality it can be very difficult to analyze data without preconception. To illustrate this fact, if you study the figure below (Figure 1.1), what do you see? A rabbit, a duck, or perhaps both?

When viewed on paper (or on a screen) the figure has black dots on a white background. So where are the duck and the rabbit? Are they

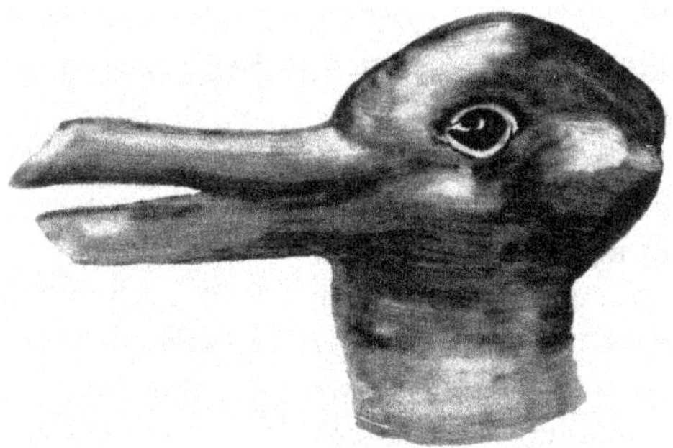

Figure 1.1 The duck/rabbit.

figments of the imagination? Are they preconceptions? Hence, to say that science is the systematic analysis of data without preconception is too restrictive; it would rule out most of what is called science today. Indeed, all observations and analyses are based on preconceptions.

Another way to differentiate science from nonscience is to say that preconception is acceptable in the context of discovery but not in the context of justification, i.e., when the data are tested. The basic idea is that the pattern of nature is neutral, and will stand out in the end. However, this does not solve the problem of preconception when the hypothesis is tested, as testing presupposes observations, and observations presuppose preconceptions.

A third classical demarcation criterion is that a scientific hypothesis or theory can be contested with possible observations, i.e., it can be falsified (or refuted).[2] However, theories are seldom really falsified,[3] and to falsify a theory presupposes that the researcher has an idea about what will happen. As scientists we are seldom ready to give up our theories, and instead add specifications or modifications.

It has been argued that scientists are preoccupied with puzzle-solving, i.e., solving puzzles within a given mode of thought (paradigm), when they should be concentrating on falsification.[3] Until the 1960s it was commonly thought that science progressed in a linear and piecemeal fashion, in which new knowledge added to existing knowledge. However, Thomas Kuhn (1922–1996) and others argued that science evolved through anomalies. Hard cases that could not be explained within the given paradigm challenged existing theories and resulted in a scientific revolution. A new set of theories established a new paradigm, and scientists turned to puzzle-solving within this new paradigm. The shift from Newton's mechanics to Einstein's theory of (special) relativity is a key

BOX 1.1 Selected Characteristics of Nonscience

1. **Belief in authority**: Some person or persons have a special ability to determine what is true or false, and others have to accept their judgment.
2. **Nonrepeatable experiments**: Reliance on experiments with outcomes that cannot be reproduced by others.
3. **Handpicked examples**: Examples that are not representative of the general category to which the investigation refers are considered decisive.
4. **Unwillingness to test**: A testable theory is not tested.
5. **Disregard of refuting information**: Ignoring or neglecting observations or experiments that conflict with a theory.
6. **Built-in subterfuge**: Arranged the testing of a theory so that it can only be verified but never falsified by the outcome.
7. **Explanations abandoned without replacement**: Giving up tenable explanations without replacing them, so that more is left unexplained in the new theory than in the previous one.[4,5]

example. However, if science is defined by a given paradigm, this potentially makes everything science, as long as scientists define it as a paradigm and are preoccupied with solving its small problems.

The above mentioned four demarcation criteria are but a few of many. None of them are flawless, and it has turned out to be quite difficult to differentiate science from nonscience. In order to remedy this, several sets of more pragmatic criteria to identify nonscience have evolved (Box 1.1).

1.4 KNOWLEDGE AND TRUTH: WHAT IS KNOWLEDGE AND WHAT CONSTITUTES A SCIENTIFIC FACT?

The great tragedy of science, the slaying of a beautiful theory by an ugly fact.

T. H. Huxley (1825–1895)

The standard definition of knowledge is: "*justified true belief.*" Hence, one key criterion of knowledge is that it can be justified. This is done all the time when researchers do experiments, present results, and publish papers. If a statement about events in the world cannot be justified (by theory, analogies, or experiments), it is not accepted as knowledge. Another criterion is *truth*. Traditionally, scientists have thought that knowledge is true. If one knows that individuals with Huntington's disease have more than 39 repetitions of the Huntington gene, one holds this to be true. But what about the third criterion of knowledge: *belief*? Many think that science and belief belong to different realms. However, if scientists say that

BOX 1.2 Statistical Methods and Advanced Technology May Muddle Truth

In a study some researchers showed photographs of people in various social settings to a salmon and measured the difference in brain activity. The salmon "was asked to determine what emotion the individual in the photo must have been experiencing." The results showed that the salmon could read human emotions. However, the salmon was dead. The point was to show that brain researchers can use complicated instruments and simple statistics to show anything they want, even meaningful brain activity in a dead salmon.[6]

they know something, they must believe that it is true, and they must be able to justify it to their peers. To say that one knows something but does not believe it is known as Moore's paradox. For example, "I know that type I diabetes results from the autoimmune destruction of the insulin-producing beta cells in the pancreas, but I do not believe it (Box 1.2)."

So how does one decide what is true? There are vast debates on whether mammography screening reduces breast cancer mortality and whether it results in overdiagnosis. How should this be decided? As with the demarcation criteria for science, there are a large number of criteria for assessing the truth of statements, theories, and hypotheses about events and objects in the world. The *correspondence theory of truth* says that something is true if it corresponds with events or things in the world; but how does one know the nature of the world if not through one's observations? If one is to decide whether a certain phenomenon can be seen in a microscope, a hypothesis must compare what one expects to see with what one actually sees through the ocular. However, as seen in Figure 1.1, observations also strongly depend on preconceptions. Hence, when correspondence is verified, only the hypothesis and the conceptions of the observations are compared. In other words, it is hard to speak of correspondence with events or things in the world without having direct access to the world (independent of our observations).

This problem is avoided by the *coherence theory of truth*, which contends that a scientific hypothesis or theory is true if it coheres with other hypotheses or theories. If it is completely out of the realm of established knowledge, it is most likely untrue. However, take the Australian scientists Barry Marshall and Robin Warren, who identified *Helicobacter pylori* in patients with chronic gastritis and gastric ulcers and hypothesized that it

BOX 1.3 Facts and Truths

Facts are simple and facts are straight.
Facts are lazy and facts are late.
Facts all come with points of view.
Facts don't do what I want them to.
Facts just twist the truth around.
Facts are living turned inside out.

David Byrne

may be a causal factor. In 2005 Barry Marshall was awarded the Nobel Prize for Physiology for this research (Chapter 4). However, their hypothesis certainly did not cohere with the established medical knowledge of the time, which held that ulcers were caused by stress, too much acid, and spicy foods. Indeed, according to the coherence theory of truth, Marshall and Warren were wrong, which shows how theory can prevent new and important insights. Therefore, it has been argued that scientific truths should be established through consensus, e.g., at consensus conferences. If the best experts in a field, after thorough analyses of the evidence, open debates, and careful deliberation, decide on something, it is the closest thing to scientific truth in this field. However, scientists can be wrong, even when they agree, as is illustrated by the case of *Helicobacter pylori*.

Another theory, called the *pragmatic theory of truth*, argues that truth is what is useful or what works. Truth can be decided through inquiry by a community of scientific investigators[7] or by whether it is trustworthy and reliable.[8] The problem with a pragmatic theory of truth is that it tends to be either circular or relativistic. It is circular when truth depends on what works, and researchers' assessment of what works depends on their conception of what is true. It is relativistic if anything can be defined as truth as long as it is part of scientists' common beliefs or decisions (Box 1.3).

Example

Sol Spiegelman (1914–1983) was a molecular biologist who developed the technique of nucleic acid hybridization. During the 1970s he worked to establish that retroviruses cause human cancers, and he identified them in human leukemia, breast cancer, lymphomas, sarcomas, brain tumors, and melanomas. His persistent work and optimistic statements impelled enthusiasm, which fueled funding, and so on, in a self-perpetuating manner. However, he saw viruses where nobody else could; nobody was able to replicate his experiments. Hence, there are many sources of facts, and we do well in reflecting critically on their existence and their emergence.

BOX 1.4 Why Most Research Results in Emerging Fields are False

1. The studies conducted in an emerging scientific field are small.
2. The effect sizes in an emerging scientific field are small.
3. The number of tested relationships is great, but the selection of tested relationships is small.
4. The flexibility in designs, definitions, outcomes, and analytical modes in an emerging scientific field are great.
5. The financial and other interests and prejudices in a scientific field are great.
6. The scientific field is hot (more scientific teams are involved).

In an article published in *PLOS Medicine*, John Ioannidis[9] argues that when it comes to emerging fields of knowledge, much of what medical researchers conclude in their studies is misleading, exaggerated, or flat-out wrong. In this article, which is one of the most frequently downloaded from *PLOS Medicine*, he concluded that in such studies it is difficult to get a positive predictive value above 50% that a research result published in a peer-reviewed journal is true. The reason for this is given in Box 1.4.

1.5 THE GLUE THAT HOLDS THE WORLD TOGETHER: CAUSATION

Another core concept of science is causation. A pivotal task of the biomedical sciences is to find the causes of phenomena such as disease, pain, and suffering. However, what is the implication of saying that something is the cause of a disease? What does it mean when one says that smoking causes lung cancer?

According to Robert Koch (1843–1910), who was awarded the Nobel Prize for Physiology or Medicine in 1905 further to his discovery of tubercle bacillus, a parasite can be considered the cause of a disease if it can be shown that its presence is not random. Such random occurrences may be excluded by satisfying what have been called the (Henle-) Koch postulates:[10,11]

1. The organism must be found in all animals suffering from the disease but not in healthy animals.
2. The organism must be isolated from a diseased animal and grown in pure culture.

3. The cultured organism should cause disease when introduced into a healthy animal.
4. The organism must be reisolated from the experimentally infected animal.

The Koch postulates require that there be no disease without presence of the parasite, and no presence of the parasite without disease. That is, the parasite is both a *necessary*[i] and *sufficient*[ii] condition for the disease. Please note that conditions are not the same as conditionals, although there are affinities.[iii] However, as Koch realized when he discovered asymptomatic carriers of cholera, the requirement of both necessary and sufficient conditions for causation is overly rigorous. Indeed, if such postulates were used as the general criteria for something to be considered a cause in the biomedical sciences, causation would be quite rare.

1.5.1 Necessary Conditions

What if only the *necessary conditions* are known, i.e., *necessary but not sufficient conditions*—can they be called causes? The cholera bacterium (*Vibrio cholera*) is necessary for one to develop cholera, but it is not sufficient, as not all people with the bacterium develop cholera. One would normally say that *Vibrio cholera* causes cholera. However, the presence of an intestinal wall (where the bacteria thrive and produce toxins) is also a necessary condition for developing cholera. If one does not have intestinal walls, one does not develop cholera. This does not mean it is correct to say that the intestinal wall causes cholera. Indeed, there may be many necessary conditions for an event that are not considered causes. However, necessary conditions are germane to health care, as without them the event (e.g., disease) will not occur. Hence, necessary conditions are relevant

[i] A necessary condition here means a condition without which an event would not occur, i.e., *sine qua non*. Without HIV, a person does not develop AIDS. Necessary conditions work through their *absence*, because if you remove a necessary condition, you also remove or prevent the event.

[ii] A sufficient condition here means a condition that is sufficient for an event to occur, i.e., the event occurs every time the condition occurs. Sufficient conditions work through their *presence*, because if you provide the condition, then you also provide the event.

[iii] A conditional is a statement of the form "If p, then q." Conditional statements are not statements of causation, as the relationship between p and q is logical, and not temporal. E.g., the conditional statement "If Barack Obama is president of the United States in 2014, then Norway is in Europe" is true, although we would not say that Barack Obama causes Norway to be in Europe.

through their absence; e.g., tuberculosis can be eradicated by eliminating one of its necessary conditions: *Mycobacterium tuberculosis.*

1.5.2 Sufficient Conditions

What then is a researcher to say when there are *sufficient but not necessary conditions*? For example, when a person develops skin burns after being exposed to ionizing radiation known to be of the sort and strength that results in burns, one would tend to claim that the ionizing radiation caused the skin burns. Of course, the burns could have been caused by other factors (sufficient conditions) in the absence of ionizing radiation. Nevertheless, if a sufficient condition is present, the researcher knows that the event (the effect) will occur. Hence, sufficient conditions for an event are usually acknowledged as causes. Contrary to necessary conditions, sufficient conditions work through their presence; however, for how many diseases are the sufficient conditions actually known? Indeed, if causes are restricted to known sufficient conditions, not many causes are known. For instance, being infected with *Yersinia pestis* is not sufficient to get the plague.

1.5.3 Combination of Conditions that Together are Necessary and Sufficient

Another situation is when there are two conditions that, individually, are insufficient for a certain event, but combined they can make an event occur. For example, an anaphylactic reaction cannot be caused simply by being stung by a bee, or simply by being hypersensitive to bee venom. However, acting jointly in certain circumstances, both of these conditions are necessary and sufficient for an anaphylactic reaction, therefore both conditions are said to cause the event. In short, each of the conditions is an *insufficient but necessary part of a necessary and sufficient condition for the event.* Again the question must be asked, for how many diseases are combined sufficient and necessary conditions known? It is more difficult to find examples than one would like to think, which may be due to the complicated characteristics of nature itself.

1.5.4 Combination of Conditions that Together are Sufficient

It has been argued that cause usually refers to an *insufficient and necessary part of an unnecessary but sufficient condition* (INUS condition),[12] and is characterized by a combination of conditions that together are sufficient but not necessary. For instance, a person who is not exposed to the sun and does not take vitamin D will have osteomalacia. Here the lack of sun exposure and lack of vitamin D are necessary but insufficient conditions. However, together they are sufficient but not necessary conditions for osteomalacia, as this disease may also result from other conditions. Correspondingly, having blood is a necessary condition for having sepsis, as no person without blood has sepsis; this does not mean one can say that blood causes sepsis.

One way to visualize INUS causation is through the use of "causal pies." Figure 1.2 illustrates the conditions that are considered to be sufficient for cardiovascular disease. Indeed, one can use this model to find and fill in various combined sufficient conditions for any disease, as there may be several conditions that are sufficient for a disease to occur (Figure 1.3).

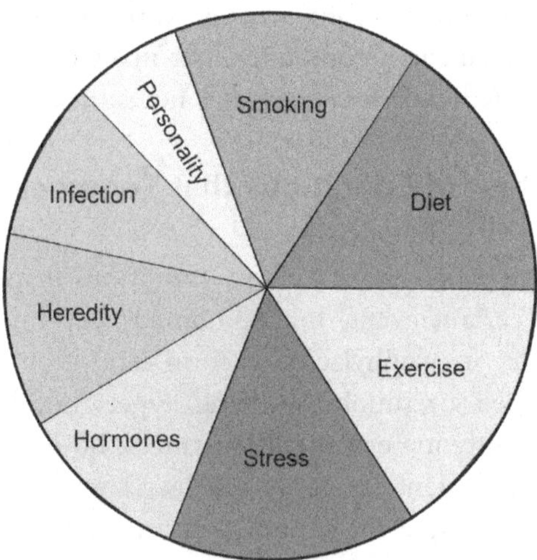

Figure 1.2 Conditions considered sufficient for cardiovascular disease. Causal pies were suggested by Kenneth Rothman in 1976.[13]

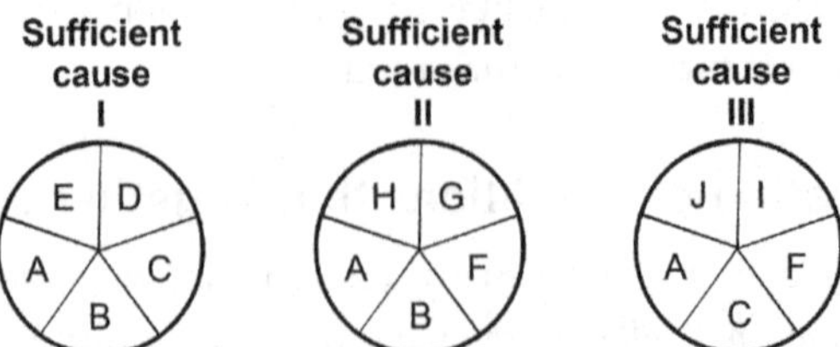

Figure 1.3 There may be several sets of sufficient conditions for a disease. Note that component A is a necessary condition because it appears in every pie.[13]

Take the example of smoking and lung cancer: are the sufficient conditions for lung cancer known, or is smoking an INUS condition for lung cancer? An INUS condition accommodates the fact that not all smokers develop lung cancer, and not all people with lung cancer have been smokers. However, it requires a concert of conditions for which lung cancer follows when smokers, but not nonsmokers, are subjected to them.

Approaches that define causation in terms of combined necessary and sufficient conditions commonly hinge on scientific determinism: the idea that complex phenomena can be reduced to simple, deterministic mechanisms, and therefore in principle can be predicted. In the case of smoking and lung cancer, all the conditions are not known. Hence, the existence of hidden conditions must be assumed in order to retain scientific determinism. The belief in unidentified conditions, as well as the difficulty in explaining a dose–response relationship, has challenged the conception of the different versions of causation based on components of sufficient conditions (sufficient condition, insufficient but necessary part of a necessary and sufficient condition, and INUS).

1.5.5 Probabilistic Causation

Rather than satisfying an INUS condition, the observation is that smokers develop lung cancer at higher rates than do nonsmokers.[14] This leads to the belief that the increased probability of lung cancer among smokers constitutes causation, i.e., probabilistic causation. The central idea in probabilistic causation is that causes raise the probability of corresponding effects.[15]

Acknowledging that overly stringent criteria for causation minimize the chances of identifying any causes of disease, the British medical statistician Austin Bradford Hill (1897–1991) outlined tenable minimal conditions germane to establishing a causal relationship between two entities.[16] Nine criteria were presented as a way to determine the causal link between a specific factor (such as cigarette smoking) and a disease (such as emphysema or lung cancer). Bradford Hill called them "viewpoints" on causation (Box 1.5). However, they are frequently referred to as criteria (see also Chapter 9).

The Bradford Hill criteria are less exacting than the Koch postulates. Nonetheless, there are many cases where one might refer to "the cause of the disease," but where the Bradford Hill criteria do not apply.

Despite their plausibility, probabilistic approaches to causation have been challenged on several levels. First, association is not causation.

BOX 1.5 Bradford Hill's "Viewpoints" on Causation

1. Strength of Association—the stronger the association, the less likely the relationship is due to chance or a confounding variable.
2. Consistency of the Observed Association—has the association been observed by different persons, in different places, circumstances, and times (similar to the replication of laboratory experiments)?
3. Specificity—if an association is limited to specific persons, sites, and types of disease, and if there is no association between the exposure and other modes of dying, then the relationship supports causation.
4. Temporality—the exposure of interest must precede the outcome by a period of time consistent with any proposed biologic mechanism.
5. Biologic Gradient—there is a gradient of risk associated with the degree of exposure (dose–response relationship).
6. Biologic Plausibility—there is a known or postulated mechanism by which the exposure might reasonably alter the risk of developing the disease.
7. Coherence—the observed data should not conflict with known facts about the natural history and biology of the disease.
8. Experiment—the strongest support for causation may be obtained through controlled experiments (clinical trials, intervention studies, animal experiments).
9. Analogy—in some cases, it is fair to judge cause–effect relationships by analogy—"With the effects of thalidomide and rubella before us, it is fair to accept slighter but similar evidence with another drug or another viral disease in pregnancy."[16]

The rooster's crow does not make the sun rise. Down syndrome is strongly associated with birth rank, but birth rank is not considered to be a cause of Down syndrome (because the effect of birth rank is mediated by the association between Down syndrome and maternal age). Correspondingly, there is a significant association between chocolate consumption in a country and the number of Nobel laureates per million inhabitants,[17] though most people would not say that chocolate consumption creates Nobel laureates.

Second, not all associations (even strong ones) are causal, so it is not easy to say how strong an association must be in order to classify it as a causal relationship. Where does one set the limit on causation? It is argued that aspirin "causes" Reye's syndrome in children, and that certain tampons "cause" toxic shock syndrome, though the probabilities are very low indeed. The limits of what can be classified as causation appear to be rooted in social values, not scientific values. Indeed, social commitments appear to play a role in probabilistic causation. For instance, the association between exposure to low-dose ionizing radiation (such as from medical X-rays) and cancer is very low for most types of cancer in adults. Nevertheless, radiation protection is organized and based on a causal relationship, as there is a strong social commitment to protect people undergoing medical investigations. If, as seems to be the case, the limits of what can be called causation depend on social values, causation diffuses from the realm of science into the realm of society.

1.5.6 Counterfactual Conditions

Another approach to causation highlights whether the presence or absence of a cause "makes a difference." If Mr. Hanson had not been exposed to the hepadnavirus, he would not have got hepatitis B. If Mrs. Jones had taken two aspirins instead of just a glass of water an hour ago, her headache would now be gone. A counterfactual draws on the contrast between one outcome (the effect) given certain conditions (the cause), and another outcome given alternative conditions. C causes E if the same condition except C would result in a condition different from E, when all other conditions are equal. The last premise is called *ceteris paribus* (all other conditions kept equal). Counterfactual conditions seem similar to necessary conditions, as without the necessary condition, the event will not occur. However, counterfactuals support or undermine suppositions, and are acceptable or not acceptable, while necessary conditions are either true or false.[12]

Counterfactual conditions are often considered to be deterministic, but they can also be probabilistic, as in the following counterfactual: "if Mrs. Jones had taken two aspirins instead of just a glass of water an hour ago, she would be much less likely to have a headache now." One of the challenges with counterfactuals is that in practice it is not easy to know or to assess what the case would have been if the situation had been different. Although epidemiological data and experiments may give clues about differences between groups, it is hard to know whether Mrs. Olsen would have avoided lung cancer if she had not smoked. Hence, it is difficult to satisfy the *ceteris paribus* condition, because the same individual cannot be observed in the exact same situation as both a smoker and a nonsmoker.

Table 1.1 sums up the deterministic and probabilistic conceptions of causation commonly referred to in the life sciences, which are discussed in this chapter.

Table 1.1 Deterministic and probabilistic conditions of causation

Deterministic conceptions of causation	Probabilistic conceptions of causation
Sufficient condition for an event *Insufficient* but *necessary* part of a *sufficient* and *necessary* condition INUS condition Counterfactual If the condition, C, had not been present, the event, E, would not have occurred, i.e., C makes a difference with respect to E.	Probabilistic Raised probability for an event Counterfactual If the condition, C, had not been present, then it is less likely that the event, E, would have occurred, i.e., C makes a difference with respect to the probability of E.

Altogether, this means that when scientists say that X causes Y, many meanings can be inferred, and much confusion could be avoided if these meanings were clearer. However, the fact that there is no one clear-cut concept of causation should not result in disappointment or despair; even in his time Aristotle identified four kinds of causes: material cause, formal cause, efficient cause, and final cause. Moreover, there are many other meanings of "cause," of which researchers must also be aware. For example, consider the following five statements:

1. The heart's capacity to pump causes the blood to circulate.
2. A gene causes Huntington's disease.
3. Smoking causes lung cancer.
4. Aspirin causes Reye's syndrome.
5. Professional interests cause (Norwegian) radiologists to be negative to expanding radiographers' tasks (towards diagnostics).

Here the word cause is used in many ways. In statement 1, "cause" denotes a mechanism, in statement 2 a sufficient condition, in statement 3 a significantly raised probability, in statement 4 a slightly raised (but socially important) probability, and in statement 5 "cause" denotes a reason.

Correspondingly, causation may have many levels. If a small child dies in Nigeria, what was the cause of his death? Was it the diarrhea, or was it that the diarrhea was not properly treated? Was it the malnutrition, or was it that his parents did not have access to clean water? Was it the low income, the lack of education of the population, the poor infrastructure, international politics, or the history of colonial exploitation? In fact, all of these factors may be relevant (Figure 1.4).

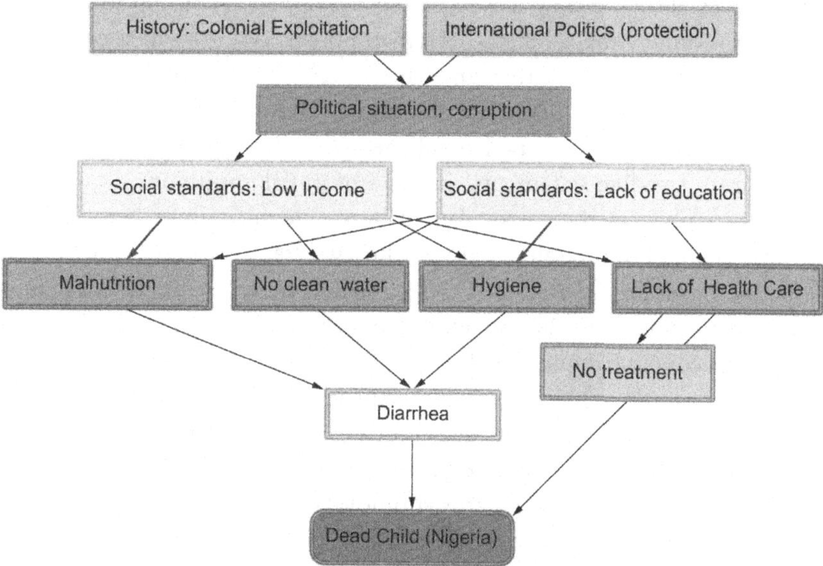

Figure 1.4 Web of causation. There are several levels to consider.

Much confusion, and also the adverse effects of hype, could be avoided if scientists were more explicit and clear when they claim that they have "solved the riddle of cancer" or that "X causes (disease) Y."

1.6 SCIENTIFIC EXPLANATION

Causation, and the various accounts of causation discussed above, are inherent in scientific explanations. From the time of Aristotle, philosophers have realized that a distinction could be made between two kinds of scientific knowledge—roughly "knowledge that" and "knowledge why." It is one matter to know that myocardial infarction is associated with certain kinds of pain (angina pectoris); it is a different matter to know why this is so. Knowledge of the former type is descriptive; knowledge of the latter type is explanatory, and it is explanatory knowledge that provides a scientific understanding of the world.[18]

How, then, can phenomena studied in the biomedical sciences be explained? For example, how does one explain the change in hematopoietic cell growth in a medium when the temperature changes? What criteria exist for something to be considered an acceptable scientific explanation? The standard answer to such questions is that something is explained by showing how it is expected to happen according to the laws of nature (nomic expectability).[19] Hematopoietic cell growth is explained by the laws that govern this growth as well as the initial conditions, including the type of medium, humidity, temperature, and pressure. Accordingly, a singular event is explained if (a description of) the event follows from law-like statements and a set of initial conditions.

When a phenomenon is explained by deducing it from laws or law-like statements, the sequence of deductive steps is said to follow a deductive-nomological model (DNM), which turns an explanation into an argument where law-like statements and initial conditions are the premises of a deductive argument (Box 1.6).

In other words, a phenomenon is explained by subsuming it in a law. For this reason DNM is often referred to as "the covering law model of explanation." One reason for the prominent position of DNM is its close relation to prediction. A deductive-nomological explanation of an event amounts to a prediction of its occurrence.

However, DNM does present some challenges. One is that DNM allows for symmetry. For instance, certain conditions of a growth medium for cells (temperature, humidity, light, etc.) can be explained by the growth rate of hematopoietic cells (in this medium) given the same laws. This is in contrast with the desire for there to be asymmetry between cause and effect; that is, what is considered to be a cause leads to an effect, and not the other way round.

Moreover, if the biomedical sciences can provide explanations only when phenomena subsume under deterministic laws of nature, it means that there are innumerable phenomena that cannot be explained. For instance, it is often stated that lung cancer can be explained by smoking, although there is no strict law stating which smokers will develop lung cancer. There is a straightforward solution to this problem, which entails replacing deterministic laws with probabilistic statements. This engenders

BOX 1.6 The DNM of Explanation

Premise 1:	Initial conditions	Type of medium, humidity, light, temperature
Premise 2:	Universal law(s)	Laws of hematopoietic cell growth
Conclusion:	Event or fact to be explained	Greater growth due to temperature increase

the deductive-statistical model (DSM) of explanation, the form of which is shown in Box 1.7.

DSM is a version of DNM that supports explanations of statistical regularities by deduction from more general statistical laws (instead of deterministic laws). However, DSM cannot explain singular events, such as Mr. Hanson recovering from bacterial meningitis after taking antibiotics. DSM can only explain why persons taking antibiotics will recover (in general). In order to explain singular events in terms of statistical laws, one can refer to the inductive-statistical model (ISM) of explanation, which explains likely events inductively from statistical models (Box 1.8).

Table 1.2 summarizes the traditional models of explanation, DNM, DSM, and ISM. Common to all these models is the idea that explanations are arguments (deductive or inductive) based on initial conditions and on

BOX 1.7 The DSM of Explanation

Premise 1:	Initial conditions	Having bacterial meningitis
Premise 2:	Statistical laws	Taking antibiotics probably leads to recovery
Conclusion:	Event or fact to be explained	Persons taking antibiotics will recover

BOX 1.8 The ISM of Explanation

Premise 1:	Initial conditions	Mr. Hanson has meningitis and takes antibiotics
Premise 2:	Probability (r) of event, given 1	The probability of recovery in such cases $= r \approx 1$
Induction:	Event or fact to be explained	Mr. Hanson will recover

Table 1.2 Models of explanations according to Salmon[18]

Laws	Singular events	General regularities
Universal laws	DNM	DNM
Statistical laws	ISM	DSM

DNM, deductive-nomological model; ISM, inductive-statistical model; DSM, deductive-statistical model.

law-like statements, be they deterministic or statistical (nomic expectancy). The standard form of each such argument is:

Premise 1: Initial conditions

Premise 2: Law-like statement

Implication: Event or fact to be explained

Most explanations in the biomedical sciences appear to fit these models.

Nevertheless, all these models of explanation present challenges. One is that arguments with true premises are not necessarily explanatory. For instance, if Mr. Hanson is a man and takes oral contraceptives (initial conditions), and if no man who takes oral contraceptives becomes pregnant (law), it leads deductively to the conclusion that Mr. Hanson will not become pregnant. According to DNM, oral contraceptive use explains why Mr. Hanson does not become pregnant, but this is intuitively wrong, as the premises are explanatorily irrelevant.

As already indicated, DNM permits symmetry. For example, DNM enables the use of plane geometry and the elevation of the sun to find the height of a flagpole from the length of its shadow, as well as to predict the length of the shadow from the height of the flagpole. However, as the length of its shadow clearly does not explain the height of the flagpole, DNM does not present a set of sufficient conditions for scientific explanation.

DNM, DSM, and ISM are the principal models relevant to the biomedical sciences, but represent only three of the many models that can be used for scientific explanation. The challenges they present in the way of relevance and symmetry have made some philosophers of science argue that explanations should be based on causation; to explain is to attribute a cause. According to a causal model of explanation, one must follow specific procedures to arrive at an explanation of a particular phenomenon or event:

1. Compile a list of statistically relevant factors.
2. Analyze the list by a variety of methods.
3. Create causal models of the statistical relationships.
4. Test the models empirically to determine which is best supported by the evidence.

However, these procedures reveal some of the core challenges of demonstrating causation. Moreover, although stating that to explain a phenomenon is to find its cause is intuitively correct, it is not necessarily the case in practice. Indeed, David Hume (1711–1776) argued that causation entails regular association between cause and effect, though the conception of causation as regularity adds nothing to an explanation of why one event precedes another. Accordingly, Bertrand Russell (1872–1970) claimed that causation "is a relic from a bygone age, surviving, like the monarchy, only because it is erroneously supposed to do no harm."[20] Indeed, defining explanation in terms of causation would enhance our ability to predict, but not to understand phenomena.[21] Accordingly, explanation entails more than referring to a cause—it invokes understanding, and thus one could argue it must include the laws of nature.

1.7 MODES OF INFERENCE

The biomedical sciences tend to employ three modes of inference first set forth in 1903 by Charles Sanders Pierce (1839–1914): deduction, induction, and abduction.

- Deduction is inference from general statements (axioms, rules) to particular statements (conclusions) via logic. If all persons with type I (insulin-dependent) diabetes are known to have deficiencies in pancreatic insulin production (rule), and Mr. D has type I diabetes (case), then Mr. D has deficiencies in pancreatic insulin production (conclusion).
- Induction is inference (to a general rule) from particular instances (cases). If all persons with deficiencies in pancreatic insulin production have symptoms of type I diabetes, and these persons are from the general population (i.e., they were not included in the study sample because of other deficiencies that could cause these symptoms), one can conclude that all persons with deficiencies in pancreatic insulin production have symptoms of type I diabetes.
- Abduction infers the best explanation. When a certain observation (case) is made, a hypothesis (rule) can be found that makes it possible to deduce a conclusion. If Mr. D has deficiencies in pancreatic insulin production, and all persons with type I diabetes have deficiencies in pancreatic insulin production, then Mr. D has type I diabetes.

The crucial question is whether these modes of inference are valid. Deductive inference is knowledge-conservative; if the axioms are true, the conclusion is true. However, the ultimate issue is whether the axioms hold. Inductive and abductive inferences are both knowledge-enhancing, and are therefore called ampliative inference. The challenge with induction and abduction is to justify the knowledge enhancement. In induction, inference is made from some cases (conclusion) to the general rule, and in

DEDUCTION	
All F's are G's (Rule)	All balls in this urn (F) are red (G)
All S's are F's (Case)	All balls in this particular sample (S) are taken from this urn (F)
All S's are G's (Conclusion)	All balls in this particular sample (S) are red (G)
INDUCTION	
All S's are G's (Conclusion)	All balls in this particular sample (S) are red (G)
All S's are F's (Case)	All balls in this particular sample (S) are taken from this urn (G)
All F's are G's (Rule)	All balls in this urn (F) are red (G)
ABDUCTION	
All F's are G's (Rule)	All balls in this urn (F) are red (G)
All S's are G's (Conclusion)	All balls in this particular sample (S) are red (G)
All S's are F's (Case)	All balls in this particular sample (S) are taken from this urn (F)

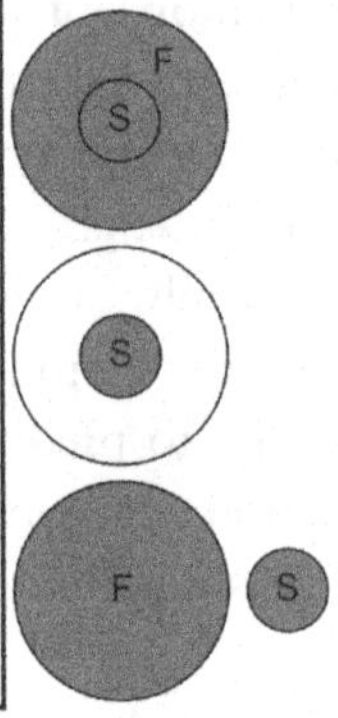

Figure 1.5 Modes of inference.

abduction there could of course be other rules that could explain what is observed even better. Figure 1.5 illustrates the differences among these three modes of inference.

1.8 WHAT SCIENCE IS ABOUT

The biomedical sciences are about this world and its biomedical phenomena. This raises the philosophical question, what is this world? Many scientists find this question odd, even irrelevant. Scientists deal with viruses, cells, molecules, and the effects of interventions, and it is clear to most researchers that cells exist and that they more or less correspond to scientific theories. However, there are innumerable examples throughout history of situations where convictions of the reality of the entities contained in different theories, such as phlogiston, proposed by the German physician and alchemist Johann Joachim Becher (1635–1682); ether; miasms; and the "cadaver poison," identified by Ignaz Semmelweiss (1818–1865) have been replaced by new convictions and new entities. Therefore, how can one be sure that the world is as science portrays it, and how can changes in theories be explained?

Scientific realists hold that successful scientific research enhances knowledge of the phenomena of the world, and that this knowledge is largely independent of theory. Furthermore, scientific realists hold that such knowledge is possible even when the relevant phenomena are not observable. According to scientific realism, there is good reason to believe what is written in a good, contemporary medical textbook, because the authors had solid scientific evidence for the (approximate) truth of the claims put forth about the existence and properties of viruses and cells and the effects of interventions. Moreover, there is good reason to think that such phenomena have the properties attributed to them in the textbook, independent of theoretical concepts in medicine.

Consequently, scientific realism can be viewed as the scientists' own philosophy of science. On the other hand, scientific antirealism holds that the knowledge of the world is not independent of the mode of investigation. A scientific antirealist might say that photons do not exist, and that theories about them are tools for thinking. These theories explain observed phenomena, such as the light beam of a surgical laser. Of course, the energy emitted from a laser exists, as well as the coagulation, but the photons are held not to exist. The point is that there is no way to know if the world is independent of our scientific investigations and theories.

There are several levels of scientific realism. A weak notion of scientific realism holds that a real world exists that is independent of scientific scrutiny, without advancing any claim about what this real world is like. A stronger notion of realism argues that not only does the real world exist independent of scientific inquiry, but it also has a structure that is independent of this inquiry. An even stronger notion of scientific realism holds that certain things, including entities in scientific theories such as photons and DNA, exist independent of humans and our scientific inquiry of the world. Accordingly, the scientific realist claims that when phenomena, such as entities, states, and processes, are correctly described by theories, they actually exist. Scientific realism is common sense, and certainly "common science," as a researcher does not doubt that the phenomena he/she studies exist independent of their investigations and theories. However, how can this intuition be justified? This is where the philosophical challenges start. Three arguments justifying scientific realism are commonly advanced: transcendental, high-level empirical, and interventionist.

The *transcendental argument* asks what the world must be like to make science possible. Its first premise is that science exists. Its second premise is that there must be a structured world independent of human knowledge of science. There is no way that science could exist, considering its complexity and extent, if the things science describes did not exist.[22] Hence, the argument reasons from what one believes exists to the preconditions for its existence. Even if science is seen as a social activity, the same question applies. How can this activity exist without the precondition that the world actually exists? Science is intelligible as an activity only if scientific realism is assumed. One premise of the transcendental argument is that science expands our knowledge of the world and corrects errors, but how does one know this? Furthermore, how can one reason from what one believes to exist to the preconditions for its existence? The answer is through thought experiments. The effects of certain microbiological events could not be pondered without the existence of DNA. Based on this one can argue that the existence of DNA is a necessary condition for microbiological events. But how can one be sure that the reason microbiological events cannot be pondered without the existence of DNA is not simply due to the limits of scientific imagination?

The high-level empirical argument contends that scientific theories are (approximately) true because they best elucidate the success of science. The best way to explain progress and success in science is to observe that (1) the terminology of mature sciences typically refers to real things in the world, and (2) the laws of mature sciences are typically approximately true.[23] However, this is an abductive inference, in which one argues from the conclusion (science has success) and the rule (if science is about real things, then it has success) to the case (science is about real things). Abductive arguments are knowledge-expanding, therefore there may be other, better explanations that have not yet been discovered.

The interventionist argument holds that one can have well-grounded beliefs about what exists based on what one can do.[24] Intervention can be employed to test whether the entities contained in scientific theories actually exist. If an intervention on an entity from a theory does not work, the entity does not exist, but "if you can spray them, then they are real."[24] Hence, one can test whether something is real. One problem with the interventionist argument is that it is not robust with respect to explanation. If one were to test whether ghosts are real by spraying them with red paint, one may conclude that ghosts are not real. However, how could a person know that this is the correct method to show that ghosts are real? Could it not be that red paint does not adhere to ghosts, whereas yellow paint does?

Scientific realism, which most scientists consider common sense, is irritatingly difficult to justify. We could, of course, dismiss the whole question about the existence of the entities in our theories by arguing that observable results are what matters, and the validity of the existence of said entities, be they photons or arthritis, does not matter. However, at a certain point a scientist may need to reflect upon the nature of being of the entities studied.

1.9 SCIENTIFIC RATIONALITY

Rationalism is the position that reason takes precedence over other ways of acquiring or justifying knowledge. Traditionally, rationalism has been contrasted with empiricism, which claims that true knowledge of the world can only be obtained through sensory experience. In antiquity, rationalism and empiricism referred to two schools of medicine, the former relying primarily on theoretical knowledge of the concealed workings of the human body, the latter relying on direct clinical experience.

One might argue that the demarcation between rationalism and empiricism has become irrelevant in science, yet remains relevant in clinical practice. There are many examples of treatments established on rationalistic grounds, such as the ligation of arteria mammaria interna as a treatment for angina pectoris, that have been revealed by empirical studies to have no effect (beyond placebo). Hormone replacement therapy in postmenopausal women was said to prevent cardiovascular disease. Laying babies on their stomach was thought to prevent sudden infant death syndrome because it would avoid suffocation from vomit that could occur if babies were put on their back. Some established treatments prescribed based on a physician's experience have also been revealed to be without effect, or even detrimental. However, modern biomedical scientists tend to rely on rationality as well as experience. For example, hypotheses may be generated on rationalistic grounds (the substance S should have the effect E because it has the characteristics X, Y, and Z), and are then tested empirically, such as in animal models or in randomized clinical trials.

Therefore, the enduring rationalism-empiricism debate still seems relevant in the biomedical sciences given the limitations of scientific methodology. There may be ethical reasons, such as reluctance to use placebo surgery, that limit empirical research; there may be lack of knowledge of mechanisms that limits a rationalistic approach, such as when one wishes to test a substance that appears to have promise in eliciting a desired effect, but for which one lacks the knowledge of how it works.

1.10 HYPOTHESIS TESTING

Whenever a theory appears to you as the only possible one, take this as a sign that you have neither understood the theory nor the problem which it was intended to solve.

Karl Popper

The author of one of the most prominent Hippocratic writings, *The Art* (of medicine), identified three challenges in medical treatment and research: (1) the obtained effects may be due to luck or accident (and not intervention), (2) the obtained effect may occur even if there is no intervention, (3) the effect may not be obtained despite intervention. In the terminology of causation, researchers are faced with the challenges that the intervention is not a necessary condition (as in case 2), nor a sufficient condition (as in case 3) for the effect, and that there may be either a probabilistic relationship between the intervention and the effect, or other (unknown) conditions for the effect (as in case 1). Almost three millennia later, scientists still struggle with the same kind of question: how can one be certain that the theories and hypotheses of the world are true, given the large variety of possible errors? The standard solution to this problem is to test the hypothesis according to the hypothetical-deductive method.

1.10.1 Hypothetical-Deductive Method

The hypothetical-deductive method[iv] is the scientific method of testing hypotheses by predicting particular observable events, then observing whether the events turn out as predicted. If so, the hypothesis is verified, and if not, the hypothesis is falsified. The steps in the hypothetical-deductive method are:

1. State a clear and experimentally testable hypothesis.
2. Deduce the empirical consequences of this hypothesis.
3. Perform empirical experiments (in order to compare the results with the deduced empirical consequences).
4. If the results concur with the deduced consequences, one can conclude that the hypothesis is verified, otherwise it is falsified.

According to the traditional interpretation of the hypothetical-deductive method, hypotheses can be verified and scientific knowledge is accumulated through the verification of ever more hypotheses (Table 1.3).

However, as Karl Popper (1902–1994) showed, the verification approach to hypothesis testing is flawed.[2] First, the verification of a hypothesis presupposes induction. However, as already pointed out, reasoning from some (group) to many (the population) is not warranted. Second, the logical form of the model is not sound (Table 1.4).

[iv] This is also frequently called the hypothetico-deductive method or model.

Table 1.3 Simplified comparisons between the structure of verification and falsification

	Verification	Falsification
1. Hypotheses	A is better than B	B is better than A
2. Deduced empirical consequences	If A is better than B, one must observe that A gives better results than B in the empirical setting	If B is better than A, one must observe that B gives better results than A in the empirical setting
3. Experiments and observations	We observe that instances where A is used obtain better results than B	We observe that instances where A is used obtain better results than B
4. Conclusion	The experiment confirms the hypothesis	The experiment refutes the hypothesis and lends support to the alternative hypothesis (A is better than B)

Table 1.4 Logical comparisons between the structure of verification and falsification

	Verification	Falsification
Logical structure	If p, then q q p (Confirming the consequent)	If p, then q Not q Not p (Modus tollens)
Mode of knowledge acquisition	Accumulation of knowledge, confirming hypotheses	Exclusion of false hypotheses, narrowing down and delimiting knowledge

Moreover, Popper was critical of the lack of standard criteria for establishing scientific truth that existed in the early twentieth century, and of the corresponding trend to employ (scientific) authority to decide what was true, which made it difficult to differentiate science from other social activities. Popper's radical solution was to avoid stating explicit (authoritative) criteria for truth and to provide stringent procedures for testing hypotheses. Furthermore, he broke with the ideal that truth could ultimately be determined, and instead strove to provide a scientific knowledge base of nontruths, or falsified hypotheses. For Popper, scientific knowledge progressed through enlarging the graveyard of falsified hypotheses. The method of falsification rather than verification makes all truth provisional, conjectural, and hypothetical. According to Popper, experiments cannot determine theory, only delimit it, nor can theories be inferred from observations. Experiments only show which theories are false, not which theories are true (Figure 1.6) (Box 1.9).

In empirical fields, the hypothetical-deductive method is used almost daily and often without a thought. The control experiment is a typical example. Can a possible effect or an absent effect have a trivial explanation? Might changes over time or in titrations of solvents produce effects, or might the cells have failed to respond? Control experiments are included to rule out such trivial explanations.

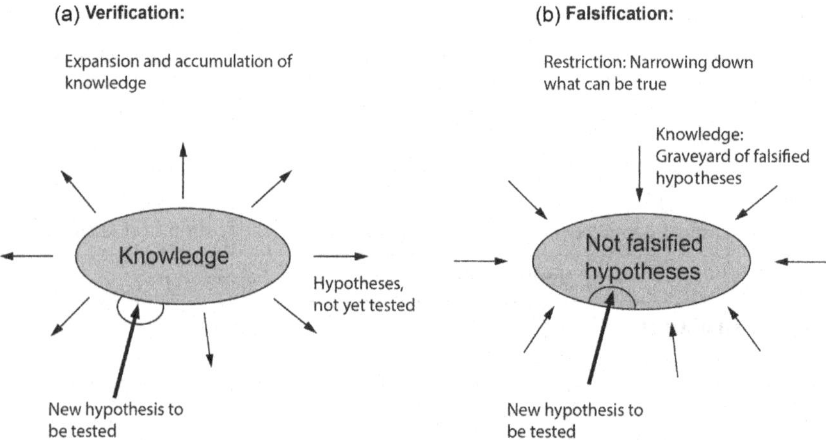

Figure 1.6 Knowledge generation in verification versus falsification.

BOX 1.9 Popper on "The Success of Refutation"

"Refutations have often been regarded as establishing the failure of a scientist, or at least of his theory. It should be stressed that this is an inductivist error. Every refutation should be regarded as a great success; not merely a success of the scientist who refuted the theory but also of the scientist who created the refuted theory and who thus in the first instance suggested, if only indirectly, the refuting experiment.

Even if a new theory (such as the theory of Bohr, Kramers, and Slater) should meet an early death, it should not be forgotten; rather its beauty should be remembered, and history should record our gratitude to it—for bequeathing to us new and perhaps still unexplained experimental facts and, with them, new problems; and for the services it has thus rendered to the progress of science during its successful but short life."[2]

In clinical trials involving new drugs, patient symptoms may be strongly influenced by the treatment situation, and a placebo may cause an effect; a placebo group is included to rule out (falsify) the placebo effect. Correspondingly, tests are conducted in a double-blind manner to falsify the hypothesis that the observed effect of a treatment is due to the expectations of the researcher.

Correspondingly, statistical tests are performed to investigate whether an obtained result is due to selection bias, or other types of bias. These tests assess whether recorded differences between groups are random. This is done by formulating a null hypothesis, denoted H_0, that states there is no difference between the groups, and thereafter assessing the probability that H_0 is true. If that probability is very small, the null hypothesis is rejected, which strengthens the principal hypothesis that there is a real, instead of a random, difference.

A hypothesis must have testable implications if it is to have scientific value. Popper contended that if a hypothesis it is not testable, and thus not falsifiable, it is not science. The lack of adequate methods often hinders scientific progress, because limited testability restricts the scope of topics that can be subjected to scientific inquiry. Therefore, great leaps in science are often made thanks to new, more powerful methods that open up new areas of research. Outstanding instances include Kary Mullis's development of the polymerase chain reaction (PCR) in molecular biology, which was recognized by a Nobel Prize in 1993, and the development of the patch clamp in neurobiology by Erwin Neher and Bert Sakmann, which was recognized by a Nobel Prize in 1991.

The development of hypotheses is closely associated with the development of models and the planning of experiments. Many hypotheses are too imprecise and ambiguous to be rejected, and consequently cannot be challenged as Popper requires. Formally, there should be two alternative hypotheses that are mutually exclusive, and an experiment should be designed to distinguish between them. If a hypothesis is falsified, this may lead us to the development of new hypotheses, which in turn can be tested (Figure 1.7).

Moreover, a scientific hypothesis should have the power to explain. It should relate to an existing, generally accepted theoretical basis of the field. There must be good grounds to reject established theories, such as an accepted law of nature. Whenever newer observations so indicate, it is advisable to modify a hypothesis. Modifications are acceptable if they make the hypothesis more testable. However, sometimes a theory that

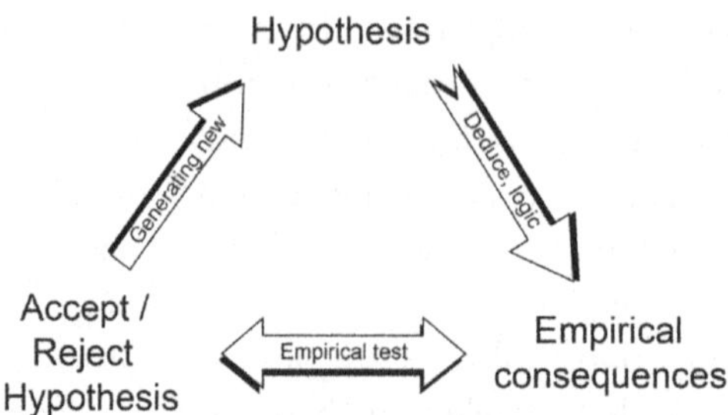

Figure 1.7 Sketch of the hypothetical-deductive method.

flounders on the grounds of falsifying experiments is defended by its remaining adherents by proposing *ad hoc* hypotheses that save the favored hypothesis by nullifying the negative observational evidence, not by making it more testable. While modified hypotheses are part of the ordinary scientific process, *ad hoc* hypotheses are not, as they hinder rather than promote scientific development.

Although falsification has become common in empirical biomedical research, its strengths and weaknesses are not always appreciated. According to Popper, a theory or hypothesis should be bold and far-reaching. Its empirical content should be high; that is, it should have great predictive power. Furthermore, the hypothesis should be testable. If the results of empirical tests support the hypothesis, it is corroborated (but not verified); if not, it is falsified. Moreover, if one can choose among several hypotheses, it is better to select the simplest, according to what has been called Ockham's razor, after the British Franciscan monk William of Ockham (1288–1347) (see also Chapter 4).

Regardless of how influential Popper's approach has been and continues to be, falsification in empirical research in the biomedical sciences has been severely criticized. Five of the most cited challenges of falsification are:

1. When one falsifies theories, their prospective robustness is not tested. They are only tested on past evidence.
2. A severe test is one which is surprising and unlikely based on present evidence. However, this means that knowledge of what is likely is used to set up a test that is unlikely, which necessitates the application of induction. Accordingly, if one really defies induction, there is no reason to act on corroborated theories or hypotheses, because doing so would be based on induction.
3. Falsification of a theory is based on empirical observations. However, observational statements should also be fallible. Hence, the falsification

of a theory may be erroneous if the observational statements are not true.

4. Popper's method can lead to falsification of robust and fruitful theories with high empirical content, e.g., due to errors in the test procedure.
5. In practice one does not falsify single theories but rather groups or whole systems of hypotheses. One reason why one cannot test single hypotheses in isolation is that a series of background assumptions is made to test the hypothesis (the Quine-Duhem thesis). Any of these background assumptions may fail when one tries to falsify a hypothesis.
6. Moreover, in practice one does not falsify a potentially fruitful theory based on a single observation or study. Instead, new experiments are conducted and *ad hoc* hypotheses generated to investigate or explain the falsifying observation.

1.11 THE AIM OF SCIENCE: REDUCING UNCERTAINTY

The primary aim of science is to increase knowledge in order to explain, understand, and intervene. Scientific knowledge is necessary to reduce our uncertainty. It is practical to differentiate among four kinds of uncertainty: risk, uncertainty, ignorance, and indeterminacy (Table 1.5).

Risk is when the system behavior is basically well known, and the chances of different outcomes can be defined and quantified by structured analysis of mechanisms and probabilities. One of the tasks of science is to find the outcomes of a given situation or intervention, as well as the probability of these outcomes. An example is the incidence of cardiovascular disease among patients with type II (noninsulin-dependent) diabetes practicing prophylactic statin use.

Uncertainty is characterized by knowledge of the important system parameters but not of the probability distributions. In this case the major outcomes of a certain intervention may be known, but not their respective probabilities. There may be many sources of uncertainty in a study,

Table 1.5 Modes of uncertainty[25]

		Outcome	
		Known	**Unknown**
	Known	*Risk*	*Indeterminacy (Ambiguity)*
Probability	**Unknown**	*Uncertainty*	*Ignorance*

including uncertainty in reasoning (i.e., how to classify a single case with regards to general categories) or uncertainty in biomedical theory (i.e., when all mechanisms in a certain field are not known in detail, or because of multifactor causation). Moreover, diseases are complicated, and it can be difficult to know and understand all their causes (see Figure 1.4 above for an example). In the case of uncertainty, the main task of science is to provide the probability distributions, to estimate risks, and to provide sensitivity analyses.

Ignorance describes a situation in which neither the outcomes nor their probabilities are known. One of the tasks of science is, of course, to find both, but when it comes to ignorance this is difficult, as a scientist does not know what he does not know. For example, the effect of thalidomide taken by pregnant women on fetuses and children (severe limb deformation) was difficult to foresee. The role of science is to reduce ignorance while being observant of and attentive to the unexpected. Indeed, it took an unnecessarily long time to discover the detrimental effects of thalidomide.

Even if scientists are successful in reducing ignorance to uncertainty and all cases of uncertainty to risk, they might still be subject to *indeterminacy*. It is not always a question of uncertainty due to imprecision (which is assumed to be narrowed by further research) but also a question of the properties or criteria used to classify things. When myocardial infarction is classified according to a set of clinical criteria, the resultant perspective will be different than if it is classified according to the level of troponin in the blood. Likewise, if pain is being investigated in terms of neural activity or according to a visual analogue scale, the risk, uncertainty, and ignorance may differ. Hence, pain may be ambiguous. Accordingly, processes may not be subject to predictable outcomes from given initial conditions, due to imprecise classification. Table 1.5 summarizes and compares the four modes of uncertainty.

1.12 THE EMPIRICAL TURN IN THE PHILOSOPHY OF SCIENCE: SCIENCE IN SOCIETY

Although many of the previously mentioned challenges in the traditional philosophy of science have been addressed and progress has been made, interesting and fruitful contributions have been fueled through empirical studies of science and scientific practices. Close empirical studies have revealed social aspects that are typical in science. In particular, the norms and activities of scientists have been shown to be basically similar to the norms and activities of other societal groups.[26] Science is but one of many socially organized activities that generates knowledge; studying science as a socially organized activity and comparing it to other such activities has given important insights into the field.

Where the philosophy of science has traditionally been theoretical and has focused principally on the products of science, that is, knowledge and its conceptual preconditions, more recent approaches are empirical and focus on the social processes of science (and its interaction with material matters). A seminal and famous study of scientific activity showed that knowledge is not accumulative and that science does not develop in a linear manner.[3] Instead, it evolves in an abrupt way (scientific revolutions) with intervening quiescent periods.

Inspired by Kuhn's paradigmatic conception of scientific progress, and by Wittgenstein's theories on rule-following and language games,[27] a series of studies of science, termed the Sociology of Knowledge (SoK) movement, emerged. The key goal was to show that science is a social activity that follows the same social patterns as other activities in society. The question of how things are in the world cannot be addressed without questioning how the social group comprising scientists conceives of and handles these things. It follows that things, be they photons or DNA, cannot be attributed a role in our world independent of symbols and meaning. Hence, while the traditional philosophy of science had procedural criteria for demarcating science from nonscience, such as Popper's criteria for falsification, the SoK movement applies social criteria. Whereas the normative aim of traditional studies of science was to free science from the authority and power inherent in the social structures among scientists and in society, the SoK movement strives to disclose power within the scientific society and to emancipate.

In many respects, the key issue in the traditional philosophy of science has been the relationship between scientific theories and nature. In the SoK movement the focus is on the relationship between theory and culture: in what way do scientific theories reflect social structures (instead of the structures of the world)? Nevertheless, what appears to be similar in both the traditional philosophy of science and the SoK movement is the focus on epistemological issues. In both cases the key question is: what do scientific theories represent? In the first case, scientific theories represent patterns in nature; in the second, they represent social structures. In the former case theories about DNA represent biomolecular entities in cells, whereas in the latter case they represent social activities. In both cases there is something behind the theories; something that the theories represent.

Later studies of science have tried to avoid this representational pattern, i.e., the belief that there is something behind a theory. Instead of investigating the relationship between theories and the social processes and structures in science, the scientific process itself was studied, including its material premises. How do scientists behave, and how do they produce the facts of science? For example, Bruno Latour studied how the daily activities and processes of scientists at the Salk Institute contributed to the establishment of scientific facts about the thyrotropin-releasing factor (TRF), e.g., that it is a peptide.[28] This can be referred to as a *processual approach*, according to which science is the change as it restructures, makes new, and stabilizes things and theories. What characterizes the social process of science is an interaction of methods, material, activities, and processes, where negotiations lead to the stabilization and generation of facts. When species of the *Helicobacter pylori* bacteria were found to be associated with gastric and peptic ulcers, scientific debate ensued on the bases of the embedded, previously accepted theories, until negotiations based on continued empirical work established *Helicobacter pylori* as a key factor.

It is not a question of what the theory represents (either nature or culture) but rather a question of negotiation between different scientific groups with regards to what will be considered fact. Hence, according to the processual approach the issue is not the relationship between theory and nature/culture (epistemological and representational) but what scientists regard and treat as real (ontological and processual). It is not a question of what is behind prions in nature or in culture but the scientific practices and processes that constitute facts about prions.

1.13 PHILOSOPHY OF THE SOCIAL SCIENCES

A significant part of the overall spectrum of health-care problems constitutes matters that are not principally biological. Should one wish to find out why patients do not take prescribed medications, why incorrect medications are sometimes administered in hospitals, or why it is difficult to attain fully informed consent for trials or treatment, one cannot search for answers in human biological research; instead one must turn to the methods of the social sciences.

For this reason it is essential to know how the philosophies of the social sciences and the biological sciences differ, so one does not erroneously use the criteria for one area to judge another. In the social sciences, many different methods are used, and there are various schools of theory.

The following comprises only a brief introduction and does not cover the broad scope of methods or schools of theory.

1.14 INTERPRETATION, UNDERSTANDING, AND EXPLANATION

The social sciences differ from the biological sciences in two respects:

1. They entail greater elements of overt interpretation that often enter into the collection of data.
2. In many cases, a research result is an understanding, not an explanation.

1.14.1 Explanation and Understanding

The principal goal in the biological sciences is to elicit causal explanations of the phenomena studied, for instance, the cause of a particular manifestation of a disease. Some projects in the social sciences also seek causal explanations of social phenomena, but many seek an understanding instead. Understanding is a form of knowledge that enables us to know why a person or a group behaves in a particular way, why and how they experience a specific situation, how they themselves understand their way of life, etc. We attain understanding through interpretation.

The distinction between explanation and understanding was first expressed by the German philosopher, psychologist, and educator Wilhelm Dilthey (1833–1911), who believed that these two ways of viewing the world were characteristic of the natural sciences and the human sciences (*Geisteswissenschaften*), respectively. However, the difference between explanation and understanding is not as distinct as many believe. Often theories of the social sciences include elements of both causal explanation and noncausal understanding. For instance, when one explains why the poor have greater morbidity than the rich, one usually refers to both causal factors, like the greater physical risks present in poor areas, and noncausal factors, like the influence of the working-class culture on lifestyle choices.

1.14.2 Interpretation

All content-bearing objects and statements can be interpreted. People express themselves not just in speech, writing, and deeds but also in architecture, garden design, clothing, etc. If, for example, one investigates where and why institutions for psychologically ill patients were built, one will find that the history reflects varying understandings of psychological illness. The architecture of the asylum is also content-bearing. However, in this chapter the focus will be on the interpretation of texts and other linguistic statements, as it is germane in the discussion of the theory of interpretation, often called hermeneutics.

Interpretation may have many goals, but in general, through interpretation one seeks to understand the information put forth in content-bearing material. The various theories of interpretation are based on differing concepts of the nature of content and how it should be identified. Is there content in a statement itself, in the thoughts of the person making the statement, in the social structure in which the statement is made, etc.? These differences are germane when analyzing the validity of specific methods of the social sciences but are of lesser importance here in our general discussion of interpretation.

The question of whether one ever obtains a true interpretation of a text is an old one. All written religions have sets of hermeneutic rules for interpreting the content of their holy texts. For example, in Christian theology, biblical exegesis concerns the interpretation of the scriptures.

In modern times, interest arose in the interpretation of secular statements, first as part of literary and historical research, and then as a part of research in the social sciences. The goals of the various hermeneutic methods that have been developed are to arrive at an understanding of content that can be defended as a valid, intersubjective understanding; that is, an understanding that can be substantiated and discussed rationally.

As Popper pointed out, the elements of interpretation enter into all observations and thereby into all forms of science. Humans lack direct access to the world "as it is" through our senses. We always view the world through a theoretical filter, and all observations are theoretically loaded. For example, to say that the Sun rises is to reflect the influence of the old geocentric worldview in which the Sun circled the Earth. And the "description" that a pathologist gives of a histological preparation seen through a microscope is to a large extent an interpretation based on theories of cells, of inflammation, etc. Biological research findings are interpreted in a similar manner; no *P*-value speaks for itself.

1.15 THE HERMENEUTIC CIRCLE, HORIZON OF UNDERSTANDING, AND "DOUBLE HERMENEUTICS"

The hermeneutic interpretation of a text rests on each individual part, as well as on the understanding of how these individual parts are related to the whole. Neither an individual part nor the whole text may be interpreted without reference to each other. Therefore, interpretation is circular; the hermeneutic circle. In principle, this circle cannot lead to certain closure, as one will never know if a deeper analysis of the text may change its interpretation. The problem of attaining valid, intersubjective interpretations has long been and continues to be discussed, and optimistic interpretation theoreticians speak of a hermeneutic spiral, which implies that interpretation gets better and better. At the pragmatic level, the problem of the hermeneutic circle is less worthy of attention, since agreement on the meaning of a text can usually be attained.

The concrete interpretation is also influenced by the interpreter's "horizon of understanding," a concept introduced in *Wahrheit und Methode*, the principal work of German philosopher Hans-Georg Gadamer (1900–2002). Gadamer argues that before one has begun a conversation with another person or begun to interpret a text, one already has preconceptions about it based on one's horizon of understanding, a collective term for one's worldview. The horizon of understanding is built throughout life and comprises one's understanding of particular words, the connotations that particular words and concepts hold, and so on. For a resident of London, the word *city* connotes financial affairs, while for people elsewhere it simply connotes an urban concentration of population. This difference in connotations means that two people engaging in a conversation may believe that they have understood each other without actually having done so. Full understanding is possible only when the two people in the conversation have acquired each other's horizons of understanding, or have "fused horizons." This explains the problem of interpretation, as in interviews in the social sciences that often are too short for the interviewer to understand the interviewee's horizon of understanding. Consequently, a vital part of the interview comprises an effort to learn how the interviewee uses and understands words and concepts in the area being discussed.

Furthermore, English sociologist Anthony Giddens (1938–) pointed out that social science research comprises "double hermeneutics."[29,30] In reality, social science research interprets interviewees' interpretations of their own understandings, and parts of their understandings arise through concepts that they have acquired from the theories of the social sciences (such as the Marxist concept of class or the incest taboo of psychology). Hence, there is a complex interaction between the interpretations of the researcher and the interviewee, which is why an additional level of interpretation may often be needed to focus on how an interviewee's self-image is affected by the theories of social science. An interviewee may be misunderstood if the interviewer does not take such reflections into account.

1.16 POWER, IDEOLOGY, AND INTERESTS

One's interpretations of the statements and deeds of others are influenced by aspects other than one's horizon of understanding. The German philosopher Jürgen Habermas (1929–) pointed out that power, ideology, and interests also play leading roles. Usually, one is not a neutral or objective observer; instead, one interprets according to the power of one's position, ideology, and the interests one wishes to further.[31]

In Habermas's view, ideology is not restricted to political ideology. An ideology is simply a set of assumptions that furthers the interests of a particular group in society. For example, the assertion that "an extensive hospital system is essential in health care" is an apolitical ideology that, in addition to safeguarding the interests of patients, furthers the interests of doctors and other health-care professionals.

One of the difficulties of ideologies is that they are often concealed, as researchers are neither aware that they have them, nor of where they came from. So ideologies can influence actions and interpretations "behind our backs," so to speak. Consequently, Habermas maintains that the principal task of the critical social sciences is to identify prevailing ideologies so one may be freed from them.

1.17 VALIDITY

In the above, widely recognized problems have been discussed: that an interpretation and the understanding attained through interpretation can never be a final truth about the meaning of a particular statement, unless the statement is extremely simple. Therefore, one is obliged to ask how one can judge the validity of a scientific interpretation. The simple answer is that if a researcher is aware of these problems and has taken the best possible steps to avoid or avert them, such as by trying to identify which ideologies and interests have influenced the various elements of the research process, there are grounds to rely on the interpretation—not because it is necessarily true, but because it constitutes a well-founded hypothesis without significant sources of error in the research process.

1.18 REDUCTIONISM AND EMERGENCE

Some biological researchers contend that there is no need for social scientific interpretation, as in the final analysis all knowledge can be reduced to facts about physical conditions. Social phenomena can be reduced to group psychology, which in turn can be reduced to individual psychology, which in turn can be reduced to neurology, which in turn can be reduced to cellular biology, and so on, until one reaches the physical level at which prevailing physical laws provide explanations for all phenomena observed at higher levels. This view, called reductionism, is in strong dispute.

It is crucial to distinguish between methodological reductionism and general reductionism. In some research projects, methodological reasons may dictate the exploration of one or more factors that can influence the phenomenon of interest, without indicating that other factors are unimportant. There are no methods that can acquire data and at the same time investigate "the whole." Our attention must be focused on something more specific. Methodological reductionism can be meaningful and necessary, whereas general reductionism has been refuted. If, for instance, one wishes to examine a biological relationship, it may be necessary to ignore an ancillary social relationship. Conversely, if one examines a social relationship, it may be necessary to ignore a biological relationship. Methodological reductionism in itself is straightforward, as long as the factors examined are sensible. It becomes problematic only when a set of factors is systematically excluded, such as the correlation between poverty, social deprivation, and disease.

There are many arguments against reducing social phenomena to physical phenomena. Two of them will be summarized here. The first problem confronting the reductionist is that it is doubtful that individual psychology can be reduced to neurophysiological processes. Dispute persists on the precise description of the relationship between psychological phenomena and cerebral activity; scientists are no closer to solving the "mind-brain" riddle today than they were a century ago. If this link in the reductionistic chain fails, reductionism as a whole cannot be carried out. The other problem for the reductionist is that many social phenomena are emergent, that is, they are socially not reducible as they occur at particular social levels and have no meaning when reduced to lower levels (individual psychology, neurology, etc.).

Paper money, for example, is an emergent social phenomenon. A £10 banknote has no value in itself (save the value of the portrait of scientist

Charles Darwin). It cannot be exchanged for gold or other objects of value at the central bank. But it is integrated in social relationships that enable it to be exchanged for goods or services worth ten pounds. Otherwise, it is just a small, rectangular scrap of paper. Emergence at the social level may also be ascribed to a particular set of social conventions or formalized laws. For instance, most societies have the institution of marriage, but the concrete implications of being married and the social effects of it vary from society to society. The human penchant to form pair relationships might be reduced to the biological level, but the concrete institution of marriage in a particular society cannot be similarly reduced. However, it is clear that the concrete, nonreducible institution of marriage affects human actions and considerations, so a full description of these actions and considerations is possible only at the social level. Similarly, in health care the professions of doctor, nurse, etc. have different roles. These roles are not naturally given but are a result of social convention. Despite this, it is important to understand these specific roles if, for instance, one is doing health services research aimed at improving the function of a particular type of hospital ward.

If the antireductionists are right, scientific effort in the social sphere is useful and may employ methods that differ from those applicable at lower levels.

1.19 GENERALIZATION

Statistics are often useful in research projects that employ quantitative methods. Whenever a study sample is selected from a well-defined source population, statistical inference to the source population is performed, for instance, by calculating the statistical confidence. Inferences to a wider population, the target population, are then made. In principle, this implies that results from research conducted in the United States may be directly applicable to treatment choices in Norway. However, it is worth noting that such generalization of results is only acceptable when there are grounds to assume that the populations are in fact similar. In the previous example this would mean assuming that there is no biological difference between Americans and Norwegians.

Generalization may be employed in much the same way in quantitative social science research, but statistical methods cannot be employed in research that is not quantitative. Does this imply that understanding attained in social science research cannot be generalized? The answer would be yes if statistical generalization were the only form of generalization available. But there is a form of generalization that is not quantitative and is frequently employed across all the sciences. It is called theoretical or conceptual generalization, sometimes called transferability. We often generalize, not in exact numbers, such as the cure rate for a particular drug, but rather within a conceptual or a theoretical frame of understanding. For instance, when teleological explanations based on the theory of evolution are used in biology, they rest upon a theoretical generalization of the theory of evolution, not upon a statistical generalization. Social scientific concepts and theories may be generalized in the same manner.

In all forms of generalization, both statistical and conceptual, it is important to keep in mind that conditions change with time. Generalizations that were once valid can be rendered invalid if there are changes in the supporting biological conditions, such as the resistance patterns in bacteria or the structures of families.

QUESTIONS TO DISCUSS

1. What makes your research scientific and subject to funding above other social knowledge-generating activities? Mention and assess four demarcation criteria for science.
2. How do you know that DNA actually exists?
3. Can you be more or less sure that RNA exists compared with that pain exists? Why?
4. Discuss four types of uncertainty mentioned in this chapter. Do they apply to your research project, and what can you as a scientist do to reduce them? What should you do if you are not able to reduce them?
5. Describe the type of causality that is relevant to your research project. What are the pros and cons of this conception of causality?
6. Which kind of scientific explanation is prominent in your discipline?
7. Give an example of abduction from your field of research.
8. What is the problem with verification, according to Karl Popper?
9. What are the challenges of falsification?
10. Mention four criteria for truth. Discuss the pros and cons of each of them. Which conception of truth is prevalent in your field of research?

ACKNOWLEDGMENTS

We thank Carina V. S. Knudsen, Institute of Basic Medical Sciences, University of Oslo, for producing the figures.

REFERENCES

1. Tranøy KE. Science and ethics. Some of the main principles and problems. In: Jones AJI, editor. *The moral import of science. Essays on normative theory, scientific activity and Wittgenstein*. Bergen: Sigma Forlag; 1988. p. 111–36.
2. Popper KR. *Conjectures and refutations, the growth of scientific knowledge*. London: Routledge and Kegan Paul; 1963.
3. Kuhn TS. *The structure of scientific revolutions*. Chicago, IL: University of Chicago Press; 1969.
4. Hansson SO. *Vetenskap och ovetenskap*. Stockholm: Tiden; 1983.
5. Hansson SO. In: Zalta EN, editor. *Science and pseudo-science*. Winter 2012 Edition The Stanford Encyclopedia of Philosophy; 2012. <http://plato.stanford.edu/archives/win2012/entries/pseudo-science>.
6. Bennett CM, Baird AA, Miller MB, Wolford GL. Neural correlates of interspecies perspective taking in the post-mortem Atlantic Salmon: an argument for multiple comparisons correction. *J Serendipitous Unexpected Results* 2010;**1**(1):1–5. <http://prefrontal.org/files/posters/Bennett-Salmon-2009.pdf>.
7. Peirce CS. In: Houser N, Kloesel C, editors. *The essential Peirce, selected philosophical writings, Volume 1 (1867–1893)*. Bloomington and Indianapolis, IN: Indiana University Press; 1992.
8. Dewey J. *The quest for certainty: a study of the relation of knowledge and action*. New York, NY: Minton, Balch, and Company; 1929.
9. Ioannidis JPA. Why most published research findings are false. *PLoS Med* 2005;**2**(8):e124.
10. Koch R. Untersuchungen über Bakterien: V. Die Ätiologie der Milzbrand-Krankheit, begründet auf die Entwicklungsgeschichte des *Bacillus anthracis* [Investigations into bacteria: V. The etiology of anthrax, based on the ontogenesis of *Bacillus anthracis*]. *Cohns Beitrage zur Biologie der Pflanzen* 1876;**2**(2):277–310.
11. Evans AS. Causation and disease: the Henle-Koch postulates revisited. *Yale J Biol Med* 1976;**49**(2):175–95.
12. Mackie J. *The cement of the universe*. Oxford: Clarendon Press; 1974.
13. Rothman KJ. Causes. *Am J Epidemiol* 1976;**104**:587–92.
14. Lipton R, Odegaard T. Causal thinking and causal language in epidemiology: it's in the details. *Epidemiol Perspect Innov* 2005;**2**:8.
15. Woodward J. Probabilistic causality, direct causes and counterfactual dependence. In: Galavotti MC, Suppes P, Costantini D, editors. *Stochastic causality*. Stanford: CSLI; 2001. p. 39e63.
16. Hill AB. The environment and disease: association or causation? *Proc R Soc Med* 1965;**58**(5):295–300.
17. Franz H, Messerli MD. Chocolate consumption, cognitive function, and Nobel laureates. *N Engl J Med* 2012;**367**:1562–4.
18. Salmon W. *Four decades of scientific explanation*. Minneapolis, MN: University of Minnesota Press; 1990.
19. Hempel CG. *Aspects of scientific explanation and other essays in the philosophy of science*. New York, NY: The Free Press; 1965. p. 331–96.
20. Russell B. *Mysticism and logic*. London: Allen & Unwin; 1959. p. 180.

21. Psillos S. *Causation and explanation.* Chesham: Acumen Publishing Limited; 2002.
22. Bhaskar R. *A realist theory of science.* London: Verso; 1997.
23. Putnam H. *Reason, truth and history.* Cambridge: Cambridge University Press; 1981.
24. Hacking I. *Representing and intervening: introductory topics in the philosophy of natural science.* Cambridge: Cambridge University Press; 1983.
25. Stirling A. Keep it complex. *Nature* 2010;**468**:1029–31.
26. Stengers I. *The invention of modern science.* Minneapolis, MN: University of Minnesota Press; 2000.
27. Wittgenstein L. *Philosophical investigation.* Oxford: Blackwell Publishing; 2001.
28. Latour B, Woolgar S. *Laboratory life: the social construction of scientific facts.* Beverly Hills, CA: Sage Publications; 1979.
29. Giddens A. *New rules of sociological method: a positive critique of interpretative sociologies.* London: Hutchinson of London; 1976.
30. Giddens A. *The consequences of modernity.* New York, NY: Polity Press; 1990.
31. Habermas J. *The theory of communicative action, reason and the rationalization of society.* London: Polity Press; 1986.

FURTHER READING

Boyd R, et al. *The philosophy of science.* Cambridge, MA: MIT Press; 1991.

Hacking I. *The social construction of what?* Boston, MA: Harvard University Press; 2001.

Hempel CG. *Philosophy of natural science.* Englewood Cliffs, NJ: Prentice Hall; 1966.

Hollis MH. *The philosophy of social science—an introduction.* Cambridge: Cambridge University Press; 1994.

Hugles J. *The philosophy of social research.* 2nd ed. Harlow: Longman; 1990.

Lynch MP. *Truth as one and many.* Oxford: Oxford University Press; 2009.

Parascandola M, Weed DL. Causation in epidemiology. *J Epidemiol Community Health* 2001;**55**:905–12.

Popper KR. *The logic of scientific discovery.* London: Hutchinson; 1959.

Resnik DB. A pragmatic approach to the demarcation problem. *Studies Hist Philos Sci Part A* 2000;**31**(2):249–67.

Williams M, May T. *Introduction to the philosophy of social research.* London: Routledge; 1996.

Wynne B. Uncertainty and environmental learning. Reconceiving science and policy in the preventive paradigm. *Glob Environ Change* 1992;111–27.

CHAPTER 2

Ethics and Scientific Conduct

Søren Holm[1], Bjørn Hofmann[2] and Petter Laake[3]
[1]Centre for Social Ethics and Policy, School of Law, The University of Manchester, England; Centre for Medical Ethics, Institute of Health and Society, University of Oslo, Blindern, Oslo, Norway and Department of Health Science and Technology, Aalborg University, Denmark
[2]Section for Health, Technology and Society, University College of Gjøvik, Gjøvik, Norway and Centre for Medical Ethics, Institute of Health and Society, University of Oslo, Blindern, Oslo, Norway
[3]Oslo Centre for Biostatistics and Epidemiology, Department of Biostatistics, Institute of Basic Medical Sciences, University of Oslo, Blindern, Oslo, Norway

2.1 WHY THE CURRENT FOCUS ON SCIENTIFIC MISCONDUCT?

Throughout history even the most prominent scientists have been accused of fraudulent work. The groundbreaking work of Gregor Mendel (1822–1884) was questioned by the famous statistician and geneticist Sir Ronald A. Fisher (1890–1962). Through Mendel's long-lasting experiments on peas he described what is now known as Mendel's Laws of Inheritance, the foundation of modern genetics. More than 50 years later Fisher claimed that Mendel's results were "too good to be true," as Mendel's reported results had too little variation to be consistent with both Mendel's law and the laws of statistics. No proof of fraud was ever presented, and regardless of how Mendel did his experiments in the monastery garden, Mendelian genetics now has a legitimate place in modern biology.

Misconduct in science has probably always taken place, but within the last 20 years the problem had received increasing attention. This is partly due to a number of high-profile national and international "scientific misconduct scandals." A number of these scandals have involved plagiarism in publications and PhD theses. Indeed, the Internet has made it easier to find sources to plagiarize, but it has also made it easier to detect plagiarism. The most recent high-profile cases of plagiarism have occurred in countries where academic degrees are of value in the political sphere, although few aspiring politicians make major contributions to research. In 2011, the German Defense Minister, Karl-Theodor zu Guttenberg, had to resign after parts of his PhD thesis were found to have been plagiarized, and his PhD was withdrawn.[1] The same happened in 2013, when the German Education Minister, Annette Schavan, resigned after her

Research in Medical and Biological Sciences
DOI: http://dx.doi.org/10.1016/B978-0-12-799943-2.00002-1

PhD was withdrawn.[2] Similar cases have involved the Iranian Minister of Education and Science, the Iranian Minister of Transport,[3] and most recently the Serbian Interior Minister.[4]

A number of other scandals have involved high-profile researchers and scientific misconduct that included more than just plagiarism. It was revealed that the Norwegian cancer researcher Jon Sudbø invented many of his "datasets," including one underpinning a paper published in the *Lancet*. The Danish researcher Milena Penkowa, who received the Danish Ministry of Research's very prestigious "Elite Researcher" prize, was found to have invented many of her "animal experiments." And the South Korean researcher Woo-suk Hwang falsely claimed to have derived human embryonic stem cells from cloned human embryos. There has also been increasing focus on the pharmaceutical industry's efforts to ensure that research on their products renders "the right results," to ensure that these results are reported in the most favorable light in the scientific literature, and its attempts to suppress negative results.[4] Retraction of scientific papers has also increased significantly. In the first decade of this millennium the retraction rate increased 10-fold, while the number of scientific papers only increased by 44%.[5] This is in part a result of an increase in scientific misconduct and increased attention to the problem.[6]

We are taught as children that cheating, lying, stealing, and deceiving are wrong, and that we should not do it. These principles also apply in a research context, so it does not take much argument to show that any kind of scientific conduct that can be properly described as cheating, lying, stealing, and deceiving is also wrong, and thus deemed scientific misconduct. However, the recognition that scientific misconduct is problematic does not mean that it does not take place.

2.2 WHAT DO WE KNOW ABOUT SCIENTIFIC MISCONDUCT?

Scientific misconduct is generally regarded as illicit, so it is difficult to assess its general prevalence, or to investigate specific instances of misconduct. That said, available evidence suggests that scientific misconduct is not rare. One might think that scientific misconduct is usually detected before the "results" are published in reputable journals. However, in the aforementioned cases of Sudbø, Penkowa, and Hwang, the falsifications and fabrications had been published in journals with a very high scientific ranking. In a 2005 report from the Committee of Publications Ethics (COPE),[7] the executive editor of the *Lancet* presented the results of a review of 212 articles submitted to or published in biomedical journals between 1997 and 2004 that had been reported to COPE by journal editors seeking advice. There was evidence of misconduct in 163 of these articles; as many as 19 were related to falsification and fabrication, and 17 cases of plagiarism were detected.

There are a number of individual studies that have tried to assess the incidence and prevalence of scientific misconduct. A survey from the United States included 1645 coordinators of clinical trial research and found that 21.5% of them had firsthand knowledge of incidents of scientific misconduct during the previous year. The definition of scientific misconduct used was that of the US Office of Research Integrity (ORI): "Fabrication, falsification, plagiarism, and other practices that seriously deviate from accepted standards when proposing, conducting, and reporting research."[8,9] A review conducted in 2000 in the United Kingdom looked at a variety of studies and indicated serious misconduct in 0.5–1% of clinical research projects.[10] But the best evidence probably comes from a meta-analysis performed by Fanelli.[11] Fanelli identified questionnaire studies in which active researchers were asked whether they had ever been party to various forms of scientific misconduct, or whether they had firsthand knowledge of colleagues who had engaged in such misconduct. He excluded studies in which the respondents were students and excluded questions about more general knowledge of cases of scientific misconduct. He found:

> *"A pooled weighted average of 1.97% (N = 7, 95% CI: 0.86%–4.45%) of scientists admitted to have fabricated, falsified or modified data or results at least once—a serious form of misconduct by any standard—and up to 33.7% admitted other questionable research practices. In surveys asking about the behaviour of colleagues, admission rates were 14.12% (N = 12, 95% CI: 9.91%–19.72%) for falsification, and up to 72% for other questionable research practices." (N refers to the number of studies included in the meta-analysis, CI: confidence interval).*[11]

The analysis showed that there was a nonsignificant trend towards scientific misconduct being more prevalent in the life sciences than in other areas of science. But what triggers scientific misconduct? In most cases, self-interest seems the most plausible explanation. By operating outside the rules it is easier to obtain interesting results and publish papers that add to one's list of publications. In some cases the psychological basis is apparently more complex, but there is no reason to believe that the perpetrators are in general mentally ill or psychologically disturbed. High pressure to publish increases not only scientists' productivity, but also their bias towards publishing "positive" results. An archetypal justification for misconduct was given by a researcher from the United States, Eric T. Poehlman, before he was sentenced to 366 days in prison "because his actions led to a loss to the government, obstruction of justice, and abuse of a position of trust" (ORI, p. 1).[8] Upon investigation by the ORI, Poehlman was found to have falsified or fabricated data in at least 12 publications and 19 funding proposals. Poehlman's explanation of his conduct is extracted in Box 2.1. In addition to his prison sentence, Poehlman was barred for life from applying for or receiving federal research grants.

It is also important to note that the pressure to engage in scientific misconduct often begins very early in one's research career. In a Norwegian survey, some PhD students reported that they had been under pressure to fabricate and/or falsify data, and over 10% reported pressure in relation to authorship.[12] Similar results have been found in other studies.

Research results must be published before they are accepted as science, and researchers accept results published in scientific journals. But since

BOX 2.1 Poehlman's Explanation

In a letter to Judge William Sessions, III, US District Court for the District of Vermont, Eric T. Poehlman said he had convinced himself that it was acceptable to falsify data for the following reasons:

> *First, I believed that because the research questions I had framed were legitimate and worthy of study, it was okay to misrepresent "minor" pieces of data to increase the odds that the grant would be awarded to UVM and the work I proposed could be done.*
>
> *Second, the structure at UVM created pressures which I should have, but was not able to stand up to. Being an academic in a medical school setting, I saw my job and my laboratory as expendable if I were not able to produce. Many aspects of my laboratory, including salaries of the technicians and lab workers, depended on my ability to obtain grants for the university. I convinced myself that the responsibility I felt for these individuals, the stress associated with that responsibility, and my passion and personal ambition justified "cutting corners."*
>
> *Third, I cannot deny that I was also motivated by my own desire to advance as a respected scientist because I wanted to be recognized as an important contributor in a field I was committed to.*

(ORI, p. 5)[8] (UVM: University of Vermont at Burlington)

scientific misconduct has become part of research, one can no longer trust all published results. *The Wall Street Journal* commissioned Thomson Reuters to go through all the retractions in the Web of Science posted from 2001 to 2011. The study revealed that while the number of published papers rose 44% in that time period, the number of retracted papers rose 1500%.[13] R. Grant Steen evaluated 788 papers that were listed as retracted in PubMed between 2000 and 2010.[14] Of these retracted papers more than 25% were fraudulent, and close to 75% were erroneous. But most importantly, more than 28,000 patients were included in the studies reported in the retracted papers, and more than 400,000 patients were included in the secondary studies citing a retracted paper. These results do not necessarily show that there is more scientific misconduct now than before; they may also be explained by better detection of scientific misconduct.

Perhaps more incredibly, a spoof paper with major scientific flaws easily identifiable by any researcher with adequate experience was sent to 304 open-access journals and was accepted by 157 of them.[15] Many, but not all open-access journals have lower thresholds for accepting papers than older, more traditional journals, so this finding cannot be generalized to all journals. Nevertheless, the evidence points to the likelihood that an increasing number of publications are flawed.

2.3 WHAT IS WRONG WITH SCIENTIFIC MISCONDUCT?

Contemplations of ethics have a long history going back (at least) to the pre-Socratic philosophers in ancient Greece, and they are an equally venerable component of all major religions. Yet there still is no agreement on the nature of the correct or best framework or theory for ethical analysis, although there is reasonable agreement concerning the core of the disagreement and the viable contenders for an acceptable ethical framework. For instance, it is relatively clear that moral nihilism and moral relativism are not viable contenders for an acceptable ethical framework. Cultural relativism about morality is a fairly commonly held position in public debate, essentially claiming that cultures differ in their morality, that these differences should be respected, and that morality cannot be discussed across cultures. Although it is clearly a true description of the state of the world that different societies have different moral commitments, cultural relativism is unsustainable as a coherent moral position. If cultural relativism were true, criticism of the moral judgments of people outside your own culture would be nonsensical, universal human rights (even the right not to be tortured) would be meaningless, and it would be impossible to make sense of a notion of moral progress (although most of us, for instance, believe that the abolition of slavery is moral progress wherever it occurs).

Cultural relativism is seen as an attractive position, in part because two distinctions are not made. The first is the distinction between the claim that moral values are universal (e.g., that harming people is wrong wherever and whenever it occurs) and the claim that they are absolute (e.g., that you can never harm another person, even for some greater good). The second is the distinction between inflexibility and context dependency in the application of moral values. Many deem cultural relativism to be attractive because the ostensible alternative is that ethical values are universal, absolute, and inflexible. But strictly speaking, a denial of cultural relativism only includes the claim that moral values are universal; they may well be nonabsolute and flexible in their application, with a sizeable scope for variation depending on the context, thereby allowing for cultural differences. We may, for instance, claim that respect for privacy is a universal value, while still recognizing that the exact contours of such a right will and must vary between societies, and between different contexts within a society. For instance, the shape of a right to privacy will depend in part on the prevalent types of living accommodation, be they communal longhouses or individual flats; similarly one can legitimately expect more privacy in one's own bedroom than in a hospital ward.

Most ethical frameworks hold that moral values are universal, but permit flexibility in their application. The principal differences in opinion in ethical theories or frameworks are about whether the basic level of ethical evaluation is an evaluation of acts, states of the world, or people (see further elaboration below), and whether the rightness or goodness of an act is essential to its evaluation. The three most frequently used ethical theories or frameworks are listed in Box 2.2.

BOX 2.2 Frequently Used Ethical Theories or Frameworks

- Consequentialism
- Deontological ethics
- Virtue ethics

2.3.1 Consequentialism

Consequentialism is the simplest possible ethical theory. It holds that goodness is primary, and it defines the right act as that which maximizes goodness. In thinking about an act, an agent should consider the various possible acts that he or she could perform and then choose the one that maximizes the good consequences. This is equivalent to choosing the act that maximizes the goodness of the state of the world. What consequentialism repudiates is that the type of act matters in itself. When deciding whether or not to perform an act, it should not matter that it involves lying, the only thing that matters is whether it has good consequences. For example, scientific misconduct is wrong because it decreases society's trust in science, decreases research funding, and decreases the social impact of science. There are various consequentialist theories, and they differ in three major ways: what they count as good consequences, whether they claim that good consequences should be maximized, and what class of entities should be included when calculating good consequences. The first two differences often do not matter for practical purposes, but the last difference can have significant implications. For instance, in animal research ethics it matters hugely for which animals good consequences should be maximized (see Chapter 3).

The main criticism of consequentialist theories is that in some situations they justify acts that most people consider wrong, for instance the sacrifice of the interests, and perhaps even the lives, of a few research participants in order to gain important scientific knowledge. This criticism has led to the development of a variant of consequentialist theory called rule consequentialism. Rule consequentialism claims that we should not consider individual acts in isolation, but instead focus on the rules that will render the best consequences if they are followed. A rule consequentialist might, for instance, contend that it makes sense to have a rule against lying, because following the rule maximizes good consequences over time.

2.3.2 Deontological Ethics

Deontological theories are in some ways the opposite of consequentialist theories. According to deontological theories of ethics, the primary concern is whether an act is right, not whether it has good consequences, i.e., one should not lie even if it has good consequences. Thus scientific dishonesty is wrong *per se*. Various deontological theories differ in how they classify acts as right or wrong and in how one should choose between two wrong acts in a situation without a right act.

Deontological theories fit well with our prereflective commitment to the belief that there are some acts that are inherently wrong (e.g., torturing newborns). The principal criticism of deontological ethics is that it fails to explain why, for instance, "white lies" should be considered as seriously wrong according to the deontological theory. Indeed, a deontologist sees any lie as seriously wrong, even a well-intentioned lie with good consequences. But for many it is difficult to see that as the case.

2.3.3 Virtue Ethics

Virtue ethics differs from consequentialism and deontological ethics in that it focuses on the person performing an act instead of on the act itself. According to virtue theories, it is possible to identify the set of character traits and motives that a morally good person should possess (character traits that are designated as "virtues"). The morally right act is the act a virtuous person would perform and that flows from his or her virtues. A good scientist does not plagiarize, falsify, or fabricate results, and does not take credit for unwarranted authorship. The main criticisms of virtue theories are the lack of one accepted list of virtues and their difficulty accounting for the fact that even morally evil people seem able to perform the occasional good act. However, an important component in any initiative to prevent scientific misconduct is to inculcate young researchers with the right scientific virtues, thus making them scientifically virtuous.

The ethical theories described above can be recognized in various rules, codes, and guidelines for appropriate research. Many professional organizations and institutions have their own. The Swedish Science Board[16] has taken the initiative to develop the eight general rules for good research listed in Box 2.3.

2.4 SCIENTIFIC CONDUCT AND MISCONDUCT

Scientists have ethical obligations in their research activities, obligations that are derived from the ethical frameworks discussed above. All systems of ethics can, for instance, explain why lying and other forms of dishonesty are problematic in most circumstances, and why deceiving others for your own gain is unacceptable. Similarly, all forms of ethical theory find the exploitation of the powerless by the powerful problematic. This clearly supports rules against well-known types of scientific misconduct, including plagiarism, the fabrication of data and results, and false or gift authorship.

BOX 2.3 Eight General Rules for Good Research

1. Tell the truth about your research.
2. Openly report your methods and results.
3. Openly disclose any commercial interests and other ties.
4. Consciously examine and present the basic assumptions underlying your studies.
5. Do not steal research results from others (e.g., from younger colleagues).
6. Conduct your research in an orderly manner (e.g., by maintaining documentation and retaining data).
7. Do not conduct your research in a way that could harm other people (e.g., subjects).
8. Be fair in your assessment of other people's research.

These requirements are summed up in words like *Honesty* (1, 5), *Openness* (2–4), *Orderliness* (6), *Consideration* (7) and *Impartiality* (8) and can be defended from the theories presented above.

But perhaps better or more specific guidance on proper scientific conduct can be derived from an analysis of the purpose and goals of the research enterprise. It is clear from the analysis presented in Chapter 1 that there is no univocal, uncontested definition of science, research, or truth. But most would accept that a core feature of science is its aim to produce publicly available, well-justified knowledge that can be achieved only through a long-term effort involving many different researchers and research groups, using open and transparent methodology. Indeed, that view is corroborated in the famous remark, "we see further because we stand on the shoulders of giants" (although, as it has been mischievously pointed out, we would also see further if we stood on the shoulders of midgets); the original remark is often attributed to Isaac Newton but seems to have originated with Bernard of Chartres in the twelfth century.

If one accepts this categorization of science as a goal-driven activity, it makes sense to ask how all the participants in the research process must act in order to achieve the goals set forth.

This question has been analyzed extensively by the philosopher Knut Erik Tranøy and by many sociologists, most notably Robert Merton. Tranøy argues that scientific work is characterized by and requires three different kinds of norms (Box 2.4).[17,18]

BOX 2.4 Norms in Science

- Internal norms
- Linkage norms
- External norms

All three of these norms guide scientific work in different ways. If scientific activity were not connected to society at large, only the internal norms would be necessary to guide researchers, but because there are a variety of interactions between science and society, linkage norms and external norms also come into play.

2.4.1 Internal Norms

Within internal norms, there is a distinction between epistemic norms that guide the activity of each individual researcher and social norms that guide the collaboration between researchers and research groups in a scientific endeavor. In epistemic norms, Tranøy includes truth seeking, testability, consistency, coherency, and simplicity; in the social norms he includes openness, open-mindedness, and honesty. He argues convincingly that unless individual researchers and research communities accept these norms and act in accordance with them, science as an activity aimed at generating knowledge cannot succeed. Consequently, these norms are mandatory. But they are not imposed from the outside; they arise through the nature of scientific activity.

2.4.2 Linkage Norms and External Norms

Linkage norms include utility, fruitfulness, and relevance, and explain why society permits scientists freedom of research, whereas external norms are the limits society places on scientific conduct (e.g., in relation to the use of human subjects and experimental animals). There are significant similarities between the results of Tranøy's analysis and those obtained by the American sociologist Robert Merton. Based on studies of scientists, Merton claimed that the scientific community was committed to a set of norms denoted by the acronym CUDOS, for "Communism of knowledge" (subsequently changed to "Communalism" when "Communism" became contentious), "Universality," "Disinterestedness," and "Organized skepticism."[19] Tranøy's analysis was significant because it provided a reason to label certain activities as problematic scientific misconduct, even if the general ethical frameworks are believed to be extrinsic to science and not applicable to science or scientific activity. Tranøy showed that there are purely internal reasons related to the epistemic claims that science makes.

2.5 SCIENTIFIC MISCONDUCT THAT AFFECTS THE TRUTH CLAIMS OF SCIENTIFIC FINDINGS

Scientific misconduct is most serious when it affects the truth claims of scientific findings, as it undermines the cumulative nature of scientific work and development, and may lead to practical applications that are harmful to patients. Scientific misconduct is a continuum from harmless errors to outright fraud.[20] Misconduct includes everything from wrong or missing citations, to wrong analysis, to plagiarism and outright fraud (Figure 2.1).

As pointed out by Nylenna and Simonsen,[20] a strict definition of scientific misconduct (falsification and fabrication) might be suitable for legal

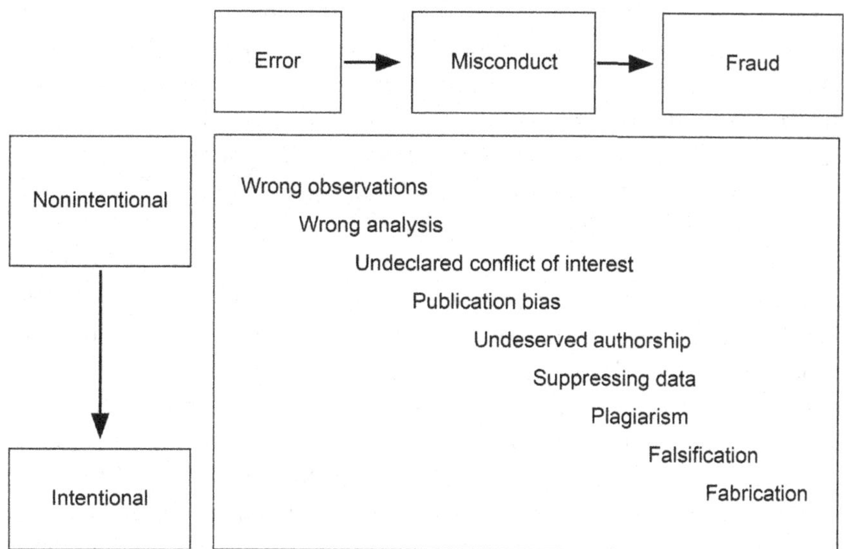

Figure 2.1 The continuum of scientific misconduct, from honest errors to intentional fraud. The horizontal axis represents the extent of deviation from acceptable scientific behavior. The vertical axis represents extent of blame, from excusable errors to willful action. *Reproduced from Nylenna and Simonsen[20] with permission.*

action, while a broader definition, i.e., all deviations from acceptable scientific behavior, should be used for prevention. The prevention of deviation from good behavior will be addressed later in this chapter.

2.5.1 Wrong Observations and Wrong Analysis

Wrong observations cover everything from a harmless error made in the laboratory, to a misunderstanding in a questionnaire, to falsification or fabrication of data. Wrong observations can be avoided, or at least kept to a minimum, by implementing routines in the laboratory or wherever data are collected. We will return to the falsification and fabrication of data later in this section.

A prevalent error is the performance of repeated statistical analyses to find a statistically significant result. A study of PhD students at all medical faculties in Norway showed that 38% did not find it inappropriate to try a variety of different methods of analysis to find a statistically significant result.[12] Wrong analysis may seem harmless, but errors in study design or data analysis can result in invalid conclusions, which might be anything but harmless. Andersen[21] listed more than 200 examples of clinical studies in which either wrong or imperfect study designs or analyses were used. These are examples of scientific misconduct in the sense that they reduce the quality of the research. In the worst-case scenario this can invalidate the corresponding conclusions, but it cannot be categorized as cheating.

Somewhere on the scale between wrong observations and fabrication one can also find wrongful framing and wrongful presentation of results. Indeed, every day the news media present sensational reports about the promising results of an experimental treatment for, say, cancer, or a new diet that may reduce the risk of diabetes or cardiovascular disease. It is the responsibility of both researchers and journalists to ensure that only valid and reproducible scientific results are presented in the news media, as in many cases preliminary results are reported or leaked to the press based on preclinical animal research, or results from Phase I or Phase II trials.

Reproducibility of scientific findings is vital. Findings should be replicated in independent experiments, and results should not be published until they are corroborated. In a paper published in *Nature* in 2012, Begley and Ellis[22] reported the results of a study in which they tried to reproduce the results of 53 "landmark" studies that introduced new approaches to targeting cancer or for clinical use of therapeutics. In only six cases (11%) could the scientific findings be reproduced, which, of course, was shocking to the authors.[3] Journal referees and editors share some responsibility for publishing results that are flawed or irreproducible. But the scientists and the academic environment around them must shoulder the lion's share of the blame, as they are ultimately responsible for the validity and the reproducibility of their results.

2.5.2 Plagiarism

Plagiarism is claiming the work of another to be one's own. The word stems from the Latin *plagium/plagiarius* meaning kidnapping, and it counts as theft. It is an old problem with modern relevance and has even been mentioned in popular, satirical songs (Box 2.5). In publications, plagiarism may be total, such as the submitting of copies or translations of papers previously published by others elsewhere (actually not rare), or more limited, such as copying and pasting other people's work into one's own papers, either in its original form or slightly paraphrased, without proper citation. Ubiquitous word processing and the Internet have made plagiarism easy and consequently commonplace. However, the ease of Internet searches has also simplified the detection of plagiarism. Several professional plagiarism detection packages are now available and are commonly used by scientific journals.

It is generally recognized that there is more plagiarism in university student essays than there used to be, and some famous plagiarizing politicians were mentioned in the introduction, but the most prolific plagiarist of all time in biomedical research was probably E. A. K. Alsabti.

BOX 2.5 Tom Lehrer on Plagiarism

A verse of "Lobachevsky," a song by Tom Lehrer (1928–):

I can never forget the day I first meet the great Lobachevsky.
In one word he told me the secret of success in mathematics:
Plagiarize!
Plagiarize,
Let no one else's work evade your eyes,
Remember why the good Lord made your eyes,
So don't shade your eyes,
But plagiarize, plagiarize, plagiarize—
Only be sure always to call it please "research."

Alsabti worked in the United States, the United Kingdom, and South Africa in the late 1970s. The complete extent of his plagiarism is not known, but he is suspected of plagiarizing at least 60 full articles.[23]

There is currently significant controversy regarding "self-plagiarism," i.e., using text, for instance a methods section, from one's own publications in a subsequent publication. Some claim that this is a form of scientific misconduct, although not strictly a form of plagiarism, whereas others claim that it is acceptable as long as the author references the previous publication. The main issue of contention is whether each new publication has to be the result of a new effort, or whether it is acceptable to recirculate one's previous effort, as long as there is no duplicate publication of results (see below on duplicate publication).

Another form of plagiarism is the theft of ideas. For instance, there have been incidents where journal referees have seen an interesting idea in a paper they are reviewing and have set their laboratory to work on the idea, while holding up the review process so that they can submit a paper based on the idea before the paper they are reviewing is published.

One of the most prominent cases concerning scientific priority (i.e., who made a particular discovery first) and potential plagiarism was the dispute between Luc Montagnier and Robert Gallo over who first isolated and identified the virus now known as HIV. Montagnier was the first to publish and submit the patent applications on tests for the virus, but Gallo nonetheless claimed scientific priority and managed to obtain the valuable US patent. Both researchers denied the other's claim, and the dispute was only resolved when the then presidents of the United States and France intervened to broker an agreement whereby the two researchers agreed to be named as codiscoverers of the virus and to share the patents. The case is well documented, and there is little doubt that Gallo used a virus sample that he obtained from Montagnier and essentially reisolated the same virus; see the papers by Prusiner, Montagnier, and Gallo in a thematic issue of *Science*.[24,25] It is also worth noting that when the Nobel Prize was awarded for the identification of HIV it went to Montagnier, not Gallo.

2.5.3 Fabrication

Fabrication of data and results is arguably the most blatant form of misconduct affecting truth claims of scientific findings. It ranges from the invention of all data and results reported to the invention of some of it (i.e., partial fabrication, for instance because of recruitment problems and time constraints). Fabrication is probably not common, but it is not exceptionally rare either, especially partial fabrication. Each year the ORI records several cases of fabrication.

The recent high-profile case of Jon Sudbø, touched on previously in this chapter, illustrates the perils of fabrication and falsification. Early in 2006, it was discovered that Sudbø, a young, prominent researcher from Norway in the field of oral premalignant lesions and carcinomas, had fabricated all the data in an article he coauthored that was published in October 2005 in the *Lancet*. Later investigations found evidence of fabrication and falsification in this and a number of Sudbø's previously published papers, including some published in the *New England Journal of Medicine*. In an article in the Norwegian newspaper *Aftenposten*,[26] *Lancet* editor Richard Horton called the case "the worst the research world has seen," but that may be an overstatement as the *Lancet* was one of the journals that had been deceived and published Sudbø's fraudulent research. All of the papers were retracted, but because of the original and high-profile nature of Sudbø's "research" it is likely that some of his "findings" have led to suboptimal treatment of patients.

Data fabrication is so common in clinical trials that statistical methods have been developed to detect it.[27] Some of these methods rely on the same idea Fisher used in his analysis of Mendel's data, since less variation is often found in fabricated data compared to real data.

Other types of misconduct in this category are the suppression of data or unwanted results, or intentionally biased analysis of the data to obtain "desired" results, both of which lead to misleading information. The same is true whenever researchers publish false or misleading accounts of their methodology in order to slow down competing research groups. Even if the research findings themselves are not affected, the deliberate introduction of falsehoods into the scientific record is tantamount to falsification.

2.6 AUTHORSHIP

Disputes about authorship are probably the most common type of conflict within research groups. Bylines on papers have several functions in the scientific community. They designate who was involved in a published work, and accordingly, who should share the honor related to the findings reported. Because of this significance, being an author plays a prominent role in employment, promotion, and the decisions of funding bodies.

Authorship thus functions as a sort of currency that can be cashed in later in a researcher's career. Hence, the frequency of disputes about authorship is understandable.

The International Committee of Medical Journal Editors (ICMJE) promulgates authorship criteria aimed at minimizing the number of disputes and ensuring that problematic forms of authorship can be clearly identified (available at http://www.icmje.org/recommendations/browse/roles-and-responsibilities/defining-the-role-of-authors-and-contributors.html). Note that these criteria concern primarily the biomedical sector; other sectors may have different rules. The 2014 version of the ICMJE authorship criteria are summarized in Box 2.6.

It is generally recognized that there are six main types of misconduct related to authorship (Box 2.7).

BOX 2.6 ICMJE Authorship Criteria

The ICMJE recommends that to be named as an author, one should meet all of the following criteria:

- Substantial contributions to the conception or design of the work; or the acquisition, analysis, or interpretation of data for the work;
- Drafting the work or revising it critically for important intellectual content;
- Final approval of the version to be published; and
- Agreement to be accountable for all aspects of the work in ensuring that questions related to the accuracy or integrity of any part of the work are appropriately investigated and resolved.

Examples of activities that alone (without other contributions) do not qualify a contributor for authorship are: acquisition of funding; general supervision of a research group or general administrative support; and writing assistance, technical editing, language editing, and proofreading.

BOX 2.7 Misconduct in Authorship

- Exclusion from authorship
- Gift authorship
- Authorship achieved by coercion
- Unsolicited authorship
- Ghost authorship
- Refusal to accept responsibility as an author when other misconduct is detected

2.6.1 Exclusion from Authorship

Exclusion from authorship happens when someone who has contributed significantly to a project and fulfills the criteria for authorship is not named in the author list, although he or she so wishes (there is no requirement stating that one must list an author who meets the criteria but does not wish to be listed). This happens most often to junior researchers, but can also happen when a research group has split before publication. Unjustified exclusion from authorship is tantamount to theft.

2.6.2 Gift Authorship

Gift authorship is defined as offering authorship to someone who has not fulfilled the criteria for authorship. There are different scenarios in which gift authorship might occur: it may be a swap, "I'll give you author status on my paper if you give me author status on yours"; it may be to add a famous name to a paper that might help it through the peer-review process; it may be a way to "improve" the CV of junior researchers in a laboratory; or it may be a way for a pharmaceutical company to get a prestigious name on a review that was essentially written by the company. Gift authorship is problematic partly because it adds authors who can take no true responsibility for the claims made in the paper, and partly because it gives those who receive the gift an unjustified advantage in the job market and other areas.

2.6.3 Authorship by Coercion

Authorship achieved by coercion commonly occurs when a senior researcher, often the head of a laboratory, demands to be an author on all publications from the laboratory, regardless of whether or not he or she fulfills the criteria for authorship.

2.6.4 Unsolicited Authorship

Unsolicited authorship is where someone is listed as an author without their knowledge or consent. An example from the anonymous case records of COPE is given in Box 2.8.

2.6.5 Ghost Authorship

Unsolicited authorship almost always involves ghost authorship; that is, the person who really wrote the paper is not listed as an author.

BOX 2.8 Paper Submitted by a PR Company Without the Knowledge of the Authors

A paper was submitted for which there were seven contributors, but no corresponding author. The only identification of who had sent the paper was an accompanying e-mail from a public relations company.

When contacted by the editorial office, the PR company confirmed that the paper was to be considered for possible publication. The named contributors were then contacted and asked whether they had given permission for their name to be attached to the paper, asked who was the corresponding author, and also if they wished to declare any conflict of interest.

This produced a very interesting flurry. One author said the paper had been produced as a result of a seminar to which he and the other contributing authors had been invited. He himself believed that he was simply giving advice to the drug company concerned, for which he had received a fee. He believed that a misunderstanding had led the PR company to send the paper for review, but had no knowledge that they had done so, and suggested that the paper be shredded.

Another author telephoned to say he could remember very little about it and certainly hadn't seen the final document. A third author telephoned in some distress, anxious that he might be accused of some form of misconduct and had never thought that his involvement would lead to a paper being submitted to a journal.

The most interesting letter of all was from the first-named author, who had subsequently written an editorial for the journal that was fairly critical of the drug concerned. The PR company who was acting for the drug company, she said, had submitted the paper on her behalf without her knowledge.

...

The same company had previously published another article to which they had put her name, but which she had not written. This author feels very abused, particularly as she wrote to the PR company requesting that they did not use her name again.

(COPE case 00/06)

However, when it comes to ghost authorship it is probably more frequent that "authors" in the author list know that they did not draft the paper, but accept authorship nonetheless. Ghost authorship has received particular attention in relation to papers that have been written by employees of pharmaceutical companies, but that name only *bona fide* researchers as authors. This hides the true nature of the paper from the reader.

2.6.6 Refusal to Accept Responsibility

In accepting that one's name appears in the author list of a paper, one also accepts responsibility for at least a part of its content. Yet, in many cases where fabrication or some other form of serious misconduct has been revealed in a jointly authored paper, people who were happy to be listed as coauthors suddenly renounce responsibility for the paper. This is either in itself a form of misconduct, or it points to earlier misconduct in accepting authorship without due care, such as not carefully revising a manuscript for publication. It might be claimed that many coauthors are not intending misconduct, e.g., they do not personally intend to fabricate data and results, or to mislead, and that they are therefore not culpable. It is probably often true that coauthors do not intend misconduct, but that does not automatically free them from all culpability; negligence also generates culpability.

2.7 SALAMI, IMALAS, AND DUPLICATE PUBLICATION

A final type of misconduct worth noting is the phenomena of salami and imalas publication. Both terms were coined by Professor Povl Riis, the first chairman of the Danish National Research Ethics Committee, and were both taken from salami, the highly salted Italian sausage that is usually served in thin slices, imalas being salami spelled backwards.[28] Salami and imalas publications seek to maximize the number of papers published from a given research project by reducing each to the "least publishable unit"; carving up the results of a project into the thinnest possible slices. Imalas publication is the sequential publishing of what are essentially the same results, but with a few new data included in the analysis each time (e.g., publishing results of a study with a planned recruitment of 100 people, after data on the first 30 and 60 people have been generated).

Salami publication constitutes misconduct, because it makes it more difficult for the users of the research results to gain an overview of the complete project. An especially problematic type of salami publication is where a large trial is published at the same time as parts of the trial are published (for instance, reporting on the patients recruited in one of the participating institutions or countries). If the link between the complete trial and the part is not made clear in the publications, it can lead to double counting of the evidence in later reviews or meta-analyses.

Imalas publication leads to the literature being cluttered with interim results, which again makes it more difficult to gain an overview of the definitive results of a project. The limiting case of imalas publication is duplicate or multiple publication of the same research results as new results. In addition to the general effects of imalas publication, it involves direct deception of the second journal, as most journals prohibit duplicate publication. Duplicate publication is generally only acceptable if the first publication is in an international journal and the second publication is in a national, native language journal, and the relationship between the two papers is made clear.

2.8 THE INVESTIGATION, PREVENTION, AND PUNISHMENT OF SCIENTIFIC MISCONDUCT

Scientific misconduct undermines the internal value system of science and tarnishes the external validity of truth claims in science, so it should be handled like other forms of crime. This means that allegations of scientific misconduct should be investigated and that those found guilty of it should be punished. But who should investigate and what punishment should be meted out?

Traditionally, scientific misconduct has been investigated by the institution employing the researcher against whom allegations are raised. But this approach is problematic, because the institution is not a neutral party. It has interests in maintaining its reputation and its relationship with funding bodies. Previously such interests often led to the suppression of allegations of scientific misconduct and the persecution of whistle-blowers (i.e., those who bring a case of possible misconduct to the attention of the relevant authorities or the public); more recently they have sometimes led to overreactions and severe sanctions against researchers accused of misconduct before a case has been fully and impartially investigated. Consequently, several countries have set up noninstitutional systems for investigating allegations of scientific misconduct.

One example is the Danish Committees on Scientific Dishonesty (DCSD). The DCSD was established by law as part of the Danish Research Agency and consists of three separate committees for health sciences, natural and technical sciences, and social and human sciences, respectively. An allegation of misconduct can be made to these committees against researchers working at Danish public institutions, including universities or organizations supported by public funds. First, an initial investigation is made; if the relevant committee thinks that there might be a case of scientific misconduct an *ad hoc* investigative committee is established, consisting of members of the DCSD and experts in the relevant area of research. At the end of the investigation, a determination is made as to whether there has been scientific misconduct, and if so its nature. This determination is sent to the researcher's employer(s) and to the journals in which the research has been published. All investigated cases are reported anonymously in annual reports. The DCSD also deals with researchers who have been publicly accused of misconduct and seek to have their names cleared. The DCSD has no formal punishments at its disposal. The ORI in the United States, however, has a range of sanctions that it can impose on researchers or institutions when misconduct has been proven. For example, it can bar them from applying for or receiving federal research grants. Most commonly this bar lasts for 1–3 years, but there have been lifetime debarments, and it can also require that the employer of the guilty researcher implement strict supervision, even of senior researchers.

In Norway, the system for the investigation of allegations of misconduct relies on a combination of institutional and national bodies. Each university has its own institutional body, and the national body, The National Commission for the Investigation of Research Misconduct, was established by law and is charged with handling more serious cases of misconduct, or cases that cannot be decided at the institutional level.[29] Some Norwegian universities also have a research ombudsman who can investigate and mediate between parties, for instance in cases involving authorship disputes between students and supervisors.

Journals are sometimes involved in the investigation of allegations of scientific misconduct, especially when the journal has the appropriate resources and no other organization is willing to undertake the investigation. The range of sanctions available to journals is obviously more limited than those available to a national body, but they include official retraction of papers, which will then also be listed as retracted in databases such as Medline, refusal to publish more papers by the same author, and publication of the finding of misconduct. However, investigation of scientific misconduct is often time-consuming, controversial, and complicated, and many journals do not have the resources to do it (or to do it well). There are many guidelines and codes of conduct for researchers, supervisors, editors, and research institutions emphasizing and promoting the internal norms described above. Nevertheless, it is important for any academic institution to build a culture of integrity. Research is a multifactorial system, with an interplay between formal and informal building blocks that ensures academic integrity.[20,30] Thus, knowledge of formal guidelines and regulations, together with research training and professional supervision, must be available to the researchers. This multifactorial system of research must also be taught to new generations of researchers. Supervisors and senior researchers play a crucial role in this education, a role that needs to be acknowledged (Box 2.9).

In line with the viewpoints put forth in Box 2.9, supervisors and senior researchers are the cornerstones of a sound and sustainable academic environment. All academic institutions should have a training program for supervisors and PhD students. At Karolinska Institutet in Sweden, a course has been established for supervisors to provide them insight into the responsibilities their role entails. An outline of the contents of this course is given in Box 2.10.

There are many strategies designed to prevent scientific dishonesty. To avoid plagiarism, different countries use clearly stated policies, sanctions, increasing knowledge about plagiarism, training, software (control), and research on plagiarism, though each country prioritizes these approaches differently.[31]

BOX 2.9 Safeguarding Against Scientific Misconduct

But most important of all, as the first scientific studies of the factors behind good conduct confirm, is the example set by senior researchers themselves. It is here in the laboratory—not in the law courts or the offices of a university administrator—that the trajectory of research conduct for the twenty-first century is being set.

(Nature *editorial*[30])

BOX 2.10 Contents of a Supervisor's Course at Karolinska Institutet

- Insight into the role of a supervisor
- Knowledge of institutional and legal requirements and procedures for supervisors and graduate students
- Responsibility and influence
- Knowledge of the ethics of research and supervisory practice
- Intercultural obstacles and possibilities

But what happens to researchers after scientific misconduct is uncovered? One study showed that traceable junior scientists who committed misconduct between 1993 and 2007 were less likely to continue to publish than other scientists. Only 11% of these junior scientists were publishing more than one paper per year after being found guilty of scientific misconduct.[32]

QUESTIONS TO DISCUSS

1. Paul collaborates with Paula on a research project. They belong to separate research units. They have planned two publications together. Paul contributes to the first publication, on which Paula is going to be the first author. Paula's supervisor is also a coauthor of this paper. Paula contributes to the second publication, on which Paul is going to be the first author. During the work Paula's supervisor says he also wants to be a coauthor on the second paper, and that he wants to have one of his external research partners as an additional coauthor. Paul does not know Paula's supervisor's research partner, who has not made any contributions so far. He does not like the situation. What is the problem? What can Paul do? What should Paul do? Why?
2. Give examples of the six authorship issues mentioned in Box 2.7. How can they be avoided?
3. What distinguishes imalas publications from salami publications?
4. Why is scientific misconduct wrong? List at least five reasons and explain the ethical framework on which each of them is based.
5. Why do researchers cheat? How can we avoid scientific dishonesty? List at least five measures to prevent scientific dishonesty.
6. The fabrication of data and results is considered to be more serious than performing repeated analysis to obtain a statistically significant result. Why is this so? Is the latter an ethical or a scientific problem? Why?
7. Give an example where the presentation of results from the statistical analysis might deceive the reader.
8. Mention some of the core internal norms in science, and explain how they are sanctioned.
9. What is the role of supervisors and senior researchers in preventing scientific misconduct?

APPENDIX 1 ICMJE RECOMMENDATIONS ON THE ROLE OF AUTHORS AND CONTRIBUTORS

Why Authorship Matters

Authorship confers credit and has important academic, social, and financial implications. Authorship also implies responsibility and accountability for published work. The following recommendations are intended to ensure that contributors who have made substantive intellectual contributions to a paper are given credit as authors, but also that contributors credited as authors understand their role in taking responsibility and being accountable for what is published.

Because authorship does not communicate what contributions qualified an individual to be an author, some journals now request and publish information about the contributions of each person named as having participated in a submitted study, at least for original research. Editors are strongly encouraged to develop and implement a contributorship policy, as well as a policy that identifies who is responsible for the integrity of the work as a whole. Such policies remove much of the ambiguity surrounding contributions but leave unresolved the question of the quantity and quality of contribution that qualify an individual for authorship. The ICMJE has thus developed criteria for authorship that can be used by all journals, including those that distinguish authors from other contributors.

Who Is an Author?

The ICMJE recommends that authorship be based on the following four criteria:

- Substantial contributions to the conception or design of the work; or the acquisition, analysis, or interpretation of data for the work;
- Drafting the work or revising it critically for important intellectual content;
- Final approval of the version to be published; and
- Agreement to be accountable for all aspects of the work in ensuring that questions related to the accuracy or integrity of any part of the work are appropriately investigated and resolved.

In addition to being accountable for the parts of the work he or she has done, an author should be able to identify which coauthors are responsible for specific other parts of the work. In addition, authors should have confidence in the integrity of the contributions of their coauthors.

All those designated as authors should meet all four criteria for authorship, and all who meet the four criteria should be identified as authors. Those who do not meet all four criteria should be acknowledged—see Section on Nonauthor Contributors below. These authorship criteria are intended to reserve the status of authorship for those who deserve credit and can take responsibility for the work. The criteria are not intended for use as a means to disqualify colleagues from authorship who otherwise meet authorship criteria by denying them the opportunity to meet criteria 2 or 3. Therefore, all individuals who meet the first criterion should have the opportunity to participate in the review, drafting, and final approval of the manuscript.

The individuals who conduct the work are responsible for identifying who meets these criteria and ideally should do so when planning the work, making modifications as appropriate as the work progresses. It is the collective responsibility of the authors, not the journal to which the work is submitted, to determine that all people named as authors meet all four criteria; it is not the role of journal editors to determine who qualifies or does not qualify for authorship or to arbitrate authorship conflicts. If agreement cannot be reached about who qualifies for authorship, the institution(s) where the work was performed, not the journal editor, should be asked to investigate. If authors request removal or addition of an author after manuscript submission or publication, journal editors should seek an explanation and signed statement of agreement for the requested change from all listed authors and from the author to be removed or added.

The corresponding author is the one individual who takes primary responsibility for communication with the journal during the manuscript submission, peer review, and publication process, and typically ensures that all the journal's administrative requirements, such as providing details of authorship, ethics committee approval, clinical trial registration documentation, and gathering conflict-of-interest forms and statements, are properly completed, although these duties may be delegated to one or more coauthors. The corresponding author should be available throughout the submission and peer review process to respond to editorial queries in a timely way, and should be available after publication to respond to critiques of the work and cooperate with any requests from the journal for data or additional information should questions about the paper arise after publication. Although the corresponding author has primary responsibility for correspondence with the journal, the ICMJE recommends that editors send copies of all correspondence to all listed authors.

When a large multi-author group has conducted the work, the group ideally should decide who will be an author before the work is started and confirm who is an author before submitting the manuscript for publication. All members of the group named as authors should meet all four criteria for authorship, including approval of the final manuscript, and they should be able to take public responsibility for the work and should have full confidence in the accuracy and integrity of the work of other group authors. They will also be expected as individuals to complete conflict-of-interest disclosure forms.

Some large multi-author groups designate authorship by a group name, with or without the names of individuals. When submitting a manuscript authored by a group, the corresponding author should specify the group name if one exists and clearly identify the group members who can take credit and responsibility for the work as authors. The byline of the article identifies who is directly responsible for the manuscript, and Medline lists as authors whichever names appear on the byline. If the byline includes a group name, Medline will list the names of individual group members who are authors or who are collaborators, sometimes called nonauthor contributors, if there is a note associated with the byline clearly stating that the individual names are elsewhere in the paper and whether those names are authors or collaborators.

Nonauthor Contributors

Contributors who meet fewer than all four of the above criteria for authorship should not be listed as authors, but they should be acknowledged. Examples of activities that alone (without other contributions) do not qualify a contributor for authorship are acquisition of funding; general supervision of a research group or general administrative support; and writing assistance, technical editing, language editing, and proofreading. Those whose contributions do not justify authorship may be acknowledged individually or together as a group under a single heading (e.g., "Clinical Investigators" or "Participating Investigators"), and their contributions should be specified (e.g., "served as scientific advisors," "critically reviewed the study proposal," "collected data," "provided and cared for study patients," "participated in writing or technical editing of the manuscript").

Because acknowledgment may imply endorsement by acknowledged individuals of a study's data and conclusions, editors are advised to require that the corresponding author obtain written permission to be acknowledged from all acknowledged individuals.

REFERENCES

1. <http://www.theguardian.com/world/2011/mar/01/german-defence-minister-resigns-plagiarism>.
2. <http://www.bbc.co.uk/news/world-europe-21395102>.
3. <http://www.nature.com/news/2009/090930/full/461578a.html>.
4. Goldacre B. *Bad Pharma: how drug companies mislead doctors and harm patients*. London: Fourth Estate; 2012.
5. Noorden RV. Science publishing: the trouble with retractions. *Nature* 2000;**478**:26–8.
6. Steen RG, Casadevall A, Fang FC. Why has the number of scientific retractions increased? *PLoS One* 2013;**8**(7):e68397.
7. COPE Commission on Publication Ethics. *Annual Reports 2005*. <http://publicationethics.org/about/annualreports>.
8. Office of Research Integrity. *Newsletter* 2006;**14**:4.
9. Broome ME, Pryor E, Habermann B, Pulley L, Kincaid H. The scientific misconduct questionnaire—revised (SMQ-R): validation and psychometric testing. *Account Res* 2005;**12**:263–80.
10. Evans SJW. How common is it? *J R Coll Physicians Edinb* 2000;**30**(1):9–12.
11. Fanelli D. How many scientists fabricate and falsify research? A systematic review and meta-analysis of survey data. *PLoS One* 2009;**4**(5):e5738.
12. Hofmann B, Myhr AI, Holm S. Scientific dishonesty—a nationwide survey of doctoral students in Norway. *BMC Med Ethics* 2013;**14**(3). <http://www.biomedcentral.com/1472-6939/14/3>.
13. The Wall Street Journal. *Mistakes in scientific studies surge*. August 10, 2011.
14. Steen RG. Retractions in the scientific literature: do authors deliberately commit research fraud? *J Med Ethics* 2011;**37**:113–17.
15. Bohannon J. Who's afraid of peer review? *Science* 2013;**342**:60–5.
16. Swedish Research Council. *Good research practice—what is it?* Stockholm: Swedish Research Council; 2006 [Report 1].
17. Tranøy KE. Science and ethics. Some of the main principles and problems. In: Jones AKI, editor. *The moral import of science. Essays on normative theory, scientific activity and Wittgenstein*. Bergen: Sigma; 1988. p. 111–36.
18. Tranøy KE. Ethical problems of scientific research: an action-theoretic approach. *Monist* 1996;**79**:183–96.
19. Merton R. *Science and democratic social structure. Social theory and social structure (enlarged edition)*. New York, NY: The Free Press; 1968. [The article was first published in 1942].
20. Nylenna M, Simonsen S. Scientific misconduct: a new approach to prevention. *Lancet* 2006;**367**:1882–4.
21. Andersen B. *Methodological errors in medical research*. Boston, MA: Blackwell Scientific Publications; 1990.
22. Begley CG, Ellis LM. Raise standards for preclinical cancer research. *Nature* 2012;**483**:531–3.
23. Lock S. Research misconduct: a résumé of recent events. In: Lock S, Wells F, Farthing M, editors. *Fraud and misconduct in medical research*. 3rd ed. London: BMJ Publishing Group; 2001.
24. Cohen J. HHS: Gallo guilty of misconduct. *Science* 1993;**259**:168–70.
25. Science. AIDS and HIV: historical essays. *Science* 2002;**298**:1726–31.
26. Aftenposten. Research scam makes waves. *Aftenposten* 2006. 17 January.
27. Pogue J, Sackett DL. Clinician-trialist rounds: 19. Faux pas or fraud? Identifying centers that have fabricated their data in your multi-center trial. *Clin Trials* 2014;**11**:128–30.

28. Riis P. Authorship and scientific dishonesty. *Committee on scientific dishonesty, annual report 1994*. Copenhagen; 1995. p. 33–39.
29. Act of 30 June 2006 No. 56 on ethics and integrity in research. <https://www.etikkom.no/en/In-English/Act-on-ethics-and-integrity-in-research/>.
30. Editorial. *Nature* 2007;**445**:229.
31. IPPHEAE Project Consortium. *Impact of policies for plagiarism in higher education across Europe. Comparison of policies for academic integrity in higher education across the European Union, Coventry;* 2013. <http://ippheae.eu/images/results/2013_12_pdf/D2-3-00%20EU%20IPPHEAE%20CU%20Survey%20EU-wide%20report.pdf>.
32. Redman BK, Merz JF. Effects of findings of scientific misconduct on postdoctoral trainees. *AJOB Primary Res* 2013;**4**:64–7.

FURTHER READING

Judson HF. *The great betrayal—fraud in science*. Orlando, FL: Harcourt, Inc; 2004.

Lock S, Wells F, Farthing M, editors. *Fraud and misconduct in medical research*. 3rd ed. London: BMJ Publishing Group; 2001.

Macrina FL. *Scientific integrity*. 4th ed. Washington, DC: ASM Press; 2014.

CHAPTER 3

Ethics in Human and Animal Studies

Søren Holm[1] and Bjorn Reino Olsen[2]
[1]Centre for Social Ethics and Policy, School of Law, The University of Manchester, England; Centre for Medical Ethics, Institute of Health and Society, University of Oslo, Blindern, Oslo, Norway and Department of Health Science and Technology, Aalborg University, Denmark
[2]Harvard School of Dental Medicine, Harvard Medical School, Boston, MA,USA

3.1 BASIC PRINCIPLES OF HUMAN BIOMEDICAL RESEARCH ETHICS

The basic principles of human biomedical research ethics were developed initially for clinical research involving patients, and these principles have influenced the way in which, for instance, the requirements of informed consent are specified. The typical clinical research project is characterized by the features listed in Box 3.1 (please note that this is not a claim that all research projects share these features, just that the typical research project exhibits them).

In a typical clinical research project there is an asymmetry in power and knowledge and the participants are in a certain sense "used" so that others (i.e., the researchers and future patients) may benefit. Using people in this way is problematic unless they understand that they are being so used and have agreed to participate (otherwise they are truly research "subjects"). This follows from the more general ethical principles of respect for self-determination and, where a project involves intervention, respect for bodily integrity. There may be circumstances in which conscription for the common good is accepted, but biomedical research is not usually seen as one of those circumstances. (For arguments that it should be, or that there is at least a moral obligation to be a research participant, see Evans[2] and Harris.[3] For an argument against conscription see Holm.[4])

This analysis of the research context yields three main principles of human biomedical research ethics, which are listed in Box 3.2.

The second of these principles is operationalized as a requirement for voluntary, informed consent, which can be parsed into the four discrete elements listed in Box 3.3.

Research in Medical and Biological Sciences
DOI: http://dx.doi.org/10.1016/B978-0-12-799943-2.00003-3

BOX 3.1 Characterization of a Typical Clinical Research Project

- The primary aim of the research is not the benefit of the actual research participants; it is to gain knowledge that will be of benefit in the future.
- It is possible the participants will be harmed by participation.
- The researchers are socially powerful, or at least more powerful than the average participant.
- The researchers are those who know the most about the project.
- The researchers have a personal interest in the success of the project.
- Participants may depend on the researchers for their continued clinical care in cases where the researchers are also the participants' physicians.

BOX 3.2 The Three Main Principles of Human Biomedical Research Ethics

- The potential harm to research participants should be minimized.
- Participation should be voluntary and based on an adequate understanding of the project.
- Participants should have an absolute right to withdraw from the study.

BOX 3.3 The Elements of Voluntary, Informed Consent

Information elements:

- full information given
- full information understood

Consent elements:

- consent ability (legal and actual)
- voluntarism (absence of coercion)

Exactly what constitutes full information is controversial, but as discussed in Section 3.2, it is specified in some detail in international regulatory documents. Because of the central role of informed consent in human biomedical research ethics, all circumstances in which participants cannot give informed consent become "problem cases" even if they are not ethically problematic in any other way.

Like other societal activities, biomedical research is subject to general considerations concerning social justice that impinge on how research participants should be selected and recruited.

3.2 INTERNATIONAL REGULATION

There are two significant and influential examples of international regulatory documents that substantiate the commonalities mentioned above and underline the aforementioned basic principles of human biomedical research ethics. These documents are the Helsinki Declaration of the World Medical Association (WMA) and the Oviedo Convention of the Council of Europe and its additional protocol on biomedical research.[5,6]

Although the Helsinki Declaration is technically binding only upon researchers who are members of a medical association that is a member of the WMA, it has nevertheless attained the status of an authoritative human rights document because it was one of the first of its kind (the first version was issued in 1965), and because it has had significant influence on how human biomedical research is regulated in most countries.

The Oviedo Convention is technically binding upon the European states that have signed and ratified it, but it is also used by the European Court of Human Rights to aid in the interpretation of specific human rights cases, even against European states that have not ratified the Convention.[7]

3.2.1 Consent

Both the Helsinki Declaration and the Oviedo Convention require that significant amounts of information be given to prospective participants and furthermore require that this information be understood by the participant. Hence, obtaining informed consent for participation in research is not a simple matter; it is best understood as a process that involves presenting both written and oral information, engaging the person in reflection concerning the project, and allowing the person sufficient time to make up his or her mind as to whether to participate. Only if this process is completed successfully is the signature on the informed consent form a valid documentation of consent. When designing the recruitment procedures for a specific project, it is necessary to consider how much time this process will take and allow adequate time for it. Unless there is some medical necessity (i.e., experimental treatment has to start now and not tomorrow), potential participants should be given ample time to make their decision; they should not be coerced into deciding on the spot (Table 3.1).

It is well known that informational materials often are written in complicated language and are thus difficult to understand. In order for

Table 3.1 Requirements for informed consent in the Helsinki Declaration and the Oviedo Convention

Helsinki Declaration—2013 version (World Medical Assembly)	**Oviedo Convention—additional protocol on biomedical research (Council of Europe)**
26. In medical research involving human subjects capable of giving informed consent, each potential subject must be adequately informed of the aims, methods, sources of funding, any possible conflicts of interest, institutional affiliations of the researcher, the anticipated benefits and potential risks of the study and the discomfort it may entail, post-study provisions and any other relevant aspects of the study. The potential subject must be informed of the right to refuse to participate in the study or to withdraw consent to participate at any time without reprisal. Special attention should be given to the specific information needs of individual potential subjects as well as to the methods used to deliver the information. After ensuring that the potential subject has understood the information, the physician or another appropriately qualified individual must then seek the potential subject's freely given informed consent, preferably in writing. If the consent cannot be expressed in writing, the non-written consent must be formally documented and witnessed. All medical research subjects should be given the option of being informed about the general outcome and results of the study.	Article 13—Information for research participants 1. The persons being asked to participate in a research project shall be given adequate information in a comprehensible form. This information shall be documented. 2. The information shall cover the purpose, the overall plan and the possible risks and benefits of the research project, and include the opinion of the ethics committee. Before being asked to consent to participate in a research project, the persons concerned shall be specifically informed, according to the nature and purpose of the research: i. of the nature, extent and duration of the procedures involved, in particular, details of any burden imposed by the research project; ii. of available preventive, diagnostic and therapeutic procedures; iii. of the arrangements for responding to adverse events or the concerns of research participants; iv. of arrangements to ensure respect for private life and ensure the confidentiality of personal data; v. of arrangements for access to information relevant to the participant arising from the research and to its overall results;

(Continued)

Table 3.1 (Continued)

Helsinki Declaration—2013 version (World Medical Assembly)	Oviedo Convention—additional protocol on biomedical research (Council of Europe)
	vi. of the arrangements for fair compensation in the case of damage; vii. of any foreseen potential further uses, including commercial uses, of the research results, data or biological materials; viii. of the source of funding of the research project. 3. In addition, the persons being asked to participate in a research project shall be informed of the rights and safeguards prescribed by law for their protection, and specifically of their right to refuse consent or to withdraw consent at any time without being subject to any form of discrimination, in particular regarding the right to medical care.

the average person to understand informed consent documents, they should be no more difficult to read than the sports pages of a popular newspaper. Only when participants are exclusively recruited among academics should academic-style informational materials be used.

3.2.2 Inability to Consent

There are many situations in which prospective research participants are legally or factually unable to consent, but here only three of these will be considered: research on children, research on permanently incapacitated adults, and research in an emergency setting where the disease process has rendered a person temporarily unable to consent (e.g., research on stroke or on cardiopulmonary resuscitation).

Research involving children is important, since children are not just small adults and information from studies in adults cannot always be safely extrapolated to children, for example, studies on pharmacokinetics and pharmacodynamics. The "gray baby" syndrome, caused by the use of chloramphenicol in neonates, and Reye's syndrome, with a plausible link to acetylsalicylic acid, are cases in point. And of course there are also many conditions that only affect children and therefore cannot be researched in adults. The research ethics system thus has to balance the protection of children with the need for research.

With regard to children who are not legally able to consent, the general rule is that their parents or guardians can give consent for them to participate in research if it is not against the interests of the child. But then no researcher should ask for a child's participation if it is not in the interest of the child. Because children are seen as a vulnerable and exploitable group, both the Helsinki Declaration and the Oviedo Convention have restrictions on the type of research in which children can participate. In the Oviedo Convention additional protocol, for instance, there are the following restrictions in relation to people who are unable to consent:

Article 15—Protection of persons not able to consent to research

1. *Research on a person without the capacity to consent to research may be undertaken only if all the following specific conditions are met:*
   i. *the results of the research have the potential to produce real and direct benefit to his or her health;*
   ii. *research of comparable effectiveness cannot be carried out on individuals capable of giving consent;*
   iii. *the person undergoing research has been informed of his or her rights and the safeguards prescribed by law for his or her protection, unless this person is not in a state to receive the information;*
   iv. *the necessary authorisation has been given specifically and in writing by the legal representative or an authority, person or body provided for by law, and after having received the information required by Article 16, taking into account the person's previously expressed wishes or objections. An adult not able to consent shall as far as possible take part in the authorisation procedure. The opinion of a minor shall be taken into consideration as an increasingly determining factor in proportion to age and degree of maturity;*
   v. *the person concerned does not object.*

2. *Exceptionally and under the protective conditions prescribed by law, where the research has not the potential to produce results of direct benefit to the health of the person concerned, such research may be authorised subject to the conditions laid down in paragraph 1, sub-paragraphs ii, iii, iv, and v above, and to the following additional conditions:*
   i. *the research has the aim of contributing, through significant improvement in the scientific understanding of the individual's condition, disease or*

disorder, to the ultimate attainment of results capable of conferring benefit to the person concerned or to other persons in the same age category or afflicted with the same disease or disorder or having the same condition;

ii. *the research entails only minimal risk and minimal burden for the individual concerned; and any consideration of additional potential benefits of the research shall not be used to justify an increased level of risk or burden.*

3. *Objection to participation, refusal to give authorisation or the withdrawal of authorisation to participate in research shall not lead to any form of discrimination against the person concerned, in particular regarding the right to medical care.*

It has been argued that these restrictions are too narrow, since children are already moral agents and we have no reason to believe that they do not want to help other people.[8] The counterargument is that if the restrictions were less narrowly defined, children would be exploited by unscrupulous researchers.

Although children are legally unable to consent, they may well have their own opinions on whether or not they want to participate in a specific research project. Even if they are so young that one may assume they cannot form valid opinions on such a complex matter, they nevertheless need to know what is going to happen to them during the research. It is therefore generally accepted that children should be informed using age-appropriate informational materials and techniques. The assent of older children should be sought and any refusal respected, even if the parents want the child to participate.

Permanently incapacitated adults differ from children in that they often have no legally defined proxy decision maker. They fall into one of two groups: those who were previously competent and those who have never been competent. With regard to consent, both groups are essentially in the same legal position as children (this does not imply that they are actually comparable to children in any other way). They cannot consent to research themselves, but they may in some instances be able to assent or refuse. Inclusion in the research can only take place if there is a proxy who is legally authorized to give proxy consent for research purposes, and this proxy must make decisions based on an understanding of the best interests of the incapacitated person. If the person was previously competent, the proxy can take into account the previously expressed wishes and values of the person and decide in accordance with these. Some countries allow competent people to execute advance directives concerning their treatment, which become operative when the person becomes incompetent; these are usually limited to treatment, but a few countries allow advance directives for research as well.

In people who are acutely incapacitated, it is unlikely that there is a legally recognized proxy, as many legal systems only allow proxies to be appointed through a legal process after a previously competent person has become incompetent. Even when there is an automatic legal presumption, for instance that the spouse becomes the proxy, there may be no time to contact the spouse before the patient has to be entered into the research protocol. Requiring proxy consent may therefore completely block emergency research among acutely incapacitated people, or make it very, very difficult. This was the case in the United States for some years during the 1990s, when the informed consent requirement was interpreted very restrictively. Nevertheless, it is clearly important to carry out research on, for instance, the improvement of cardiopulmonary resuscitation.

It is important that research ethics do not become so protective of a specific group that they make it impossible to do research on the condition of a specific group or the corresponding treatment. Doing this would leave the group in what the Danish research ethicist and codrafter of the 1975 version of the Helsinki Declaration, Professor Povl Riis, called a "golden ghetto." This situation has been increasingly recognized in recent years, and exemptions from informed consent requirements for emergency research can now be found in both the Helsinki Declaration and the Oviedo Convention. In the Helsinki Declaration, for instance, the following exemption is written (the italics represent our emphasis):

> 30. Research involving subjects who are physically or mentally incapable of giving consent, for example, unconscious patients, may be done only if the physical or mental condition that prevents giving informed consent is a necessary characteristic of the research group. In such circumstances the physician must seek informed consent from the legally authorised representative. *If no such representative is available and if the research cannot be delayed, the study may proceed without informed consent provided that the specific reasons for involving subjects with a condition that renders them unable to give informed consent have been stated in the research protocol and the study has been approved by a research ethics committee.* Consent to remain in the research must be obtained as soon as possible from the subject or a legally authorised representative.

3.2.3 Randomized Controlled Trials

Specific problems arise in randomized controlled trials, especially with regard to the use of placebo in the control group. If there is an effective treatment for the condition under study that is not given to the participants in the control group but that they would have received if they were not in the trial, they are in fact worse off by their participation (and we would know this in advance for everyone in the control group).

A small detriment to their well-being is probably something a research participant can consent to (i.e., whether my ingrown toenail is treated this week, or in 2 weeks' time if the randomization places me in the control group), but whether it should be possible to consent to a potentially large detriment to well-being is more controversial.

The Helsinki Declaration previously seemed to prohibit the use of placebo completely if there was an existing treatment, but the 2013 version allows the use of placebo in such circumstances, with certain restrictions:

> *33. The benefits, risks, burdens and effectiveness of a new intervention must be tested against those of the best proven intervention(s), except in the following circumstances:*
>
> *Where no proven intervention exists, the use of placebo, or no intervention, is acceptable; or where for compelling and scientifically sound methodological reasons the use of any intervention less effective than the best proven one, the use of placebo, or no intervention, is necessary to determine the efficacy or safety of an intervention and the patients who receive any intervention less effective than the best proven one, placebo, or no intervention will not be subject to additional risks of serious or irreversible harm as a result of not receiving the best proven intervention.*
>
> *Extreme care must be taken to avoid abuse of this option.*

The exact interpretation of "compelling and scientifically sound methodological reasons" is, however, still a matter of controversy.

A somewhat similar issue arises in some kinds of psychiatric research, in which it is methodologically advantageous to have a long "washout period" where the patient's ordinary therapy is stopped before the patient commences the experimental treatment. This practice is ethically problematic if there is a significant risk of deterioration during the washout period.

A considerably more complicated situation arises in those cases where trials using placebo in the control group are carried out in a developing country by a firm or research institute from a developed country. In this case, there may be an effective treatment for the condition under study that is available to everyone in the developed country where the research institute is located, but that is so expensive that no one can get it in the developed country where the research is taking place (this has been the case in a number of HIV/AIDS trials). In these circumstances, would using placebo in the control group entail a breach of the Helsinki Declaration or the Oviedo Convention? (Note that the Oviedo Convention applies to this research only if the research institute is European, according to Article 29 of the additional protocol.) The answer to this question hinges on the interpretation of whether the formulation in the Oviedo Convention, "there are no methods of proven effectiveness," is interpreted on a global scale or a local one. Is this condition satisfied if the treatment method available is not used in the study setting, or is it only satisfied if there is no treatment method available anywhere?

The reason this question gets complicated and murky, and the discussion often heated, is that it intersects with other controversial questions concerning the relationship between rich and poor countries, e.g., Do rich countries exploit the poor ones? Do rich countries have an obligation to help poor countries? Is it hypocritical or expressive of double standards to carry out research in another country that you would not allow at home? Keep in mind that no one in the control group is actually worse off when developed countries carry out this kind of research in developing countries. Patients in the control group do not get effective treatment as research participants, but they would not have received it as regular patients either. It has been suggested that this makes the research acceptable either on its own, or if the research results will further benefit the community in which the research is carried out. There is, however, still no agreement on this issue.

3.2.4 Vulnerable Research Participants

There are groups of research participants that are vulnerable to exploitation, not because they are unable to give consent, but because they are institutionalized, socially powerless, or dependent on the researchers in some way. The voluntarism of the consent of such groups is an issue, and research ethics committees (RECs) will often require specific justification as to why such a group has been chosen as the research population and/or assurance that it will be made even more clear than usual that participation is purely voluntary, that it is acceptable to refuse, and that refusal will have no negative consequences.

3.2.5 Epidemiological Research, "Big Data," and Biobank Research

It is evident that the model of research covered by human biomedical research ethics is derived from typical clinical research, but large-scale epidemiological research is gaining in importance. This research often uses existing health data recorded in relation to treatment, and it is increasingly combined with the analysis of biological samples from large, structured sample collections (biobanks). These projects are often of very long duration, they may use data and biological samples collected for unrelated purposes, and the focus of the research may change during the life of the project.

This means that achieving the ideal situation of full informed consent at the beginning of the project becomes problematic. Suggestions for how this problem can be solved include "broad" or "unspecific" consent, and some form of recurrent consent. A requirement for recurrent or renewed consent for new uses of old data and biological samples clearly creates a significant burden on researchers, and such consents may be impossible to obtain if the participants are untraceable or dead. This idea has thus been strongly resisted by the research community. However, broad consent is not unproblematic. Indeed, the broader the consent, the less valid it is. Has one really consented to a given use of one's biological samples in the ethical sense of the term "consent," if one has consented to "any future use," but that use refers to one that neither the participant nor the researchers have envisaged at the time of consent? In research ethics there is great resistance to the idea that consent creates some form of contract between researcher and research participant, but considering consent as constituting a contract is probably the only way to justify a reliance on broad consent. It has also been suggested that epidemiological research on nonsensitive topics may not require consent, because there is no possibility of harming the research participants. However, this argument has not been universally accepted.

There is, however, a consistent trend towards reducing the amount and stringency of regulation of research that only uses existing data sources. In many cases it is possible to use these data sources in such a way that the individual subjects remain anonymous/nonidentifiable to the researcher, for instance when a data source containing outcome data is linked to a data source containing exposure data by a third party, and the researcher is only given access to a de-identified dataset. In some jurisdictions research on fully anonymized datasets of this kind does not require approval by an REC.

A slightly different issue occurs when the researcher works with identifiable data, and where there might be a presumption of obtaining informed consent, but where it would be very difficult to trace the original subjects. Here most countries follow the Helsinki Declaration and allow the research to proceed without consent if contacting the subjects whose data will be used is deemed "impossible or impracticable."

3.2.6 The Role of RECs

To ensure that research ethics regulations are adhered to, researchers are required to submit a research protocol including the relevant patient information and informed consent forms to an REC (known in some countries as a research ethics board or institutional review board) for approval before the project commences. The REC will typically have both professional and lay members, who will assess the scientific validity of the proposal, any ethical problems that the project contains, and the informational material and informed consent documents. Based on this information they will decide whether the project should be allowed to begin, or whether elements of it need to be modified.

REC approval is a legal requirement in many countries and it is reinforced by peer-reviewed journals, as most will only publish research that has REC approval. However, the exact structure of the REC system varies considerably across countries. In some countries, RECs also have control functions concerning the actual conduct of the research, that is, whether it is conducted in accordance with the approved protocol.

The journal requirement for REC approval can sometimes create problems, because RECs in many countries may decline to assess a project if they do not see it as research, but for instance as an audit. If the researchers still want to publish the results of their "audit," they may have to explain the situation to the journal, and provide documentation that the project was submitted to an REC, and that it was the REC that declined to review.

3.2.7 Data Protection, Good Clinical Practice, and Other Regulations Influencing Biomedical Research Ethics

Specific research ethics regulation is not the only kind of regulation that influences the ethical conduct of biomedical research and the work of RECs. Most countries have data protection laws that independently require consent for the collection and processing of person-identifiable health data, except for official statistical purposes. It may also be necessary to obtain permission from the national data protection agency to store and process person-identifiable health data, if such an agency exists. Moreover, some countries also have specific rules concerning the storage and use of human tissue, and the use of this material in research also requires explicit consent in many contexts.

The international Good Clinical Practice guidelines that apply to all research performed to support registration of drugs or medical devices in the United States, the European Union, and Japan also require explicit, informed consent from all research participants and approval by an independent REC.

The presence of all these national ethical committees, safeguards, and regulations means that even if the Helsinki Declaration and the Oviedo Convention suddenly disappeared, little would change in the basic ethical requirements for the conduct of biomedical research on competent, adult human beings. Only the more specific rules concerning specific cases would be affected.

3.3 THE ETHICS OF ANIMAL RESEARCH

The two basic problems in the ethical analysis of animal research as it applies to biomedical research on experimental animals are: (i) whether the use of animals in research is justifiable at all, and (ii) if so, the conditions under which it is justifiable. It is immediately obvious that the standard justification for human research does not apply in the case of experimental animals. Animals cannot consent to participation in research; they cannot voluntarily take upon themselves the burden of being a research participant (even in veterinary research aimed at benefiting a specific species). Some argue that the use of animals in research cannot be justified at all, since such research infringes on the animal's right to life or causes the animal suffering. The question of whether animals have a right to life, and if so who would have the obligation to protect that right, is highly controversial. It is, however, generally accepted that animals can suffer and that their well-being can be negatively affected; from a moral standpoint it is very difficult to argue that animal suffering should not matter at all. The main reason it is wrong to kick a dog is simply that it causes the dog pain, not, as Immanuel Kant seemed to believe, that it desensitizes the kicker. To the degree that animal research causes pain and other kinds of suffering, one must justify that causing this pain is necessary. But how can one qualify the good that outweighs the suffering?

The answer is that the good is the scientific knowledge and the future medical benefits that flow from animal research. Although it is sometimes denied by animal rights activists, there is no doubt that animal research has been, and continues to be, a necessary component of the long and complicated research processes that lead to medical progress. But this is not a compelling justification for the practice of animal research. A further argument is needed to show that the amount of animal suffering caused by research is justified by the beneficial outcome (medical or veterinary progress). This assessment is extremely complicated, partly because there is no good way to estimate the magnitude of the various kinds of suffering involved, including both the animal suffering caused and the human and animal suffering averted, and partly because it is difficult to estimate the exact contribution of animal research to medical benefits.

So far, this argument has been taking place on the general level of biomedical research as one comprehensive activity. This means that even if one accepts that animal research is justified in general, one does not necessarily have to accept that every project involving animals is justified. There may well be individual projects where the expected benefits do not outweigh the predicted suffering of the experimental animals.

Therefore, the official regulation of animal research has two aims. First, to ensure that the general level of animal suffering caused by animal research is minimized; this is for instance the justification for requiring training and certification of researchers and laboratory technicians, and good housing conditions for all experimental animals. The second aim is to evaluate individual projects or broader research plans to ensure that there is an acceptable balance between suffering and benefit in each case.

Over the years, a consensus has emerged that a useful tool for reaching these aims is the "Three Rs" approach (replacement, reduction, and refinement), which is based on concepts initially developed by Russell and Burch[9] (Box 3.4).

Some countries now require researchers to specify how they have taken the three Rs into account in their project planning.

Because the consideration of animal suffering is so central in the standard approach to animal biomedical research ethics, the most problematic category of research is usually experiments where the animal survives for a long time in a state of some suffering. In contrast, experiments where the animal is anesthetized, experimented upon, and then painlessly killed are deemed less problematic. Here it is important to note that a significant

BOX 3.4 The "Three Rs"

- Replacement: The use of methods that permit a given scientific purpose to be achieved without conducting experiments or other scientific procedures on living animals.
- Reduction: The use of fewer animals in each experiment without compromising scientific output and the quality of biomedical research and testing, and without compromising animal welfare.
- Refinement: Improvement of all aspects of the lifetime experience of animals to reduce suffering and improve welfare.[10]

proportion of short- and long-term toxicological studies that fall into the most problematic category are legally mandated as part of premarket testing of new chemicals and pharmaceuticals.

A question that still causes controversy is whether there should be different levels of protection for different animal species, and what the rational basis for increased protection should be. In the past, some countries had special categories for animals that are traditionally kept as pets, so one could carry out experiments in pigs that one could not carry out in dogs. Such special protection for animals that humans happen to like does not seem rationally warranted. The existing evidence suggests that pigs can suffer to the same extent as dogs, and that they are probably more intelligent than dogs. A more rational approach is to consider what kind of suffering an animal can experience based on its cognitive abilities and species-specific lifestyle. Sentience is the basic criterion for any kind of suffering, but there are other kinds of suffering besides pain. Animals that usually live in social groupings may suffer if housed in single cages; animals that have higher cognitive capacities (like humans) may suffer from boredom if housed in an unstimulating environment or may suffer because of the anticipation of pain to come. Different countries have operationalized these considerations differently in their regulations, but a typical classification of increasing protection follows.

Not protected:

- invertebrates (no specific protection because not believed to be sentient)

Increasingly protected:

- vertebrates (a few countries also protect cephalopods)
- mammals
- nonhuman primates

- great apes (many countries now prohibit or strongly discourage research on great apes)

A hierarchical list like this is clearly not sufficiently sensitive to the many differences between animal species, and it is therefore important that both researchers and laboratory technicians are knowledgeable about the specific species they are working with so that they can minimize species-specific suffering (e.g., animals that have rooting as a normal behavior should have rooting materials in their cages, and animals that live in social groups should not be housed individually) (see also Chapter 6).

3.3.1 Animal Research and Other Human Uses of Animals

In most industrialized countries the rules concerning the use of animals in biomedical research are quite stringent, as are the rules about the breeding and housing conditions of these animals. There is often a glaring inconsistency between these rules and the rules, or absence of rules, concerning other human uses of animals. Whereas all countries prohibit cruelty to animals, there is often very little enforcement of these rules. Exotic vertebrates can be kept as pets without their owners knowing anything about the requirements for their well-being. Mammals are used in dangerous sports where many of them will suffer. One mammal can be used to hunt and kill another mammal (this is not a reference to fox hunting in the United Kingdom, but the activities of domestic cats all over the world). And, perhaps most importantly in terms of numbers, farm mammals are kept in conditions that are much worse than the conditions required for experimental animals. It is worth considering whether animal rights activists should not first target these other, much more problematic uses of animals, before targeting animal research. Even vegans may want to benefit from the progress "bought" through the suffering of animals in biomedical research.

QUESTIONS TO DISCUSS

1. When does research on human tissue specimens, cells, cell lines, or data involve human subjects?
2. Are you proposing research on human subjects if your studies will only use cell lines?
3. Who should receive the required education on the protection of human subjects?
4. What does it mean to "obtain" identifiable private information or identifiable specimens?

5. When is it appropriate to consider a waiver of written informed consent?
6. Are you proposing research on human subjects if your studies will use only cadaver samples?
7. Does your research involve human subjects if you collect blood from a patient in order to create a cell line?
8. Do research studies involving privately owned animals, such as pets, need institutional approval in your country?
9. Do the animal welfare policies in your country apply to larval forms of amphibians and fish?
10. What factor determines whether activities involving privately owned animals constitute veterinary clinical research or care?

REFERENCES

1. Deleted in review.
2. Evans HM. Should patients be allowed to veto their participation in clinical research? *J Med Ethics* 2004;**30**:198–203.
3. Harris J. Scientific research is a moral duty. *J Med Ethics* 2005;**31**:242–8.
4. Holm S. Conscription to biobank research? In: Solbakk JH, Holm S, Hofmann B, editors. *The ethics of research biobanking*. New York, NY: Springer; 2009. p. 255–62.
5. WMA Declaration of Helsinki—ethical principles for medical research involving human subjects. <http://www.wma.net/en/30publications/10policies/b3/>.
6. Convention for the protection of human rights and dignity of the human being with regard to the application of biology and medicine: convention on human rights and biomedicine, ETS 164. <http://conventions.coe.int/Treaty/en/Treaties/Html/164.htm>.
7. Plomer A. *The law and ethics of medical research: international bioethics and human rights*. London: Cavendish Publishing; 2005.
8. Harris J, Holm S. Should we presume moral turpitude in our children? Small children and consent to medical research. *Theor Med* 2003;**24**:121–9.
9. Russell WMS, Burch RL. *The principles of humane experimental technique*. London: Methuen; 1959. Available at: <http://altweb.jhsph.edu/publications/humane_exp/hettoc.htm>.
10. Nuffield Council on Bioethics. *The ethics of research involving animals—a guide to the report*. London: Nuffield Council on Bioethics; 2005.

FURTHER READING

Brody BA. *The ethics of biomedical research—an international perspective*. New York, NY: Oxford University Press; 1998.

Emanuel EJ, et al. *Ethical and regulatory aspects of clinical research—readings and commentary*. Baltimore, MD: Johns Hopkins University Press; 2003.

Emanuel EJ, et al. *The Oxford textbook of clinical research ethics*. Oxford: Oxford University Press; 2010.

Monamy V. *Animal experimentation: a guide to the issues*. 2nd ed. Cambridge: Cambridge University Press; 2009.

CHAPTER 4

Research Strategies, Planning, and Analysis

Haakon Breien Benestad[1] and Petter Laake[2]
[1]Department of Physiology, Institute of Basic Medical Sciences, University of Oslo, Blindern, Oslo, Norway
[2]Oslo Centre for Biostatistics and Epidemiology, Department of Biostatistics, Institute of Basic Medical Sciences, University of Oslo, Blindern, Oslo, Norway

4.1 INTRODUCTION

In science, it is the achievements of individuals that drive progress. The most famous recognition of this fact is embodied by the Nobel Prizes, which have been awarded each year since 1901 for outstanding achievements in chemistry, physics, physiology or medicine, literature, and efforts for peace. Since ancient times the debate has ensued on just what constitutes achievement; in science, an achievement could arguably be described as a discovery or an improvement, with a discovery classified as the finding of something previously unknown, such as Roentgen's discovery of X-rays or the Curies' discovery of radium, and an improvement as the production of something better. One example of an improvement is the invention of dynamite: in 1866 Alfred Nobel mixed kieselguhr, an absorbent, in highly unstable nitroglycerine to make this stable explosive. With today's level of scientific knowledge, spectacular discoveries are few. By the same token, few improvements today have consequences as rewarding as Nobel's invention of dynamite, which underpinned the business he used to build the fortune that now funds the Nobel Prizes. Instead, today most scientists achieve by contributing to the greater body of scientific knowledge, providing support for a view or hypothesis that is generally accepted, but not proven. In concert, these contributions can be significant, as illustrated by the example of neurotransmitters, which is covered in Chapter 7 of this book. Indeed, in the early twentieth century, the hypothesis that signals propagated between nerve cells by chemical, not electrical means was accepted, though unproven; it was decades before the first neurotransmitter was identified.

Scientific developments can be classified as neither discoveries nor improvements, but they bring about progress nonetheless. Mathematical

Research in Medical and Biological Sciences
DOI: http://dx.doi.org/10.1016/B978-0-12-799943-2.00004-5

BOX 4.1 Reflections from Others on Scientific Problems

Unfortunately, solving problems is not all there is to the scientific endeavour. Even more important than solving problems is finding relevant problems—formulating questions that really matter. The world offers us an infinite number of problems to solve, of which we select some and disregard others. Much of the art of doing science is then deciding which problems to concentrate on and which to ignore.

...

... The ability to solve problems requires logical thinking, and hence a rational mind, whereas the ability to identify consequential problems is only in part based on logic; mostly it is based on instinct, intuition, subconscious perception, a sixth sense, inborn proclivity, talent, irrational impulse or whatever you might want to call it.

There is plenty of evidence from the history of science for an independent assortment of these two traits.

(Klein[1])

modeling can identify weaknesses in scientific data or help determine when there are insufficient data on which to base prevailing opinion. Compiling, improving, and testing laboratory methods can also provide significant contributions. In these endeavors, entrepreneurship is a special skill and a prerequisite for coordinating the work of large teams, be it in basic research, applied research, or other "big science," as well as for successfully appealing to funding bodies. Many personality types are needed; "it takes all sorts to make the [scientific] world" (Box 4.1). In all the aforementioned cases, it helps to have thorough basic knowledge of the relevant field, to be prepared for the unexpected, and to be able to identify analogies between dissimilar scientific fields, even if it takes some imagination, as exemplified by Pasteur's inventions. Pasteur believed that "chance favors the prepared mind," so serendipity—making fortunate discoveries by accident—counts. Indeed, Horace Walpole's entertaining story about the three princes from Serendip, now Sri Lanka, is often quoted in scientific writings.[2]

4.2 IDENTIFYING SCIENTIFIC PROBLEMS

First and foremost, a scientific problem must have significance, and its solution must have value; not simply hold fascination for the researcher.

BOX 4.2 On Untrodden Paths and Unexpected Finds

In basic research, everything is just the opposite. What you need at the outset is a high degree of uncertainty; otherwise it isn't likely to be an important problem. You start with an incomplete roster of facts, characterised by their ambiguity; often the problem consists of discovering the connections between unrelated pieces of information. You must plan experiments on the basis of probability, even bare possibility, rather than certainty. If an experiment turns out precisely as predicted, this can be very nice, but it is only a great event if at the same time it is a surprise. You can measure the quality of the work by the intensity of astonishment. The surprise can be because it did turn out as predicted (in some lines of research, 1 percent is accepted as a high yield), or it can be confoundment because the prediction was wrong and something totally unexpected turned up, changing the look of the problem and requiring a new kind of protocol. Either way, you win.

(Thomas[3])

Attaining the goal of a solution may require time and effort, so avoid trivial scientific problems and be sure to plan thoroughly in advance!

Many scientific problems have existed for a long time, but have not been tackled as the means to solve them were not available. For example, greater portions of molecular biology—genetic material structure, the principles of protein synthesis, the regulation of gene activity, etc.—had to be elucidated in order to understand how a fertilized egg develops into a complete, multicellular organism, before the problems of developmental biology could be addressed. It is often said that technology builds on basic research, but often information also flows in the other direction. A new technique, method, or apparatus can often be well suited to problem-oriented research, aiding scientists in the solution of older, mature problems in biology. Conversely, although it can spawn many publications, learning a new technique and exploiting it repetitively constitutes technically oriented, not problem-oriented research (Box 4.2).

Sir Peter Medawar, the philosopher of science who shared the 1960 Nobel Prize for Physiology or Medicine with Sir Frank MacFarlane Burnet, maintains that "science is the art of the soluble." Accordingly, researchers need to conduct definitive inquiries that either support or falsify their working hypotheses. A favored hypothesis should be looked at from many angles; the best is often to try to falsify it so one does not end up with a "so-what" paper. It must be possible to falsify the hypothesis.

A thought-provoking example of this is the research work of Barry Marshall, who questioned the then-accepted hypothesis that peptic ulcers were caused by stress or too much acid. Through clinical studies, in which he even experimented on himself, he succeeded in showing that *Helicobacter pylori* is in fact the main cause of peptic ulcers. Initially there was noticeable skepticism towards the idea, but Marshall's perseverance prevailed, and he was awarded the Nobel Prize for Physiology or Medicine in 2005.

When selecting a scientific problem, it is best to heed the principle of Ockham's razor; that is, select the simple. It is preferable to choose the simplest hypothesis or explanation that is compatible with the facts that require interpretation. Complex assumptions can have many reservations, which lead to exceptions, which require modification of the hypothesis and ultimately produce a hypothesis that is difficult to falsify. The experimental procedures need not be overly complex lest the risk of error increase. A research project should be planned in an unhurried manner, step by step, so that the corresponding potential outcomes and consequences can be easily determined.

Reproducibility is as vital as relevance, testability, and congruence. A potentially publishable finding needs to be replicated in one, or preferably two, new, independent experiments before it is published. In the same manner, skepticism is warranted when considering sensational published developments, until they have been confirmed by other research groups.

4.3 THE EXPERIMENTAL DESIGN

All of science builds on observations, and causal explanation of observed phenomena is routine in the natural sciences. Causality can be straightforward when there is only one cause, but often several causal factors act simultaneously to produce an observed effect. Causal explanations represent a natural starting point for understanding or explaining an observed phenomenon. This does not discount the importance of studies that demonstrate interesting associations or findings that cannot be defined as cause-effect relationships. Regardless of the starting point, when it comes to science, causal thought enters into the explanation of empirical associations and is essential to the hypothetical-deductive method. Understanding is enhanced through hypotheses that can be falsified by empirical studies. The goal of all empirical studies is to arrive at valid, reproducible conclusions concerning associations. This is achieved through well-planned studies, with valid inference and valid analyses.

To falsify a hypothesis, the collection of observations is crucial. Any observation is a combination of "signal" and "noise." The hypothesis is related to the signal; the goal is to detect differences in the signal from one experimental condition, characterized by the experimental intervention, to another. One example might be the observation of the effect of an experimental intervention to determine whether the signal is different when the intervention is active. Generally, the noise will confound the signal, and too much noise might make it difficult to differentiate the signal from the noise. A well-designed experiment is one where the signal stands out, or equivalently, the noise is under control. The concepts of signal and noise are formalized in statistics, and will be elaborated on in Chapter 11.

The design of the experimental or observational study dictates how observations are collected (Box 4.3).

Interventions play an important role in experimental studies, as they do in basic medical research. The experiment or intervention must employ the requisite controls. Indeed, method checks and other aspects of quality assurance included in the design of an experimental study are the best markers of a researcher's skill. The use of appropriate controls is vital; controls should differ from tests in a single aspect: the variable under investigation, be it a chemical or a training scenario. No matter the situation, controls should be congruent with tests: control patients should correspond with test patients, control animals with test animals, and control phases with test phases in trials involving humans or animals that serve as their own controls (baseline versus experimental phase). As an alternative,

BOX 4.3 Study Designs

Experimental studies:

- Basic medical research
- Intervention studies
- Randomized controlled trials (RCTs)

Observational studies:

- Ecological studies
- Cross-sectional studies and surveys
- Cohort studies, longitudinal studies
- Case-control studies

between-individual variations can be eliminated by arranging for patients, experimental animals, organ preparations, or cell cultures to be their own controls, for example by successively applying test and control conditions.

A randomized controlled trial (RCT) is an experiment carried out in a clinical setting, with the aim to study different treatments. RCTs commonly compare the effect of two or more alternative treatments, or two or more doses of the same treatment. Randomization of patients to different treatment groups is crucial to ensure that any differences observed between the groups are due solely to the treatment, not to any confounding factors. See Chapter 8 for more on RCTs.

The variable of main interest is called the outcome variable (also known as the effect variable, the response variable, or the dependent variable). The variable used to explain the outcome is called the exposure variable (also known as the causal variable, the explanatory variable, the independent variable, or the covariate). A research hypothesis contends that there is either an association or a cause-effect relationship between a given exposure variable and a given outcome variable. The goal of a research project is to conclude whether the research hypothesis is true, i.e., whether there is in fact an exposure-outcome association. Once data have been collected and statistical inference has been done, the goal is to support the research hypothesis, now called the alternative hypothesis, by rejecting the null hypothesis, which says that there is no association between the outcome variable and the exposure variable. Usually the null hypothesis is denoted H_0 and the alternative hypothesis is denoted H_1. When statistical inference is performed, a significance level, i.e., the probability below which the null hypothesis can be rejected, must be chosen. This threshold is usually set at 5%.

4.4 DATA TYPES

Observations provide data, and data must be analyzed to provide a basis for valid and reproducible conclusions. In the natural sciences, statistics such as mean, median, variation, and measures of association are the starting point for further analyses. The manner in which these statistics are computed depends on the type of data included in an analysis, and statistics are expressed on measurement scales that suit their nature. As computations are numerical, the scales are represented as numbers. In turn, the scales affect the presentation of data, the effect estimates, and the statistical analyses. This will be discussed further in Section 4.9.

BOX 4.4 Data Types

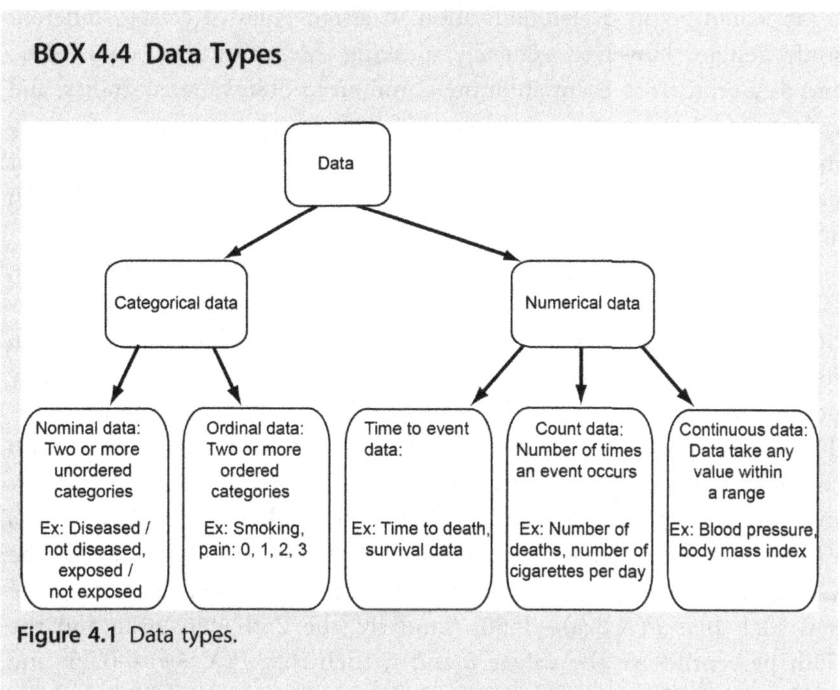

Figure 4.1 Data types.

Box 4.4 lists the different types of data.

There are two principal types of data, categorical and numerical. Categorical data are classified into nominal and ordinal variables. Smoking is an example of a nominal variable, as it is divided into two categories: nonsmoker (category value 0) and smoker (category value 1). Smoking habits are a good example of an ordinal variable, as this variable can be divided into many categories with an underlying ranking: nonsmoker (category value 0), former smoker (category value 1), occasional smoker (category value 2), and daily smoker (category value 3). Numerical data include count data, data on time to event, or continuous data. Data acquired over time or space may be expressed in discrete numbers, such as count data. The number of times an event occurs, measured as cases per year, is a typical example. Data on time to event is often used for survival analysis. Continuous data are characterized by the measurement of ranges of variables. Typical examples include weight, hormone concentration, blood pressure, and age, for which the data are the actual observed values.

It would be an oversimplification to assign types of data to different study designs. However, generally speaking categorical data, count data, and data on time to event are more common in observational studies, and continuous data are more common in experimental studies, particularly in basic medical research. As the normal distribution plays a prominent part in continuous data, the basic properties of the normal distribution are covered below.

The normal distribution function, also called the Gaussian function, is symmetrical and is described by its location (μ) and its standard deviation (SD) (σ). Its variance is the square of the SD (σ^2). The distribution function is bell-shaped and its location is at the same time the symmetry point and the value where the function reaches its maximum value. The size of the SD gives the variability of the distribution and is also the inflection point of the Gaussian curve. When a variable X is normally distributed, the probability that X falls below a specified value a can be computed. This is written $P(X \leq a)$. The entire area under the Gaussian curve is 1. The median is the 50th percentile, that is, the value a is such that $P(X \leq a) = 0.50$. Similarly, the 25th percentile and the 75th percentile are the values b and c, such that $P(X \leq b) = 0.25$, and $P(X \leq c) = 0.75$, respectively. The difference between the 75th and the 25th percentile, called the interquartile distance, is often used as a measure of variation.

For a normal distribution with a location parameter μ and an SD σ, it follows that $P(\mu - 1.96\sigma \leq X \leq \mu + 1.96\sigma) = 0.95$ (95%). This gives a neat interpretation of σ, namely that there is a 95% probability that a normally distributed variable lies within 1.96 times the SD of its location. On the other hand, if an interval within which the measurements will fall can be specified with a probability of 95%, the length of that size is about four times the SD. This will render an impression of the variability of the measurements.

The concepts of signal and noise mentioned in Section 4.3 can be mathematically expressed by introducing the simple statistical model $X = \mu + \varepsilon$. Here μ represents the signal and ε the noise, often called random noise. For a normally distributed variable the SD of ε is σ.

Once the data have been collected, the location and the SD can be estimated. If it is assumed that there are n observations $X_1, X_2, \ldots, X_n$ that follow a normal distribution having a location μ and an SD σ, the mean of the observation $\overline{X}$ is the estimate of the location. Furthermore, the empirical variance is given by $s^2 = 1/(n-1)\sum_{i=1}^{n}(X_i - \overline{X})^2$, which is the estimate of the variance σ^2. Then $\mathrm{SD} = \sqrt{1/(n-1)\sum_{i=1}^{n}(X_i - \overline{X})^2}$ is the estimate of σ. Note that SD here denotes the empirical standard deviation.

4.5 METHODS AND EXPERIMENTAL TECHNIQUES

The body of scientific literature is extensive and overwhelming. It cannot be completely covered, but it is important to have a solid and diverse basic knowledge of it in order to see what is and what is not biologically plausible. Moreover, new research projects should not replace work that has already been carried out by others. For this reason new research should be based on thoroughly-phrased literature searches in relevant databases. The most important bibliographical databases are Medline/PubMed, the Cochrane Library, EMBASE, and the Web of Science. For more on literature searches and bibliographical databases, see Chapter 5. Getting started with practical work also is essential lest too much time be spent in theoretical preparation. For this reason it is advantageous to be a member of a scientific group. Indeed, it is tough being the Lone Ranger; it is more beneficial if the group can divide reading among its members and hold regular literature seminars.

Of course, choosing the optimal materials and methods is decisive. In experimental studies, the scientific problem can be approached using human subjects, experimental animals, organ preparations, or a cell culture or subcellular system. The best procedure often consists of conducting experiments within a hierarchy of materials, from fully developed humans to cellular constituents, preferably employing the procedure that is the simplest and easiest to control first. The higher in this test hierarchy a research project is started, the greater the chance for physiologically meaningful results may be, but the chance of misleading results may also be greater. For example, if a new hormone is tested in an experimental animal, there is a nonnegligible risk that the hormone concentration at the locus of action and the time-effect profile will be above and beyond the physiological. Many years ago, when critiquing a study, a member of an academic appraisal committee wrote that the estrogen doses given to mice would have been more suitable for an elephant. On the other hand, the injected agent may metabolize, or become deactivated in some other way, so rapidly that no effects will be observed at all; a compensatory mechanism might take over, or the mechanism under investigation might manifest itself only under special conditions, as often has been the disappointing experience with transgenic animals that otherwise seem to have no flaws. Indeed, an injected substance A can provoke the secretion of hormone B, which in turn yields a measured response C. Thereafter it is difficult to assert that A causes C. Response C may not be observed at all because physiological, homeostatic reactions (negative feedback) contribute to maintaining a steady state, and thereby reduce C to an unmeasurable level. This is analogous to the fact that removing one cause of depression does not guarantee improvement of a patient's self-esteem.

Misleading results or artifacts can also be deceiving when working with systems that are simpler than fully developed organisms, and it is easy to be deceived. Indeed, many new cancer treatments were considered promising after reports of encouraging results on experimental animals, or *in vitro* results on cancer cells. Unfortunately these treatments rendered disappointing results when they were finally tested in RCTs. Perhaps the best precaution is to meticulously chart the experimental setup (Box 4.5).

BOX 4.5 Experimental Research Checklist

Methods:

- Precision: standard error, quartile interval, coefficient of variation (CV)?
- Accuracy: compare to a "gold standard"?
- Dose-response?
- Time course?
- Physical-chemical conditions: pH, osmolarity, etc.?
- Positive controls? Sensitivity controls? Controls with known answers?
- Negative controls, and possibly other specificity controls as well?
- Milieu control?
- Statistics: Sample size? Test power? Uncertainty measure: 95% confidence interval, standard error of the mean? *P*-value? Parametric or non-parametric test? Regression or correlation?

Study setup:

- Test = control plus exposure? Multifactorial relations? Observational study?
- Scoring with a blind or double-blind setup?
- Selection criteria, exclusion criteria, drop-outs, randomization (simple, blocked, stratified), source population?
- Validity? Inadequacies in design or analysis? Several routes to the conclusion?
- Repeatability? How many replicate studies? Reproducibility?

Interpretation:

- Biologically likely?
- Statistically as well as biologically significant? (The interocular impact test.)

The scientific community is often most interested in conclusions that are valid in humans. Nonetheless, nonhuman subjects are often used. For instance, the earliest experiments on nerve cells were conducted using the giant axon of a squid, as sufficiently precise instruments were not yet available to conduct similar experiments on human nerve fibers. Indeed, experiments conducted on animals can reveal mechanisms and connections that are not as readily observable in humans. The lack of understanding of some phenomena, such as tickling, stitches caused by strenuous exercise, migraines, and fibromyalgia, is due in part to an absence of sufficiently comparable animal models and in part to the fact that these phenomena are not amenable to *in vitro* study. Consequently, newer transgenic animals hold great promise. Examples include a mouse model of anemia, which is similar in many respects to thalassemia, or Mediterranean anemia, and another mouse model that ostensibly simulates schizophrenia. It is less expensive and less emotionally stressing for researchers to use small animals, such as mice and rats, rather than larger mammals. Moreover, small rodents can be selectively bred to be as alike as identical twins, whereas it is very difficult, and requires extensive ethical evaluation, to undertake research on apes, livestock, cats, or dogs—many of which are no longer used in scientific research.

Although a study conclusion is most convincing when it has been checked in many ways, such as by addressing a scientific problem at several levels, this ideal may become audacious in practice. Particularly if a researcher is new in a research group, it is a risky business to attempt to change its routines or expand its armamentarium of methods. Time is usually the most limited resource. And as an old adage advises: "Never change a winning team!" The group's experimental procedures are probably well documented, and there is likely a lot of data already available for further experimentation. That said, a reluctance to bring in new procedures and models can easily result in stagnation. So it is advantageous if the group as a whole innovates occasionally, and it is important that everyone is allowed to take part in that effort. It is wise to first gain knowledge by pursuing the literature on materials and methods, but it is not advisable to postpone going to laboratories where the desired procedures are routine. Indeed, in doing so, one can save effort and gain new contacts.

4.6 EXPERIMENTAL RESEARCH CHECKLIST

Correct design and execution of an experiment is essential to derive valid conclusions (see Box 4.5).

Technical proficiency is essential, as it enables an experimental researcher to rely upon results and reproduce findings, and can only be had with practice. It is important that researchers try the experimental procedures for themselves, for example, perform a series of replicate measurements and record the spread of the results, preferably expressed as the coefficient of variation (CV). For data that are continuous (or preferably normally distributed) the CV is defined by $CV = \sigma/\mu$, which is estimated by $CV = SD/\overline{X}$. Expressed in simple terms, the CV is the size of the SD relative to the mean. It is common to present CV as a percentage. A blood hemoglobin level might render a precise measurement with an analytical CV of 1.3%, whereas this would not be the case with a corresponding serum folate analysis with a CV of 16%. This determines the number of replications that have to be performed in the test and control groups. The more precise the analysis, the fewer parallel measurements are needed to find a mean value with a low degree of uncertainty, or a significant difference between two groups of data. At this point, it may be wise to perform a power calculation, either with or without the help of a statistician. A power calculation determines the amount of data required to, for instance, assume with 80% probability that a study can reveal a 30% difference between two groups of data at the 5% level, when the analytical plus the biological CV (which permits the estimation of the SD) are known.

The basic idea of sample size calculations can be illustrated by the classic situation of comparing two means. There are data from two groups, for instance of patients, randomized into two groups. One group is exposed to a new drug regimen (test group), the other to placebo treatment (control group). The alternative hypothesis is that the two regimens are different with respect to a measured outcome. It is assumed that the outcome variable is normally distributed. The measurements from the two groups follow distributions located at μ_1 and μ_2, respectively. There is a shift towards positive or negative outcome measurements if the locations are different. Thus, the null hypothesis to be tested is

H_0: there is no difference in location between the groups, that is, $\mu_1 = \mu_2$,

versus the alternative hypothesis

H_1: the locations of the groups are different, that is, $\mu_1 \neq \mu_2$.

Table 4.1 Decisions and probabilities of the decisions in statistical testing

Result of the significance test	The truth: The null hypothesis is true	The truth: The null hypothesis is false
The null hypothesis is rejected	Type I error Probability = significance level	Correct Probability = power
The null hypothesis is not rejected	Correct Probability = 1 − significance level	Type II error Probability = 1 − power

The decision whether the null hypothesis should be rejected is based on a significance test, and the corresponding *P*-value is computed from the sampled data. For details on the significance test and the *P*-value for this specific comparison, please see Section 11.7. Possible decisions and the related probabilities are summarized in Table 4.1.

The significance level, which is the probability of rejecting a null hypothesis that is actually true, should be low, and is usually set to 5%. Rejecting a null hypothesis that is true is called a type I error. The power is the probability of rejecting a null hypothesis that is false. As this is the desired outcome, the power needs to be high. Usually a minimum power is 80%. Accepting a null hypothesis that is false is called a type II error. One minus the power is the probability of type II error.

The size of the effect that the new drug regimen will have over placebo now has to be quantified. Let the size of the effect be denoted by Δ. Furthermore, the power and the significance level have to be quantified. Finally, the SD of the measurements has to be specified. If the CV is known, SD can be calculated, since SD is equal to CV times the mean. The significance level is set at 5% and the power to 80%. The required number of observations in each group is then given by:

$$n = 2 \cdot \left(\frac{\mathrm{SD}}{\Delta}\right)^2 \cdot 7.9. \tag{4.1}$$

The number 7.9 is related to a required significance level of 5% and a power of 80%. If a higher power of 90% is needed, 7.9 must be changed to 10.5.

Example

A mean of 80 units in the control group is expected and the goal is to detect an effect of 30% in the test group. Thus $\Delta = 24$, and the mean of the test group 104. If $CV = 16\%$, the SD is 17 ($104 \cdot 16\%$) in the test group. Different SDs are expected for the control group and the test group. The highest value of SD is used, since that will give the highest value of n to ensure the required power. The desired power is 80%, and a significance test at a level of 5% is used. Then a sample size of $n = 2 \cdot (17/24)^2 \cdot 7.9 = 8$ is needed in each group, or a total of 16 patients. The reason for such a low study sample is evidently that the expected effect size is quite large compared to the SD.

Avoiding systematic error is essential; all work must be accurate. Precision and accuracy are often confused, but may be distinguished using an example. When shooting at a target using a rifle with a sight that is skewed to the right, the shots may fall close to each other (high precision), but they will systematically fall to the right of the bull's eye (low accuracy). Precision is related to standard error, and accuracy to bias. High precision means low SD, and an accurate instrument is said to be unbiased. Accuracy requires that measuring instruments be calibrated against a recognized standard, which may or may not be commercially available. The nature of the procedure will determine how often measuring instruments need to be recalibrated.

The principal aspects of quality assurance at a large hospital laboratory can be used as a paradigm for a research laboratory. In this paradigm two types of references are used: calibrators with known, stated values, and control material used daily. Calibrators are used to establish working characteristics or standard curves (the functional connection between instrument response and analytical concentration). Some calibrators are sold by independent companies, but for many analyses, the only calibrators available are those supplied by the instrument manufacturer as part of the reagent kit.

Control material is used to monitor analytical quality for routine, internal analytical quality assurance. This consists of including control material in various concentrations in all analyses. Repetitions of control material are often included in long series of analyses, usually at the start, in the middle, and at the end, to monitor for instrument drift. Control material should be included at a minimum of two levels, one low and one high, when the working characteristic or standard curve is linear, and at three levels when the characteristic is nonlinear. For diagnostic applications, the levels of control material should represent the extremes of the range in which clinical intervention is indicated. But for the pure technical quality assurance needed in a research project, there should be a control concentration in the middle as well as at either end of the working range where standard curves deflect, such as at ED_{20} and ED_{80} in a radio-immune assay, where ED means effective dose and ED_{20} and ED_{80} are 20% and 80% of maximum response.

In principle, control material and calibrators other than those provided by the instrument manufacturer are preferable, but in practice this is often not possible, such as in hematological analyses. Control material can be made in house for many analyses by freezing portions of the analyte that can be defrosted just before use. Control material should have the same matrix, that is, the same composition of proteins, ions, etc., as that of the test material.

Moreover, external organizations (LabQuality, Murex, etc.) send out control materials for analysis at different intervals, from every 2 weeks to six times a year, allowing individual laboratories to report back with their findings. This procedure is called external quality assessment.

Some experimental procedures may require exceptional vigilance to spot systematic errors. Indeed, recorded observations may change if an experimental animal moves, or its breathing might have an effect. Subjective choices may also play a role, for example, if only information from the largest nerve cells is recorded; if blood is collected only from the most accessible arteries; or if cells are counted only in areas of a preparation that are presupposed to be rich in the type of cell under investigation. In such cases, those recording the data should be blinded.

Establishing the dose-response relationship for an analysis is part of the rules of craftsmanship of a study design, and may be essential for proper performance of an experiment. Moreover, this may allow for future experiments in which a particular dose can be selected as a continuation of a previous research project. Either saturation doses or half-maximum effective doses (ED_{50}) may be chosen, and in such cases the strength of influence of the dose-response relationship must be known.

Time course also is significant, because in a subsequent research project data can be recorded at the time when the maximum response was previously observed. The measuring instrument may have a long response time that can be altered, or at least that must be kept in mind. A lethargic mechanical part might be replaced by an electronic component, or other causes of unacceptable latency may be removed. For example, when non-invasive ultrasonic blood flow measurements replaced invasive measurements on experimental animals, it became possible to measure blood flow in humans by recording changes in the course of a couple of heartbeats, and to corroborate that those changes in the bloodstream were triggered by nerves, not hormones, which act more slowly. It is also essential to determine the period during which measurements are stable, in order to keep instrument recalibration to a minimum. If a stable base level, or baseline, can be documented, followed by an observable outcome resulting from an experimental intervention, and finally a return to base level when the intervention ceases, the experimental setup is splendid. In such cases, it is advantageous to compare the observations of the measured variables with the means of the two nearly identical control phases.

For *in vitro* studies, physical-chemical conditions must be monitored, such as pH, osmolarity, temperature, gas pressure, etc. Positive controls are also essential, particularly with negative findings. Positive controls are related to calibrators, or control material in various concentrations, and to dose-response studies, but are most often used at higher, more complex levels. For example, there are many possible interpretations when a given intervention does not render the expected effect: (i) the working hypothesis may have been falsified, (ii) the researcher may have been misled by a compensatory reaction in a human subject, experimental animal, or organ preparation that more or less masked the expected response, (iii) the experimental procedure may not have actually intervened in the expected manner because a chemical agent was no longer active ("past its *best by* date"), or (iv) cells that had lost their ability to respond were used. This puts the researcher in a position where he or she must document that the intervention did indeed proceed as it should have. For example, if a researcher finds that a graft-versus-host reaction does not affect blood formation as assumed, then the occurrence of a genuine graft-versus-host reaction must be confirmed, for example by demonstrating enormous spleen growth in an experimental animal.

Sensitivity control goes one step further in that an intervention must not only have an effect, the researcher must also document a strength-effect relationship, as discussed previously for individual analyses. If the strength of the stimulus required to elicit the desired effect or outcome is physiologically unreasonable, the results can be considered tenuous at best. Supra-physiological stimuli can cause various artifacts. In intact animals, unspecific stress reactions or alerting responses may trigger the observed effects. Large drug doses or other factors may have nonspecific effects. Sensitive cells and tissue that are roughly handled *in vitro* may not exhibit the same reactions as *in situ*. Indeed, cell lines comprise transformed cells that do not necessarily respond the same way *in vitro* as normal varieties do *in vivo*. It also seems that infection with mycoplasma or viruses, and genetic or epigenetic transformation, can cause spontaneous changes in continuous cell lines.

These examples underline the importance of specificity controls or negative controls. Before the media is informed that some remarkable substance kills cancer cells, it is of the utmost importance to be sure that it does not also kill normal cells. Moreover, it is important not to publish scientific results in the general media before the publication of the scientific article. If a factor has been isolated that believably specifically stimulates something like a hormone-secreting cell, it must be proven that an actual secretion process was observed, and that the secretion was not simply the result of cell damage (as indicated by release of a cytoplasmic enzyme like lactate dehydrogenase). Furthermore, specificity should be proven by confirming that the effect occurs via a particular receptor. This can be done by blocking the effect with an antagonist or antibody to the receptor. Alternatively, a ribozyme or inhibitory RNA molecule that cleaves or inactivates mRNA for the receptor protein can be used, or a "knock-out" mouse that lacks the receptor and hence also the observed hormone secretion in wild-type cells can be created. On the other hand, specificity can occur in an intact organism for "geographical reasons," though this cannot be demonstrated *in vitro*: autocrine, juxtacrine, neurocrine, and paracrine agents are site-specific in an organism and have no remote effects, as they cannot diffuse over any distance.

Finally, quality assurance of study conditions is needed for milieu controls. Subjects should usually be at rest, balanced, fasting, comfortable, and properly informed as to the nature of the study. Please see Chapter 3 for more information. Experimental animals should be handled according to regulations, and should have adapted to their new environment in the experimental animal center. Proper consideration must be given to the diurnal physiological variation of variables, preferably by recording data at the same time each day. Anesthesia, analgesia, circulatory and respiratory parameters, and fluid and electrolyte concentrations should all be controlled and standardized. These precautions parallel those of the physical-chemical conditions for *in vitro* tests.

4.7 REPEATABILITY, REPRODUCIBILITY, AND RELIABILITY

When creating a study design, it is important to remember that the signal must be distinguished from the noise. The observation is the sum of signal and noise, but the signal contains the information that pertains to the hypothesis. Noise gives rise to variation, and occurs because test entities are observed with measuring instruments that are never completely precise, or because there are other sources of random error. When variation is small, the signal is readily observable and the null hypothesis can be rejected. Large variation, however, can complicate this task.

In assessing the variation or the random error of observations, one must distinguish between repeatability and reproducibility. Repeatability is understood to be the degree to which the same results are observed when measurements are repeated under identical experimental conditions, which may be identical methods or identical observers. The model for the *i*th observation can be described as follows:

$$X_i = \mu + \varepsilon_i \quad \text{for } i = 1, 2, \ldots, n, \tag{4.2}$$

where μ is the population mean, and ε_i the noise. The noise is random, in the sense that the variance is the same for all observations, that is, $\text{Var}(\varepsilon_i) = \sigma^2$. Repeatability is expressed by the variation coefficient σ/μ, which, of course, can be estimated by $\text{CV} = \text{SD}/\overline{X}$, where $\overline{X}$ is the mean, and $\text{SD} = \sqrt{s^2}$ (see Section 4.4). To understand the concept of repeatability, note that whenever data follow a normal distribution, about 95% of the observations will fall within $\overline{X} \pm 1.96$ SD.

Example

When performing immunoblotting, a semiquantitative technique for protein analysis, loading equal amounts of protein is crucial. Therefore, the absolute quantity of protein in each sample is quantified using a special kit and a spectrophotometer. Table 4.2 shows the output from a spectrophotometer for six hearts, from each of which five tissues samples were taken. The measurements across the tissue samples for each heart are considered repetitions.

Table 4.2 Quantity of protein in the tissue samples of six hearts[a]

	Sample 1	Sample 2	Sample 3	Sample 4	Sample 5
Heart 1	0.386	0.397	0.398	0.399	0.400
Heart 2	0.300	0.305	0.306	0.306	0.307
Heart 3	0.409	0.415	0.416	0.417	0.418
Heart 4	0.286	0.290	0.291	0.292	0.296
Heart 5	0.281	0.287	0.288	0.288	0.289
Heart 6	0.307	0.313	0.314	0.315	0.315

[a]We are grateful to Lars Henrik Mariero, Department of Physiology, Institute of Basic Medical Sciences, University of Oslo, for providing us with us with these data.

If the CV for heart 1 is considered, the mean over the six tissue samples is $\overline{X} = 0.396$, and SD = 0.0057. The SD gives the analytical variation in the protein levels within one heart. That gives CV = 0.0057/0.396 = 0.014 (= 1.4%), which is very low.

The example above can easily be extended to a more general and more interesting situation in which there are observations for different experimental conditions. This makes it possible to estimate the variation within experiments, as well as the variation between experiments. When the experiments are performed on human subjects or organ preparations, the variation between experiments is called biological variation, and the variation within experiments is called analytical variation.

Assuming I experimental settings, and n_i repetitions for the ith setting, the observations are modeled by:

$$X_{ij} = \mu + \varepsilon_{1i} + \varepsilon_{ij} \quad \text{for } j = 1, 2, \ldots, n_i \text{ and } i = 1, 2, \ldots, I. \qquad (4.3)$$

Here $\sigma_1^2 = \text{Var}(\varepsilon_{1i})$ is the variation between experiments and $\sigma^2 = \text{Var}(\varepsilon_{ij})$ is the variation within experiments. The total variation in the experiment is $\sigma_1^2 + \sigma^2$.

The CV, as a measure of repeatability, is given by σ/μ. The estimation of σ_1^2 and σ^2 is not as straightforward as that for the simpler model above. It must be calculated using a statistical software program for repeated measurements, or for so-called mixed models. The estimates of σ_1^2 are denoted by S_B^2 and the estimate of σ^2 by S_W^2. The subscripts relate to variation between (B) and within (W) experimental settings. The analytical CV is now defined by $CV_W = SD_W/\overline{X}$, and the biological CV by $CV_B = SD_B/\overline{X}$. These will be used in the following example.

Example

When the data in Table 4.2 are analyzed for all six hearts, they show that $S_B^2 = 0.003158$ and $S_W^2 = 0.000015$. Since S_W^2 gives the size of the analytical variation, the analytical CV is $CV_W = SD_W/\overline{X} = \sqrt{0.000015}/0.396 = 0.010$ (=1.0%), and the biological CV is $CV_B = SD_B/\overline{X} = \sqrt{0.003158}/0.396 = 0.142$ (=14.2%). This shows that the biological variation between hearts is small, but the variation within tissue samples is much smaller. Consequently the analytical CV is also very small.

Reproducibility is understood to be the degree of variation observed when experimental conditions change. This may depend on the measurement method or on the observer. Hence, when planning a study, it is wise to consider any factors that may cause variation. For instance, in a study of changes in blood pressure, the aspects of the measurements that may cause variation must be assessed. These may include body position (whether the subjects are in a supine or sitting position), the measurement instruments used, and the time of day measurements are taken. After all such matters have been considered, the aspects that are vital and create variation can be assessed, and perhaps mitigated in the design of the experiment.

Whenever the variation of a measuring instrument is unknown, it is wise to start with a preliminary study that assesses the repeatability and reproducibility of the instrument's measurements. Reliability measures are used as a means to identify and estimate reproducibility, such as that over measuring instruments or positions. For instance, an assessment of the degree to which a supine or sitting position affects blood pressure measurement should be made before a blood pressure study starts. This can be done by performing a pilot study, taking blood pressure measurements from subjects while they are in a supine and a sitting position. This will give two measurements that will show the variation between measurements taken in these two positions, as well as the variation between subjects. Such measures of reliability are calculated as the intraclass correlation coefficient (ICC) for continuous data, and as Cohen's kappa coefficient for categorical data. Cohen's kappa coefficient is the counterpart of ICC for categorical data. Briefly, ICC is a measure of reliability since it quantifies how different methods or different observers measure the same quantity. It can also be interpreted as the correlation between the paired measurements.

Consider an assessment of the reliability of an experimental procedure. A reproducibility study is conducted by taking the same measurements from the same subjects using different measuring methods. The model for these observations is as in Eq. (4.2) above, and the two variances σ_1^2 and σ^2 are given as that between and within experiments. Then the ICC is defined as $\sigma_1^2/\sigma_1^2 + \sigma^2$.

ICC is the ratio of the between-subject variation and the total variation. Accordingly, ICC is a measure that varies between 0 and 1, and the closer it is to 1, the higher the reliability. High reliability is attained when the within-subject variation, and hence the variation between measurement methods, is small. As noted above, the estimation of σ_1^2 and σ^2 must be done using a statistical software program for mixed models. This renders estimates S_B^2 of σ_1^2 and S_B^2 of σ^2. Then the estimate of the ICC is given by:

$$\mathrm{ICC} = \frac{S_B^2}{S_B^2 + S_W^2}. \tag{4.4}$$

Example

The aim is to assess the reproducibility of blood pressure measurements in a supine or sitting position. This is done by taking blood pressure measurements from all subjects in both a supine (measurement 2) and a sitting (measurement 1) position with the same measuring instrument, in order to estimate variations both within and between subjects. The study measurements are listed in Table 4.3.

Table 4.3 Blood pressure measurements of 10 subjects in both a supine (measurement 2) and a sitting (measurement 1) position

Subject	Measurement 1	Measurement 2
1	93	91
2	78	79
3	84	81
4	72	69
5	83	85
6	90	91
7	80	80
8	91	89
9	82	79
10	73	75

By using a program for mixed models, such as SPSS, the result is $S_W^2 = 2.25$, and $S_B^2 = 49.39$. The observed variation between subjects is far greater than that within subjects, so high reliability can be anticipated. ICC is calculated as $\text{ICC} = 49.39/(49.39 + 2.25) = 0.96$.

Of course, there is no one size a reliability measure must be to ensure agreement between two sets of measurements, but a guideline is given in Box 4.6.

Based on the example above, there is a very good agreement between the blood pressure measurements taken in supine and sitting positions.

Reliability computations are particularly necessary when testing the reproducibility of different measurement methods, or of repeated

BOX 4.6 Reliability: Degree of Agreement

- 0.00–0.20: Poor
- 0.20–0.40: Moderate
- 0.40–0.60: Good
- 0.60–0.90: Very good

measurements in the same subjects. Another example might be the analysis of the inter-rater reliability, i.e., the agreement of data recorded for the same subjects by different observers. For more on reliability measures please see chapter 14 in Veierød et al.[4]

4.8 MULTIFACTORIAL RELATIONSHIPS AND OBSERVATIONAL STUDIES

An experiment, which often involves an intervention, is a typical study of the association between an outcome and an exposure, such as a study investigating the effect of a new drug where the outcomes of a test group and a control group are assessed. The equivalent situation exists in laboratory experiments exploring the effect of a hormone on growth rate in a cell culture.

Causal relationships often are multifactorial, but even then the effect of a single exposure variable can be studied by randomizing test units into two groups that differ by that variable alone. This is the case when test subjects are randomized into treatment and control groups. Randomization ensures that the two groups are "identical," save for the exposure, i.e., the drug regimen or other intervention to which the test group is subjected. An experiment that tests a hypothesis on a single exposure variable also simplifies the statistical analysis, as only the difference between groups needs to be statistically tested.

Observational studies differ from experimental studies in that they are multifactorial, as randomization into test or control groups in these studies is impossible. Although the effect of a particular exposure variable on the difference between the test group and the control group is often of interest, this effect cannot be directly assessed in an observational study, as controls for other exposure variables are not included in the study design through randomization. Consequently, the exposure variable of interest must be studied through statistical analysis using multivariable statistical methods. For more information, please see Chapter 11.

Confounding arises when the outcome variable of interest is intermixed with the effects of other variables. In experimental studies such as RCTs, randomization is used in an effort to eliminate the effects of confounding. In observational studies, confounding is more severe, so multivariable models and appropriate study designs are used to reduce its effects. In multivariable statistical methods, control of exposure variables other than the exposure variable of interest is accomplished through the use of a statistical model that includes all the main exposure variables. The effect of the exposure variable of interest can then be estimated. Examples of such analyses include linear regression analysis, logistic regression, Poisson regression, and Cox regression. See Sections 11.9 to 11.13 or chapters 3–6 in Veierød et al.[4] for more on regression analysis.

The types of observational studies used in the medical and biological sciences were listed in Box 4.3. Please see also Chapter 9 on epidemiology.

Ecological studies are studies where the outcome and the exposure are measured at the group level. Since the outcome is not measured at an individual level, and control for confounding factors is not possible, valid conclusions regarding the exposure-outcome relationship cannot be drawn. Nevertheless, ecological studies are easy to perform and can give valuable information.

Cross-sectional studies and surveys are conducted within a source population at a particular point in time. At that time, the population studied may include people of various ages, at various phases of disease and with various exposure variables for that disease. The outcome and the exposure variables are not tracked over time, so a cross-sectional study is not well suited to exploring the relationship between cause and effect. A cross-sectional study is suited to study the prevalence, but not the incidence of a disease.

Cohort studies are prospective studies or follow-up studies of a source population. In studying the association between exposure variables and a disease (as the outcome variable), a healthy population is observed until either the disease occurs or the study is stopped. Key examples of such studies include progress studies and survival studies, which are often long in duration. The information on the association between the exposure variables and the disease lies in the number of cases and the length of follow-up before the disease occurs. Age, other exposure variables, and the outcome variable are monitored over time, so this sort of study is well suited to examining the relationship between cause and effect, and it allows for the investigation of both the incidence and prevalence of a disease. Typical follow-up studies include those of the association between nutrition or diet and cancer incidence.

Longitudinal studies are conducted on test subjects or experimental animals through repeated monitoring over time. At the same time, the progress of the outcome variable and the exposure variables are studied. Data are acquired through repeated measurements, so the parameters do not include follow-up time; instead they are limited to disease observations and changes in the exposure variables during the study period. In cohort studies, on the other hand, one usually follows the study sample until a given event occurs, with different measurements recorded during follow-up. Longitudinal studies are useful whenever there are relatively rapid changes in diseases or exposure variables. They are well suited to studying the associations between cause and effect. Examples of longitudinal studies include investigations of the effects of hormones on growth in animals or humans, using monthly growth measurements. Such studies may be conducted over months or years.

A drawback of cohort and longitudinal studies is that they are lengthy and costly. Moreover, a long-term study often suffers loss of some subjects during follow-up. The challenge of conducting this type of study is to ensure that any decline in the number of subjects does not threaten study validity or conclusions.

Case-control studies are characterized by the inclusion of subjects with the outcome variable as the starting point. In studies of disease, the outcome variable is characterized by ill/healthy. These in turn become cases (ill) and controls (healthy). Exposure variables can be analyzed retrospectively to examine the effects over time. Case-control studies can be conducted by matching cases and controls by variables other than the exposure variable under investigation, such as age and gender. In this manner, cases and controls are made equal with respect to the matching variables. The most classic example is when two groups are similar with respect to all variables except the exposure variable. In cancer epidemiology, associations with diet or sun exposure are often investigated using case-control studies.

4.9 VALIDITY, EFFECT ESTIMATE, AND CHOICE OF STATISTICAL TEST

When it comes to validity, it seems appropriate to define some concepts of population. The target population is the population to which conclusions are to be generalized. The source population is defined as a subset of the target population, which is obtained by applying certain given selection criteria, and will in most cases differ from the target population. The study sample is drawn from the source population. By proper analysis, valid conclusions from the study sample can be drawn that also apply to the source population. Finally, generalizations to the more general target population must be addressed.

BOX 4.7 Validity

- Concept validity
- Internal validity
- External validity

BOX 4.8 Internal Validity Can Be Threatened by:

- Sample selection bias
- Information bias
- Statistical confounding

Validity relates to the generalization to the source population (internal validity), and then to the target population (external validity). Three types of validity can be distinguished (Box 4.7).

As the name implies, concept validity is associated with the concept being studied, while internal and external validity concern the conclusions drawn from a study. Validity entails the consideration of whether the associations or differences observed are real, which is germane in all research, and also applies to conclusions; a large gap between conclusions drawn and data acquired may indicate poor validity. If a chemotherapeutic agent kills the cold virus in a cell culture, it cannot be concluded that it is a good treatment for the common cold; even though professors report an average work week of 55 h, it is not certain that they actually work that much.

Concept validity assesses the degree to which the data reflect the variables under investigation that cannot be recorded directly. The problem studied should be operationalized to be suitable and adequate, and accordingly several variables or tests may be used to measure the concept. This will often be the case when new methods are used, from methods used to measure neurotransmitters to the use of a questionnaire in psychiatry. A new method has to be validated against a well-used or well-known method before it can be accepted as an alternative method. A questionnaire should be validated by two or more raters to evaluate their agreement before being put into use. The measures of reliability discussed in Section 4.7 are used to assess agreement, and validity is ensured when the agreement is high.

Internal validity is associated with valid inferences about the source population, and consequently it has potential weaknesses (Box 4.8).

Bias can occur when the study sample is not representative of the source population. Nonresponse or exclusion from a study may also induce bias. Loss to follow-up can seriously threaten the validity of a study and is especially problematic in long-term studies and in study designs where loss to follow-up is not uniform across the study groups being compared. Results from clinical trials may be biased when the test group and the control group differ.

Information bias occurs when the information recorded in a study is flawed. This may occur in the laboratory but is more likely to happen if the study subjects themselves report information, and do so incorrectly. This results in misclassification of variables, which is particularly serious when misclassification differs between study groups, such as in clinical trials with test and control groups, or in case-control studies.

Research in the medical and biological sciences employs statistical methods and statistical testing of hypotheses. Internal validity is weakened when improper methods are used. For example, statistical validity depends on using the right effect measures and statistical tests, and using a sufficient sample size to avoid type I and type II errors (see Section 4.6 and Chapter 8).

The effect measure is the quantity that provides the statistical description of the hypothesis in a study and is derived from the outcome variable. For example, there may be a mean difference in the outcome variable between a test group and a control group. Based on these observations the effect estimate is calculated, which is simply the sample analogue of the effect measure. The effect estimate and its uncertainty, presented in the form of a confidence interval, are used to assess the significance of study results. The *P*-value provides information on the degree to which the result of a sample can be generalized to the source population. Many choose 5% as a threshold for statistical significance, and consequently only present results with *P*-values less than 0.05. In general, there is no reason to reduce the information contained in the *P*-value to a statement of significant or not significant, so presenting the *P*-value is recommended. That said, *P*-values higher than 0.05 may contain considerable information. For more on effect estimates, confidence intervals, and *P*-values, please see Chapter 11.

A valid statistical analysis comprises all the parameters listed in Box 4.9.

Most studies have more than one effect estimate and entail more than one statistical test. If the limit for rejecting a test is set at 0.05, the

BOX 4.9 Presentation of Main Statistical Results

- Effect estimate, possibly adjusted for confounding.
- Uncertainty of the effect estimate, expressed as a confidence interval (preferably a 95% confidence interval).
- Results of one or more statistical tests, expressed as *P*-values.

probability of rejecting at least one of several tests will be greater than 0.05. So the significance levels of the individual tests must be adjusted to preclude excessive testing, such as by applying the Bonferroni multiple comparison correction.

Of course, the choice of a statistical method is closely associated with the effect measure used, which in turn is closely related to the data chosen to measure a given variable. Table 4.4 comprises an overview of effect measures and statistical tests used with various types of data. The relevant details are covered in Chapter 11.

External validity concerns generalization, i.e., the ability to apply a conclusion from a source population to the more general target population. It is a complex challenge, as any conclusion depends on the design, population, and statistical model used in the study. The relationship between the source population and the target population is particularly important, but regardless of how meticulous one is in the planning and analysis phases of a study, a discussion of external validity is marked to a great degree by assessments and speculations and poorly defined criteria. That said, conclusions should be explicit, so that the reader can properly assess the external validity.

4.10 RESEARCH PROTOCOL

A research protocol usually is required as part of a funding proposal. It is absolutely necessary to think meticulously through all the possibilities and consequences when choosing a certain experimental procedure. It takes perseverance to do this instead of rushing to the laboratory or beginning data collection to get the work started, but its importance cannot be overstated. Ethics committees, experimental animal panels, database authorities, and others also require a research protocol before even preliminary

Table 4.4 Statistical tests for independent and dependent samples

Number of groups	Effect measure	Statistical test: Independent sample	Statistical test: Dependent sample (repeated measurements)
Nominal data			
Two or more groups	Relative risk/ odds ratio	Chi-squared test	McNemar's test
Continuous exposure	Odds ratio	Logistic regression	Conditional logistic regression
Ordinal data			
Two or more groups	Difference in medians	Wilcoxon-Mann-Whitney test/ Kruskal-Wallis test	Wilcoxon sign test/ Friedman's test
Time to event data			
Two or more groups or continuous exposure	Hazard ratio	Kaplan-Meier plot, Cox regression	
Count data			
Two or more groups	Incidence rate ratio	Poisson regression	
Continuous data			
Two or more groups	Difference in means Difference in medians	*t*-test/ANOVA/ Wilcoxon-Mann-Whitney test/ Kruskal-Wallis test	One-sample *t*-test for paired data/ Wilcoxon sign test/Friedman's test
Continuous exposure	Regression coefficient	Regression analysis	Repeated measurements/ mixed models

research can be undertaken. It is always wise to think ahead to minimize the chance of failure in practical work (or of replicating work that has already been done), as well as to minimize the number of unproductive

BOX 4.10 Basic Research Protocol Checklist

- Project title?
- Problem and goal: the what and why of the project. What is the hypothesis?
- Background: what is now known? Give references to published work.
- The setup: the how of the project.
- Design: how will observations be collected?
- The size of the study.
- Material: what will be included in the study (data from medical registries, human subjects, experimental animals, organ preparations, cell cultures, subcellular systems, chemicals)?
- Variables: what will be measured—and when?
- Analysis: how will data be analyzed? Statistical methods?
- Presentation of results: scientific journal or report or perhaps also a popular press article? Order of authors' names?
- Financing and administration: who is paying; who is responsible for what; what are the deadlines?
- Time plan: when will the project finish?

If a funding proposal is involved, this list might be supplemented with:

- Why should the project be conducted; why it is important?
- Indication of expertise: are all procedures to be used routine in the primary group or co-operating groups, or do you intend to devote time and resources to establishing special procedures?

pilot studies. A good rule of thumb is to count on the project taking three times longer than originally anticipated.

You may supplement the basic research protocol checklist in Box 4.10 with aspects discussed elsewhere in this chapter, including the experimental research checklist in Box 4.5 and the advice of Box 4.11 to establish the logical relationship between the hypothesis and the specific prediction following from it. Be sure to consult others, perhaps by holding a seminar with colleagues from different disciplines, including technicians who may have valuable input. Medawar[5] believed that "...technicians are colleagues in collaborative research: they must be kept fully in the picture about what an experiment is intended to evaluate and about the way in which the procedures decided upon by mutual consultation might conduce to 'the sum of the business' (Bacon)."

Describe the scientific problem or the goal of the project, clarify the materials and methods that will be used to find the answers, explain what has

BOX 4.11 On Scientific Conclusions

...to establish firmly the logical relationship between the hypothesis and the specific prediction following from the hypothesis. It is no use spending a great deal of time, effort and money for the purposes of designing and executing a series of experiments to answer a particular question if, at the end of all this, answering that question is not a decisive test of the hypothesis. For example, suppose the hypothesis is that a high-fat diet can result in premature death. Suppose also that it is known that diets rich in fat can cause obesity. To show that obese subjects have a decreased life expectancy in comparison with non-obese controls does not necessarily support the original hypothesis. Obese individuals might differ from controls not only in the size of their fat stores but also in the intake of some other foodstuffs. A decisive test of the hypothesis would require the two groups of individuals to be identical in all factors (including degree of obesity and dietary habits) except for fat intake....

(Scott and Waterhouse[6])

been done previously in the field, and analyze the consequences and possible results of the suggested study design. After doing this, the study's reliability and validity can be substantiated (as discussed in Boxes 4.6 and 4.7). Validity and reliability must be considered, even if they are not specifically included in the research protocol.

How extensive or numerous should experiments, trials, or data collection be? Answering that question often entails compromising study goals, as dictated by the time, funds, and resources available. No matter what, the research material must be as large as the power calculation requires to preclude the risk of type II error, i.e., when the amount of data is inadequate to reveal a real and significant difference between test and control groups in an experiment, or between test and control patients in an RCT.

Systematic error must be avoided, which entails the consideration of a few questions: Will the essential variables change over time, e.g., with diurnal cycles, the seasons of the year, apparatus wear, lapse of biological material and chemicals, or the effects of repeated measurements on biological preparations? Observer bias must also be avoided, as discussed above, and in Box 4.12. Subjective evaluations, such as differential counts of white blood cells, must be blinded, i.e., the person evaluating must not be able to identify preparations. Studies with human subjects should be

BOX 4.12 Example of Observer Bias ("the Experimenter Effect")

The experimenter effect is particularly prominent in behavioural research, where people exchange signals unintentionally, without speaking. These signals may be transmitted by gestures, by auditory or visual channels, by touch or even by smell. In every experimental situation the experimenter may thus convey to the subjects his (or her) feelings without even knowing that he has done so.

...

Rosenthal and his collaborators carried out experiments designed to detect the experimenter effect on rats. ... They had to train rats in seven different tasks. The experimenters were deliberately biased by having been provided with false information that some of the rats were "bright" and the others were "dull" while in fact all the rats were from the same colony, were of the same age and sex and had performed similarly. The "intelligence" of the rats was said to have been determined in previous maze running experiments. Eight teams were given rats described as "bright", and six teams were told their rats were "dull". At the end of the experiment, the experimenters had to rate themselves, as well as the rats. It turned out that the experimenters believing their subjects to be generally "bright" observed better performance on the part of the rats and rated themselves as more "enthusiastic, friendly, encouraging, pleasant and interested" in connection with the performance of their rats, than the experimenters working with "dull" rats. The differences between the two groups were statistically significant. The explanation given to the experimenter effect in rats was that the rats defined as "bright" and supposedly performing better, were liked better by their experimenters and were therefore touched more. Indeed, Bernstein showed in 1957 that rats learned better when they were handled more by the experimenters. If mere physical contact could affect the learning behaviour of rats surely more dramatic effects may be expected in human experimentation.

(Kohn[7])

double-blind so that neither the subjects nor the researchers know whether a subject is in the test group or the control group.

It is important to state in the research protocol how the results will be processed. Specify which statistical test will be used (see Table 4.4), and whether the methods will be parametric or nonparametric, whether regression analysis will be used to control for confounders, whether multiple significance testing will be controlled using the Bonferroni multiple comparison correction or another correction. Mention any special test used that checks whether statistical significance has been attained while experiments are ongoing before accumulating new data (see Chapter 8).

It is wise for collaborators to agree in advance on their respective responsibilities, and on deadlines. It is a good idea to determine who shall compile the first draft of a manuscript for publication, who will be a coauthor according to the Vancouver guidelines (www.icmje.org, see also Chapters 2 and 13), and the order of authors named. However, it is important to allow for unanticipated changes during the process, which may necessitate rearrangement of these matters.

Studies on human subjects should be cleared by the relevant ethics committee, and often they must be insured. Likewise, experiments on animals must be authorized by the relevant agency or committee (see Chapter 3). All researchers concerned should have been approved, and have a copy of the research protocol.

As the research protocol is constructed, it is wise to also compile a register form for the individual experiments. The form should include spaces for the trial or experiment number, title, data, goal of the study, and a reference to a detailed description of the method. Moreover, it should include routine data that are recorded each day of the experiment, such as the time at which the procedures were performed, cell concentrations, room temperature, etc. Finally, there should be space for conclusions, so that one is obligated to learn from each experiment. Instrument printouts can be stapled to the register form. The form and its attachments should be collected in a research protocol with fixed pages. The research protocol should have a table of contents in which trial numbers and titles are entered sequentially. The following guidelines, based on those prepared by Newcastle University, illustrate how strict the rules for research protocols can be.

1. Records of primary experimental data and results should always be made using indelible materials. Pencils or other easily erasable materials must not be used. Where primary research data and results are recorded on audio or video tape (e.g., interviews), the tape housing should be labeled as set out in paragraph 4.
2. Complete and accurate records of experimental data and results should be made on the day they are obtained and the date should be indicated clearly in the record. When possible, records should be made in a hard-backed, bound notebook in which the pages have been numbered consecutively.
3. Pages should never be removed from notebooks containing records of research data. If any alterations are made to records at a later date they should be noted clearly as such, date of the alteration stated, and the alteration signed by the person making it.

4. Machine printouts, photographs, tapes and other such records should always be labeled with the date and with an identifying reference number. This reference number should be clearly recorded in the notebook referred to above, along with other relevant details, on the day the record is obtained. If possible, printouts, photographs, tapes, and other such records should be affixed to the notebook. When this is not possible (e.g., for reasons of size or bulk), such records should be maintained in a secure location in the University for future reference. When a "hard copy" of computer-generated primary data is not practicable, the data should be maintained in two separate locations within the University, on disk, tape, or other format.
5. When photographs and other such records have been affixed to the notebook, their removal at a later date for the purpose of preparing copies or figures for a thesis or other publication should be avoided. If likely to be needed, two copies of such records should be made on the day the record is generated. If this is not practicable, then the reason for removing the original copy and the date on which this is done should be recorded in the notebook, together with a replacement copy or the original if this can be re-affixed to the notebook.
6. Custody of all original records of primary research data must be retained by the principal investigator, who will normally be the supervisor of the research group, laboratory, or other forum in which the research is conducted, and who shall follow any instructions on confidentiality issued by an appropriate ethics committee. An investigator may make copies of the primary records for his or her own use, but the original records should not be removed from the custody of the principal investigator. The principal investigator is responsible for the preservation of these records for as long as there is any reasonable need to refer to them, and in any event for a minimum period of 10 years.

Some scientists recommend carrying a notebook, in which ideas that come at unexpected moments can be quickly noted. It is said that Loewi dreamed of how he could prove the existence of the postulated "Vagus-stuff" (now known as acetylcholine) that was believed to be exuded from the vagus nerve to the heart, causing it to beat more slowly. He woke, made a note of a research plan, and fell asleep again. The next morning, he could not understand what he had written. Then the dream came again. He woke, went to his laboratory, and set up two frog heart preparations. Both hearts beat in a Ringer saline solution, one with and one without the vagus nerve. He electrically stimulated the vagus nerve of one heart for a couple of minutes, so its beat slowed. Then he transferred the saline solution from the heart with the slower beat to the other heart, which then also began to beat more slowly, as if it had been vagus stimulated, thus proving the existence of a Vagus-stuff.[8]

4.11 EXPERIMENTAL ROUTINES

When conducting laboratory experiments or collecting data, it is important to try to establish workable routines that further efficient work, and promote cleanliness and a physically and socially pleasant working environment. Laws, regulations, and common sense should regulate the storage, use, and disposal of drugs, poisons, radioactive materials, microbes, and other chemicals. When doing any of the aforementioned, it pays to be meticulous. An otherwise worthy experiment is worthless if it is based on the use of unstable chemicals, or biological materials that have lost their potency due to improper storage or handling (storage at high temperature, repeated freezing and thawing of proteins, failure to use desiccants, or the premature opening of deep frozen containers so that moisture condenses on their contents). If the age of reagents is in doubt, don't hesitate to buy new. The working time lost when using defunked reagents probably costs more.

Cleaning can be critical. Traces of aluminum or detergents can ruin the contents of a reagent tube or Petri dish. Some plastics adsorb proteins and peptides, as well as cells if the medium has no proteins. Abnormal loss of key ingredients during cleaning or separation procedures should be monitored. Biologically active substances such as mycoplasma or endotoxin can cause misleading results in many types of studies. It is important to routinely test for the absence of these "robbers." If sterilization of the experimental material is necessary, autoclaving, dry heat, X-ray radiation, boiling, or ethylene-oxide gas treatment can be considered; some solutions that cannot be bought sterile can be filtered.

The responsibility for routine controls or calibration, as discussed in Section 4.6, can be distributed among the staff in the group. Even electronic scales should be checked twice a year, preferably with old-fashioned physical weights.

Ensure proper documentation of results and raw data, as well as detailed procedure descriptions, so that experiments can be accurately re-established at a later date. This is important so that everything can be recalled when it comes time to write the final paper, which may happen much later.

ACKNOWLEDGMENTS

We thank Carina V. S. Knudsen, Institute of Basic Medical Sciences, University of Oslo, for producing Figure 4.1.

QUESTIONS TO DISCUSS

1. What is required for an association (or positive correlation) to suggest causality?
2. Why is randomization important for animal or human test and control groups?
3. Explain why it is important to know and minimize experimental variation. Are there situations where the size of the variation does not matter?
4. What is the difference between precision and accuracy?
5. Why are statistically significant differences not always biologically significant?
6. Discuss the main differences between an experimental and observational study. Give two study examples: one where an experimental study design should be applied, and one where an observational study design should be applied.
7. In a case-control study, exposure-disease effects may be revealed that cannot be reproduced in a cohort study. Why is that? Discuss the advantages and disadvantages of those two study designs.
8. You are on a team that will study quality of life in cancer patients with fatigue. You have to design a questionnaire that you will send out to the cancer patients. How would you validate your questionnaire?
9. In a funding proposal, what issues should be addressed?
10. It has been suggested that there might be an interaction between grapefruit juice intake and ACE inhibitors. ACE inhibitors are a group of pharmaceuticals primarily used in the treatment of hypertension and congestive heart failure, and they are in some cases first-line treatments. You are invited to plan a study on a possible interaction between grapefruit juice intake and ACE inhibitors.

    Define the search terms and the databases to use for a literature search on the interaction between grapefruit juice intake and ACE inhibitors.

    Discuss the pros and cons of two experimental study designs for this purpose: basic medical research and an RCT.

    Make a sketch of a research protocol for such a project.

REFERENCES

1. Klein J. Hegemony of mediocrity in contemporary sciences, particularly in immunology. *Lymphology* 1985;**18**:122–31.
2. Austin JH. *Chase, chance and creativity, the lucky art of novelty.* New York, NY: Columbia University Press; 1978.
3. Thomas L. *The lives of a cell: notes of a biology watcher.* New York, NY: Viking Press; 1974.
4. Veierød MB, Lydersen S, Laake P. *Statistical methods in clinical and epidemiological research.* Oslo: Gyldendal Akademisk; 2012.
5. Medawar PB. *Advice to a young scientist. Human action wisely undertaken.* New York, NY: Harper & Row; 1979.
6. Scott EM, Waterhouse JM. *Physiology and the scientific method.* Manchester: Manchester University Press; 1986.
7. Kohn A. *False prophets: fraud and error in science and medicine.* Oxford: Basil Blackwell; 1986.
8. Mazarello P. What dreams may come? *Nature* 2000;**408**:523.

FURTHER READING

Beveridge WIB. *The art of scientific investigation.* London: Heineman; 1974.

Kirkwood BR, Sterne AC. *Essential medical statistics.* 2nd ed. Malden, MA: Blackwell Science Ltd; 2006.

Lang TA, Secic M. *How to report statistics in medicine.* 2nd ed. Philadelphia, PA: American College of Physicians; 2006.

Medawar PB. *Induction and intuition in scientific thought. Jayne lectures for 1968.* Philadelphia, PA: American Philosophical Society; 1969.

Meinert C. *Clinical trials: design, conduct, and analysis.* 2nd ed. Oxford: Oxford University Press; 2012.

Rothman KJ, Greenland S, Lash TJ. *Modern epidemiology.* 3rd ed. Philadelphia, PA: Lippincott, Williams & Wilkins; 2008.

Strike PW. Quality control in laboratory medicine. In: Armitage P, Colton P, editors. *Encyclopedia of biostatistics.* 2nd ed. Chichester: John Wiley & Sons Ltd; 2005. p. 4339–48.

Strike PW. *Measurement in laboratory medicine: a primer on control and interpretation.* Oxford: Butterworth-Heineman Ltd; 1996.

White VP. *Handbook of research laboratory management.* Philadelphia, PA: ISI Press; 1988.

CHAPTER 5

Literature Searches and Reference Management

Anne-Marie B. Haraldstad and Ellen Christophersen
University of Oslo Library, Medical Library, Blindern, Oslo, Norway

If I have seen further, it is by standing on the shoulders of giants

***Isaac Newton*[1]**

5.1 INFORMATION LITERACY

As exemplified in this quote by Isaac Newton, research does not occur in a vacuum. Much of it is built on discoveries painstakingly gleaned by other scientists over many years. In order to truly know one's research topic, one must be familiar with past discoveries and keep abreast of current ones. This is why literature searches are the foundation on which new research is built, and why the handling of information is a vital skill that should be developed along with professional competence. Indeed, the body of scientific literature is complex, voluminous, and constantly growing; special knowledge and expertise are required to sort out which publications are relevant and reliable. Information literacy is the ability to identify the information required, locate suitable resources of information (e.g., bibliographic databases), and critically assess their contents, their organization, and their efficient use. It concerns the use of all available resources, be they publications (i.e., books, journals, newspapers), radio and TV, or communications with colleagues, and the establishment of a plan and a structure to avoid random and fortuitous research results.

5.2 LITERATURE SEARCHES

A literature search can be systematic or nonsystematic (traditional). A systematic literature search is both methodical and reasoned; it is based on a systematic approach that permeates the entire process. But not all situations call for a systematic literature search. A traditional literature search

Research in Medical and Biological Sciences
DOI: http://dx.doi.org/10.1016/B978-0-12-799943-2.00005-7

may well be performed with less planning and less knowledge of both search techniques and bibliographic databases.

Most literature searches in basic research are conducted by content on specific topics, for instance biological subjects, diseases, or drugs. During a literature search by content, search terms are put together in various logical combinations. Literature searches by content do not differentiate by publication type (i.e., editorials, clinical trials, guidelines, original research articles, or reviews). The search results are displayed at random, e.g., by date of publication, and will include studies on human subjects, experimental animals, laboratory tests, and case studies alike. Furthermore, these results do not discriminate between protocols, primary and secondary sources of information, or basic and clinical research. A literature search by content may be a workable approach if the aim is to find "everything" (Box 5.1).

A literature search by methodology is mostly used in clinical research. Such a search will, as the name implies, filter the results by methodology. For example, if the intention is to find literature for clinical practice, the methodology filter for clinical research must be applied, which will sift out relevant literature that fulfills this criterion. Methodology filters have been developed for various categories of clinical queries, such as diagnosis, treatment, prognosis, etiology, and clinical prediction guides. The filters

BOX 5.1 Overview of a Literature Search

- Determine the research question. A focused research question will clarify the components of the search.
- The subject of the research will determine which sources of information are the most suitable, in order of relevance.
- Determine the appropriate time frame: how far back is far enough?
- List the relevant print-based sources (check your local academic library's catalogue).
- Fully exploit the chosen resources: knowledge of scope and search skills.
- Assess the quality of the search results. Try a variety of approaches to the research question.
- Both during the research project and after its completion, repeat the search to capture the latest information and to identify possible retracted studies.
- Store or file the retrieved publications using reference management software or compile a reference list.

BOX 5.2 Ancillary Details of Literature Searches in Clinical Research

- Determine the type of clinical question (diagnosis, treatment, prognosis, etiology, lived experience, health-care costs).
- Determine which study design best answers this question; the study design determines the best bibliographic databases for the literature search.
- List possible sources of information—bibliographic databases included—in prioritized order, best source first.
- Prefer secondary studies to primary studies when available.

comprise lists of key concepts of methods and statistics that together reflect the research design corresponding to the clinical query involved (Box 5.2).

As with other research procedures, literature searches should be documented and reproducible. Indeed, as there is no set method for performing literature searches, the documentary procedure is germane.

5.2.1 Bibliographic Databases

Literature searches take advantage of the existence of a vast number of bibliographic databases, which compile the data from a selection of journals. Some examples of bibliographic databases include Medline and EMBASE. The principal bibliographic biomedical databases are described in Section 5.8. Most bibliographic databases have their own thesaurus, i.e., a list of controlled vocabulary; for example, the thesaurus used in Medline is called Medical Subject Headings (MeSH), whereas in EMBASE the thesaurus used is EMTREE. These thesauri contain what are called "index terms," and publications included in bibliographic databases are "indexed" (i.e., assigned index terms) according to their topic. Publications are usually indexed by author, text words, and keywords. Index terms can be used as search terms in a literature search.

5.2.2 Sources of Information

Literature searches may be carried out using both electronic and print sources. Indeed, no single bibliographic database provides the answers to all medical queries. Medline is the best known biomedical bibliographic database; but other, potentially better resources should not be excluded in order to avoid systematic distortion (Tables 5.1 and 5.2).

Table 5.1 Queries

Query		Study design	Systematic reviews available?	Database/source
How many ...	Prevalence	Cross-sectional study	Very few/none	1. For systematic reviews and studies: Medline and/or other general databases[a] 2. National registers and statistics
How can we determine ...	Diagnosis	Cross-sectional study (with a gold standard)	Some	1. UpToDate, Best Practice 2. Other reviews and technology assessments (Cochrane Library) 3. Medline and/or other general databases[a]
How will it turn out ...	Prognosis	Cohort study	Very few/none	1. UpToDate, Best Practice 2. Medline and/or other general databases[a]
Why ...	Etiology	1. Cohort study 2. Case-control study 3. Patient series	Some	For queries on the side effects of treatments (as with specific drugs): 1. UpToDate, Best Practice 2. Other reviews (Cochrane Library), Evidence-based... journals 3. Medline and/or other general databases[a] Queries on other etiologies: 1. Medline and/or other general databases[a]
What can be done ...	Effect of intervention (treatment, prevention, rehabilitation)	1. Randomized controlled trial 2. Controlled trial without randomization 3. Cohort study 4. Case-control study	Many	1. UpToDate, Best Practice 2. Cochrane Reviews, other reviews and technology assessments (Cochrane Library) 3. Trials (Cochrane Library) 4. Medline and/or other general databases[a]
How is it experienced ...	Experience	Qualitative studies	Very few/none	1. CINAHL (good coverage in qualitative research) 2. Medline and/or other general databases[a]

[a]In this context, "general databases" means those not limited to specific study designs.

Table 5.2 Overview of the principal scientific bibliographic databases

	Medline	EMBASE	Web of Science	BIOSIS Previews	PsycINFO	CINAHL	Cochrane Library
Time span	1946–	1947–	1900–	1926–	1806–	1981–	Aims at comprehensiveness
Extent	>5600 journals from 70 countries Pre-Medline: new articles not yet indexed	>7600 journals from 70 countries	>12,000 journals and series, of which >3000 biomedical from 80 countries	About 5000 journals from 100 countries	About 2500 journals from 45 countries	About 3000 journals	>5600 systematic reviews >763,800 controlled trials
Scope	General medical database Covers medicine, nursing, dentistry, veterinary medicine, public health, allied health	Biomedical and pharmacological database. Human medicine, clinical and experimental Broad coverage of pharmacology, toxicology, drug therapy	Most medical disciplines, psychology, psychiatry, public health	Biology and biomedical science Pharmacology 80% medically oriented material 70% global coverage, 30% United States	Psychology, criminology, psychological aspects of education, sociology, health care, nursing, nutrition	Nursing Physiotherapy Occupational therapy Nutrition	Cochrane Reviews Other Reviews (DARE) Trials Methods Studies Technology assessments Economic evaluations
Focus	Biomedicine Clinical medicine	Medicine Pharmacology Drug research	Multidisciplinary research	Preclinical research in biology and biomedicine	Psychology Psychological aspects of disease	Nursing	Clinical medicine EBM
Updating	Varies according to vendor: daily, weekly, quarterly	Weekly. New entries indexed earlier than in Medline	<1 week after publication	Weekly	Weekly	Weekly	PWR (Publish When Ready)
Thesaurus	MeSH	EMTREE	Key Words Plus	BIOSIS Codes, MeSH, CAS Registry Nos	Thesaurus of Psychological Index Terms	12,000 CINAHL subject headings organized like MeSH	Partly MeSH

(*Continued*)

Table 5.2 (Continued)

	Medline	EMBASE	Web of Science	BIOSIS Previews	PsycINFO	CINAHL	Cochrane Library
	27,000 subject headings	>68,000 subject terms of which >30,000 cover drugs and chemicals	No thesaurus, but lists author keywords and automatically generated keywords		>7900 subject terms		
Abstracts	>83%	>80%	Yes	About 90%	Yes	Yes	Yes
Benefits	Free access in PubMed interface. Pre-Medline	Best database for pharmacological topics	Cited reference searching	In addition references to books, US patents, meeting reports and conference proceedings	Starting 1987 also references to English books and book chapters	Qualitative studies	Updated systematic reviews in full text
	Daily updates	Strong on European research	Provides information on impact factor		Unique source for older material	Includes references to books and book chapters, dissertations, videos and educational software	Best database for controlled trials
	General database for all healthcare workers	More subject terms assigned than in Medline					
References	>23 million	>22 million	>36 million	About 15.6 million	About 3.6 million	About 1 million	>521,000
Overlapping	60% overlapping at journal level versus EMBASE	60% overlapping at journal level versus Medline 10–75% overlapping depending on topic searched	75% overlapping at journal level versus Medline	About 30% overlapping at journal level versus Medline, about 3000 unique journals versus Medline, EMBASE likewise		1/3 unique material versus Medline	All controlled trials from Medline included, likewise from other sources
	Covers South America and Japan, better than EMBASE	Covers Africa and Asia, better than Medline					Systematic reviews indexed in Medline

5.2.2.1 Primary Sources of Information and Primary Studies

Primary sources of information are bibliographic databases or other resources that contain the bulk of individual studies (e.g., Medline and EMBASE). The individual studies contained in primary sources of information will be referred to in this chapter as primary studies. A primary study may present skilled, methodical research, but its conclusions apply to that specific study only.

5.2.2.2 Secondary Sources of Information and Secondary Studies

Secondary sources of information are bibliographic databases or other resources that provide access to secondary studies. The Cochrane Library, UpToDate, Best Practice, and Clinical Evidence, as well as secondary journals such as *ACP Journal Club* and several evidence-based journals (*Evidence-Based Medicine/Evidence-Based Dentistry/Evidence-Based Nursing*, etc.), are examples of secondary resources of information.

Secondary studies are compilations of primary studies; they include narrative reviews, systematic reviews, meta-analyses, clinical guidelines, and economic evaluations. In secondary studies data are taken from selected primary studies, sometimes including unpublished data; the data are then pooled, synthesized, analyzed, and conclusions drawn. The results of any review are never better than those of the primary studies on which it is based, but in general the conclusions drawn from secondary studies will have greater significance and will give an understanding of a topic that is beyond what is possible in a single study.

A narrative review usually presents the results and conclusions of two or more publications on a particular topic. They are often solicited by journals from experts in the field. For narrative reviews in which the authors fail to explain their search strategy and inclusion criteria, the reader should question whether essential studies have been omitted, and treat the conclusions accordingly. Indeed, due to their nature, narrative reviews are open to manipulation, as the authors may promote their own theories by deliberately selecting studies that support them. On the other hand, narrative reviews can offer a good, broad introduction to a new topic.

Systematic reviews often cover a more limited subject area. They aim to provide an overview of the literature on a specific topic and to draw conclusions from the publications they include. The methods section of a systematic review should include a description of the procedures used to collate the primary studies on a specific, relevant topic from the bibliographic databases exploited (see also Chapter 12).

Meta-analyses employ statistical methods to analyze and compile the results of several studies as though they were a single study. This method requires that the studies have comparable research designs and data. If this is not the case, the publication will be a systematic review, not a meta-analysis. Not all meta-analyses are based on systematic literature searches, so as a reader it is important to rule out any selection bias. Indeed, meta-analyses compiled from deliberately chosen studies may produce mathematically correct results, but may be clinically misleading. Again, an examination of the methods section should reveal the breadth and depth of the literature search.

5.2.3 The Evidence Pyramid

Critical appraisal is an essential part of evidence-based clinical practice, in which research evidence is examined for validity and relevance. It is therefore rational and time-saving to include critically appraised secondary studies when available. The hierarchy of different resources is often illustrated as a pyramid—the evidence pyramid—in which systems make up the top of the hierarchy, and single studies represent the bottom of the hierarchy. Figure 5.1 shows this hierarchy along with examples of corresponding resources. A systematic literature review starts at the top of the pyramid and drills down until an answer is found.

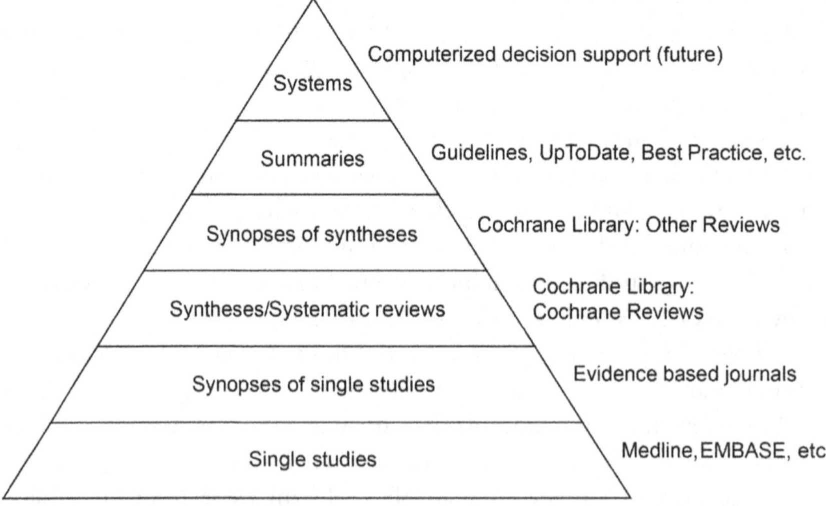

Figure 5.1 The evidence pyramid.

5.2.4 Search Strategy

Search strategies form the basis of literature searches and differ for basic and clinical research. Well-structured search strategies will retrieve more relevant publications, whereas more broadly phrased searches may generate endless lists of publications. Search strategies comprise all the steps that lead up to the actual literature search, i.e., establishment of a search query, selection of the most relevant sources of information, determination of the dates to be covered in the search, the search terms to be used, and the use of different search techniques. All of these steps are integral to a successful literature search.

5.3 ESTABLISHMENT OF A SEARCH QUERY

The preparation of a search strategy begins with the very important step of formulating a concise search query. New ideas on search queries often come to mind during a search session, but preparing a search query will save time in the actual literature search and, equally importantly, ensure the quality of the search strategy. A literature search based on a search strategy that starts with a vague search query will likely produce vague results. Literature searches must reflect the thesaurus of the bibliographic database being targeted. A search query comprising several components will make literature searches easier to perform, and search progress easier to control. Begin by dividing the search query into groups of keywords and synonyms, and use the thesaurus of each targeted bibliographic database to find relevant search terms.

5.3.1 PICO: A Focused Approach

The PICO approach (Population/patient/problem (P), Intervention/exposure (I), Comparison (C), and Outcome (O)) is often used in the formulation of focused research questions. However, PICO is also a useful tool for developing a search query based on that research question. Although search queries may not include all PICO components, most will have two or three of them.

For example, if the research question is "Does urban air pollution increase mortality in a population?" the search query is formulated by breaking this question down into the PICO components and entering them into the PICO table (Figure 5.2). Each column is an independent PICO component, as shown. The different PICO components are

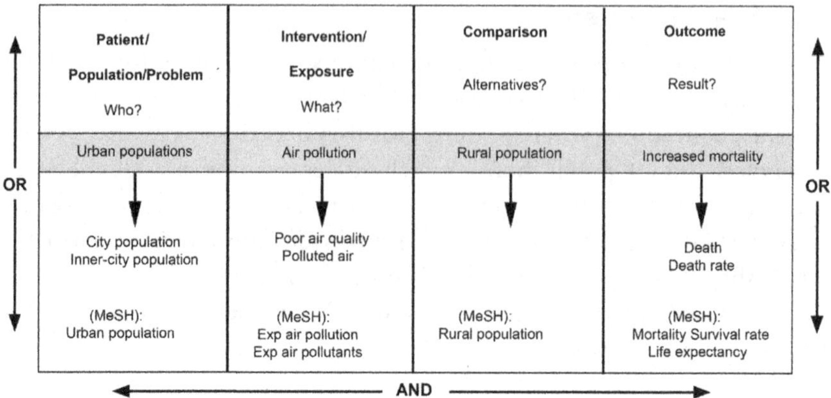

Figure 5.2 PICO table.

identified, and adequate search terms and synonyms are listed in the relevant columns of the PICO table, including text words and index terms from the thesaurus of each targeted database. Thereafter, the search terms are collated with OR (vertical list) and finally the search terms are combined with AND across all columns.

The search terms and combinations gleaned from the PICO table are then transferred to the search strategy. Though not all search terms are included, the structure of the search is evident (Figure 5.3).

5.3.2 Boolean Operator

Boolean operators define the relationships between words or groups of words. The AND, OR, and NOT operators are the most frequently used. The AND operator narrows the search, stipulating that all terms must be present at the same time to be included in the results. The OR operator is used to broaden the search, for instance to include synonymous search terms, stipulating that either term may be present in the results. The NOT operator is used to exclude a search term. Use NOT cautiously. It is a powerful operator that can eliminate valuable references. As shown in Figure 5.4, the various combinations of search results can be represented by circles. The black indicates the search results using the AND, OR, and NOT operators.

The choice of bibliographic database and the applications it offers determine the search functionalities available in that database. The search strategy can be set up in the form: search term, operator, search term (Table 5.3).

Number	Parameter	Hits	
1	Urban population (kw)	2810	Collect all concepts that characterize the patient group/population/problem
2	Urban population (tw)	1705	
3	1 OR 2	29,002	
4	Exp air pollution (kw)	35,114	Collect all concepts that characterzie exposure or intervention
5	Exp air pollutants (kw)	31,757	
6	Poor air quality (tw)	26	
7	4 OR 5 OR 6	58,760	
8	Mortality (kw)	28,799	Collect all concepts that characterize the outcome/ result
9	Life expectancy (kw)	8130	
10	Death rate (tw)	4890	
11	8 OR 9 OR 10	38,940	
12	3 AND 7 AND 11	75	Final combination

kw = keyword; tw = textword; exp = exploded/expanded term (Ovid)

Figure 5.3 Search strategy.

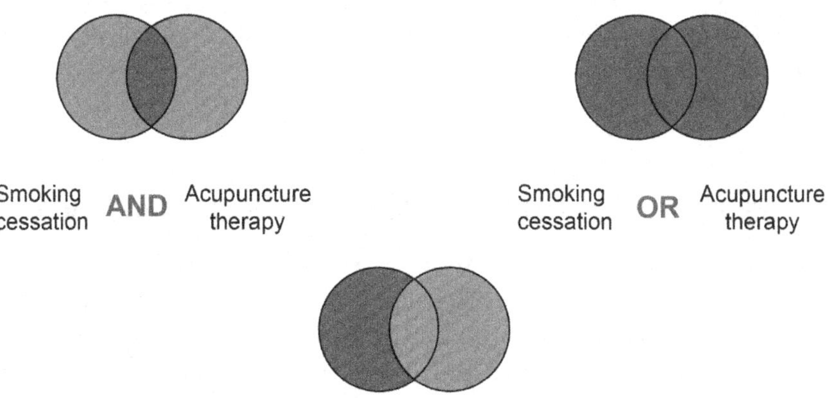

Figure 5.4 Boolean operators.

Table 5.3 Examples of uses of search terms and operators

Search term and operator	Result
Birth	Finds all documents containing that term
birth OR labor	Finds all documents containing either or both of the terms
birth AND forceps	Finds all documents containing both terms, but not necessarily adjoining
forceps ADJ delivery	The terms are adjacent to each other in the document
delivery NEAR birth	The terms are located near, within *n* words of each other, independent of order
birth WITH pain	The terms are in the same field, e.g., both are in the abstract
back NEXT pain	The terms are adjacent to each other in the order entered
injur*	Searches the word stem and the wildcard (*) yields injury, injuries, injured, etc.

Table 5.4 Operators supported by the principal database vendors

Operator	OVID	PubMed	Cochrane	ProQuest	EBSCO
AND	x	x	x	x	x
OR	x	x	x	x	x
NOT	x	x	x	x	x
ADJ	x				
NEXT			x		
NEAR/N			x	x	

The databases with the greatest number of available operators offer the greatest flexibility. For example, a Medline search via ProQuest, with its register of operators, allows for a more refined literature search than does a Medline search via PubMed. The operators supported by the principal scientific bibliographic databases are listed in Table 5.4.

5.4 SELECTION OF RELEVANT SOURCES OF INFORMATION

There is no cover-all database, so be sure to search several bibliographic databases when carrying out a literature search. Use various primary and secondary resources of information (e.g., PubMed's Related articles function, which is accessible from the abstract view), and do not limit the search to journal publications only. Check sources like the US National

Library of Medicine's Locator Plus (http://locatorplus.gov) for books and other materials that are pertinent to the research subject. The scope of journals or other media included is often a crucial criterion when selecting target databases. In addition, the research subject and the strength/weakness of the individual bibliographic databases should be considered. See Table 5.2 for an overview of the principal scientific bibliographic databases.

5.5 SEARCH TECHNIQUES

Knowledge of search techniques is important if one is to plan a successful search strategy. Indeed, scientific bibliographic databases comprise huge amounts of information, which can present a challenge. A literature search can turn into a search for a needle in a haystack if proper search techniques are not applied. Most bibliographic databases offer two types of searches: a Basic search and an Advanced search. Advanced searches are recommended to fully exploit the features of a given bibliographic database.

5.5.1 Thesaurus Mapping

Thesauri automatically try to map, or translate, plain-language phrases to their respective index terms. For instance, if one types in "kidney stones" in EMBASE, its thesaurus EMTREE will map that term to "nephrolithiasis," whereas the Medline thesaurus MeSH will map to "kidney calculi." This feature is of great help when building a search strategy, as it automatically maps to adequate index terms and synonyms. For instance, when searching Medline via PubMed, be aware that any truncation (i.e., word-stem search) of search terms plus the wildcard (*) will switch off the mapping feature. To illustrate this point, searching for "kidney stone*" with the aim to retrieve both singular and plural forms will not map to the corresponding MeSH term. So while the intention is to broaden the search, the result in fact narrows it considerably.

5.5.2 Federated Search

A federated search is a search of several databases in one operation. Indeed, some bibliographic database vendors sell packages of databases and support federated searches (e.g., in OVID, with the option "select more than one database to search"). When starting a research project, a

BOX 5.3 General Search Tips

- When creating a search strategy, be as topic-specific as possible. Use the thesaurus of the bibliographic database to find index terms and their interrelation.
- Remember that index terms can have their own "histories," meaning they may have changed over time. Check Scope Note (in Medline via OVID) or the MeSH Browser in PubMed, and formulate the search strategy accordingly. For instance, "Mongolism" has been changed to "Down's syndrome."
- As the indexing of publications is never completely correct in all contexts, consider searching by combinations of index terms and text words. Think of alternative terms and synonyms. Do not rely on one search term alone; add an index term or text word.
- In addition, perform an author search.

federated search can give a bird's-eye view of the relevant publications and identify interesting sources of information. However, federated searches should be executed with care to avoid the pitfalls that can result when the databases used have different structures, thesauri, publication types, etc. Indeed, surveys have shown that a large number of publications are indexed in different ways in different bibliographic databases. Hence, federated searches should be avoided when the goal is to find all available information on a particular topic. In those cases the different bibliographic databases should be searched individually for best results. However, advanced search techniques can be used to eliminate duplicate entries in bibliographic databases provided by the same vendor, resulting in a single set of publications. Such advanced searches require in-depth knowledge of literature searches, and cooperation with your local academic library may be productive (Box 5.3).

Regardless of how well prepared and structured search strategies may be, it is not uncommon that the search results do not render what one hoped. Sometimes search results may retrieve no publications. This might mean that the research problem in question is still relatively unexplored, but it is more likely that the search strategy needs to be recalibrated. See Box 5.4 for tips on expanding search strategies.

On the other hand, when search results render far too many publications, the search strategy should be revised as well. See Box 5.5 for tips on narrowing search strategies.

BOX 5.4 Tips on Expanding Search Strategies

- The members of a research team should search independently, as each member may find relevant, unique publications.
- Check the index terms assigned to relevant articles and use them to initiate complementary searches.
- Track down more recent articles that cite an important, older article by using "Cited ref search," which is available through the Web of Science (Thomson Reuters, subscription required). Note that a similar service is available free of charge through Google Scholar's "Cited by."
- Expand the search using related terms (mortality/survival rate).
- Sometimes a search strategy can be augmented using contrasting terms, e.g., entering "rural population" in a search strategy that concerns urban populations.
- Consider broadening the time scale of the search, and check to see if the index terms entered have changed over time. MeSH terms, for example, have scope notes with definitions for every entry.
- Be aware of how various databases handle expanded search commands like "Explode." In PubMed, all search terms are automatically expanded whenever possible. However, in Medline via OVID, search terms are expanded only upon command.

5.5.3 Free-Text Search

Free-text searches in bibliographic databases retrieve articles with those words in the title, abstract, or keyword fields. This is also sometimes called a text-word search. Although text words will map to the corresponding terms in the thesaurus of a bibliographic database when possible, only indexed publications will be retrieved that way. When a comprehensive literature search is required, consider adding text words to the search strategy. This will help retrieve the latest publications, including those that have not yet been fully indexed. Even when the index terms in thesauri are dynamic, reflecting the developments in the fields they cover, it takes some time before new index terms are established for scientific breakthroughs.

Hence, the search strategy must take free-text searches into consideration (Box 5.6).

5.5.4 Searching by Index Term Versus Publication Type

A search by index term searches content, not publication type; therefore, searching by index term and by publication type renders very different

BOX 5.5 Tips on Narrowing Search Strategies

- Consider whether the search term should restrict hits to articles in which the term describes principal content. Limit by using "Focus of the article" (Medline via OVID)/"Major topic" (PubMed). In literature searches on a single topic, limit to the major topic. In literature searches on more than one topic, do not limit to the major topic.
- Be cautious in limiting to subheadings due to inconsistent indexing. Instead, try to limit the search in other ways. For example, by "human," or by publication type (review, randomized controlled trial, etc.), methodology filters, etc.
- For studies concerning humans only, use the "human" limit. Since many articles are tagged with both "human" and "animal," limit the search with "animal" as well, and then exclude them with the NOT operator. This will retrieve articles on humans only.
- To preclude language bias, be cautious in limiting to a specific language, as it may lead to the loss of valuable information.
- The delineations of age groups differ between databases; be sure to use the proper designation.
- Limiting to age groups, human or animal studies, language, and publication type is carried out as the last step in the search strategy.

BOX 5.6 Free-Text Search Tips

- Use synonyms, such as those arising from differing terminology between countries or disciplines (e.g., bed sores/decubitus ulcer; SIDS/cot death).
- Truncation (with wildcard): includes singular and plural forms, and other orthographic endings. For instance "injur$" will search for injury, injuries, or injured. Truncation symbols (e.g., * $: ?) may differ from database to database.
- Include British and American spellings (tumour/tumor). Search both: "tumo?r".
- Abbreviations of terms (AIDS/acquired immunodeficiency syndrome).
- Trademarked and generic names (Tamiflu/oseltamivir phosphate).
- A free-text search also elicits false hits, as the text words may have connotations other than those relevant to the search strategy at hand.

results. It is possible to limit a search strategy by publication type, as stated under "Limits" in individual bibliographic databases. For example, in Medline via PubMed the subject heading "randomized controlled trials (plural) as topic" applies to articles that describe methods or discuss applications, while "randomized controlled trial" (singular) will be treated as a publication type, and thus only articles that in fact are randomized controlled trials will be retrieved. Since publication indexing is not always correct in every detail, it is recommended to simply perform a free-text search for "randomized controlled trial," which will retrieve all publications, including irrelevant ones. In Medline via PubMed the same applies for "meta-analysis as topic" (MeSH) and "meta-analysis" (publication type), and "review literature as topic" (MeSH) and "review" (publication type). The "as topic" terms are searched in the same way as for any other subject. A search may be limited by publication type using the database's "Limits/Limit to" function.

Lastly, when the literature search is complete, print and save the search history to document the literature search method used. State the databases used and time spans covered. Create a personal account with the database, which can be used to store search strategies (see Section 5.9).

5.5.5 Methodology Filters

Each year, millions of scientific publications are released, but their quality varies. High-quality research must have a solid basis and must be carried out to a high standard. One way of sifting out studies of a presumed higher quality is by formulating a search strategy that also focuses on the methods described. This is particularly important in clinical research and practice, known as evidence-based medicine (Box 5.7).

Methodology filters are based on previous systematic literature searches for a specific time span, in which the goal is to find search strategies that yield optimal results for each of the research designs. Methodology filters are formulated to obtain a search result that closely approaches the gold standard of a completely successful search. The filters may consist of comprehensive search strategies, covering the words and concepts that together delineate the research design. Consequently they are suitable for

BOX 5.7 Methodology Filters in Brief

- One filter for each study design.
- Developed and tested to retrieve high-quality research papers.
- Reduce the number of hits in a search result.
- Elicit quality rather than quantity.
- Adapted to the structures of individual databases.

combination with any subject search (i.e., a search consisting of both index terms and text words) in which a specific research design is sought. In short, a therapy filter includes terms like "random allocation" and "randomized controlled trial," while a diagnosis filter includes the obvious pair of words "sensitivity and specificity."

There is no standard logical structure for bibliographic databases. Because of this, methodology filters must be developed for each individual database and must suit the search terms and publication types used by that bibliographic database. In databases, methodology filters are available via clinical queries or through "Limit to" facets.

5.5.5.1 PubMed Filters—Clinical Queries

In PubMed, methodology filters are implemented as automatic searches accessible using the search entry "Clinical Queries," making it easy to search by clinical study category. The filter tables at http://www.ncbi.nlm.nih.gov/books/NBK3827/#pubmedhelp.Clinical_Queries_Filters list the search parameters of the individual filters.

PubMed has filters for etiology, diagnosis, therapy, prognosis, clinical prediction guides, systematic reviews, and medical genetics. In addition to the methodology filter, "Sensitive/Broad" or "Specific/Narrow" may be checked. A literature search that focuses on sensitivity will elicit a greater number of references than a corresponding search focused on specificity. In this context, sensitive is defined as a broad search that captures as many relevant references as possible, and specific is defined as a narrow search that weeds out as many irrelevant references as possible.

The "Systematic reviews" filter is matched to systematic reviews and meta-analyses and is searchable either from the Clinical Queries entry, or as a limitation by "Article types" in the general search pane.

The "Medical Genetics" filter finds citations related to various topics in medical genetics.

5.5.5.2 Searching for Qualitative Research

There are relatively few qualitative studies compared to the vast number of quantitative publications in the scientific field. Searching for qualitative studies may be a challenge due to lack of abstracts, inconsistent indexing across databases, etc. Hence, search strategies should be created to include the terms relevant to this methodology. It is not considered wise to rely on index terms alone, but rather to search for a combination of index terms and text words.

The Cumulative Index on Nursing and Allied Health Literature (CINAHL) is the preferred bibliographic database for literature searches on qualitative research, because it has a wider range of index terms that reflect this type of research. In addition, CINAHL offers well-structured methodology filters on three levels through the database's "Limit to" facet: High sensitivity—High specificity—Best balance.

Although there are tools in PubMed that allow one to search for different types of quantitative study designs, there are no such tools as predefined search filters that allow one to search for the different categories of qualitative research. The qualitative research filter available in PubMed is not intended for clinical research, but rather for health-care quality and cost. Therefore, it is listed under "Health Services Research Queries." No methodological criteria are included in this filter.

5.6 CRITICAL ASSESSMENT

Upon completion of the literature search, the researcher must evaluate the quality of the retrieved publications. The list of references should be examined to decide which papers, books, or other media should be read in detail. At this point, one may often contact the local academic library to order any publications that cannot be found in the research institution's collection. There is no simple way to determine whether a publication fulfills quality requirements, but there are a few key things to keep in mind when deciding whether a more thorough evaluation is required. The researcher must be sure that there are no retracted articles among the publications retrieved, and that information is not taken from the abstract of a publication alone. After this, each publication must be critically assessed: is it of sufficient quality to be included in the dossier? Visit http://www.cebm.net/ for a description of relevant procedures on critical appraisal (see also Chapter 12).

5.6.1 Abstracts

Most references in bibliographic databases include abstracts. It may be tempting to use information from abstracts only, as this requires less time (and money). However, as shown convincingly by Pitkin et al.[2] and confirmed more recently by others, it is inadvisable to rely upon abstracts alone. Pitkin et al.'s study aimed to assess the degree to which the contents of abstracts were consistent with the contents of the corresponding publications. They found that "the proportion of deficient abstracts varied widely (18−68%) and to a significant degree..." Even in articles published by the renowned "Big Five +" medical journals included in the survey (*BMJ, Lancet, JAMA, New England Journal of Medicine, Annals of Internal Medicine, Canadian Medical Association Journal*), the conclusion was that the content of an abstract was often inconsistent, or even in contrast with the content of the corresponding publication. This indicates that basing work on information gleaned from abstracts alone is risky; many might even maintain that it constitutes unethical research.

Researchers should only cite publications they have read in full. Regrettably, this practice is not adopted by all, as evidenced by the many incorrect references that are widely used by authors who simply copy the publications cited by other authors, without making a critical check of the original publications. Whenever such chicanery is exposed, the author's credibility is tarnished and the resulting publication has to be reassessed accordingly. Surveys have shown that citing erroneous references may be a widespread practice, which creates extra work for those who strive to find original publications to cite.[3]

5.6.2 Errata and Retracted Publications

Scientists are concerned with keeping abreast of developments in their respective fields. There are several ways to stay informed using current awareness services and e-mail alerts, which will be described in detail in Section 5.9. However, if one acquires publications in this manner only, one may miss articles that have been subsequently retracted. There are many possible reasons for retracting a publication. Sometimes the authors themselves have reasons for retracting a publication; other times employers may not accept responsibility for the publication. In the worst case, the work may have been exposed as scientific misconduct or fraud. Regardless of the severity of the reason for retraction, studies are retracted every year. However these studies remain listed in bibliographic databases with accompanying notes such as "retracted publication," "retraction of publication," or "erratum."

Searches have shown that bibliographic databases differ in the speed with which they trace and register changes such as retractions and errata. Consequently, a paper may be listed in Medline as a valid publication, but simultaneously tagged with an erratum in EMBASE. This is yet another reason to search more than one bibliographic database. Indeed, each scientist is obligated to ensure that incomplete or flawed studies are not included in their background material. Studies have shown that even though a paper has been retracted and marked accordingly in bibliographic databases, it may appear in the reference list of publications. Scientists who glean publications from reference lists published by others, instead of from their own, well-formulated literature searches in bibliographic databases, can overlook retracted publications. Moreover, such practice is unethical. To identify possible retracted articles and eliminate them, search for "erratum" or "retracted publication" as a text word, and use the databases' limiting features when available. Consider finishing all searches with "limits: retracted publications" (Medline via OVID). Alternatively, use the search filter "Retracted Publication," which is available through PubMed's "Topic-Specific Queries."

5.7 BIBLIOMETRIC MEASURES

5.7.1 Impact Factors

Starting in 1961, the ISI Science Citation Index provided a basis for the impact factor (IF), which is a measure of the frequency with which an "average article" in a journal has been cited in a given year. The IF for a specific year is calculated as the ratio of citations over 1 year generated by articles published in that journal in the previous 2 years, calculated as follows:

$$\text{Impact factor} = \frac{\text{Total number of citations}}{\text{Total number of citable articles}} \text{ in that specific journal.}$$

The IF was never intended to be a universal criterion for journal quality, even though it is frequently regarded as such. For example, scientists may consult the IF when deciding to which journals they should submit papers. Employers and funders may use the IF as a criterion in hiring or appointing scientific staff. The IF is also used by institutions when evaluating journal subscriptions, and by funding bodies when allocating research funds. Quite naturally, funding bodies seek assurance that funds are channeled to the scientific communities in which they will have the greatest benefit, and employers are interested in hiring the best people. It is also natural that scientists wish to disseminate their works as widely as possible. Nevertheless, it is uncertain whether the IF should be a decisive criterion.[4] A thought-provoking study even pointed out an apparent correlation between IF and the probability of retraction, reporting that the higher the journal's IF, the greater the chance that its articles will be retracted at a later date.[5] Indeed, the IF has too many weaknesses to justify the position it has attained when it comes to the assessment of publication quality.[6] It is preferable to assess the quality of publications by the number of citations, and who cites the publications, rather than only considering the journals in which articles are published. In short, there is no easy way to assess quality.

"A true appreciation of the validity of quality judgments will, therefore, require a knowledge of the quality of the judges, for a committee of camels will never approve a horse."[7]

The IF should be used with care for the following reasons:

- It does not reflect the quality of an individual article, only the average measure of the use of a journal as a whole for the 2 years included in the calculation. Usually, only a few articles generate half of the citations that can be attributed to the journal. Hence, articles not cited also enjoy the full weight of the IF.
- The IF includes self-citations (graphs of self- and nonself-citations are included for each journal).
- Preclinical journals are cited more often than clinical journals, and clinical researchers often cite preclinical articles. Understandably, the reverse is less frequent, as knowledge must be acquired before it is put to clinical use.
- The IF cannot be compared across disciplines or specialties due to differing citation practices.
- Dynamic disciplines that continually produce new findings that supplant old knowledge (such as biochemistry and molecular biology) are favored due to the time limit used in the IF calculation. Work of greater longevity is disfavored.
- Journals with supplements are at a disadvantage, because supplement series usually generate fewer citations and lower citation frequency.
- Review articles are more often cited, and thus journals that publish many such articles are favored.
- Journals with extensive correspondence sections have an advantage, because although letters to the editor are not taken into account as the basis of calculation (the denominator) for the IF, citations to them are included.
- IF is calculated for selected indexed journals, predominantly English-language journals published in the United States. Estimating coverage for various disciplines is difficult, but the Journal Citation Reports database includes more than 12,000 titles in the subject areas of

medicine and the natural sciences, of which some 3000 are medical journals. In any case, the selection of journals indexed in the Journal Citation Reports database is small in the overall scheme of journals published. According to the Ulrichweb Global Serials Directory, there are some 33,900 journals published in the fields of medicine and health.

- Might unfavorable mention be preferable to no mention at all? If, for instance, many researchers wish to comment on and reject a work exposed as scientific misconduct, this will result in many citations of that work, which thus increases the corresponding journal's IF.

IFs can be found in the Journal Citation Reports database, which is available by subscription. Another approach is to use a general internet search engine and enter "impact factor" and the journal name as search terms. The home pages of journals also often list their IF.

5.7.2 The Highly-Cited Index

In 2005, J. E. Hirsch[8] introduced an alternative bibliometric measure, the Highly-cited index (H-index), which focuses on the individual researcher's impact on his or her academic environment. Hirsch's definition is as follows: "A scientist has an H-index h if his or her N_p publications have at least h citations each, and the other $(N_p - h)$ publications have $\leq h$ citations each."

One's own H-index can be calculated if the number of citations for each publication is known; Table 5.5 gives an example of an author with an H-index of 7.

There are H-index calculators available online, which can be found by performing a general Internet search.

5.8 PRINCIPAL SCIENTIFIC BIBLIOGRAPHIC DATABASES

An overview of the principal scientific bibliographic databases was given in Table 5.2, and a narrative overview is given in this section. Most

Table 5.5 H-index calculation for an author with an H-index of 7

Number of publications	1	2	3	4	5	6	7	8	9	10
Citations	45	32	17	15	11	10	8	3	2	0

bibliographic databases require a subscription. Contact your local academic library to find out which databases are available through local networks, and to obtain training in how to search them effectively.

5.8.1 Medline

Medline was developed and is maintained by the US National Library of Medicine. It is the largest and most widely used biomedical bibliographic database. It contains records of more than 23 million articles published in more than 5600 journals and covers the fields of medicine, nursing, dentistry, veterinary medicine, health-care systems, and preclinical sciences, as well as associated disciplines. It is international, but most of the references (~91%) are from journals published in English. Medline contains records of articles from 1946 to the present, and approximately 83% of these records include abstracts in English. Medline is available through many interfaces, both by subscription through database vendors, such as OVID and EBSCO, and for free via PubMed.

Medline's thesaurus MeSH contains approximately 27,000 subject headings (in 2014), which are organized both alphabetically and hierarchically in a tree structure that is searchable at various levels, from the narrow (more specific) to the broader (more general), and the reverse (Figure 5.5). At the general hierarchical level are terms such as "anatomy,"

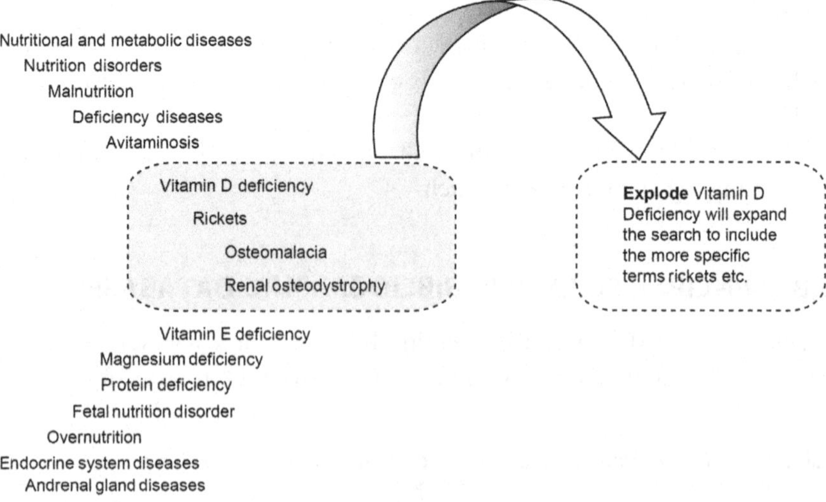

Figure 5.5 Section of a tree structure: hierarchical index.

"nutrition disorders," and "enzymes and coenzymes." At the more specific level are terms such as "ankle," "avitaminosis," and "calcineurin." MeSH is updated once a year to reflect changes in medicine and medical terminology.

MeSH terms are used to index the content of articles included in Medline. There may be 10–20 MeSH terms assigned to any one article. By examining these MeSH terms, one can get additional information about an article's content beyond reading the abstract. MeSH terms may also be a starting point to find other relevant search terms.

A search by MeSH terms involves synonym control and yields precise search results. Indeed, one MeSH term spans many concepts, so a single term retrieves relevant references regardless of which words or concepts the author(s) of the article may have used in the title, abstract, or keywords. For example, a search using the MeSH term "cerebrovascular accident" will cover all words and concepts associated with stroke. Applying the Explode function will include all subordinate aspects of a general subject term. A subheading may be associated with a subject heading in the form "subject heading/subheading" in order to search for one or more aspects of a subject, such as:

- molecular biology/methods
- stem cell transplantation/ethics
- low back pain/radiography
- myocardial infarction/drug therapy

5.8.2 PubMed

PubMed (www.pubmed.gov), developed by the National Center for Biotechnology Information (NCBI) at the US National Library of Medicine, affords free access to several major databases, including PubMed, Nucleotide, Protein, Structure, Genome, Taxonomy, OMIM, and others.

PubMed provides access to all of Medline and is updated daily with reference data supplied directly by publishers, often before a journal issue is released. These records are often available with abstracts, but they are not immediately indexed. PubMed records are linked to full-text electronic articles whenever publishers have an agreement with the National Library of Medicine. Individual users may access the electronic journals to which their institutions subscribe. Contact your local academic library for an overview (Box 5.8).

BOX 5.8 Differences Between PubMed and Medline

- From general science to chemistry journals, only articles on the life sciences are indexed for Medline. PubMed, however, includes all articles, including those outside the Medline scope (e.g., articles on astrophysics, plate tectonics).
- PubMed includes in-process citations that provide a record for an article before it is indexed.
- PubMed includes links to full-text articles published by PubMed Central (the digital archive of biomedical and life sciences journals published by the US National Institutes of Health).

5.8.3 EMBASE

EMBASE (www.embase.com) is published by Elsevier Science. It is a European-oriented database with records from 1947 onwards on more than 22 million articles from more than 7600 scientific journals in pharmacology and medicine, including about 2000 journals not indexed in Medline. More than half of the journals included in EMBASE are published in Europe. The topics covered concern drug research, toxicology, dependence and abuse, clinical and experimental medicine, health care, psychiatry, forensic medicine, and biotechnology. More than 80% of the newer references have abstracts, and approximately 75% are from journals published in English. EMTREE is the EMBASE thesaurus, which includes approximately 68,000 index terms, including MeSH terms, organized in a tree structure. The overlap between EMBASE and Medline is 10–75% depending on the topic.

5.8.4 Web of Science

The Web of Science is published by Thomson Reuters and is an interdisciplinary database with records from several bibliographic databases, among them Science Citation Index Expanded (SCI-EXPANDED) and Social Sciences Citation Index (SSCI). SCI-EXPANDED includes records from most of the medical disciplines. SSCI coverage includes public health, psychology, and psychiatry. The Web of Science contains records of publications from 1900 to the present.

In addition to searching by topic, it is also possible to carry out a citation search in the Web of Science, i.e., to find to what degree a particular author or work has been cited. A citation search in Google Scholar is described in Section 5.10. Further information is available at www.wokinfo.com.

5.8.5 BIOSIS Previews

BIOSIS Previews (www.thomsonreuters.com/biosis-previews) is published by Thomson Reuters and comprises references to articles published in 5000 journals in the fields of biology and the biomedical sciences. The database also includes references to books and patents held in the United States, as well as reports from meetings and conferences. BIOSIS Previews contains references from 1926 to the present. Approximately 90% of the references have abstracts, and approximately 30% of the journals included therein are also included in Medline and EMBASE.

5.8.6 PsycINFO

PsycINFO (www.apa.org/pubs/databases/psycinfo) is published by the American Psychological Association and comprises references to articles published in approximately 2500 scientific journals, as well as records of research reports, doctoral dissertations, and conference reports. The database covers psychology and related topics, including education, sociology, medicine, nursing, physiology, nutrition, and psychiatry. PsycINFO contains records from 1806 to the present and books and book chapters from 1987 onwards published in English.

5.8.7 Cumulative Index to Nursing and Allied Health Literature

CINAHL is published by EBSCO and includes records on articles, book chapters, theses, etc., within the fields of nursing, physiotherapy, occupational therapy, and other allied health sciences. More than 3000 journals are included in CINAHL, mainly in the English language; depending on the subscription chosen they date back to 1981. CINAHL should be the first choice when searching for qualitative research. Further information is available at www.ebscohost.com/academic/the-cinahl-database.

5.8.8 Cochrane Library

The Cochrane Library (www.thecochranelibrary.com) is the best source for finding systematic reviews and randomized controlled trials concerning the outcomes of health-care initiatives. It is updated on a published-when-ready basis, and comprises several databases, as follows.

5.8.8.1 Cochrane Reviews

Cochrane Reviews are full-text systematic reviews compiled by The Cochrane Collaboration. Cochrane Reviews are updated whenever new scientific documentation is available. The database also contains protocols of systematic reviews that are in the process of compilation.

5.8.8.2 Other Reviews

The Cochrane Library also comprises published abstracts of high-quality systematic reviews from other sources. The abstracts are compiled by the Centre for Reviews and Dissemination at the University of York, United Kingdom, by the American College of Physicians' Journal Club, and by the journal *Evidence-Based Medicine.*

5.8.8.3 Trials

Trials is a database of controlled trials compiled by various Cochrane Review Groups. The references are gathered from several bibliographic databases worldwide, as well as through manual searches in journals. Trials is the best single source for controlled trials.

5.8.8.4 Economic Evaluation

This database comprises structured abstracts concerning economic evaluations of various measures in health care.

5.9 STAYING UP TO DATE

Staying up to date in the medical field is challenging, and over the years professionals have developed different strategies to stay informed. Alert services are one of the strategies that are constantly being developed and improved. However, it is important to remember the previous warnings in this chapter against limiting information to the most recent resources only, which can lead to missing information regarding retractions or errata issued at a later date.

5.9.1 Saving Search Strategies

Well-structured search strategies should be saved for future use. A search strategy that is limited to, for example, the last 30 days before being saved can be re-run once a month in order to stay updated. This is a current awareness service, formerly known as Selective Dissemination of Information.

Some bibliographical databases offer personal accounts (My Profile, My NCBI in PubMed, etc.) that can be used as efficient tools to store reference lists and save search strategies for later use. However, different bibliographic databases do this differently. For example, PubMed search strategies can be saved and retrieved in a manner that is clearer and easier to read than in the "Save in My NCBI" option on the search set number's drop-down menu. In Save in My NCBI, the search strategy is saved as one single search string that is quite challenging to read. When the purpose is to present the search strategy as a part of the documentation and methods, it is better to use the "Download history" option, which saves the history as a comma-separated values (cvs) file. This method leaves a search strategy nicely organized with hits shown for every stage in the process. One can open this file in Excel, mark one column at a time, and follow the instructions in the "Convert to columns" wizard. On the data tab click "Text to columns," choose file type "Delimited," choose the delimiter "Comma," and select Finish. Finally, adjust the column width according to the contents.

5.9.2 Alert Services

Alert services are time-saving and efficient services offered by both bibliographic databases and electronic journals. These services deliver new references or articles by e-mail or Really Simple Syndication (RSS) feed on the basis of a saved search strategy. RSS is an Internet standard for the delivery of news and other frequently updated content provided by websites. Some electronic journals also have their own alert services, like table of contents alerts via e-mail or RSS. RSS feeds are available via an RSS reader (free Internet download) or an e-mail client.

5.9.3 Awareness Tools: Apps for Mobile Devices

Consumer electronics are constantly evolving, and new equipment and services continue to be launched for different platforms. Current mobile apps (applications) include awareness tools that allow researchers to access, read, and store journal articles.

The following apps can be downloaded for free, but the journals available through these apps are dependent on an institution's subscriptions, with the exception of open access journals. All the apps below offer similar services and choosing one or the other might be a matter of taste. Testing them can help to determine which app best meets one's individual functionality requirements.

5.9.3.1 BrowZine (for iOS and Android Devices)

This app presents personalized favorite journals in a virtual bookshelf, organized either alphabetically or by subject. One can browse tables of contents and select interesting articles to read in full text. Articles can also be stored for reading offline, or uploaded to bibliographic software like Mendeley or RefWorks. Uploading to EndNote Web will be a later feature. A large number of universities support BrowZine (http://thirdiron.com/browzine/).

5.9.3.2 Docphin (for iOS and Android Devices)

Using Docphin, one can select journals to follow. When an institution is selected, subscription-based and open access journals are available. Specific articles can be tracked down using the search feature. In addition to traditional search criteria, articles can be filtered to retrieve particular publication types, like meta-analyses, practice guidelines, reviews, etc. The Medstream entry lists the current highly read highlights of medical topics in the news. Docphin can be downloaded from the Apple App Store or Google Play.

5.9.3.3 Read by QxMD (for iOS Devices, Coming on Android)

This app introduces itself as a "Flipboard for medical journals." It is like an interactive magazine in which the journals to be followed are selected either individually or by category. The PDFs of key articles allow interactivity, like highlighting text, adding notes, etc. Articles can be easily shared with colleagues either by e-mail or via social media, thus encouraging professional discussion. Calculate by QxMD is a separate app, with a large number of calculators supporting the different subspecialties of the medical field.

More information on this app is available at http://www.qxmd.com/apps/read-by-qxmd-app; it can be downloaded from the App Store.

5.10 MEDICAL AND SCIENTIFIC INTERNET SEARCH ENGINES

Although a great deal of information is available for free via the Internet, in many situations its scope is insufficient, as you "get what you pay for." Depending only on information freely available on the Internet is a risky business, as chance, not system, dictates the results. Instead, after exhausting the search possibilities found in bibliographic databases, consider using Internet search engines to find supplementary information. Several dedicated search engines are available for this purpose, but as none of these search the entire Internet, several should be used in order to make thorough searches. Three examples of medical and scientific Internet search engines follow.

5.10.1 Google Scholar

Google Scholar is a search engine for academic literature with links to free material on the Internet. It is stronger in biomedicine than in other disciplines. It harvests information from PubMed and a number of publishers and scientific associations, but its coverage is incomplete. A general Internet search is suited to rapidly gain a perspective over the material available. A general Internet search can be a useful source for gray literature, such as unpublished conference material.

"Cited by" in Google Scholar is an interesting function that links to theses and papers that have cited original publications. Google Scholar is available at http://scholar.google.com/.

5.10.2 MacPLUS

MacPLUS is a medical and scientific search engine designed "to find the best evidence-based answer to your clinical questions by simultaneously searching the leading evidence-driven medical publications and high quality clinical literature." In fact, federated searches are performed as the search engine takes into consideration the different structures of the databases covered. No complicated search strategy is needed; just type in one or two words to cover the topic. The search results are listed according to the hierarchy of sources shown in the evidence pyramid (see Figure 5.1). MacPLUS is free and available to medical and health-care professionals through ACCESSSS (http://plus.mcmaster.ca/ACCESSSS, registration is mandatory), although full text may still involve traditional interlibrary lending services.

5.10.3 SUMSearch2

SUMSearch2 simultaneously searches various medical Internet resources and databases, mainly the US National Library of Medicine, DARE, and the National Guideline Clearinghouse, presenting results in groups ranked

BOX 5.9 Evaluating Quality on the Internet

- Source description: who is the publisher?
- Extent: what is covered?
- Objective or just advertising?
- Updating?
- Is the content plausible?

according to form and relevance. It is principally for clinicians, and was developed for evidence-based medicine. SumSearch2 is available at http://sumsearch.org.

5.10.4 General Guidelines for Internet Searches

Internet searches, performed either by surfing or by using search engines, require a high degree of awareness in order to assess the quality of the sources. Critical vigilance is essential, as anyone can put anything on the Internet. Some guidelines to keep in mind are listed in Box 5.9.

5.11 FINDING RESEARCH PROTOCOLS AND ONGOING PROJECTS

In the initial stages of a research project it is important to track down ongoing research activity in the subject area, as planning research projects that have already been undertaken by others is a waste of time and resources. All clinical trials should be registered as part of a researcher's responsibility to carry out ethical scientific research.

The World Health Organization's International Clinical Trials Registry Platform is a portal that harvests clinical trials listed in registries around the world. The search portal is available at http://apps.who.int/trialsearch/. However, since search interfaces may be poor, it is wise to search additional trial registries:

- ClinicalTrials.gov Publisher: The US National Institutes of Health, http://www.clinicaltrials.gov. EU Clinical Trials Register, available at: http://www.clinicaltrialsregister.eu. Cochrane Database of Systematic Reviews Protocols are tagged.
- PROSPERO—International prospective register of systematic reviews.
  Since systematic reviews are especially time-consuming and resource-demanding, it is important to register protocols at their inception to avoid

unplanned duplications. PROSPERO covers research protocols for systematic reviews in health and social care. PROSPERO is available at: http://www.crd.york.ac.uk/PROSPERO/.

5.12 REFERENCE MANAGEMENT

It is wise to keep a record of all publications accessed. Reference management software can be helpful to keep track of publications, as it can record them in a personal literature database created by the researcher. It is better to include too many publications in the literature database than too few, as in the early stages of work it is not always clear which publications will finally be cited in the research report or journal article.

5.12.1 Efficient Handling of References, Publications, and Manuscripts

Building a personal literature database is an essential part of a research effort. Reference management software is specifically designed to record and utilize bibliographic records, and to link them to full text/PDFs whenever possible. The software can be integrated with word processors, and reference lists can be produced in the formats required by different publishers and academic journals. References can also be imported from bibliographic databases. EndNote is an example of a reference management program often used in academic settings, due to the flexibility and capacity it offers, but there are many more packages available on different platforms. Details can be found in this comparison table: http://en.wikipedia.org/wiki/Comparison_of_reference_management_software.

Contact your local academic library to find out which reference manager programs are available through your institution and to obtain training in how to use them efficiently.

5.12.2 EndNote

EndNote (www.endnote.com) is produced by Thomson Reuters and has been specially developed to manage bibliographic references. Its principal functions are described below; see the program manual for further details.

EndNote is an easy-to-follow program that handles nearly 50 reference types (articles, books, chapters in books, conference proceedings, websites, etc.). Each reference contains a particular number of fields, according to its type, and a literature database in EndNote (an EndNote library) can store an unlimited number of references. Each single reference registered receives a unique ID number, and can have several files attached, for instance PDFs, figures, and videos. Even though it is possible to set up an unlimited number of libraries, it is recommended that all references be registered in one main library. This eases the preparation of manuscripts, the generating of reference lists, and the moving of files. The "Group" function can be used to sort references by topic within an EndNote library.

5.12.2.1 Settings for an EndNote Library

Several settings options may be chosen from the Preferences menu. One may specify how thoroughly references are to be compared for duplicates, activate OpenURL to use an institution's server to access full text, determine how to handle PDFs, modify reference types, and so on.

5.12.2.2 Journal Index

The installation package for EndNote includes text files with lists of journal titles in full and in standard abbreviated form within a range of disciplines, for instance medicine. It is advantageous to import this list into a newly established library before references are registered. The lists are useful aids in compiling reference lists with journal names, either full or abbreviated.

5.12.2.3 Entering References

There are several ways to register references in a library:

1. Add the references manually.
2. Directly export references from a bibliographic database or an electronic journal, or import text files with references.
3. Directly import from bibliographic databases by using EndNote as a search engine.
4. Import PDF files to create references.

Adding references manually should be the last resort as it is time-consuming and prone to error. However, if the source is a book or journal article that is old, or otherwise not registered in a bibliographic database, manual entry is necessary.

Many interfaces, such as PubMed, OVID, and Web of Science, offer a function that enables direct export from their websites into an EndNote library. Increasingly more electronic journals also offer this type of function for direct export of references. In this way the extra steps of downloading references into bibliographic software via text files are eliminated.

Importing text files with references downloaded from a search in a bibliographic database is a good alternative to direct export, but in doing so one must be sure the references are downloaded in a format EndNote can read. EndNote has numerous import filters, which make it possible to read and import references from various databases.

It is also possible to import references directly from bibliographic databases by using EndNote as a search engine. Performing a search of bibliographic databases via EndNote is a simple and effective way of collecting references, but this method can be recommended only for very basic searches, for instance in tracing the complete bibliographic details for one or more articles where the author, title, etc., are already known. Subject searches should be performed directly in the appropriate bibliographic database (for instance Medline via PubMed or OVID), where it is possible to take advantage of the interface designed for that specific database, with its own special features. The results can then be transferred to EndNote. Trials have shown that subject searches in Medline via EndNote produce fewer results and incidental references than similar searches using the same strategies in the native database interface.

Full-text articles in PDF format can also be imported, both single files and folders containing several files, to create references in EndNote.

5.12.2.4 Organizing References

The "Group" function offers many options to organize references within a library. References can be sorted into single groups, group sets, and smart groups based on a search strategy.

5.12.2.5 Cite While You Write

It is often advantageous to cite references during the writing of a manuscript. It is therefore sensible to use reference management software from the very beginning, as it saves time when the manuscript and its reference list are being finalized for submission. Cite While You Write is a function that supports the insertion of citations in a manuscript during its drafting. Embedded in each citation is a link between the manuscript and the database from which the reference has been taken. Citations are entered in the text either via an application available in Microsoft Word or by highlighting references in the EndNote library and inserting them in the manuscript. The example in Box 5.10 illustrates the most common construction of an unformatted citation: the first author, year of publication, and ID/sequence number, all enclosed in brackets. If instant formatting is

BOX 5.10 Appearance of Unformatted and Formatted Citations in a Manuscript

Unformatted

Several assertions have now been put forward which must be documented with references. Here is the first assertion which must be supported by the following {Pitkin, 1999 11 /id} {Seglen, 1997 4 /id}. The article continues with a discussion and a statement of the approach to the problem and concludes with a reference to the following {Minozzi, 2000 1 /id}.

Formatted in the Vancouver Reference Style

Several assertions have now been put forward which must be documented with references. Here is the first assertion which must be supported by the following (1,2). The article continues with a discussion and a statement of the approach to the problem and concludes with a reference to the following (3).

Reference List

1. Pitkin RM, Branagan MA, Burmeister LF. Accuracy of data in abstracts of published research articles. JAMA 1999; 281(12): 1110–1111.
2. Seglen PD. Why the impact factor of journals should not be used for evaluating research. BMJ 1997; 314(7079): 498–502.
3. Minozzi S, Pisotti V, Forni M. Searching for rehabilitation articles on MEDLINE and EMBASE. An example with cross-over design. Arch Phys Med Rehabil 2000; 81(6): 720–722.

turned on, the references will automatically be presented in the chosen style. If not, the software will scan the document for the citations so that a reference list can be compiled and formatted automatically in a specified bibliographic format.

5.12.2.6 Reference Lists—"Output Styles"

"Output styles" determine the presentation of citations and the corresponding reference list in a manuscript, such as whether the in-text citations should be numbered and refer to a numbered reference list, or alternatively whether in-text citations should consist of authors and years and refer to an alphabetical reference list. Output styles also specify punctuation and the fields of a reference that are to be included. There are many predefined output styles included in EndNote, including specific styles for a range of medical journals, as well as standard bibliographic formats, such as Vancouver, Harvard, and APA. If none of the existing styles is suitable, a new one can be created, either by copying and editing an existing style or by devising a new one.

5.13 OPEN ACCESS PUBLICATION, COPYRIGHT, AND SELF-ARCHIVING

5.13.1 Open Access Publication

Open access publication means that scientific publications are available on the Internet for the end user to read, download, copy, search, and link to the full text without payment. Research institutions such as the European Research Area Board and the Wellcome Trust have passed guidelines claiming that scientific publications funded by these institutions should be made available through open access publishing channels. The European Union's funding program "Horizon 2020" embraces open access policies. Likewise, publications resulting from research funded by the US National Institutes of Health must be open access publications, according to "The Public-Access Law" passed in December 2007. More and more countries are adhering to the philosophy that outputs from publicly funded research should be published in a manner that makes them freely available to the general public. There is also an important ethical aspect to the democratization of information. For those who cannot afford subscription fees, be they countries, research groups, or individuals, unrestricted online access to research publications is important if they are to stay up to date—provided they have an Internet connection.

So who is paying the bill for open access publication? This is a big question, involving national authorities, publishers, libraries, and scholars. The issue is under debate; however, it is evident that the future will embrace open access in one way or another. There are two established routes to open access, the "gold route" and the "green route" (Figure 5.6). The gold route involves peer-reviewed open access journals funded by article processing charges, implying that all publication costs are paid before publication. Articles from such journals are fully accessible on the Internet as soon as they are published.

The green route implies publication through traditional peer-reviewed channels; however, the authors reserve the right to self-archive a copy in an institutional repository. In this way no researchers need to compromise the freedom to publish wherever they think best, and the publication process retains the standard quality control. It is the authors' responsibility to deposit their papers in institutional repositories at either a pre- or

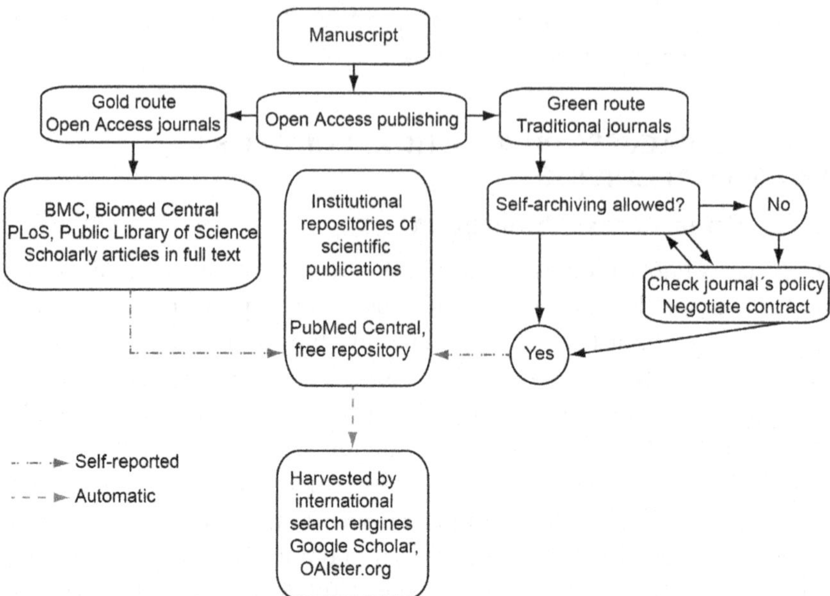

Figure 5.6 Open access publishing routes.

post-publication stage, according to the journal's open access policy. Check out local academic funding bodies for support on article processing charges, and trust their know-how on open access publishing.

Over the years the quality of open access journals has been debated, and rightly so. The Internet has made it easier for scientists who so wish to publish their papers directly. Although new Internet-based journals were launched, some turned out to be short-lived. Their quality was questioned, as was their peer-review process. However, this situation altered considerably when the traditional subscription-based journals went open access. BMJ was one of the earliest journals to make its contents electronically available. Waves of scientific journals followed. The next shift came when online-only journals were launched by the two major open access publishers in medicine and biology, namely BioMed Central (http://www.biomedcentral.com/), and Public Library of Science (PLoS, http://www.plos.org/). The Directory of Open Access Journals (http://doaj.org/) provides an index to all peer-reviewed open access journals. Full-text published material is freely available from these resources.

Of course all that glitters is not gold. There is every reason to take a close look at calls for submission from open access journals, especially when the publishers are unknown. Among the serious publishers there are also swindlers going after easy money. Some publishers have chosen a hybrid strategy, requesting extra fees from authors if they want open access publication for their article, and keeping the remainder of the articles they publish available by subscription only. Such publishers should be avoided, since no institutions will pay extra for open access when already subscribing to the journal. No one should support double payment.

5.13.2 Retaining Copyright and Self-Archiving

Authors have both economic and ideal rights to their intellectual property. The economic rights consist of controlling the publications, and the ideal rights the authors' right to be correctly cited and protected against any misuse of their publications. Among scientific authors there is a growing awareness that one need not transfer all rights to the publishers. Before signing a copyright transfer agreement, make sure it contains a passage stating the author's right to deposit an electronic copy in an institutional archive.

Following the green open access route, self-archiving is not a parallel publication, it is a deposit of a copy of a published paper in an institutional repository. Self-archiving implies that authors themselves are responsible for this last step in the publication process, and it is free of charge. In the medical field only post-prints or publishers' original versions are of interest. The majority of publishers accept self-archiving of pre- or post-prints in open institutional repositories when explicitly agreed upon in the copyright transfer agreement. Sometimes an embargo period is required. Check out the publishers' copyright and self-archiving policies on the SHERPA/RoMEO site (http://www.sherpa.ac.uk/romeo.php).

PubMed Central is a free digital archive of literature from biomedical and life sciences journals (http://www.ncbi.nlm.nih.gov/pmc/).

Institutional repositories are harvested by OAIster (OAI = Open Archives Initiative protocol, http://oaister.worldcat.org/), a catalogue of digital resources.

5.13.3 Citing

So what is the verdict on open access versus traditional publishing channels? There are scientific studies implying that the scientific impact of open access publications can be seen in terms of their citations.[9] Indeed, open access publications are cited more often, due to their availability, than articles published in subscription-based journals. However, nothing should be taken at face value. The quality issue of open access articles still merits attention.

QUESTIONS TO DISCUSS

1. Phrase your research question—be as specific as possible.
2. Formulate your search query. Use the PICO table for structure.
3. Consider the type of research question:
   a. Basic research?
      – List databases according to relevance.
   b. Clinical research?
      – What kind of question? List relevant databases/resources according to relevance. Are methodology filters available?
4. Prepare your search strategy:
   a. List adequate index terms and text words. Fill in the PICO table.
   b. Decide upon Boolean operators.
5. Perform the literature search in the targeted bibliographic databases and compare the results for:
   a. Hits, relevance, unique records.
   b. Pick an interesting article and look at how it is indexed. Do you get some ideas on how to rephrase your search strategy?
6. Do an author search for your name in different databases and compare the results:
   a. Hits. Are your papers cited? By whom?
   b. Calculate your own H-index.
7. Check out the publisher's copyright and self-archiving policy of your favorite journal.

REFERENCES

1. Newton I. *Quote. The Oxford dictionary of quotations*. 4th ed. Oxford: Oxford University Press; 1992.
2. Pitkin RM, Branagan MA, Burmeister LF. Accuracy of data in abstracts of published research articles. *JAMA* 1999;**281**:1110–11.
3. Rekdal OB. Academic citation practice: a sinking sheep? *Portal: Libr Acad* 2014;**14** (4):567–85.
4. Garfield E. The history and meaning of the journal impact factor. *JAMA* 2006;**295**:90–3.
5. Fang FC, Casadevall A. Retracted science and the retraction index. *Infect Immun* 2011;**79**:3855–9.
6. Seglen PO. Why the impact factor of journals should not be used for evaluating research. *BMJ* 1997;**314**:498–502.

7. Chargaff E. Triviality in science: a brief meditation on fashions. *Perspect Biol Med* 1976;**19**:324–33.
8. Hirsch JE. An index to quantify an individual's scientific research output. *Proc Natl Acad Sci USA* 2005;**102**:16569–72.
9. Björk B-C, Solomon D. Open access versus subscription journals: a comparison of scientific impact. *BMC Med* 2012;**10**:73.

FURTHER READING

PhD on track. Available from: http://www.phdontrack.net/.

CHAPTER 6

Basic Medical Science

Bjorn Reino Olsen[1] and Haakon Breien Benestad[2]
[1]Harvard School of Dental Medicine, Harvard Medical School, Boston, MA, USA
[2]Department of Physiology, Institute of Basic Medical Sciences, University of Oslo, Blindern, Oslo, Norway

6.1 INTRODUCTION

What is basic medical research? How does it differ from basic research in general? How can it be distinguished from translational and clinical medical research? A satisfactory introduction to the strategies and methods used in basic medical research must clearly answer these questions. However, these answers need not be precise, and there are currently no easy or generally accepted answers. Indeed, the rapid progress of biomedical research during the past 50 years has led to a gradual erasing of the historical boundaries between scientific disciplines and research fields, making it almost impossible to distinguish between basic medical research and basic biological research. In addition, the introduction of new, more powerful molecular biological, genetic, biochemical, and morphological methods has created the possibility to address fundamental questions of cellular and biochemical mechanisms using patient-derived biological materials, making the distinction between clinical and basic medical research less obvious. For these reasons, the definition of basic medical research presented in this chapter will be a practical one, i.e., research carried out in medical/dental/veterinary schools or institutes, with a major goal of enhancing the understanding of fundamental normal or pathological processes at the organismic, organ, cellular, and molecular levels.

This definition emphasizes that basic medical research seeks understanding and intellectual insights at many different levels. The strategies and methods to be used to achieve this goal must therefore also be effective at many different levels. Obviously, the methods must be able to generate data, but since the goal is to obtain intellectual insight, the data must also contribute to the understanding of the biological and pathological logic, which usually means that the results allow the researcher to test one or more structured ideas or hypotheses. For this reason, the gathering of biological facts alone has been considered of relatively little value in basic medical research and is given low priority. Today, after the human genome project and revolutionary

Research in Medical and Biological Sciences
DOI: http://dx.doi.org/10.1016/B978-0-12-799943-2.00006-9

advances in genetic, proteomic, and imaging technologies, such fact-gathering projects are now pursued vigorously. These fact-gathering projects are often called hypothesis-generating research, though it could be argued that there are always underlying hypotheses, which may sometimes not be clearly defined, that guide data collection. One example of hypothesis generating is the gigantic brain research and modeling studies taking place in the United States and the European Union.

Successful basic medical research projects are characterized by good research questions and clearly stated research problems that prompt the use of adequate and effective methodology; not the other way around (Boxes 6.1 and 6.2). To allow the research question to be defined or influenced by one's knowledge of or experience with a particular method is a poor strategy if one wishes to perform outstanding research.

6.2 LONG-TERM GOALS AND SPECIFIC AIMS

Large research projects require sustained efforts over a long period of time, and it is useful to remind oneself that science is a marathon, not a sprint. It is therefore essential to define clear long-term goals from the very start of a research project (Box 6.3).

The long-term goals serve as distant reminders of where the daily research efforts are heading. However, whether their realization is 3–5 years into the future (for a PhD dissertation), or longer (5–10 years for a junior faculty member who is establishing her or his own independent laboratory; 10–20 years for a professor who wants to make a significant impact on her or his chosen field of research), long-term goals, no matter how well they are defined, are too distant to serve as practical guideposts for daily, weekly, or monthly research efforts. Indeed, for ongoing efforts to be effective, several specific aims should be generated.

BOX 6.1 Basic Requirements of a Research Problem

- A research problem should allow a clear description of overall goals and specific aims.
- Any hypotheses (structured ideas) should be testable.
- Research problems should allow for the collection of data; the most efficient methods for testing the hypotheses should be selected.

BOX 6.2 The Importance of Important Research Problems

The purpose of scientific enquiry is not to compile an inventory of factual information, nor to build up a totalitarian world picture of Natural Laws in which every event that is not compulsory is forbidden. We should think of it rather as a logically articulated structure of justifiable beliefs about nature. It begins as a story about a Possible World—a story which we invent and criticise and modify as we go along, so that it ends by being, as nearly as we can make it, a story about real life...

... The scientific method is a potentiation of common sense, exercised with a specially firm determination not to persist in error if any exertion of hand or mind can deliver us from it. Like other exploratory processes, it can be resolved into a dialogue between fact and fancy, the actual and the possible; between what could be true and what is in fact the case.

Medawar[1]

It can be said with complete confidence that any scientist of any age who wants to make important discoveries must study important problems. Dull or piffling problems yield dull or piffling answers. It is not enough that a problem should be "interesting"—almost any problem is interesting if it is studied in sufficient depth. A problem must be such that it matters *what the answer is—whether to science generally or to mankind.*

Medawar[2]

BOX 6.3 Long-Term Goals, Specific Aims, and Specific Objectives

- *Long-term goals* reflect a vision of what the research is trying to accomplish in the long run.
- *Specific aims* serve as practical guideposts for the daily/weekly/monthly research efforts.
- *Specific objectives* are the more immediate outcomes of the study.

The specific aims serve as short-term goals, chosen in such a way that if successfully maneuvered, they will enable investigators to reach specific objectives as they work towards their long-term goals. Specific aims should be interrelated, but not sequentially interdependent. For example, successful completion of one specific aim should not be an essential prerequisite for starting work on the next one. In fact, it should be possible to work on two or three specific aims at the same time. Also, one should avoid defining specific aims in such a way that a specific outcome of one aim makes work on another aim impossible or irrelevant.

For example, in a project on the pathogenic mechanisms of infantile cutaneous hemangiomas, benign tumors of capillaries that grow rapidly for a few weeks and months after birth and then slowly regress over the next few years, it would be a mistake to have a specific aim to generate a mouse model of hemangiomas with loss-of-function mutation in the targeted gene in parallel to a specific aim to identify gene mutations responsible for hemangioma formation. Indeed, generating a mouse model of hemangiomas would only make sense if it is first demonstrated that hemangiomas are in fact caused by loss-of-function mutations.

The specific aims implicitly prescribe the most effective experimental methods for their accomplishment. Methods and detailed procedures should be seen as logical consequences of the specific aims; not chosen simply because the investigator knows how to perform them. For these reasons, the specific aims are best described in statements that both imply the hypotheses to be tested, as well as the best experimental methods to test them, without being too focused or method-oriented. For example, in a research project on the molecular mechanisms responsible for the initial rapid growth and subsequent slow regression of infantile hemangiomas, the statement "To screen for genetic loci in familial hemangiomas by linkage mapping and loss-of-heterozygosity analyses" is a better description of an aim than "To sequence genes *x*, *y*, *z* with DNA from tissues of individuals with hemangiomas." Although the work to be done will involve DNA sequencing, the first statement implies the hypotheses that inherited germ-line mutations are responsible for hemangiomas in some families (and therefore can be mapped), and that second-hit somatic mutations (detectable as loss-of-heterozygosity) may be present in hemangiomas. The statement also prescribes the technical approach to address the hypotheses, i.e., the use of linkage mapping and loss-of-heterozygosity analyses.

There are two categories of specific aims. A hypothesis-driven specific aim gives the description of a specific aim around a testable hypothesis and works well in cases where hypotheses are, in fact, being tested. Descriptive specific aims state the specific aims as a series of questions and may be better in more descriptive studies. For example, if a research project were focused on the natural history of infantile hemangiomas rather than the molecular mechanisms responsible for their formation, one may ask: "What is the phenotypic variability among individuals with sporadic and familial hemangiomas?" A hypothesis-driven specific aim is not necessarily better than a descriptive specific aim; it is simply different, but it is useful to be very clear about the category of specific aim one is pursuing, since they lead to different specific objectives.

Hypothesis-driven specific aims can lead to mechanistic insights; descriptive specific aims can lead to new hypotheses for subsequent testing (i.e., hypothesis-generating research). More and more, descriptive studies are considered important in biomedical research. Indeed, hypothesis-generating research is essential for progress in some fields, such as endeavors to understand how genomes and the brain work. Knowledge of the natural history of disease is essential for judging the effects of treatment interventions, and descriptive analyses of gene expression profiles during development and growth are necessary to understanding tissue-specific effects of mutations and new drugs that are designed to interfere with gene function.

The time period to be covered by a specific aim depends on the overall scope of the research project. For research projects funded by multiyear grants from funding bodies or foundations in many countries, it is not uncommon to have many specific aims, for example, a 3-year grant would require three specific aims, each covering 1 year of work, whereas for projects that are funded by 1-year grants, it is useful to have specific aims that cover shorter (3—4 months) time periods.

6.3 BACKGROUND AND SIGNIFICANCE

In a good research project, the long-term goals and specific aims should address a significant need for better insight and understanding in the area of investigation. Identification of such a need must be based on a comprehensive, in-depth, and critical analysis of the current level of understanding as it is reported in the scientific literature. This preparatory phase of a research project is frequently performed too hastily and postponed until the experimental phase is completed and the corresponding manuscript is being drafted.

Too often, studies are initiated based on a quick reading of recent papers in a "hot" area. If the papers are exciting and lead to obvious follow-up research, someone with the appropriate technical experience may decide to plunge ahead without much thought. In general, this approach of letting the work of others define the focus of one's research project does not generate sustained research of a high level of quality and originality. One reason for this is that if an exciting paper prompts obvious and important questions, it is very likely that the authors of the paper have already extended the work to answer those questions by the time the paper is published. Therefore, any study that is initiated based on the published results of others is likely to be too little, too late. A second reason is that papers may appear to break new ground upon initial examination but frequently turn out to be less important in light of subsequent work and more careful analysis.

It is also advisable to avoid choosing a research project based entirely on funding opportunities. Good research funding is essential for sustained research efforts, but to allow the research area, the long-term goals, and the specific aims be completely shaped by funding opportunities in a specific area is a huge mistake. At best these areas or themes are decided by committees of scientists, and at worst by governmental bureaucrats, based on considerations that may be unrelated to scientific needs.

Before research projects are started, it is essential to give serious thought to what is intellectually intriguing and interesting at the time in one's field. Focus on areas that are less well developed and therefore more intellectually challenging, and address questions that have multiple and significant biological and medical ramifications. In short, a good research project must start with thorough reflection about its background and significance. Moreover, a section on the background and significance of a research project is required in most funding proposals and research reports. Therefore, preparing a background and significance document is highly recommended, even if the research project has already been funded (Box 6.4).

When reading the background literature pertinent to a research project, it is important to extract elements that will contribute to the logical foundation,

BOX 6.4 Points to Keep in Mind During the Literature Review That Establishes the Background for a Research Project

- Do not limit the review to papers that are available electronically.
- If necessary, go back 100 years—important discoveries in anatomy, histology, and physiology were made that long ago.
- Put your ideas and thoughts concerning the rationale for the research and the arguments for selecting specific methods in writing, even if you are not preparing a funding proposal for the work.
- Think broadly—avoid choosing a project that is primarily a follow-up of specific results published by another laboratory.

or rationale, for the planned research. This requires an in-depth study of individual studies on the topic, referred to as primary studies; a quick reading of recent reviews is not sufficient. Also, avoid the common practice of restricting the literature search to publications accessible through electronic databases. Indeed, much of the medical research of the past 100 years predates Medline, PubMed, and the desktop computer! If at all possible, one should avoid trying to rediscover what others may have already reported 50–80 years ago, by including older primary studies in the background reading.

When a strong overall rationale for the research project has been built and the significance of the research is apparent, logical arguments must be structured for the specific aims. The specific aims should define the experimental steps that will most effectively advance understanding in the chosen field of research in the context of the available resources, the experience of collaborators, the past contributions by the investigator, and the state of the art in the field.

6.4 EXPERIMENTAL STRATEGIES AND METHODS

Selecting the optimal experimental strategies to address the hypotheses is crucial for the ultimate success of a research project. Inadequate experimental methods can destroy a project with outstanding and innovative hypotheses, just as innovative and powerful methods can enhance a relatively insignificant project and lead to serendipitous findings of outstanding importance. Many investigators are most comfortable with methods they know well, and tend to use these methods even in situations where they are inadequate. To avoid this, it is useful to think about experimental methods as broadly as possible, seek expert advice on unfamiliar methods, learn new techniques as necessary, and collaborate with appropriate experts as needed. Above all, it is important not to be afraid of new and technically challenging approaches, and to remember that with sufficient time and effort any experimental method can be learned and effectively applied.

Before experiments are done and detailed research protocols are written, it is useful to think through the anticipated results and decide on further experiments to perform in the case where the results come out as expected and in the case where they do not. If a major new finding is made, it is imperative that it be confirmed and supported in several different ways and by different experimental methods before a research report is submitted for publication. Finally, once the methodological possibilities have been analyzed, or even while the most efficient ways of addressing specific aims are being considered, it is useful to think of some simple and quick preliminary experiments that will provide preliminary data that may affect the direction of the research project, not the least of which is to determine the size of the experimental material needed, including a power calculation (see Chapter 4). For preliminary experiments to be meaningful, they must be carried out in the same rigorous manner as any experiment destined for publication. A preliminary experiment should set the stage for future studies and must therefore be complete, reliable, and include proper controls (positive and negative).

6.5 LEVELS OF RESEARCH—FROM ORGANISMS TO CELLS

At the beginning of this chapter a practical definition of basic medical research was given as research carried out in medical/dental/veterinary schools or institutes with a major goal of enhancing understanding of fundamental normal or pathological processes at the molecular, cellular, organ, or organismic levels. This definition can accommodate a variety of research problems utilizing a wide range of experimental methods (Box 6.5).

Specific advice that would be helpful in promoting high quality research across this broad spectrum is beyond the scope of this chapter. However, when considering a research project at the start of one's academic career, it may be useful to remember a few general rules. The first and second of these rules emphasize the need to focus on real problems of biological systems, i.e., problems related to *in vivo* processes as compared to problems related exclusively to *in vitro* models (Box 6.6). For example, details of gene regulation in cultured cell lines may not reflect gene regulation *in vivo* and should therefore be coupled with experiments demonstrating *in vivo* relevance. The third rule emphasizes the need to be courageous when choosing a research problem. The fourth recommendation is to divide challenging research projects into smaller steps without losing sight of the final goals. The fifth and final recommendation is to adopt an open, collaborative attitude in dealing with ideas and research projects. Secrecy does not produce great basic medical science; openness, intellectual generosity, and free exchange of ideas, information, and reagents are required for exciting, innovative science.

BOX 6.5 Are the Old Views of Experimental Studies Still Valid?

The *History of the Royal Society*, published in 1667 by one of its founders, Thomas Sprat (1635–1713), included a durable definition of experimental studies:

> *But they are to know, that in so large, and so various an art as this of experiments*
> *there are many degrees of usefulness: some may serve for real and plain benefit,*
> *without much delight,*
> *some for teaching without apparent profit;*
> *some for light now, and for use hereafter;*
> *some only for ornament and curiosity.*
> *If they will persist in condemning all experiments, except those which bring with them*
> *immediate gain and a present harvest;*
> *they may as well cavil at the Providence of God, that He has not made all seasons of the year,*
> *to be times of mowing, reaping, and vintage.*

BOX 6.6 Points to Keep in Mind When Choosing a Research Problem

- Select biological problems that apply to organisms, organs, or cells *in vivo*.
- Stay away from research questions that have emerged primarily from *in vitro* studies, unless those questions clearly apply to *in vivo* situations as well.
- Do not be afraid of choosing big, challenging research problems—why waste time and effort on minor questions?
- Think strategically—break big research problems into a series of smaller steps; stay focused on the long-term goals. At the same time, always keep eyes and ears open, since a serendipitous observation may take the research in a direction that is more productive than the one that was originally planned.
- Do not be so afraid of having ideas "stolen" that it prevents a productive exchange of ideas and collaborations—the best ideas are usually the products of intense discussions among groups of individuals, and usually result in the most significant research projects.

6.6 RESEARCH ON EXPERIMENTAL ANIMALS

Research on experimental animals is indispensable to basic medical research. For example, one has to work on living animals to study the integrated regulation of body function (Box 6.7). However, such research should be reduced to a minimum.

Experimental animals have been crucial to the understanding of a large number of diseases and conditions. Unfortunately, some human disorders and conditions are not easy to reproduce in animal models (Box 6.8). As a result, such conditions often remain poorly understood.

Before one embarks on *in vivo* research on experimental animals, a solid base of data obtained from work with isolated organs, tissue slices, cells, or subcellular systems may be needed. This is particularly true in cases where one needs to interpret results from experimental animal work. In such cases knowledge about component mechanisms from *in vitro* work is required. For example, if the bacterial component endotoxin is injected intravenously into a rabbit and a febrile response is observed, this

BOX 6.7 Research on Experimental Animals Represents the Best Approach to Many Research Problems

- Studies of mechanisms of disease.
- Studies of integrated regulation of body function.
- Studies of vaccination against human or animal infectious diseases.
- Studies of experimental therapies in disease models.
- Intervention studies that are judged to be crucial to the understanding of a large number of diseases and conditions.

BOX 6.8 Lack of Knowledge or Insight May Be Due to Lack of Appropriate Animal Model

- Some psychosocial disorders.
- Chronic fatigue disorder.
- Nocturnal leg cramps.
- The phenomenon of tickling.
- Some kinds of tinnitus.
- Infantile colic.
- Migraine.

does not mean (as was the widely held view for a long time) that endotoxin directly affects the temperature-regulating centers of the brain; instead, it stimulates macrophages to secrete cytokines that do have this effect. Noradrenaline injected into an experimental animal may slow the heart rate, but when applied to an isolated heart *in vitro* it will speed up the rate. In this case, the *in vivo* contraction of small arteries raises blood pressure, which evokes a baroreceptor response and a nervous reflex that slows the heart rate in order to lower the blood pressure, overriding the direct effect of noradrenaline on the heart. Thus, the former example leads to a misleading conclusion, since it relies solely on animal experiments, whereas the latter example shows that animal experiments may ultimately be required to identify the *in vivo* effects and side effects of an agent or procedure.

Considerable *in vitro* experiments are not a prerequisite to the use of experimental animals. For example, in studies of genetic mechanisms, it is often the examination of the broad consequences of altered gene functions *in vivo* that guides subsequent studies of specific cellular and tissue mechanisms *in vitro*. However, this "exploratory" use of experimental animals is in conflict with the animal rights credo, which states that animals should be treated like humans, and research on experimental animals should be postponed until it can be performed safely and painlessly. Despite the conflict, exploratory use of experimental animals often results in insights that, in the long term, are critical for the detection and treatment of human diseases. Certainly, much human suffering has been averted thanks to research on experimental animals conducted in the past that led to innovations like vaccines, modern drug treatment, etc., which would have been impossible to develop in the absence of such research.

6.7 THE THREE Rs

In their 1959 book *The Principles of Humane Experimental Technique*, Russell and Burch[3] introduced the "three R concept" (replacement, reduction, and refinement) as a main guideline for the responsible use of experimental animals. Replacement refers to the use of *in vitro* and other methods instead of experimental animals. Examples of replacements include organs and cell cultures (Box 6.9), computerized models and video, DVD, or Internet programs. Reduction refers to a decrease in the number of experimental animals required for a given experiment. This can be achieved by choosing suitable experimental procedures, by controlling environmental factors, by standardization of the animal population, and by statistical power analyses.

BOX 6.9 Examples of Replacements for Experimental Animals

Research on:

- Perfused organs (e.g., heart; lungs)
- Tissue slices (e.g., of the hippocampus)
- Cell cultures (primary cultures or cell lines)
- Subcellular systems (biochemistry; molecular biology)

BOX 6.10 The "Three R Concept" Incorporated in Legislation

- *Replacement* through the use of *in vitro* and other methods.
- *Reduction* of the number of complex animals used: standardization and optimization of procedures; replace warm-blooded animals with organisms like chicken embryos, zebra fish, or the round-worm *Caenorhabditis elegans.*
- *Refinement* of procedures that can potentially cause pain or distress: appropriate sedation, anesthesia, analgesia, and method of euthanasia.

Refinement refers to a decrease in the incidence or severity of painful or distressing procedures applied to experimental animals. Legislation mandating the incorporation of the three R concept into animal research and testing has been passed in the United States and Europe.

The Animal Research: Reporting *In Vivo* Experiments (ARRIVE)[4] guidelines for *in vivo* preclinical research are very important in research on experimental animals (www.nc3rs.org.uk/ARRIVE, January 16, 2014).

It is always important to consider carefully whether the research project has really advanced to the stage where *in vivo* experiments are justified or indispensable. If not, *in vitro* culture of organs, tissues, cells, or subcellular systems and computer simulations may represent good alternatives or supplementary approaches. Moreover, *in vivo* bioassays and toxicological tests can, and often must be, replaced by physicochemical, biochemical, and immunological analyses, or by fascinating techniques developed by molecular biologists. Remember also that research on experimental animals can mean very different things (Box 6.10). Complex biological phenomena may best be analyzed in simpler organisms, like insects or worms (*C. elegans*), at least at an early stage. However, the results of such analyses cannot necessarily be extrapolated to all animals, let alone to humans, making some research on complex animals unavoidable when it comes to human physiology and disease.

In addition, some fields of research have traditionally used a certain type of experimental animal in order to construct a solid experimental database for comparison. It may therefore be preferable to choose a warm-blooded animal, like the mouse, for some experiments. In any case, experimental animals should be treated like precious instruments in regard to control and calibration. Inherent in research on experimental animals is the diversity of variables that need calibration. These variables should be known and controlled as much as possible.

6.8 ANIMAL MODELS

The significance of results obtained from research on experimental animals depends on the selection of a suitable animal model. The extent to which the results can be extrapolated to humans depends upon the type of animal and the nature of the research, but there are no general rules for the choice of the most suitable animal model or the extrapolation procedure. Animal models can be divided into four groups: induced models, spontaneous models, negative models, and orphan models. The induced and spontaneous models are the most important.

In induced animal models, a disease, disorder, or other change is induced experimentally to obtain an intended similarity with humans, e.g., with regard to disease manifestations or etiology. The change can be induced through surgery; transfer, alteration, or removal of genetic material; or by the administration of biologically active substances. Animal models have been created for a great number of diseases and malfunctions by interfering with the environmental, dietary, endocrine, immunological, infectious, or genetic state of experimental animals. Genetically modified animals are one of the most important groups of induced animal models, as they are essential for investigating the underlying pathogenic mechanisms of human genetic disorders.

Spontaneous animal models are those that have arisen by natural genetic variation. Hundreds of strains/stocks of animals have been analyzed and categorized as displaying spontaneous diseases that resemble those in humans.

Negative models include species, breeds, or strains in which a certain disease does not develop. This term may also be applied to a model that is insensitive to a stimulus that usually has a specific effect in other species or strains. The underlying mechanisms of insensitivity in these negative models can be studied to provide further insight.

Orphan animal models refer to those in which a disease or condition is initially recognized and studied in an animal species, with the assumption that a human counterpart might be identified at a later stage. Malignant epithelial tumors caused by papillomavirus in rabbits, and Marek's disease, a highly contagious lymphoproliferative disease caused by herpes virus in chickens, are two examples of orphan animal models.

Only rarely do animal models fully mirror human health conditions or diseases. Very often, the selection of an animal model is based on a similarity between humans and animals in only one aspect of the phenomenon under study. Therefore, it is preferable to use a variety of spontaneous and induced animal models (Box 6.11).

BOX 6.11 Common Causes of Reduced External Validity in Research on Experimental Animals

- The use of young, otherwise healthy animals, when patients with the disease under study are usually elderly people with comorbidities.
- The assessment of treatment effect in a homogeneous group of animals versus a heterogeneous group of patients.
- The use of either male or female animals only, whereas the disease occurs in male and female patients alike.
- The use of models for inducing a disease or injury that is not sufficiently similar to the human condition.
- Delays in the start of treatment that are unrealistic in clinical settings; the use of doses that are toxic or not tolerated by humans.
- Differences in outcome measures and the timing of outcome assessment between research on experimental animals and clinical trials.

. . .a systematic review and meta-analysis of all available evidence from pre-clinical studies should be performed before clinical trials are started. Evidence of benefit from a single laboratory or obtained in a single model or species is probably not sufficient.

van der Worp et al.[5]

6.9 MOUSE MODELS FOR STUDIES OF MAMMALIAN DEVELOPMENT AND DISEASE

Research on experimental animals has been crucial in elucidating the basic mechanisms of embryonic development and postnatal homeostasis. To date, *C. elegans*, fruit flies, zebra fish, and mice have been the preferred species because of their rapid development and postnatal growth, and the ease with which their genomes can be manipulated, although other species—including rats—have also received considerable attention. Mice continue to be the animals of choice for mechanistic studies of mammalian development and disease. The development of methods to isolate and culture mouse embryonic stem cells, insert pieces of DNA into the mouse genome, delete pieces of DNA from the mouse genome, or edit the mouse genome in various ways has allowed investigators to carry out precise and in-depth studies of gene function and the effects of mutations.

By injecting DNA into fertilized eggs, followed by implantation into pseudopregnant females, it is possible to generate transgenic mice that express a gene under the control of a specific promoter/enhancer. The generation of such mice initially relied on the unpredictable, random insertion of the injected DNA into the mouse genome. This necessitated the establishment of several transgenic lines and careful analyses of transgene expression, as well as other control experiments, to ensure that the transgene could be expressed and that its insertion into the genome did not affect the expression of other genes. However, with new technology it is now possible to obtain highly efficient targeted insertion of one copy of the transgene into a defined locus or intergenic region. For example, based on the finding that a gene on chromosome 6, named *Rosa26*, can be disrupted in mice without producing any abnormality, there are several methods that can be used to insert a promoter and a transgene into the first intron of the *Rosa26* locus. This eliminates the possibility for the transgene to be inserted at a random site in the genome, disrupting the function of genes that happen to be located there. Investigators have also taken advantage of the properties of enzymes called integrases to insert transgenic DNA into specific sites in the genome.

One example is the ΦC31 integrase protein, which gets its name from the fact that it is encoded by the genome of the *Streptomyces* bacteriophage ΦC31. The ΦC31 integrase protein is a sequence-specific recombinase that mediates recombination between two 34-base pair sequences called attachment (att) sites. One att site (attB) is in the phage genome; the second att site (attP) is present in the DNA of the bacterial host. Although mammalian cell DNA does not contain attP sites, the ΦC31 integrase protein has been found to work in mammalian cells; therefore, if an attP-like sequence is inserted into a region of mouse genomic DNA that is not important for gene expression or regulation, one can use the ΦC31 integrase protein to catalyze insertion of DNA containing an attB sequence into that site. When used to generate transgenic mice, a circular plasmid DNA containing the transgene

BOX 6.12 Diagrams Showing the Use of Sequence-Specific Integrase to Target Transgenes to a Specific Site in the Genome of a Mouse Oocyte for the Generation of Transgenic Mice (Figure 6.1)

Step 1: A circular DNA construct containing the transgene of interest and an integrase-specific att site (attB) is prepared. Mice that carry an integrase-specific att site (attP) in their genome are used for the preparation of fertilized eggs. Step 2: Injection of the transgene construct and integrase protein (or messenger RNA, mRNA, that can be translated into protein) into eggs results in integrase-dependent binding of the construct to the genomic att site. Step 3: Integrase-mediated recombination occurs. Step 4: Transgene with flanking sequences is inserted into the mouse genome.

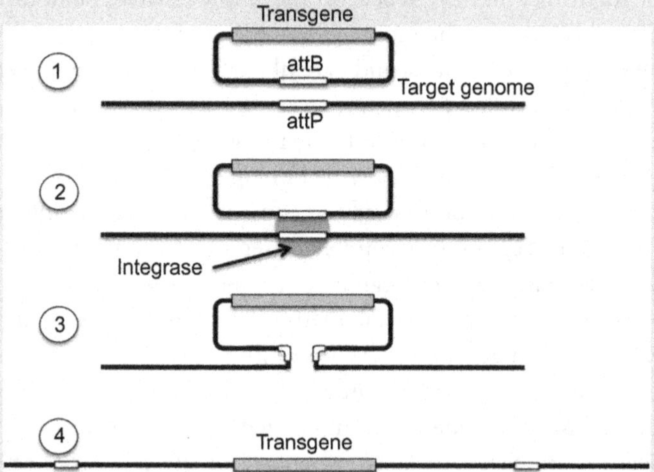

Figure 6.1 Action of integrase.

and an attB sequence is constructed and amplified in a bacterial host. This circular DNA and the recombinant ΦC31 integrase protein are injected into fertilized eggs collected from mice that have been engineered to carry an attP-like sequence at a specific genomic site. The result is an irreversible integration of the transgene construct into the attP-containing region (Box 6.12).

In cases where the transgene encodes a mouse protein and is under the control of a promoter that drives the expression of the endogenous gene, transgenic mice provide an opportunity to study the effects of overexpression of that protein. In cases where the transgene is a biologically inactive reporter that gives a visible (such as fluorescence) or measurable product (such as beta-galactosidase activity) when the promoter is activated, the mice can be used in studies of the tissue-specific expression of the promoter or its transcription factor. For example, nuclear factor-kappa B (NF-κB) is a multipotent transcription factor involved in hematopoiesis and inflammation. The activity of NF-κB in various mouse organs and cells can be studied in mice expressing a luciferase reporter transgene whose expression is dependent on activation of NF-κB. The mice carry a transgene promoter with NF-κB binding sites, coupled to the luciferase transgene. Luciferase expression is then measured after contact with luciferin plus ATP as light emission, from cell extracts, isolated cells, tissues—or even intact, living mice.

Moreover, the transgene may code for a recombinase enzyme that will be expressed in cells with a promoter for the transgene. Recombinases (such as cre recombinase from P1 bacteriophage and Flp from *Saccharomyces*) can remove DNA segments flanked by specific nucleotide recognition sites (loxP sites for cre recombinase and Frt sites for Flp) in the genome and paste the sequences on each side of the segment together. The use of recombinase enzymes in genetic engineering makes it possible to generate transgenic mice that do not express a transgene unless it is "turned on" by recombinase action. One way to do this is to insert a strong signal for stopping transcription between the promoter and the translated region of the transgene. To allow controlled removal of this STOP signal, one would place loxP or Frt sites on each side of the STOP signal. A transgenic mouse carrying this kind of "conditional" transgene can be mated with a mouse that carries a transgene encoding the cre or Flp recombinase. In their offspring, cells that express the recombinase will lose the STOP signal and the transgene will be expressed if the promoter that drives transgene expression is active in those cells. This approach allows for useful experimental strategies. For example, if the promoters for the transgene and the recombinase exhibit cell-specific expression, the offspring of the cross will only express the transgene in cells that express both promoters (Box 6.13).

By transfecting cloned DNA "constructs" into embryonic stem cells and allowing homologous recombination to take place, one can obtain stem cells in which a portion of the construct DNA has replaced the homologous genomic DNA. The cloned DNA may result in (i) a loss-of-function mutation of exon nucleotide sequences, (ii) an altered amino acid codon within an exon, or (iii) exon-flanking sequences (loxP or Frt sites, as detailed previously) that are not normally found in mice but are

BOX 6.13 Use of Recombinase to Induce Expression of a Conditional Transgene, Tomato (Figure 6.2)

Diagram illustrating how crossing a transgenic mouse (transgenic 1) carrying a transgene encoding cre recombinase under the control of a promoter (promoter 1) with another mouse (transgenic 2) carrying a conditional transgene encoding a fluorescent protein, Tomato, under the control of another promoter (promoter 2) generates mice that will exhibit red Tomato fluorescence in cells that express both promoters. In transgenic 2, a transcriptional STOP signal flanked by loxP sites prevents expression of the red fluorescent Tomato protein. However, in cells that express cre recombinase, the STOP signal is removed and Tomato expression can be induced.

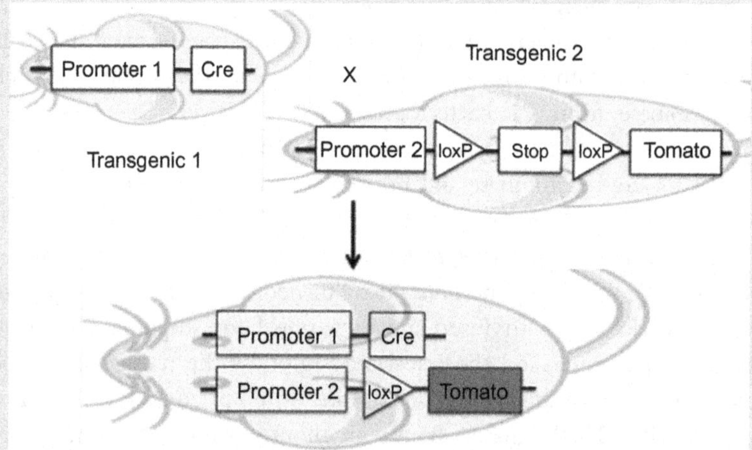

Figure 6.2 Expression of Tomato.

recognized by cre and Flp recombinases. If the screening of such cells reveals some with cloned DNA with the desired sequence change in the gene that the DNA construct was designed to target (targeting DNA construct), these cells can be injected into blastocysts so that they can become part of the adult animal.

If the targeted gene is expressed in the germ line, one can generate "knock-out" mice with loss-of-function mutation in the targeted gene, or "knock-in" mice with specific disease-causing mutations knocked into the mouse genome. If the targeting DNA construct contains loxP or Frt sites flanking an exon, the mice containing the targeted cells would be normal, but will produce offspring called "conditional knock-out" mice;

BOX 6.14 The Effects of Tamoxifen-Inducible creER(T2) Recombinase (Figure 6.3)

An example of what can happen in the cells of the offspring of a female mouse carrying a gene in which exon 2 is flanked by loxP sites in the upstream and downstream introns and a male mouse carrying a creER(T2) transgene under the control of a promoter expressed in all cells. The mouse inherited the loxP-flanked (floxed) gene from the mother and the creER(T2) gene from the father. The mouse will express the creER(T2) protein in all cells, but is normal because the recombinase did not enter cell nuclei. Administration of tamoxifen will induce translocation of the creER(T2) into the nuclei of cells, then exon 2 in the floxed gene will be deleted and, if loss of exon 2 results in the loss of the translational reading frame (the number of bases in exon 2 is not a multiple of 3), the protein encoded by the gene will not be produced. Therefore this is a conditional knock-out mouse.

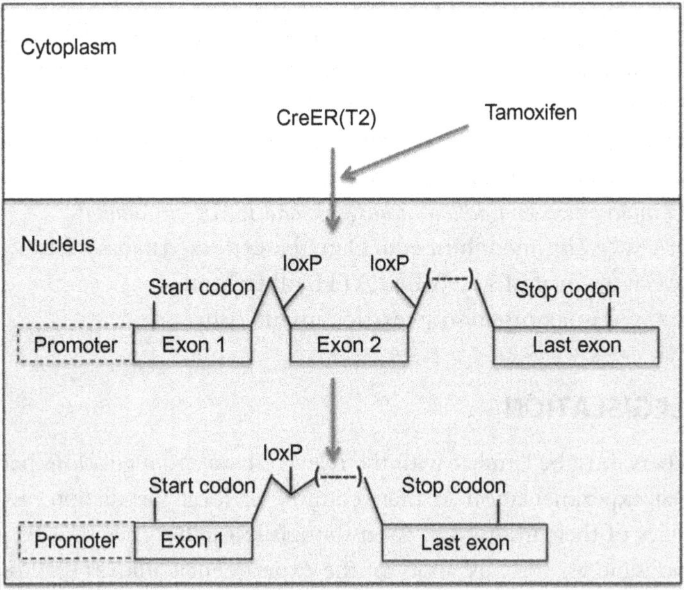

Figure 6.3 Tamoxifen-inducible creER(T2).

these mice will have lost the flanked exon when crossed with mice carrying a transgene for the relevant recombinase (Box 6.14). By using cre or Flp transgenic mice that express the recombinase in some cells but not others, conditional knock-out mice can be used to study the consequences of cell/tissue-specific loss of gene function.

Using mice carrying a cre transgene encoding a cre/estrogen receptor (modified to make it more specific to a drug that binds to the estrogen receptor, e.g., tamoxifen) fusion protein (creER(T2)), one can also control the timing of gene inactivation. creER(T2) does not enter the nucleus, and cre cannot act on the genomic DNA as a recombinase unless the estrogen receptor is stimulated by tamoxifen (see Box 6.14). Tamoxifen-inducible cre mice have made it possible to study the postnatal functions of genes that are critical for embryonic development, since induction during the embryonic stage would have killed the embryo. They have also been helpful in studies of genes that affect cell and tissue functions in a time-dependent manner.

Finally, the properties of zinc finger nucleases, transcription activator-like effector nucleases, and so-called clustered regulatory interspaced short palindromic repeats in combination with a bacterial DNA endonuclease have been utilized to develop techniques that allow site-specific modification of genomes in organisms and cells, including mice and human cells. These techniques have opened the door not only to generating mutant knock-out and knock-in mice faster and cheaper than before, but also to developing models of diseases that may be caused by a combination of mutations/variations at different sites in the genome.

More information and data on transgenic and knock-out animals:
http://www.bis.med.jhmi.edu/Dan/tbase/docs/databases.html
http://www.ncifcrf.gov/VETPATH/nihtg.html
http://www.jax.org/resources/documents/imr/

6.10 LEGISLATION

Researchers must be familiar with the relevant laws and regulations pertaining to animal experimentation in their country or local jurisdiction, as well as the policies of their institution. Even though researchers may not be personally responsible for the husbandry of the experimental animals they use, they should know the elementary rules of animal welfare (see suggestions for further reading).

The use of live animals for experimental purposes and other scientific purposes in Europe is regulated both at the international and the national levels. The international bodies include both the Council of Europe and the European Union, both of which have adopted regulations on the topic. The aims of these initiatives are to assist the European Union member states in their efforts to modernize legislation in this particular field, and of course to harmonize such legislation. The enforcement of legislation on experimental animals is the responsibility of each independent country. In the United States all animal research is covered by the Federal Animal Welfare Act of 1966 and its subsequent Amendments, all of which are enforced by the US Department of Agriculture. All institutions receiving federal funds for research purposes must have an assurance of compliance on file with the National Institutes of Health's Office for Laboratory Animal Welfare, and must establish stringent animal use programs to ensure that researchers are compliant with all rules and regulations in their research on experimental animals.

A sentence stating that the present work has been carried out in accordance with the national laws and regulations should be a natural part of the "Material and Methods" section of research reports of experiments conducted on animals.

6.11 NOTES ON THE USE OF EXPERIMENTAL ANIMALS

Uniform experimental situations during a given experimental period are important; therefore, experimental animals should be obtained from an accredited supplier running a high-quality breeding station, who can specify and document the properties of the animals, unless the home institution can guarantee the same kind of quality product.

If animal uniformity is essential and mice or rats are to be used, inbred animals should be chosen. They are more expensive and often less robust than outbred animals, but for all practical purposes they are isogenic, i.e., like identical twins. Hybrid animals that are products of a cross between members of two inbred strains are also isogenic, but are heterozygous and therefore often more robust than their parents. Special animal types, such as mutant mice—of which there are hundreds of types simulating different human diseases—and rats, which differ genetically from the parent strain at only one histocompatibility chromosome locus, have offered important scientific advantages. The development of new techniques in genetic engineering has expanded the scientific and technological possibilities enormously.

Concerning the microbiology of experimental animals, there are four choices: (i) conventional animals, (ii) specified pathogen-free (SPF) eggs or animals (Caesarean section or hysterectomy-derived), (iii) germfree or axenic gnotobiotic animals, and (iv) gnotobiotic animals associated with one or more defined type of microorganism. Gnotobiotic animals must be reared and maintained in isolators, and with strict regimes of handling, sanitation, food preparation, etc., all of which require specialist veterinary supervision. In comparison with SPF and conventional animals, gnotobiotic animals have a less-developed immune system, slower passage of intestinal contents, lower turnover of intestinal epithelium, and, at least in rats, a lower basal metabolic rate and cardiac output.

There are elaborate procedures for the breeding of SPF animals, which are similar to those used for the gnotobiotic animals, including strict personal hygiene rules for the investigator (see *Guide for the Care and Use of Laboratory Animals* in the "Further Reading" section). Compared with conventional animals, the SPF varieties are better standardized, more uniform, have a lower morbidity and mortality, and a higher resistance to some agents (endotoxin), drugs (e.g., cortisone), and physical insults (e.g., ionizing radiation). During infection they may produce less pus than conventional animals. They also have a steeper growth curve and longer life span. On the other hand, their breeding and transport are more complicated. Taken together, however, SPF animals should be preferred to conventional ones. Even if one has to maintain them under conventional conditions in the animal house, their SPF status can remain intact for weeks or months. Moreover, the researcher can be confident about what has been received, and minimize the risk of introducing an epizootic.

6.12 COMMONLY USED EXPERIMENTAL ANIMALS

The majority of experimental animals are rodents, especially mice. Mice are convenient experimental objects in studies of immunology, experimental hematology, pharmacology, embryology, neurobiology, skeletal biology, vascular biology, and oncology. They are cheaper and require less space than other mammals. Moreover, they are better genetically characterized and tailored to various applications than most other species, due to the existence of so many inbred, congenic, and recombinant strains.

The stocks and strains of experimental rats derive from the brown rat, or Norway rat (*Rattus norvegicus*; less-used synonyms are *Mus norvegicus*, the Hanovarian rat, or in German "Wanderratte"). A more proper name would have been *Rattus asiaticus*, since this rat migrated to Europe from the east about a quarter of a millennium ago and is now ubiquitous, like the mouse. The rat is one of man's close companions, though it is unpopular and even feared. Rats destroy one-fifth of the world's crops each year, can carry microbes that are pathogenic to humans, and may even attack and kill small children. Nevertheless, among researchers who have gotten to know and appreciate the domesticated experimental rat, the attitude is very different. These researchers know that when treated with calm and patience these animals are more peaceful than mice and easy to manipulate. Their intelligence is high enough to make them suitable for behavioral research; and due to the many inbred, congenic, and recombinant strains that have been developed and genetically characterized in recent years, rats have become a major experimental animal in studies of pathology, immunology, genetics, oncology, pharmacology, and physiology (e.g., endocrinology and nutrition research). The rat may represent a good compromise between economic considerations (size and space) and size considerations (possibility of surgical procedures or multiple blood or organ sampling).

As an experimental animal the rabbit (*Oryctolagus cuniculus*) has been used for the production of immune sera, diagnosis of infections, tests for pyrogens and toxicity of agents and drugs, but also for various kinds of physiological studies. Similarly to rodents, rabbits have very long, continuously growing incisor teeth, an almost exclusively vegetarian diet, and a large cecum. Even though some strains of rabbit may manifest anger and aggressive behavior, rabbits are generally silent and timid creatures. When frightened, rabbits may let out heart-rending screams that are often quite out of proportion with the damage inflicted. Worse, if frightened while restrained or sitting on a smooth support, violent contraction of the muscles in its back may break its weak lumbar spine. A quiet and careful approach to rabbits is therefore strongly recommended (Figure 6.4).

The cat (*Felis catus*) and the dog (*Canis familiaris*) were domesticated during prehistoric times. There are numerous strains or breeds of these animals, though variability is less marked for the cat than for the dog. Dogs have been used as experimental animals in physiology, pharmacology, and experimental surgery. The dog is a very expensive experimental animal, due to slow reproduction and growth, big size, and special requirements concerning husbandry. There are strong objections to their use in research among scientists and nonscientists alike. If they are indispensable for a given research project, they should be provided by an accredited vendor or breeder. The Beagle is often a convenient choice, as they are small (adult weight 10–17 kg, shoulder height 33–38 cm), smooth-haired, fertile, and friendly. However, if the Beagle is too small, the Labrador Retriever has many of the same pleasant characteristics, with a weight of 25–35 kg and a shoulder height of 55–60 cm. Physiologically, the cat is quite a labile animal. It has been used in anatomy, physiology, pharmacology, and toxicology, and a solid database has been accumulated. However, it is just as

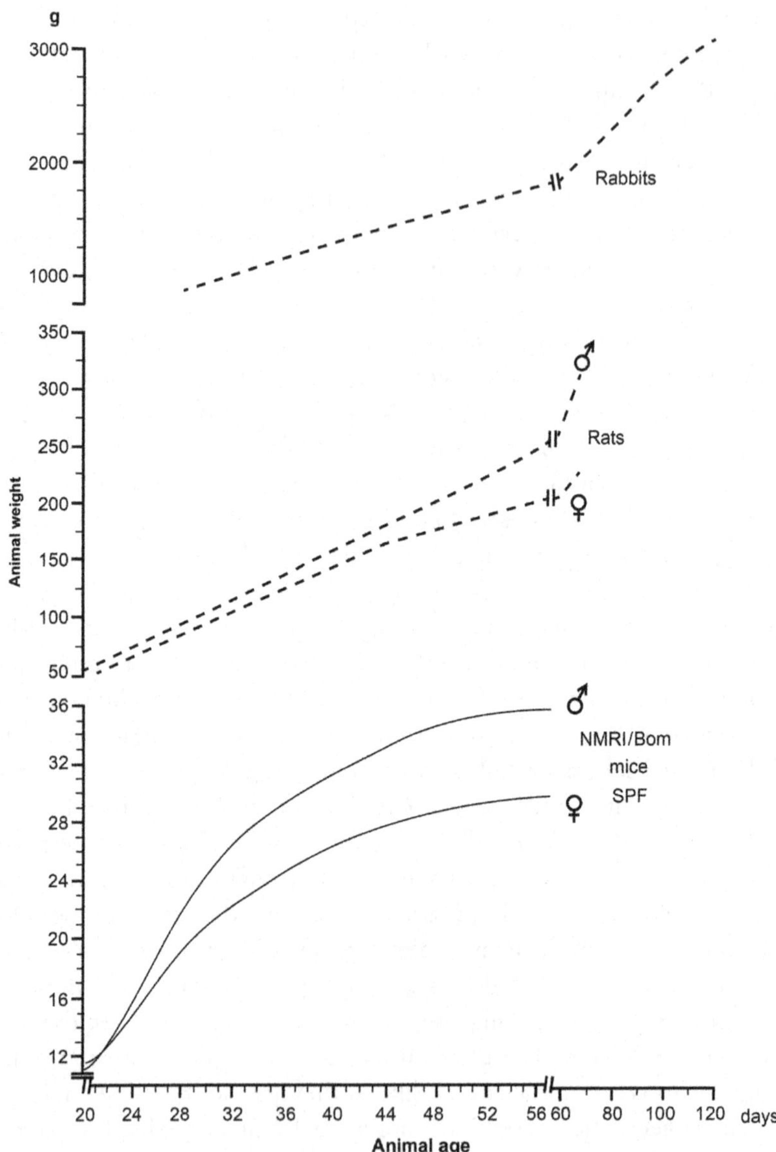

Figure 6.4 Examples of growth curves for commonly used experimental animals. It is important to establish the normal or optimal curves for strain(s) used in each research project, in order to establish whether the experimental procedures used in the project affect the weight gain of the animals.

difficult to obtain cats in a safe and reliable way as it is for dogs. The mini-pig may represent a good alternative to these larger experimental animals.

In earlier times biologists engaged in research on the taxonomy and diversity of different types of experimental animals. Similarities among different species were also noted, and it became increasingly clear that results obtained from a certain kind of animal—or even a certain bacterium or plant—could be relevant to other species, including humans. Therefore, because of accessibility and size, the large ova of the toad (*Xenopus laevis*) became a favored object of study for developmental biologists, and the giant axon of the squid became a favorite of neurobiologists. Bacteria (*Escherichia coli*) and their viruses (phages)—and more recently yeasts (*Saccharomyces cerevisiae* and *Schizosaccharomyces pombe*) as well—have been chosen by molecular biologists. Simple metazoa like *C. elegans* have been studied by neurobiologists, developmental biologists, and cell biologists, as the cells of these organisms are visible under a light microscope, and their genetics and development can be precisely defined. Modern experimental biomedicine abounds with similar examples. A fuller account of most of these examples—including the use of the fruit fly (*Drosophila melanogaster*), plants, primates, and humans as experimental systems—is given in a special issue of *Science* (1988; vol. 240, no. 4858) (Box 6.15).

6.13 CELL AND TISSUE CULTURE

During different stages of a research project, the researcher may choose to work with intact or reasonably intact organisms (see the previous section), or may have to isolate organs, tissues, cells, or subcellular components for investigation. Organisms can be studied over their entire life spans; organs removed and perfused with blood or a blood substitute for hours to days; tissue slices incubated in artificial extracellular fluids (culture media) perhaps for hours; and isolated cells for hours to months, depending on their nature and the culture conditions. Subcellular components, like enzymes, their substrates, hormones, RNA, or DNA, may be stored for long periods of time either freeze-dried (lyophilized) or frozen (at minus 20–196°C; the latter being the boiling point of liquid nitrogen). Some cell types can also be stored at low temperatures.

A mature mammalian organism is made up of cells that can be roughly classified into one of three categories: (i) cells that cannot divide (e.g., nerve and striated muscle cells), (ii) cells that constitutively divide (i.e., cells of renewal tissues, such as bone-forming cells, bone marrow cells, and surface

BOX 6.15 The Mouse

The house mouse (*Mus musculus*) is a social, nocturnal creature. Mice in a cage tend to congregate and sleep during the day and swarm actively at night.

The origin of the mouse was probably in Asia; migration made it ubiquitous. The wild-type animal is very temperamental and excitable. It will immediately try to escape or bite if one tries to catch it by its tail. The domesticated varieties may have similar characteristics, but experimental mice are mostly docile and easy to handle. Furthermore, they are lively, clean, and tidy.

Except for a sparse supply on the legs, the mouse lacks sweat glands and regulates its body temperature by varying the blood flow to its tail. Temperature-regulating centers in the hypothalamus control the blood flow—which may be large—through the one artery and three veins in the tail. Under conditions of extreme heat mice will lick themselves and their neighbors, thus cooling themselves through the evaporation of the saliva deposited on their coats.

All rodents have four very long, curved cutting (incisor) teeth, two in the upper and two in the lower jaw. They are rootless, which means that the enamel organ persists, producing continuous growth throughout life at a speed of about 2–3 mm a week. Constant length is maintained by gnawing of upper and lower incisors against each other.

The rodent diet is mainly vegetarian, with a variable meat component. The most conspicuous feature of the rodent digestive tract is the huge cecum. This is a storage, fermentation, and sorting site; during the night some of its contents are sorted out, and passed on through the large bowel; the resulting excrement is eaten (i.e., coprophagy). This process of passing food through parts of the digestive tract twice, known as cecotrophy, is presumably necessary because some essential nutrients produced by microbial activities in the cecum cannot be absorbed by the large bowel (B-vitamins, vitamin K, essential fatty acids).

Rodents are very fertile and give birth to large litters. This is probably an adaptation to naturally high mortality, due to numerous predators and few effective defense mechanisms in their natural habitat. Mouse estrous cycles and gestation times are both relatively short (Box 6.16), which is an advantage for experimental animal husbandry.

Thousands of mutant mice have been generated. These are valuable models of human diseases or physiological conditions, for example the diabetic mouse, the athymic (and "nude") mouse, the muscular dystrophic mouse, the obese mouse, various types of anemic mice, etc.

Mice may have viral infections that can change the function of their immune system or other organ systems—usually with no clinical signs. Examples include mouse hepatitis virus (MHV), minute virus of mice, and lactic dehydrogenase virus. These and other microorganisms can influence experimental results even when signs of disease are not observed. MHV, for example, is extremely contagious and can contaminate transplantable tumors and cell lines, thus altering experimental results. As health monitoring techniques have been improved, more and more microorganisms have been found to be latently present in experimental animals.

BOX 6.16 Useful Data on Some Commonly Used Experimental Animals

	Mouse	Rat	Rabbit	Dog	Cat
Biological variables					
Typical body weights	20–30 g	200–400 g	2.5–4 kg	10–35 kg	2.5–4 kg
Life span	390–420 days	2–3 years	7–8 years	13–17 years	13–20 years
Body temperature (°C)	35.5–39.5	35.5–39.5	37.5–39.5	37.0–39.0	38.5–39.5
Min. av. heart rate (min^{-1})	470	350	260	110	150
Min. av. respiratory rate (min^{-1})	138	92	40	19	26
Av. blood volume[a]	8	5	7	9	7.5
Breeding, reproduction, and husbandry					
Sexual maturity	6–8 weeks	7–11 weeks	4–5 months	9–15 months	5–12 months
Sexual cycle length (days)	4–5	4–5	14–16	21	14–24
Av. gestation period (days)	20	21	31	63	62
Litter size (range)	6–12	6–14	1–16	3–10[b]	1–8
Blood volume available (mL)					
From adults, routine sampling	0.1	0.25	1.0	2.0	1.0
Maximum inject. volumes (mL)					
Intravenously, one rapid inject.	0.5	1.0	10		
Intraperitoneally	1	4	10		

[a]Measured in mL/100 g of body weight. Note that cage life tends to make animals fat, if they are fed *ad libitum*. Then the blood volume given as a percentage of the body weight will be reduced.
[b]Breed-dependent.
min., minimum; av., average; inject., injection.

epithelial cells), (iii) cells that conditionally divide rapidly to repair cell loss (e.g., glandular cells in the liver, kidneys, and salivary glands; as well as many types of stromal cells). With current techniques, cultures of cells from all three of these categories have characteristic and limited life spans; some cells survive such a short time that even the best culture conditions could be referred to simply as incubation. Nerve, muscle, surface epithelial, and glandular cells can often be grown for days only; bone-forming, bone marrow, and stromal cells for hours to months.

Indeed, cells of renewal tissues such as bone marrow cells consist of stem cells as well as maturing and mature transit cells. The definition of a stem cell is a cell that can give rise to daughter cells that have two options: either to remain a stem cell, or to differentiate and start a suicidal course, maturing to a functional end cell with a limited life span (i.e., a transit cell). It is now known that most or all organs contain, and may be dependent on, stem cells—perhaps even cancers. So if one isolates functional end cells, one can perform meaningful experiments with cultures for periods that vary tremendously by cell type. Extended cell culture time would be permitted if the stem cells could be converted to a permanent cell line. The same would apply to induced pluripotent stem cells (iPSCs) generated by genetic manipulation of differentiated tissue cells.

6.14 AN OVERVIEW OF SELECTED TECHNICAL DEVELOPMENTS

Tissue culture was first successfully attempted at the beginning of the twentieth century. At that time small pieces of tissue were explanted from a cold- or warm-blooded animal to a culture vessel. Cells sometimes migrated out from the tissue explant, and cell division was only seen among those cells. In about 1950, a rapid expansion of the field took place in the field of tissue and cell culture (Box 6.17). Culturing of dispersed cells (single-celled suspensions) was introduced; refined culture media and culture technology appeared (antibiotics, disposable plastic vessels, clean air equipment, etc.); and it was discovered that most cells need specific polypeptide or glycoprotein growth factors for optimal survival and growth *in vitro.*

The history of hematopoietic cell culturing is particularly illuminating. Until the mid-1960s, attempts to extend the culturing of bone marrow cells had been unsuccessful. The art of cell culturing was by then not much more than 50 years old; it was called an art because no one really understood what was going on in the cultures. The exercise—at least as it started—resembled alchemy more than science.

It was of course known that when evolution produced multicellular organisms, the unicellular forerunners brought a portion of their surrounding sea with them into the new assembly, to constitute the extracellular fluid of man, animals, and plants. Normal life functions depend upon the homeostasis of this extracellular fluid, a task which falls to several of our organ systems: lungs, heart, blood, bowels, kidneys, etc. Osmolality, alkalinity (pH), ionic composition, gas pressure (pO_2 and pCO_2), nutrients, vitamins,

BOX 6.17 Some Landmarks in the Development of Tissue and Cell Culture

1885	Embryonic chick cells could be maintained alive in a saline solution outside the animal's body.
1907	An amphibian spinal cord was cultivated in a lymph clot, demonstrating that axons are extensions of single nerve cells.
1952	A continuous cell line from human cervical carcinoma was established, which later became the well-known HeLa cell line.
1954	Nerve growth factor (NGF) was shown to stimulate the growth of axons in tissue culture.
1955	The first systematic investigation of the requirements of cells in culture was made, and it was found that animal cells could be propagated in a defined mixture of small molecules supplemented with a small proportion of serum proteins.
1958–1960s	The characteristics of viral culture and transformation were established.
1961	It was shown that human fibroblasts die after a finite number of divisions in culture.
1965	The first heterokaryons of mammalian cells were produced by virus-induced fusion of human and mouse cells.
1975	The first monoclonal antibody-secreting hybridoma cell lines were produced.
1976	The first of a series of papers showing that different cell lines require different mixtures of hormones and growth factors to grow in serum-free medium were published.
1977	An efficient method for introducing single-copy mammalian genes into cultured cells was developed, adapting an earlier method.
1989	The first knock-out mouse was created.
1998	A technique was developed to isolate and grow human embryonic stem cells in culture.
2006	iPSCs were generated.

hormone concentrations, and temperature must be kept within strict limits. It is therefore only natural that the first culture media were tissue extracts. Over time, refined methods of analysis and new materials and products paved the road for the advent of standardized cell and tissue culture media of various types and defined composition. Even then some extra material of unknown but definite significance, present in sera or tissue extracts, had to be added, a practice that has been necessary until the most recent decades.

Fetal calf serum is often used in a cell culture medium, although it is very expensive and cheaper substitutes can sometimes be used. We now know that fetal calf serum is devoid of potentially harmful antibodies and contains adhesive proteins, which are important to anchor cultured cells to their plastic or connective tissue support. However, the most decisive characteristic is the serum's content of relevant growth factors, most of which are local hormones or paracrine agents that are needed for cell culture. In hemopoietic cell cultures, growth factors are called colony-stimulating factors, interleukines, or cytokines. These growth factors control not only cell division in culture, but also cell maturation and survival. Conditioned medium, which may replace or supplement serum in a culture, is mostly harvested from a different kind of cell culture. The success of bone marrow culturing, for example, depended on the production of media "conditioned" by lymphoid, stromal, or other cell populations to produce relevant growth factors; and on the later purification, characterization, amino acid sequencing, and gene cloning of these factors (Box 6.18). Adhesive proteins (dissolved or integral parts of cell membranes), as well as growth factors, are often involved when cells need "feeder cells" to grow in culture, or when cell-to-cell interactions are observed.

Another important event—which is particularly relevant to bone marrow culturing—was the production of viscous, semi-solid media, created by the addition of polysaccharide agar or methylcellulose to the culture media. This prevented daughter cells from leaving their birthplace, so that cells stemming from a single progenitor (or stem) cell (i.e., a cellular clone) would stay together in a colony. In a week, mouse bone marrow

BOX 6.18 Some Growth Factors and Their Actions

Factor	Composition	Representative activities
Epidermal growth factor (EGF)	53 amino acids (aa)	Stimulates proliferation of many cell types.
Insulin-like growth factor II (IGF-II)	73 aa	Collaborates with i.a. EGF, stimulates proliferation of fat cells and fibroblasts.
Transforming growth factor β (TGF-β)	Two chains, each 112 aa	Potentiates or inhibits response of many cell types to other growth factors. Affects differentiation of some cell types.
Fibroblast growth factor (FGF)	Acidic: 140 aa Basic: 146 aa	Stimulates proliferation of many cell types, including fibroblasts and endothelial cells.
Interleukin 2 (IL-2)	153 aa	Stimulates proliferation of T lymphocytes.
Nerve growth factor (NGF)	Two chains, each 118 aa	Promotes axon growth and survival of sympathetic and some sensory and central nervous system neurons.
Vascular endothelial growth factor A (VEGFA)	Two chains, each 120–188 aa	Promotes growth and survival of vascular endothelial cells, as well as other cells.

cells would generate colonies visible to the naked eye (minimum cell generation times about 8 h); human cell colonies would need about twice that time (minimum generation times about 24 h). A special property of blood cells is that they can grow in colonies without a solid support. Many other normal cells cannot; they need a glass, plastic, or other solid surface to adhere to. Cartilage-forming cells, chondrocytes, and cancer cells, on the other hand, can often grow in suspension and form colonies in the same manner as blood cells can. Finally, it should be mentioned that the most immature bone marrow stem cells apparently do not thrive in the types of semi-solid media presently available, nor do they thrive on plastic surfaces. It should be borne in mind that, even if some cell types survive and grow on a plastic surface, other types may perform quite differently and more naturally when present in a cellular niche that is similar to their natural one. For bone marrow stem cells, that might mean blood vessels, as well as certain types of connective tissue cells, collagen fibers, proteoglycans, etc., to make up a defined three-dimensional scaffold.

In this way, with current technology, hemopoietic stem cells can produce hemic cells for weeks in culture. Ultimately, however, this hemopoiesis ceases. Even cultures of fibroblasts, which are quite easy to culture, tend to die out. With fibroblasts, it seems that a kind of cellular senescence is programed into the cells, so that fibroblasts from embryos will divide for a longer period of time than fibroblasts from young animals, which in turn will divide longer than fibroblasts from aged animals, also known as the Hayflick limit (Figure 6.5). The biological basis of the Hayflick limit seems to be the loss of the chromosomes' telomeres. Telomeres are stretches of repeated polydeoxynucleotides forming the extremities of chromosomes. The standard DNA replicatory mechanism cannot duplicate telomere DNA; it takes special RNA primers and the telomerase enzyme to do the job. In the absence of telomerase, a segment of each chromosome's DNA will be lost in each cell division. When a critical telomere reduction has occurred, the cell can no longer divide. Significantly, the telomerase enzyme is active in cancer cells, and possibly also in stem cells.

6.15 APPLICATIONS OF CELL CULTURE

Cell culture offers the possibility to obtain homogeneous cell suspensions and increase cell multiplication, thus increasing the amount of cellular material available for analysis. Cell culture is easy to set up and manipulate, and the resulting cultured cells and cell interaction systems have been

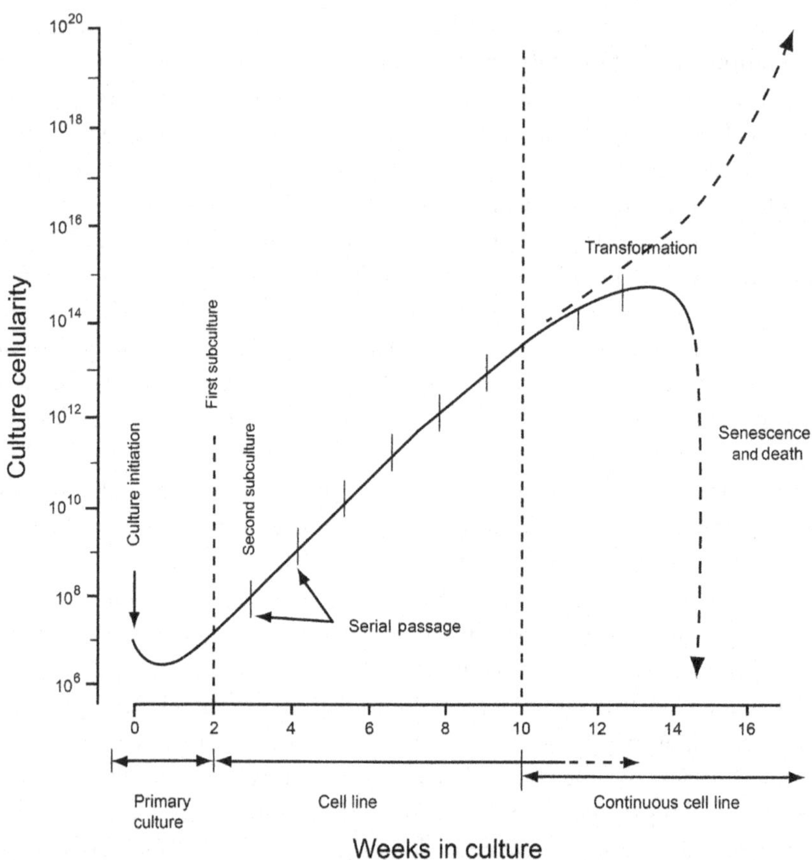

Figure 6.5 Various fates of cultured cells. Evolution of a hypothetical cell culture. The *y*-axis shows the accumulated cell yield, assuming no cell loss at passage, on a log scale. A continuous cell line can arise at any time, not necessarily after 10 weeks, as depicted here. Likewise, senescence may occur at any time—here after about 12 weeks—but for human diploid fibroblasts most likely after 30–60 cell doublings (or 10–20 weeks, depending on the doubling time)—the "Hayflick limit." *Based on Freshney.*[6]

invaluable to the great advances made in the field of cell biology during the last decades (Figure 6.6). The challenges posed by cancer and the desire to understand, diagnose, prevent, and treat viral diseases have meant that cells from warm-blooded animals and man have been the most commonly used. However, cells from lower species have increasingly been used in cell culture to solve problems in fields such as developmental biology, agriculture, and fish farming.

A primary cell culture is started from cells taken directly from an organism; when these cells are transferred to a new vessel (also referred to as

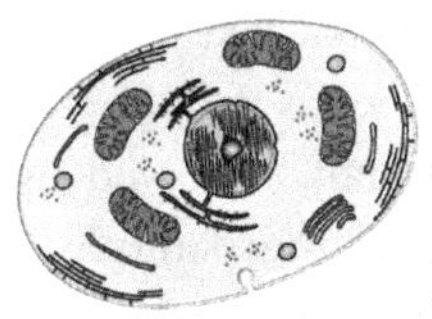

Cellular biochemistry and molecular biology

(Intermediary metabolism; DNA/RNA synthesis; cell transfection and gene therapy; anti-sense RNA; S/N/W blotting; polymerase chain reactions; gene knock-out; site-directed mutagenesis; screening for genetic diseases; production of recombinant proteins; proteomics, metabolomics, etc.)

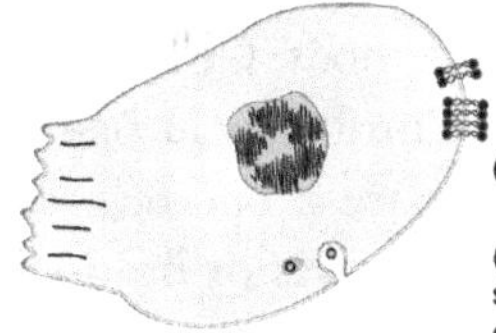

Cell biology

(Movement; secretion, endocytosis; cell adhesion; induced pluripotent stem cells (iPSC); cell division, differentiation and maturation; carcinogenesis; membrane fluxes and transport; membrane potentials; etc.)

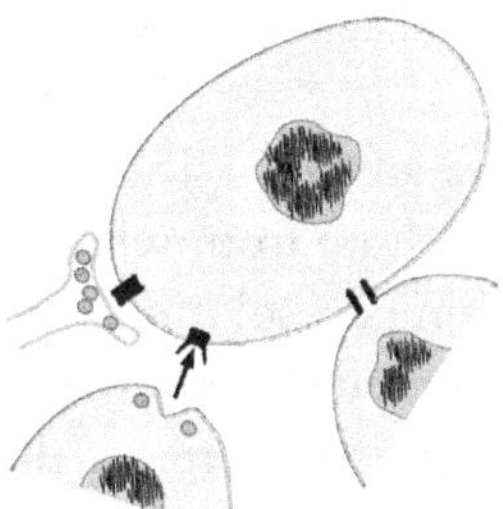

Cell communication

(Hormones; paracrine agents; neurocrine transmitters; cell-to-cell (metabolic) cooperation and signaling; cell-to-matrix interactions; receptors; etc.)

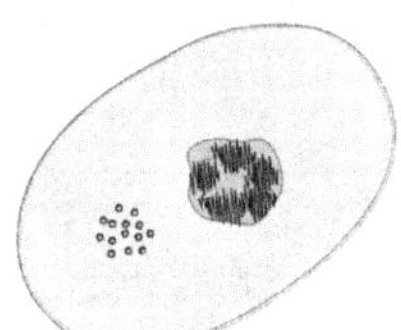

Miscellaneous

(Cell fusion (with e.g., hybridoma formation); virus culture and other infections; drug actions; toxicity studies; development of vaccines; etc.)

Figure 6.6 Survey of various ways to use cell cultures.

subculturing) they become a cell line. Cultured normal cells may become immortal either spontaneously, or through experimental manipulation (exposure to chemicals, irradiation, or viruses) (see Figure 6.5). They then form a continuous cell line, and the normal cells are considered transformed. Transformed cells may be oncogenic, i.e., able to produce tumors when injected into animals, or not. Oncogenic cell lines may give rise to benign or malignant (i.e., invading and metastatic) growths. Transformed cells seem to exhibit a genetic change, leading to the bypassing or the extinction of regulatory feedback loops, so that the cells are no longer dependent on the addition of extracellular factors for their growth or survival.

An experimental system consisting of a single type of cell can now be created by various approaches. A reasonably homogeneous primary culture can be made by dispersing the cells in a tissue fragment, followed by cell isolation/separation procedures. Cells can be dispersed mechanically (e.g., by pipetting or whorl-mixing), enzymatically (e.g., by trypsin digestion of extracellular matrix), chemically (e.g., by ethylenediaminetetraacetic acid complexing Ca^{2+} and Mg^{2+} ions, which are often instrumental in cell adhesion), or by various combinations of these three methods. Cell isolation may be based on cell size (e.g., cell sedimentation or centrifugal elutriation), cell density (e.g., density gradient centrifugation), surface properties (e.g., cell adherence to plastic or to antibody-coated vessel surfaces or beads), or on combinations thereof (e.g., fluorescence-activated cell sorting, where cells above or below a certain size threshold, or with or without attached fluorescent antibody against a cell surface antigen, are isolated by flow cytometry) (Figure 6.7).

Alternatively, homogeneous cell cultures can be obtained by serial subculturing, starting with a heterogeneous primary culture from which a certain cell type may outgrow the others. An even simpler solution is to use a continuous cell line. One great advantage of continuous cell lines is that the researcher can take experiments performed elsewhere in the world and reproduce them exactly. It is also possible to start clonal growth from a single (stem) cell, using a 96-well culture plate.

6.16 MANIPULATION OF CULTURED CELLS

Somatic cell genetics received a strong impetus from the discovery that mitosis could be induced in normal blood lymphocytes during cell culture by the plant lectin phytohemagglutinin (which had already been used for decades to aggregate, and thereby sediment, red blood cells). Immunology benefited as well, in that proliferating and differentiating lymphocytes could be studied. Over time, improved cell culture techniques facilitated reproducible studies of interactions in cell culture, such as those between T and B lymphocytes, and between lymphocytes and antigen-presenting cells, mimicking *in vivo* interactions during antibody production.

Another major advance was the technique of cell fusion. With one of the fused cells enucleated, important information on the interplay between cytoplasm and nucleus was gained. A cell with two nuclei is called a heterokaryon; in these cells a hybrid cell with one large nucleus is formed after

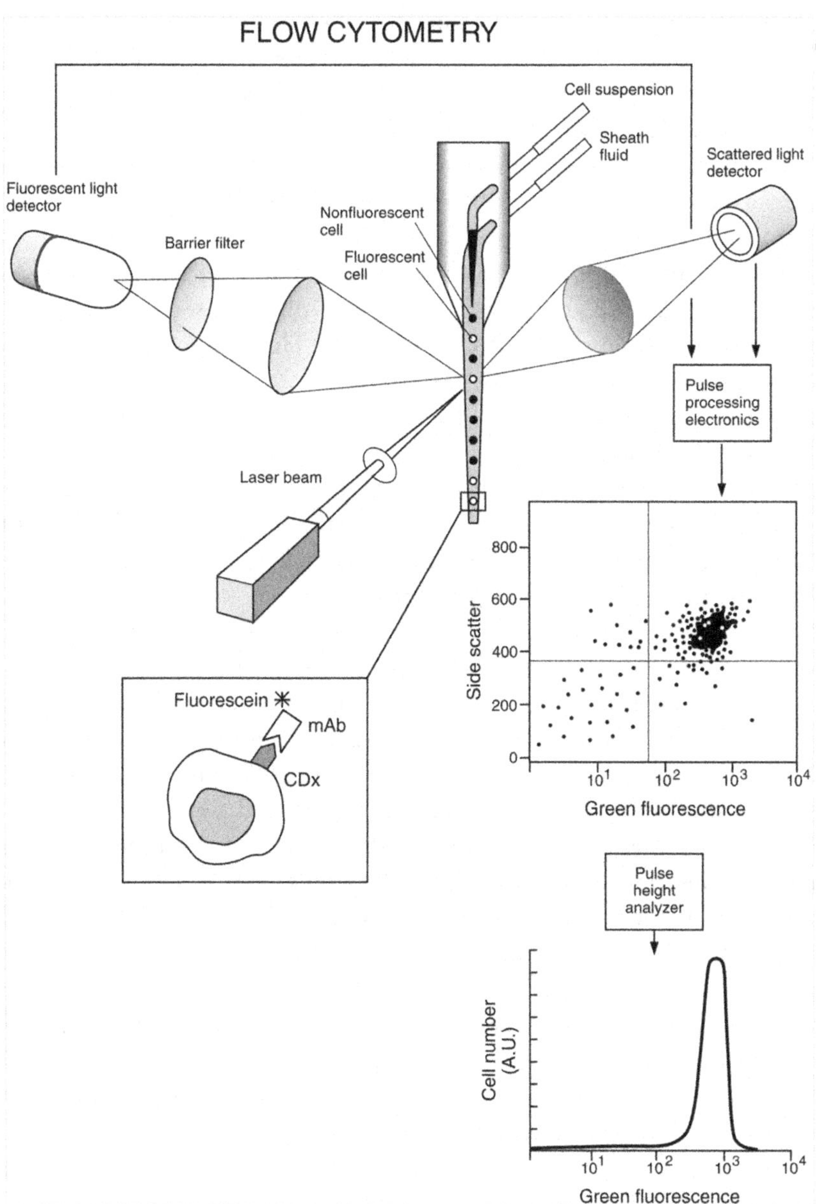

Figure 6.7 Principle of flow cytometry.

mitosis. When hybrids between human and mouse cells are created, sometimes their human chromosomes are lost. By correlating the changes in cell phenotype to the random loss of chromosomes during mitosis, it has been possible to assign certain genes to specific chromosomes.

Of immense practical importance was the discovery that the hybrid cells formed by an immune cell (i.e., a cell which produces a specific antibody) and a cell from a plasmacytoma cell line (i.e., a transformed, antibody-forming cell line) may form a continuous cell line—originating from the hybrid cell and therefore a clone—that will secrete the specific antibody into the medium of cultured cells. This antibody is a so-called monoclonal antibody and has become an indispensable tool in cellular and molecular biology, as well as in the diagnosis and therapy of human and animal diseases (Box 6.19).

A huge number of cell lines are available for purchase, or can be obtained more or less for free. The American Type Culture Collection is a unique, private, nonprofit resource dedicated *inter alia* to the distribution of living cell lines (http://www.atcc.org/).

The most recent developments in cell culture are based on molecular biology. Genes can be introduced into cells to transform them, to manipulate them in specific ways, or to make them produce protein factors or hormones, such as human interferon, insulin, or erythropoietin. Addition of an anti-sense RNA can disclose whether a certain protein (encoded by the relevant "sense" mRNA) plays a hypothesized functional role. The polymerase chain reaction (PCR) technique can be used to detect an actively transcribed gene in a single cell! It is easy to comprehend the enormous developmental potential such techniques may hold for both basic biological research and areas of applied research, from medicine to agriculture. Sometimes, the post-translational modification of proteins, like glycosylation, is not absolutely necessary for some aspects of their biological function. In such cases, prokaryotic organisms that are easier to culture can be used to host the producer gene, rather than eukaryotic cells (yeast cells or, e.g., mammalian cells).

The further and varied uses of cell culture, for example, to perform chromosome analysis on fetal cells obtained by amniocentesis, to diagnose viral infections and make viral vaccines, to assess the toxicity of pharmaceutical compounds and environmental pollutants, and perhaps form tissues suitable for transplantation purposes or reconstructive surgery, lie

BOX 6.19 Monoclonal Antibodies

Antibodies can distinguish with exquisite specificity among an amazingly large number of different antigens, a property that has made antibodies a versatile tool in biomedical research and clinical medicine.

In the past, research laboratories had no option but to use polyclonal antibodies. These are obtained by simply injecting the antigen into an animal (e.g., rabbit, sheep, or horse), usually by repeated injections several weeks apart, and then bleeding the animal to get serum. However, this serum always contains a blend of different antibodies to the antigen, each binding to or "seeing" different sites (epitopes) on the same antigen, or seeing the same epitope in different ways with dissimilar affinities. Because different antibodies are produced by different antibody-forming cells, this blend is referred to as polyclonal antibodies. A major drawback with this method is that unless the injected antigen is completely pure, the resulting serum also contains antibodies against the impurities in the antigen. Another drawback with the traditional method of producing antibodies is that each individual animal responds to an antigen in a unique way, producing different sets of antibodies with dissimilar titers, in addition to variation among different bleedings from the same animal. This means that it is often difficult to compare results from similar tests made with antibodies from different batches of serum.

These shortcomings could be overcome if single cells producing an antibody to a given epitope could be cultured to produce the same antibody. Unfortunately, normal antibody-producing cells do not grow in culture but instead die within a few days. In 1975 this obstacle was circumvented when Köhler and Milstein got the idea to fuse antibody-producing cells isolated from the spleen of an immunized mouse with malignant mouse cells. The hybrid cells retained the ability to produce antibodies from the normal, antibody-producing fusion partner and the power to survive and divide infinitely in culture from the malignant fusion partner. The latter fusion partner was selected with care. It was a myeloma cell, originally derived from a normal antibody-producing cell, but transformed to a cancer cell. Furthermore, it was deficient in a certain enzyme so that it could not survive in a selective medium, called HAT medium. This meant that after fusing antibody-producing cells with myeloma cells and culturing the cells in HAT medium, only hybrids of normal cells and myeloma cells survived.

Another advantage with this technique is that the order of events can be completely reversed, i.e., the antibody can be produced first and the antigen detected and characterized afterwards, with the antibody as a tool.

Monoclonal antibodies have numerous applications in clinical medicine and biomedical research. A well-known example is modern pregnancy tests, sold as kits in pharmacies. As early as 2 weeks after fertilization the urine of the pregnant woman contains traces of the hormone human choriogonadotropin (hCG),

(*Continued*)

BOX 6.19 Monoclonal Antibodies—cont'd

which can be detected by a monoclonal antibody against hCG and latex particles coated with the hormone. If hCG is absent from the urine, the antibodies will crosslink the particles to easily visible clusters, whereas free hCG in the urine will compete for the binding sites and thereby inhibit cluster formation. Variants of this test are based on direct detection of hCG with antibody. Similar principles, whether competitive or direct, lie behind a large array of other tests, such as radioimmunoassay, which is based on scintillation counting of radiolabeled ligands, and enzyme-linked immunosorbent assay, which is based on photometric measurements of converted substrates catalyzed by enzymes coupled to antibodies. Note that these tests are also performed with polyclonal antibodies, but the monoclonal antibody technique has greatly extended their use.

Antibodies, in particular monoclonal antibodies, can also be used as staining agents to visualize the presence and distribution of molecules on cells (immunocytochemistry) or in tissues (immunohistochemistry). The antibodies must then be tagged with a tracer that can be made visible, such as a fluorochrome that can be detected by fluorescence microscopy, a radioactive marker that can be visualized by autoradiography, or an enzyme that catalyzes the deposition of an insoluble, colored end product.

There are also interesting applications of monoclonal antibodies in clinical medicine, either in what is referred to as immuno-targeting or as drugs. In the first case the antibodies are used as vectors to carry attached groups—like cytotoxic agents or isotopes—to selected cells in the body, e.g., cancer cells. Monoclonal antibodies used as drugs neutralize the cytokine tumor necrosis factor in rheumatoid arthritis or weaken the "brakes" of an immune attack on cancer cells (anti-CTLA-4 antibodies).

New methods of making monoclonal antibodies are currently being explored based on advances in genetic engineering. In these techniques cDNA libraries from antigen-stimulated spleen cells are ligated into expression vectors and used to transform bacteria. It is therefore likely that bacteria (prokaryotes) or yeast cells (eukaryotes) will supplement hybridomas in the production of monoclonal antibodies.

The monoclonal antibody technique has developed into a large industry that has greatly facilitated biomedical research and radically transformed laboratory medicine. Moreover, it has important applications in clinical medicine.

outside the scope of this chapter. This is also the case for some routine procedures, such as techniques of asepsis, preparation and sterilization of culture media, cell separation and quantification, control of mycoplasma and other infections, and control and assessment of endotoxin contamination and the function of incubators.

6.17 POSSIBILITIES PROVIDED BY MOLECULAR BIOLOGICAL TECHNIQUES

As indicated in Figure 6.8, methods exist to measure the amounts and activities of all possible cellular constituents, even from a single cell, such as transcriptomics, proteomics, and metabolomics, which concern RNAs, proteins, and metabolic intermediates of cells, respectively. If the researcher can obtain a single-cell suspension, flow cytometry is a versatile technique (see Figure 6.7) that scores thousands of suspended cells in half a minute. With fluorescently tagged antibodies, the density of cell surface protein can be quantified, and cell size, granularity, DNA content, intracellular proteins, specific mRNAs, and viability can be determined.

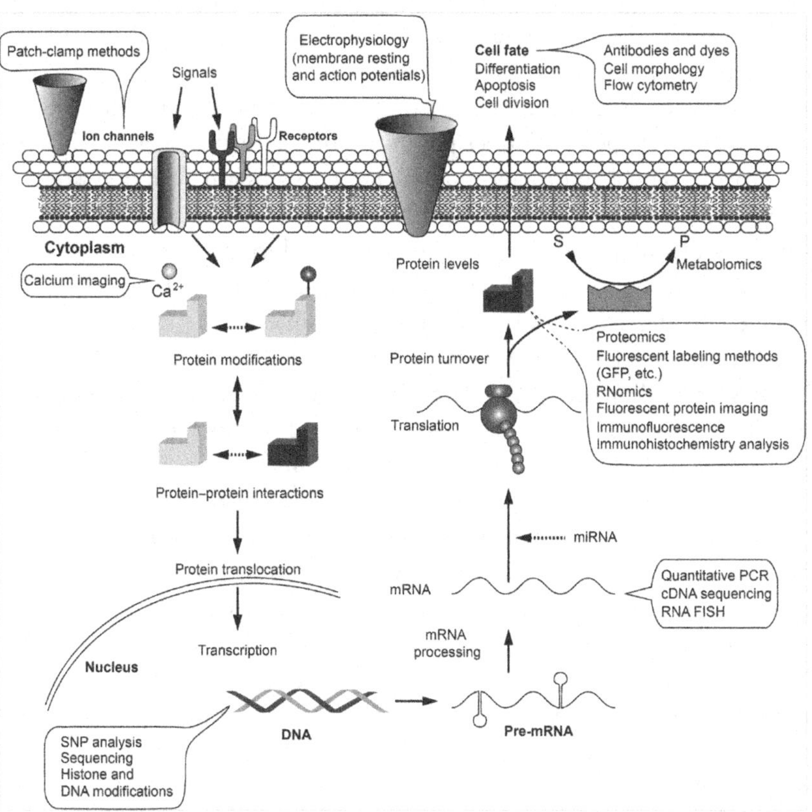

Figure 6.8 Schematic view of possibilities to analyze constituents in and function of a cell—genomics, proteomics, metabolomics, and electrophysiology/cell membrane channels.

Discoveries during the last decade have opened the doors to exciting experimental possibilities, such as the genetic engineering of cells (replacing or removing their genes; see the previous section), rapid DNA and RNA sequencing, the generation of iPSCs from differentiated cells, the analysis of epigenetic regulatory mechanisms in cells, as well as the interference from noncoding RNA species (microRNAs, etc.) with mRNA-dependent protein synthesis. Current DNA sequencing methods differ from the classical Sanger dideoxynucleotide-chain-termination method in that they read the addition of labeled nucleotides as DNA chains are copied in a massively parallel fashion. This makes it possible to quickly obtain large amounts of sequence data and deep coverage (each nucleotide is read multiple times, ensuring increased accuracy). For example, some of the available systems have outputs of multiple 1-million nucleotides per second per machine; one-time coverage of one human genome of DNA can be performed in about an hour, and the equivalent of the human genome can be resequenced 30 times per day. Using such fast methods for sequencing cDNA copies of RNA transcripts makes it possible to identify and count the relative numbers of specific transcripts in a sample. In principle, such results can provide digital information about gene expression in the entire transcriptome, even with RNA isolated from single cells. In addition, methods that allow identification of methylated residues during deep DNA sequencing have made genome-wide studies of changes in methylation-dependent epigenetic regulation (DNA methylomes) possible.

These molecular biology technologies have revolutionized the field of genetics, including the identification of gene mutations responsible for inheritable diseases in humans and animals, and of sporadic diseases caused by somatic mutations. Furthermore, by amplifying genomic DNA from single cells, one can sequence and compare the entire genomes of different cells to obtain systematic assessment of genomic diversity within an organ such as the brain.[7] Finally, the ability of the current technologies to sequence an enormous number of different DNA strands at the same time makes it possible to tease out the sequence of a minority of specific DNA strands in samples that are otherwise contaminated with large amounts of unrelated DNA. As a consequence, obtaining DNA sequences from fossilized ancient animals and plants is now possible, and has resulted in groundbreaking studies in phylogeny, biogeography, and molecular evolution. For example, portions of genomes from fossilized remains of Neanderthals can now be compared with the genomes of modern humans. How far one can go back in time depends on how well the remains were preserved; the genome of a prehistoric horse was recently determined using DNA isolated from bone fragments that were preserved in permafrost for 700,000 years.[8]

Another advance in molecular biology has led to the use of modified viruses to deliver genetic material into cells during cell culture (as well as into living organisms). Using such techniques, it is now possible to perform genome-wide, cellular signaling pathway-specific and single-gene functional analyses. Transduction of DNA using a viral vector is much more efficient than traditional methods of plasmid-mediated transfection of cells. Although it requires the extra work of packaging the DNA into infectious viral particles, core facilities at many research institutions or companies provide a low-cost virus production service, thus making available many different types of viruses suitable for different needs. Replication-defective retroviruses can carry DNA inserts up to about 10,000 base pairs long. The disadvantage is that they will only transduce DNA into dividing cells. Nevertheless their ability to integrate into the host genome is clearly a plus, since the DNA is passed on to the progeny when the cells divide. In contrast, adenoviral vectors do not integrate into the host cell genome, but they do remain in the cell nucleus and provide transient and high-level gene expression. In addition, they efficiently infect many types of mammalian cells. Adenovirus delivery of DNA is therefore a very good choice for short-term cell culture experiments. Lentiviral vectors have the advantage of infecting practically all nondividing cells, such as stem cells, primary cells, and neurons, and like retroviral vectors, they integrate into the host cell genome. Lentiviruses, having obtained their name from their long incubation periods, belong to the family of retroviruses. They cause diseases including immunodeficiency in primates, sheep and goats, horses, cats, and cattle. When used as vectors for the delivery of DNA into cells, they are made replication-incompetent for safety, and have also been engineered in other ways to reduce potential hazard. However, the use of lentiviral vectors poses a risk of biohazard, since they can transduce primary human cells. Therefore, the use of lentiviral vectors must follow Biosafety Level 2 guidelines with proper waste decontamination.

Using lentiviral vectors to transduce gene-specific short-hairpin (sh) RNAs into human or mouse cells provides an interesting method for probing gene functions in cells. If the shRNA is specific to one particular gene, the transcript levels for that gene will be suppressed, and the resulting alterations in cell function can be studied. If one is interested in finding out what the functional impact may be of suppressing any one of a number of genes that appear to work together in a signaling pathway, cells can be transduced with a collection (called a library) of different lentiviral particles, each particle encoding shRNA for a different gene transcript. After treating cells with such a library, the cell population will no longer be homogeneous, but will consist of a mixture of cells with different suppressed gene transcripts. Finally, if one would like to screen for genes that may be important for a specific cellular process, for example, differentiation of mesenchymal stem cells to osteoblast-like cells, one can use a genome-wide library of lentiviral shRNA (a mixture of virus particles in which shRNAs for all genes in the genome are represented) to infect mesenchymal stem cells. After growing the transduced cells under conditions that promote osteoblast differentiation, the cells may be stained for alkaline phosphatase (expressed in osteoblast-like cells, but not in mesenchymal stem cells). Many cells will be positively stained (not transduced or transduced with shRNA for genes that have no effect on osteoblastic differentiation); some cells may be stained more intensely because they express shRNA for a gene that normally suppresses osteoblast differentiation; negative cells may express shRNA for a gene that is needed for osteoblast differentiation. Isolation and amplification of such positive and negative cells, followed by identification of their lentiviral gene-specific constructs, provides a basis for further validation studies aimed at excluding false positives and negatives and narrowing the screen to identify genes that may really be important. Although this example uses staining for alkaline phosphatase as readout ("target") for differentiation, the transduced cells could be used to explore many processes or functions; for example, up- or down-regulation of a key bone cell transcription factor, the secretion of a bone-specific extracellular matrix molecule, or the cell-surface expression of a specific matrix-binding receptor. Selecting and subculturing cells that have up- or down-regulated expression of the target protein as a result of the lentiviral infection, followed by PCR amplification of the shRNAs in the selected cells and RNA sequencing, will identify genes that potentially regulate the expression of the target protein. Although the potential regulators must be validated by further studies, the lentiviral library screening approach can provide an important assumption-free first step in defining regulatory pathways.

The use of cell-penetrating peptides for intracellular protein delivery to cells in culture is an interesting alternative approach to plasmid transfection and viral transduction of DNA. The technique is based on the discovery that some peptide sequences called protein transduction domains (PTDs) frequently contain a cluster of basic amino acid residues and have the ability to freely cross the cell membrane and accumulate in the cytoplasm. Examples of such PTDs include a domain within the *Drosophila* antennapedia protein and a domain in the HIV TAT protein. Several PTDs are not only able to cross cell membranes on their own, but when fused to larger peptides or proteins they can mediate the transduction of the whole complex into cells. Coupled with the properties of nuclear translocation sequences (NLSs), which mediate the transport of intracellular proteins into nuclei through nuclear pore complexes, PTD technology makes it possible to modify proteins to cause them to enter cell nuclei when added to the medium of cultured cells. A good example of the usefulness of this method is the very efficient nuclear uptake and recombination efficiency of a cre recombinase-PTD-NLS fusion protein when added to cultured cells carrying a gene with floxed alleles.

Thus, there are several experimental approaches that can be utilized to study the differentiation and functions of cultured cells. Combined with the discovery of methods to induce dedifferentiation of differentiated cells into iPSCs, many cellular aspects of embryonic development can now be studied in culture in detail. Initially, iPSCs were generated by the introduction of four genes into differentiated cells such as fibroblasts. The introduced genes stripped cells of their epigenetic restrictions, so that cells became sufficiently immature and able to differentiate in virtually all directions allowed by a fertilized ovum of the species and also renew themselves by cell division without maturation (the definition of stem cell property). Although many improvements have been made in iPSC methods, the yield of iPSCs continues to be relatively low. Also, a high potential for cancer development as the cells are grown through multiple passages makes current iPSC technology unsuitable for clinical regeneration of damaged organs. In addition, technologies need to be developed that can allow cells to grow and differentiate in three dimensions before one can fully investigate the processes of tissue formation and organogenesis. In such investigations, techniques and insights from molecular biology, cellular and developmental biology, extracellular matrix biology, and tissue engineering will undoubtedly be needed.

QUESTIONS TO DISCUSS

1. Given the goals of your own research project, what are the pros and cons regarding the use of experimental animals?
2. If you were to write a research protocol that included the use of mice for review by your institutional animal care and use committee, what do you think would be the most important factors on which the committee would examine and base their decision (approval or disapproval)?
3. What are the alternatives to using experimental animals in your own research project?
4. What aspects of cell culture do you find to be most useful for the solution of a scientific problem(s)?
5. Give examples of what you find to be the most fruitful use of current cell culture technology.
6. Can you think of a cell culture technology that you wish someone would develop to help your own research project progress faster?
7. In your opinion, which are the most critical factors to keep in mind when one interprets results obtained with cell, tissue, or organ cultures and attempts to extrapolate the results to human physiology or pathology?
8. iPSCs isolated from patients with various inherited disorders are frequently described as allowing studies of human diseases in a culture dish. In your opinion, which are the most important aspects of human disease mechanisms that limit the conclusions that can be drawn from such culture experiments?
9. What do you think are the most important limiting factors in the use of genetically modified mouse models for studies of osteoporosis and osteoarthritis?
10. What is the difference between somatic and germ-line mutations? Explain how the current deep DNA sequencing methods can be used to identify somatic mutations in diseased tissues.

REFERENCES

1. Medawar PB. *Induction and intuition in scientific thought. Jayne lectures for 1968.* Philadelphia, PA: American Philosophical Society; 1969.
2. Medawar PB. *Advice to a young scientist. Human action wisely undertaken.* New York, NY: Harper & Row; 1979.
3. Russell WMS, Burch RL. *The principles of humane experimental technique.* London: Methuen & Co. Ltd; 1959.

4. Kilkenny C, Browne WJ, Cuthill IC, Emerson M, Altman DG. Improving bioscience research reporting: the ARRIVE guidelines for reporting animal research. *PLoS Biol* 2010;**8**:e1000412 1–5.
5. van der Worp HB, Howells DW, Sena ES, Porritt MJ, Rewell S, O'Collins V, et al. Can animal models of disease reliably inform human studies? *PLoS Med* 2010;**7**: e1000245 1–8.
6. Freshney RI. *Culture of animal cells. A manual of basic technique*. 2nd ed. New York, NY: Alan R. Liss; 1987.
7. Evrony GD, Cai X, Lee E, Hills LB, Elhosary PC, Lehmann HS, et al. Single-neuron sequencing analysis of L1 retrotransposition and somatic mutation in the human brain. *Cell* 2012;**151**:483–96.
8. Orlando L, Ginolhac A, Zhang G, Froese D, Albrechtsen A, Stiller M, et al. Recalibrating *Equus* evolution using the genome sequence of an early Middle Pleistocene horse. *Nature* 2013;**499**:74–8.

FURTHER READING

Davis JM, editor. *Basic cell culture: a practical approach*. Oxford: Oxford University Press; 2002.

Lanza R, Langer RP, Vacanti JP, editors. *Principles of tissue engineering*. 4th ed. Burlington, MA: Elsevier Academic Press; 2013.

National Research Council. *Guide for the care and use of laboratory animals*. 8th ed. Washington, DC: The National Academies Press; 2011.

Schaeffer WI. Proposed usage of animal tissue culture terms. (Revised 1978.) Usage of vertebrate cell, tissue and organ culture terminology. *In Vitro* 1979;**15**:649–53.

van Zutphen LFM, Baumans V, Beynen AC. *Principles of laboratory animal science, revised edition. A contribution to the humane use and care of animals and to the quality of experimental results*. Amsterdam: Elsevier; 2001.

CHAPTER 7

Translational Medical Research

Bjorn Reino Olsen[1], Mahmood Amiry-Moghaddam[2] and Ole Petter Ottersen[2]
[1]Harvard School of Dental Medicine, Harvard Medical School, Boston, MA, USA
[2]Department of Anatomy, Institute of Basic Medical Sciences, University of Oslo, Blindern, Oslo, Norway

7.1 INTRODUCTION

Translational medical research is often described as an activity that covers what is called "from bench to bedside." As the description suggests, translational medical research builds on the results of basic medical research and is aimed at developing new therapies and diagnostic procedures across the entire spectrum of human disease. Thus, the success of translational medical research is critically dependent on progress in basic medical research. However, great advances in basic medical research do not guarantee corresponding advances in drug development, the discovery of novel methods for disease detection, or in measuring the effects of medical care. Indeed, the boom in basic medical research during the past 50 years has yet to lead to an avalanche of new drugs for the most common human disorders, despite the enormous efforts made by the pharmaceutical industry to take advantage of new discoveries to develop and test new drugs. What are the reasons for this? Is there an intrinsic difficulty in drug discovery that makes the process slow and inefficient? Is drug development easier now than it was in the past? Are there unnecessary obstacles in translational medical research? Is translational medical research best left to the experts—i.e., the pharmaceutical industry?

In the following sections specific examples will be used to address these questions, and to explain the differences between translational medical research at academic institutions and in drug research and development in the pharmaceutical industry. The need to integrate translational medical research into the academic clinical research environment, and what this means for biomedical research training, will also be discussed.

7.2 AN OLD SUCCESS STORY—ASPIRIN

Aspirin is one of the most cost-effective and widely used drugs in medicine. With about 100 billion tablets swallowed worldwide every year, it is

Research in Medical and Biological Sciences
DOI: http://dx.doi.org/10.1016/B978-0-12-799943-2.00007-0

remarkably effective in treating pain, fever, and inflammation. It prevents blood clotting and thus heart attacks, strokes, fetal growth retardation, and pre-eclampsia, and it appears to be able to reduce the risk of certain types of cancer. How was aspirin discovered, what were the scientific insights that led to its current status as a superdrug, and what can one learn from the history of its development? To answer these questions one must go far back in time and follow a track that is by no means logical or linear.

Medicinal use of salicylic acid derivatives to treat pain and fever dates back at least 6000 years.[1] Assyrian clay tablets describe the use of willow bark extract to treat fever, inflammation, and musculoskeletal disorders. In Egypt, willow bark extract was used to treat wounds, and in China it was also used to treat colds, hemorrhage, goiter, and rheumatic fever, and as an antiseptic for wounds and abscesses. In Greece, at the time of Hippocrates, willow bark extract was prescribed for pain, headaches, and fever. About 200 years later, the native peoples of North America discovered that salicylate-containing extracts of birch bark were effective in treating pain. After a long period of continued use, but no real progress in the understanding of how willow bark extract worked, an English priest named Edward Stone was perhaps the first to conduct a clinical trial. He gave 50 of his parishioners willow bark extract, and in 1763 reported to the Royal Society that the medication provided relief from malaria and "intermitting disorders." One would think that this successful "experiment" would have accelerated efforts to discover the active ingredients in the extract, but progress was remarkably slow. Only 60–70 years later did German chemists purify the active substance, salicin: a glycoside that is oxidized to salicylic acid following ingestion. This discovery resulted in an increased use of salicin, but as use increased, it became clear that the drug had several drawbacks, such as severe irritation of the gastric mucosa, resulting in bleeding and ulceration.

In 1853 a French chemist, Charles Frederic Gerhardt, reported that the carboxyl group in salicylic acid could be acetylated.[2] Unfortunately, he did not do any follow-up studies, and it was almost another 50 years before chemists at Bayer, a dye and pharmaceutical company, pursued the acetylation of salicylic acid in an effort to reduce its irritating side effects.[3]

Because of Gerhardt's report, Bayer did not obtain patent protection for the acetylated drug. In addition, as salicylic acid was reputed to "weaken the heart," probably due to the large doses that were used to treat rheumatism patients, Heinrich Dreser, head of the pharmacology group at Bayer, delayed clinical testing and production of acetylsalicylic acid. Instead, testing and production of another drug was started, diacetyl morphine (heroin), which Felix Hoffmann, who acetylated salicylic acid under the direction of former university chemist Arthur Eichengrün, had chemically modified. Convinced of the potential of acetylated salicylic acid, Eichengrün tested it on himself and gave samples away to physicians for patient use. As the excellent results became known, Dreser was finally convinced to subject acetylsalicylic acid to further clinical tests. Based on these tests, he wrote a report on the new drug, which was marketed in 1899 under the name aspirin about 2 years after Hoffmann synthesized it.

Although Bayer was unable to obtain patent protection for aspirin in Germany, the company received British and American patents and set up a subsidiary in the United States for the American market to avoid paying import duties. With the outbreak of World War I, the company was soon forced to cut production of aspirin due to an increased demand for phenol, the base chemical for salicylic acid and for explosives (trinitrophenol). For a time, Thomas Edison, who needed phenol to produce phonograph records, came to the rescue. Faced with the prospect of running out of raw material for his highly popular invention, Edison started a phenol company that was able to produce 12 tons of phenol per day. This was considerably more than was required for his phonograph record production, and was quickly seized on by the Germans, who set up a front company that bought Edison's excess phenol and sent it to the American subsidiary of the German-owned Chemische Fabrik von Heyden for the production of salicylic acid. This salicylic acid was then shipped to Bayer in Germany and used for aspirin synthesis. After a few months, a Secret Service agent discovered the plot and leaked documents describing the activities to the anti-German newspaper *New York World*. The public pressure that followed forced Edison to end the deal and start sending his excess phenol to the US military. However, by the time the Edison phenol source dried up, sufficient amounts of phenol/salicylic acid had been obtained by Bayer to continue aspirin production. Unfortunately, after the United States declared war on Germany in 1917 all of Bayer's American holdings were seized and auctioned off, together with all its American patents and trademarks, the Bayer brand name, and the Bayer cross logo. It was not until 1994 that the rights to the Bayer name and trademarks were sold back to Bayer AG for 1 billion US dollars.

After World War I, Bayer tried to diversify the aspirin market; one of the results was Alka-Seltzer, a mixture of aspirin and sodium bicarbonate. The company also became part of a conglomerate of former dye companies, IG Farben, which played a most unfortunate role during the Nazi regime in Germany. After World War II, the British-owned Bayer Ltd developed Excedrin, based on aspirin, and started searching for new pain relievers. This search led to the development of acetaminophen (Tylenol, Paracet) and ibuprofen (Advil, Motrin). These new drugs led to reduced aspirin use. In addition, the connection between the potentially fatal disease Reye's syndrome and aspirin use in children further reduced enthusiasm for aspirin.

The reason it took about 6000 years to develop aspirin is many-faceted. First, for much of this period there was a lack of fundamental knowledge about the human body and how it functions. Second, the development of aspirin required chemical insights and could not happen before practical methods of chemical engineering were available. Third, the making, testing, and marketing of aspirin required the combined efforts of excellent pharmaceutical chemists and a company with the necessary chemical production and marketing resources. Had Arthur Eichengrün worked in a university instead of a company like Bayer, it is unlikely that he would have contributed to the discovery of aspirin. However, it is also possible that the 2-year delay in testing and marketing aspirin could have been prevented if not for the company's financial and risk assessments in favor of heroin. The financial incentive was clearly a strong factor in the initial push to develop the drug; at the same time the for-profit element may have delayed the necessary work to get aspirin on the market.

7.3 RESEARCH PROVIDES A PATH TO UNDERSTANDING MECHANISMS AND NEW DRUGS

Remarkably, 60 years passed between the beginning of the mass production and sale of aspirin and serious investigations into the mechanisms by which it works.[1] It is interesting to note that such investigations were not initiated by Bayer, but by a chemist, Harry Collier, who worked at the pharmaceutical company Parke Davis. However, although Collier's studies led to insights into many of the properties of aspirin, they did not reveal its fundamental mechanisms. This was accomplished by the studies of John Vane, a pharmacologist at the Royal College of Surgeons in London.[4] In 1971, Vane discovered that aspirin blocks the production of prostaglandins, and later research showed that it does this by inhibiting cyclooxygenase (COX), the enzyme that converts arachidonic acid into prostaglandin. At the same time, scientists started to examine the effects of aspirin on platelet function and reported that aspirin acetylates COX. This results in reduced formation of thromboxane in platelets, which is part of the mechanism by which a low-dose aspirin regimen exerts a cardioprotective effect. This is the basis for the use of aspirin in the treatment of patients at risk for atherosclerotic events, such as myocardial infarction, stroke, and blood clots.

The discovery that aspirin inhibits prostaglandin synthesis is an outstanding example of what research in an academic environment, where graduate students, postdoctoral trainees, and faculty members work together, can contribute to clinical progress in the treatment of disease. Harry Collier studied the effects of aspirin on animals and animal tissues for several years at a pharmaceutical company, but it was John Vane who unlocked the secret to how it works. Vane also discovered prostacyclin and shared the Nobel Prize for Medicine in 1982 with Bengt Samuelsson and Sune Bergstrom for these discoveries. In trying to explain what led him to the idea that aspirin may be an inhibitor of prostaglandin synthesis, John Vane had this to say: "I suppose in the end it also comes down to luck, or serendipity or chance. Call it whatever you like. But it plays a big part and the knack of being a good scientist is to recognize it when it comes along. To say, that's strange, that's funny and then to follow it up. I think I was lucky."[1]

7.4 THE IMPORTANCE OF UNDERSTANDING DRUG METABOLISM—THE RISE AND FALL OF CLOPIDOGREL (PLAVIX)

The introduction of aspirin as an anti-platelet drug stimulated drug companies to search for compounds that would be even more powerful in the treatment of patients with atherosclerotic disease. After a systematic 25-year-long search for new anti-inflammatory drugs, Sanofi-Aventis and Bristol-Myers Squibb Co. were able to launch an anti-platelet drug, clopidogrel, named Plavix, in 1997.[5] In initial clinical trials it appeared that Plavix was more effective than aspirin in reducing cardiovascular events in patients with a recent stroke, myocardial infarction, or peripheral artery disease. As a result, it quickly became the dominant drug for such patients, though it was 100 times more expensive than aspirin. By 2010 it was the second most prescribed drug in the world, with sales of over 9 billion US dollars. However, over time reports of serious complications and side effects, and increasing doubts about whether it really was that much of an improvement over aspirin, reduced the enthusiasm for the drug. In fact, research comparing the two drugs showed in the end that Plavix was not more effective and led to more complications and deaths. It was also discovered that the effectiveness varied in patients based on genetic variations in how the drug was metabolized.[6,7]

Some of the problems with clopidogrel resulted from the mechanism by which it worked.[5] The drug was inactive *in vitro* but produced an active drug *in vivo* after being metabolized in the liver. Loss-of-function polymorphisms in the genes encoding liver enzymes, which were required to turn the inactive "prodrug" into its active form, made some patients resistant to the drug. The active metabolite is unstable and irreversibly inactivates the platelet target, an ADP receptor, and may irreversibly react with other potential molecular and cellular targets as well. These properties have serious consequences. First, the active form of the drug cannot be synthesized, and second, if patients show adverse reactions to treatment there is no way to quickly reverse the effect of the drug.

7.5 THE NEED FOR MORE TRANSLATIONAL MEDICAL RESEARCH

The history of aspirin and Plavix provides several lessons for translational medical research. First, although chemistry, genetics, molecular and cellular biology, genomics, proteomics, and related technologies have come a long way in the past 70 years, searching for new drugs and testing candidate drugs remains a long and difficult task. Second, high-throughput molecular target-based screens of compounds must be complemented by mechanistic studies using cells, tissues, and experimental animals before beginning clinical trials in human subjects. Negative as well as positive effects need to be rigorously pursued. Drug candidates that show negative effects in certain screens but positive effects in others should not be rejected too quickly, since they may represent valuable starting compounds for continued modification and testing. Third, the costs of research and testing are high, and since over 95% of candidate drugs that make it to the clinical trial stage ultimately fail, drug companies can end up investing up to 1 billion US dollars per drug that makes it to the market.

These considerations have led to the conclusion that major changes in the drug discovery process are required.[8] The way molecular and cellular drug targets are selected must be based on scientifically sound hypotheses and research findings that are validated by being challenged and systematically confirmed by multiple investigators. Better technologies that screen for side effects in different cells and tissues at an early stage in the process must also be developed. In addition, there is a need for better education of translational medical researchers in such a manner that they can truly bridge the gap between basic medical research and treatment of disease; this requires strengthening of translational drug-discovery research at academic institutions.

Based on these considerations, in 2011 the National Institutes of Health (NIH) in the United States reorganized part of their structure to create the National Center for Advancing Translational Sciences (NCATS). The goal of this center is to get more successful medicines to more patients more quickly. To reach its goals NCATS has provided Clinical and Translational Science Awards to 62 medical research institutions in 31 states and the District of Columbia, with the aim of strengthening the country's infrastructure and training environment for translational medical research.

Within NCATS, the Office of Rare Diseases Research, in collaboration with the National Genome Research Institute and other NIH branches, provides information on rare disorders and/or genetic diseases, support for collaborative clinical research projects on rare disorders, training of clinical investigators, pilot and demonstration projects, and rare disorders workshops. NCATS is working with the Defense Advanced Research Projects Agency and the Food and Drug Administration (FDA) to support funding for projects aimed at growing several different human tissues on a chip so that potential drugs can be rapidly screened for toxic effects. Another collaboration between NCATS, the National Toxicology Program at the National Institute of Environmental Health Sciences, and the US Environmental Protection Agency involves the screening of 10,000 environmental chemicals and approved drugs against every known human signaling pathway to identify molecules that may have toxic effects.

Prior to the establishment of NCATS, many wondered if researchers with federal government support would be able to do anything that was not already being done by pharmaceutical companies. Given the initial success of many programs supported by NCATS, it is now clear that federal support for translational medical research allows investigators to address problems that could not be financially justified in the for-profit pharmaceutical industry. For example, the majority of candidate drugs that make it to the clinical trial stage are ultimately rejected, either because they do not have the desired effect or because of business reasons. By providing

NIH-funded scientists access to these rejected drugs, NCATS is stimulating translational drug-discovery research that complements industrial drug development. A drug that is rejected due to a lack of therapeutic effect on the condition it was developed for may turn out to be effective for a completely different disorder. Furthermore, existing drugs which have been developed to treat one condition may turn out to be effective for other conditions. This was the case for aspirin; a more recent success story is the development of phosphodiesterase 5 inhibitors (Viagra, Levitra, and Cialis) for treatment of both male erectile dysfunction and pulmonary arterial hypertension.[9]

7.6 TRANSLATIONAL MEDICAL RESEARCH AND THE DEVELOPMENT OF ORPHAN DRUGS

Translational medical research has a particularly important role to play in the development of treatments for rare disorders, also known as orphan diseases.[10] In the United States a disease is considered rare if it affects fewer than 200,000 Americans; in Europe it is considered rare if fewer than 1 in 2000 are affected. However, when the approximately 7000 different rare disorders are considered together, they represent a large disease burden. About 80% of rare disorders have a genetic etiology. Many of these genetic disorders are associated with phenotypes that only develop after birth; this provides a postnatal window of opportunity for intervention with appropriate drugs. Other disorders can be diagnosed *in utero*, so prenatal treatment could be considered if the right drugs were available. In a large proportion of rare disorders one should be able to treat the pathophysiology without correcting the underlying gene mutation. What is required is a combination of detailed knowledge of the molecular and cellular mechanisms by which the mutant gene causes the disease phenotype, and resources allowing one to screen a large number of compounds for their activity against the genes, proteins, and signaling pathways involved.

In some cases, elucidating the cellular and molecular mechanisms of a disorder is sufficient to identify the properties a candidate drug should have. A good example is hypophosphatasia, caused by mutations in the gene encoding tissue-nonspecific alkaline phosphatase (TNSALP).[11] TNSALP is a homodimeric ectoenzyme anchored to cell surfaces in bone, liver, kidney, and other tissues.[12] Patients with mutations in this gene have low levels of alkaline phosphatase activity, resulting in defects in calcium and phosphate levels. This leads to rickets or osteomalacia, which can often cause death in severely affected babies. Treatment by infusion of soluble alkaline phosphatase has not been successful, but a recombinant fusion protein containing the TNSALP ectodomain, the Fc domain of human IgG1 (serving as a dimer-forming anchor), and a tail consisting of 10 aspartate residues (for targeting bone) has been found to be effective in preventing hypophosphatasia in alkaline phosphatase knock-out mice[13] and to improve pulmonary function and skeletal development in infants and young children with life-threatening hypophosphatasia.[14]

In other cases, elucidating cellular and molecular mechanisms of a genetic disorder may lead to the testing of an existing drug for a different application. The autosomal dominant disease known as Marfan syndrome is an outstanding example of such a case. Named after the French pediatrician Antoine Marfan, who first described the condition in 1896, Marfan syndrome is caused by mutations in the gene (*FBN1*) that encodes fibrillin-1, a major structural component of microfibrils in connective tissues.[15,16] Given the role of fibrillin-1 in the maintenance of microfibrils and elastic fibers,[17] it is not surprising that patients with the syndrome have signs and symptoms in the skeletal and cardiovascular systems, eyes, lungs, and central nervous system. The most serious complications are defects in the heart valves and the wall of the aorta that can result in aortic aneurysms or dissections.

The discovery of mutations in a structural component of the extracellular matrix initially led to the idea that structural insufficiency was the basis for the pathological mechanisms in Marfan syndrome. However, the finding that fibrillin contains several domains that are homologous to latent transforming growth factor β-binding proteins (LTBPs) led to studies showing that microfibrils bind LTBP-TGF-β and that fibrillin mutations in Marfan syndrome result in the dysregulation of TGF-β.[17] In fact, increased TGF-β activity in the lung and aorta of a mouse model of Marfan syndrome was found to be a significant contributor to the lung and aortic defects in the mice, as blocking TGF-β activity rescued key aspects of the mutant phenotype.[18] Based on these findings, investigators turned to losartan, an FDA-approved angiotensin II type 1(AT1) receptor blocker that lowers blood pressure and has been demonstrated to inhibit TGF-β signaling in animal models. A long-term trial of losartan in a mouse model of Marfan syndrome showed that the drug normalized aortic root growth and histological features of the aortic wall and prevented defects in other organ systems as well.[19] These animal data provided the rationale for clinical trials of losartan in the treatment of Marfan syndrome in both children and adults and randomized trials are ongoing.[20] The results of these trials are unlikely to demonstrate that losartan is a miracle drug that can "rescue" Marfan patients from all the consequences of FBN1 mutations, but they will undoubtedly stimulate further efforts to develop or identify drugs that may be effective in minimizing the multi-organ clinical problems caused by the disorder.

The pharmaceutical industry may well become involved in such efforts. In the past, for business reasons, the pharmaceutical industry has not invested the resources needed to develop drugs that would prevent or minimize the developmental and physiological consequences of rare gene mutations. However, several factors have contributed to a change in development of drugs for rare disorders, or "orphan" drugs. First, legislative and regulatory incentives have been introduced to encourage companies to develop drugs for diseases that have a small market. In the United States this started with the establishment of the Office of Orphan Products Development at the FDA, followed by the Orphan Drug Act in 1983, which allows companies that develop an orphan drug to receive clinical trial tax incentives and to market the drug for 7 years without competition. The Rare Diseases Act of 2002 led to the establishment of the Office of Rare Diseases at the NIH and increased funding for drug development for patients with rare disorders. A similar program was enacted in Japan in 1993, Australia in 1998, and the European Union in 2002. In the European Union companies receive 10 years of marketing rights for orphan drugs. In 2007, the FDA and the European Medicines Agency agreed on a common application from companies for orphan drug designation. Similar programs have also been implemented in Korea, Thailand, Taiwan, and Singapore, and are being developed in Russia and China. These programs have clearly been successful.[21] For example, only 38 orphan drugs were approved in the United States prior to the passage of the Orphan Drug Act in 1983; in the period between 1983 and early 2010, 353 orphan drugs were approved by the FDA and over 2000 compounds were granted orphan drug designations. In 2011 the orphan drug market was reported to be about 50 billion US dollars. This success is due in no small part to an increase in venture capital investments and the willingness of health insurance companies to pay for these drugs. Indeed, the costs of orphan drugs can be as much as US$100,000–200,000 per year for a single individual, providing pharmaceutical companies a good return on investment even for a small market. Not surprisingly, in addition to an increasing number of small pharmaceutical companies entirely focused on orphan drug development, several of the industry giants are also getting into the orphan drug business.

7.7 FROM RARE DISORDERS TO COMMON DISORDERS

One of the most significant outcomes of research on rare genetic disorders is the insight that is gained into developmental and homeostatic mechanisms. Such insights can sometimes directly lead to the testing of new drugs for common disorders. An excellent and relatively recent example is how the discovery of gene mutations responsible for rare genetic disorders of either too little or too much bone in the skeletal system is leading to what may be the next powerful treatment of age-dependent osteoporosis.

Genetic studies of the recessive disorder osteoporosis-pseudoglioma, characterized by severe juvenile-onset osteoporosis and congenital or juvenile-onset blindness, led to the demonstration in 2001 that patients with the syndrome had homozygous loss-of-function mutations in the gene encoding a Wnt-binding receptor known as LRP5.[22] At about the same time, other investigators discovered that patients with dominantly inherited high bone mass had gain-of-function mutations in LRP5.[23,24] Based on these data, it became clear that Wnt signaling through LRP5 represented a powerful regulator of bone mass accrual.

In addition, it was discovered that rare, more severe forms of high bone mass in patients with Van Buchem disease or osteosclerosis were caused by loss-of-function mutations in sclerostin, a secreted protein largely expressed by osteocytes.[25,26] It turns out that sclerostin acts as a negative regulator of bone formation by binding to LRP5, preventing Wnt signaling.[27] Based on these discoveries, major drug companies have developed humanized antibodies against sclerostin for the treatment of age-dependent osteoporosis, and the results of ongoing clinical trials are extremely promising.

7.8 TARGET-BASED DRUG DEVELOPMENT—AQUAPORINS

In the case of aspirin, its introduction as a drug started with the observed effect of a medicinal plant—the willow tree. The effective compound was isolated much later, and the molecular mechanisms for the effects of aspirin later still. In contrast, most modern drug development projects start with identification of a specific drug target. In the case of rare disorders, identification of genetic mechanisms provides information about potential targets. As discussed above, such drug targets can also be for more common disorders. However, in most cases drug target selection is based on a combination of careful analyses of cell and molecular biological, biochemical, physiological, and pathophysiological data, as well as the chemical structure of potential candidates. Before a decision is made to focus on a particular molecular drug target, its detailed chemical structure and the likelihood that it is capable of binding to chemical compounds that could act as drugs—a property called "druggability"[28]—are carefully analyzed. It is a sobering fact that fewer than 500 molecules have proved to be successful drug targets in humans.[29] The number of druggable submolecular domains is even lower, possibly around 130. These figures are in stark contrast to the tens of thousands of proteins or protein variants in the human body and attest to the difficulties involved in designing new drugs. Another, more positive view is that there is ample potential for identifying new drug targets and for developing more efficient approaches to drug design. With the continued discovery of new, potentially druggable proteins and protein families, the universe of drug targets is growing rapidly. As the tertiary chemical structure is determined for an increasing number of proteins and protein subdomains, the opportunities for rational drug design are increasing. Given the enormous resources that are required for drug development, putative new drug targets must be selected with care. It has been estimated that more than 50% of small-molecule drug-discovery projects fail because the target is not druggable.[30]

Several routines have been proposed for predicting the druggability of potential small-molecule binding sites. It has been hypothesized that known drug-binding sites contain particular physicochemical properties and that the identification of such "druggable microenvironments" might have predictive value.[30] The size of the target, the presence of binding pockets, and the overall charge of the interaction surface are key determinants of druggability. Undruggable sites are typically strongly hydrophilic, require covalent binding, and are very small or very shallow.[30–32] Obviously, while high-affinity binding to a target is a prerequisite for the drug to become active, it also needs to be demonstrated that the ligand, once bound, exerts physiologically relevant effects. Most drugs bind to pockets that normally bind endogenous molecules that affect the target's physiological function. It is beyond the scope of the present chapter to discuss bioavailability and the other criteria a compound has to satisfy in order to qualify as a drug for clinical use.

An excellent example of how translational medical research is inspired by the discovery of new and potentially druggable proteins is the story of a family of water channels called aquaporins. Aquaporins were hailed as potential drug targets soon after they were discovered. The first aquaporin was characterized in 1992 by Peter Agre and collaborators,[33] and in 2003, Peter Agre was awarded the Nobel Prize in Chemistry for his work on aquaporin water channels. As Peter Agre pointed out in his Nobel lecture, the discovery of the first aquaporin was a case of serendipity.[34] While studying Rh blood group antigens, Agre was trying to raise antibodies to an Rh polypeptide. The immunized rabbits generated antibodies that reacted strongly with a polypeptide of ~30 kDa. This polypeptide turned out to be a contaminant unrelated to the Rh antigen. The distribution of this new polypeptide—being prevalent in renal tubules as well as red blood cells—suggested that it could be involved in water transport. This turned out to be the case. When the polypeptide was expressed in frog oocytes, the oocytes became highly permeable to water and "exploded like popcorn" when exposed to hypo-osmotic stress. The polypeptide that contaminated Agre's Rh preparations is now known as aquaporin 1—the first functionally defined water channel protein.

Since the discovery of aquaporin 1, 12 additional aquaporin water channels have been identified in mammals. They are expressed in many organs such as the kidneys, liver, brain, skin, eyes, and muscles,[35,36] and there is evidence to support their involvement in several diseases, including brain edema, epilepsy, hypertension, obesity, and cystic fibrosis. Mutations in aquaporins are also associated with conditions such as nephrogenic diabetes insipidus and cataract. The brain contains several aquaporins, including aquaporin 1, 4, and 9. Of these, only aquaporin 4 is present throughout the brain. Aquaporin 4 is concentrated in the endfeet of astrocytes that encircle brain capillaries. Knocking out aquaporin 4 or selectively depleting this aquaporin in the astrocytes of mice significantly reduced brain edema following experimentally induced ischemia.[37,38] These data offered compelling evidence that aquaporin 4 controls water influx to the brain in pathophysiological conditions such as stroke and suggested that aquaporin blockers may help forestall brain edema, e.g., in the wake of an ischemic brain attack. Since about 130,000 individuals die of ischemic stroke in the United States every year, and a fulminant, treatment-resistant edema is the most common immediate cause of death, an effective aquaporin blocker would have a significant clinical impact.

In considering aquaporin 4 as a drug target, the first question raised was whether this channel is subject to regulation. Specifically, is there a gating mechanism that could be pharmacologically exploited? Data based on investigations of plant aquaporins offered promise in this regard. Structural and other analyses indicated that a spinach aquaporin is closed when an intracellular loop of the molecule caps the channel and occludes the pore.[39] Aquaporin 4 is structurally similar and is endowed with a similar intracellular loop. These findings, combined with data showing that aquaporin 4 is sensitive to a number of metal ions,[40] raised the possibility that the channel could be closed by pharmacological means. However, a screening of drug libraries has so far not led to the identification of a promising candidate drug. In parallel, attempts have been made to develop aquaporin 4 blockers through rational drug design. One potential drug target is a binding site responsible for coupling aquaporin 4 molecules face to face at sites where the channels occur in opposing membranes. Such configurations are found in just a few places in the brain (mainly in osmosensitive areas, such as the supraoptic nucleus),[41] implying that the binding site is normally unoccupied and free for drug binding.[42] A cyclical oligopeptide designed to fit into this binding site is now being tested for affinity to aquaporin 4.[43] The idea is to use the 3D structure of any successful peptide to synthesize a compound that fills the same chemical space and that has properties conducive to uptake in the brain. The future drug must be bulky enough to prevent water entry into the pore of aquaporin 4.

About 20 years have elapsed since aquaporin 4 was identified, and about 15 years have passed since it became clear that aquaporin 4 is an obvious target for new and much-needed drugs against brain edema. But there is still no drug or even a promising drug candidate for this target.[44] This story is not unique. Translational medical research is fraught with difficulties. The path to a new drug is arduous and slow. Polio serves as another example: About 50 years separated the discovery of the poliovirus and the development of the first useful vaccine. For this reason, when it comes to translational medical research, scientists must be careful to avoid unrealistic promises in their grant applications, and funding bodies must be patient and honor quality without requiring or expecting short-term successes.

The lessons to be learned from the stories in this chapter are many. First of all, it is essential that translational medical research be based on a sound platform of high-quality basic science. In hindsight, the scientific premises for the hunt for aquaporin 4 blockers were solid enough: the original studies on the location and properties of aquaporins have withstood the test of time. But essential data regarding their regulation have been challenged. It is still unclear whether mammalian aquaporin 4 is subject to gating, and several reports have failed to find changes in the structure of aquaporin 4 or water permeability following manipulations of phosphorylation sites that could be potential sites for protein activation or deactivation. Even the original results in plants are being questioned.[45]

Another lesson is that translational medical research must be based on robust model systems and bioassays. In the case of aquaporin 4, animal models were essential in substantiating the hypothesis—based on immunolocalization studies—that aquaporin 4 constitutes an influx route for water in brain edema. Several complementary models were used, and the results were consistent. Specifically, the extent of edema formation was reduced by depletion of aquaporin 4, irrespective of the experimental approach. Depletion of aquaporin 4 by constitutive or glial-conditional *Aqp4* gene knockout offered protection, as did interference with the mechanisms that anchor AQP4 to glial membranes at the brain–blood interface.[46,47] Similarly, a protective effect of AQP4 depletion was observed regardless of whether edema was produced by arterial occlusion or hyponatremia.[37,38,46]

Typically, translational medical research integrates the use of animal experiments with bioassays in reduced model systems, such as cell cultures or oocytes transfected with cDNA encoding the target protein, or liposomes containing the target protein following its reconstitution in the membrane. High-throughput screening is costly and time consuming, and it is essential that the bioassays live up to the highest standards in terms of sensitivity and specificity. In the case of aquaporins, bioassays are routinely carried out on cultured astrocytes or on frog oocytes. One would assume that recordings of water permeability would be straightforward, but unfortunately this is not the case. Current bioassays based on astrocytes or oocytes pose methodological challenges that have stirred controversy regarding the results. Among other things, a rapid change from an iso-osmolar to a hypo- or hyperosmolar buffer is required to ensure that transmembrane water flux, rather than buffer-mixing time, limits the measured rates of swelling or shrinking.[48]

If we assume for a moment that a lead compound for aquaporin 4 blocking had been found, following high-throughput drug screening or rational drug design, major hurdles would still remain. Drug targets in the brain pose particular challenges, given the fact that any drug will have to cross the blood–brain barrier. Also, history has taught us that drug candidates targeting the central nervous system may be promising at an early phase, but often do not get FDA approval due to inadvertent side effects. In the case of aquaporin 4, major side effects would seem unlikely, as aquaporin 4 deletions produce only minor neurological deficiencies, linked primarily to sensory organs;[49] the effects on cognition are subtle.[50] But timing of treatment would be essential if an aquaporin 4 blocker were made available. This is because water movement through aquaporin 4 is bidirectional:[38] aquaporin 4 mediates water influx in the build-up phase of an edema, but also facilitates water efflux in the resolution phase. Thus, an irreversible blockade of aquaporin 4 could prove unhelpful or even harmful.

The final lesson to be learned is that translational medical research rewards science and society at large, even if successes in the form of new drugs are few and far between. The clinical potential provides an invaluable motivational factor, and the dividends in terms of new insight are often substantial. A case in point is that the search for aquaporin 4 blockers led to the discovery of an aquaporin 4-associated protein—TRPV4—that appears to serve as an osmosensor in the brain and might be a drug target in its own right.[51] Thus, although identifying and testing new drugs usually takes a long time, science has few dead ends, making translational medical research a rewarding career option for students, postdoctoral fellows, and junior scientists. There are several training options for such a career.

7.9 TRAINING OF INVESTIGATORS FOR TRANSLATIONAL MEDICAL RESEARCH

The initiatives to strengthen translational medical research have led many countries to expand efforts to train new investigators in the field. These efforts seek to enhance training programs that are already in place, improve coordination of programs in different countries, and support the establishment of new training mechanisms.

MD-PhD programs, also known as Medical Scientist Training Programs (MSTPs), provide training in both medicine and research and have been offered by medical schools in the United States since the first program was started at Case Western Reserve University in 1956. After the NIH created a mechanism for funding such dual training in 1964, a number of medical schools established similar programs. By integrating the research training into medical training, these programs have allowed students to graduate with both MD and PhD degrees in a shorter period of time than the traditional pathway of obtaining the MD first, completing a residency in the specialty of choice, and fulfilling the requirements for a PhD degree at the end.

In other countries, including the United Kingdom and other countries in Europe, the traditional pathway has been the only option until recently. However, stimulated by the successful outcomes of MSTPs in the United States, similar dual degree programs were initiated in Switzerland and the United Kingdom (by the University of Cambridge and University College London) about 20 years ago. Outcome analyses of the Swiss and University College London programs resulted in positive evaluations and recommendations that such programs be increased across Europe. Based on these recommendations, a European MD/PhD Network was established in 2012 during an MD/PhD Conference held at University College London. The Network organizes annual meetings for MD/PhD students to present data, share experiences, and promote the combination of research and medical training. In 2012 the European Science Foundation also released a Science Policy Briefing on Medical Research Education that is likely to help enhance the rigor and quality of traditional PhD training programs, with increased emphasis on stimulating PhD candidates to link and contribute to medical research. Efforts to offer PhD students translational medical research experience are being expanded in the United States in several ways. Some programs offer 1-year clinical research training for PhD students; others provide clinical research training leading to a Master of Science, and some are PhD programs in clinical research. Career options for graduates of such programs are rapidly expanding because of evolving innovative partnerships between academic institutions and pharmaceutical companies.[52,53]

In Norway, an Industrial PhD has been established to promote interaction between academic institutions and the pharmaceutical industry. This initiative aims to strengthen translational medical research and innovation. While the requirements in regard to scientific quality and format are the same as for regular PhD programs, an Industrial PhD candidate has one tutor based in academia and one tutor in the pharmaceutical industry. A research institute or another third party could also be involved. This new PhD program is a promising tool to boost translational medical research, but has yet to be fully embraced by the scientific community. This is likely to change once it is appreciated that the Industrial PhD program helps foster the two-way communication required to motivate basic medical research and to provide a sound evidence base for innovation in the private sector. Thus, at the heart of translational medical research lies the realization that the pharmaceutical industry needs to be exposed to the potential and limitations of basic medical science, while basic medical science needs to be challenged and inspired by the unmet needs to which the private sector caters.

7.10 COLLABORATION BETWEEN ACADEMIA AND THE PHARMACEUTICAL INDUSTRY

As illustrated by the history of aspirin and Plavix, market needs and return on investment have always been strong drivers of research and development (R&D) within the pharmaceutical industry. In the past, in-house company research, as in the case of aspirin and Plavix, provided sufficient innovation to ensure a relatively high likelihood of success and a good return on the investment. In fact, until about 25 years ago top pharmaceutical companies made about three times more in sales than what they invested in capitalized R&D.[54] However, 25 years later, the return on capitalized R&D is only about 80%.[54] During the same period, in the case of major companies such as Merck, Lilly, and Roche, the rate at which new approved drugs were produced did not change.[54] Thus, the classical pharmaceutical business model is not sustainable, and innovations by the industry are lagging. Based on such statistics, the continued need to develop drugs for multiple serious diseases, and the remarkable progress in basic biological and medical research at academic institutions and nonprofit research institutes, intense efforts are currently being made to facilitate partnerships between academia and the pharmaceutical industry. These efforts include assessments of the state of drug discovery in academia and the pharmaceutical industry,[55] developing the legal framework for pharmaceutical public–private partnerships aimed at moving discovery from "bench to bedside,"[52] and the establishment of innovative centers that bring academic researchers and industry scientists together.[56] For example, Pfizer has established Centers for Therapeutic Innovation in Boston, New York City, San Francisco, and San Diego with research staff including both Pfizer employees and academic scientists. Similarly, Merck has established the nonprofit California Institute for Biomedical Research; this will provide academic collaborators with industrial technical support so that therapeutic projects can be advanced to the proof-of-principle stage. Another project, aimed at developing drugs for patients with autism, includes pharmaceutical companies from the European Federation of Pharmaceutical Industry Associations, the organization Autism Speaks, and King's College, London, and is leading an academic partnership of 14 European centers of excellence. In Berlin, The Charite-Universitätsmedizin and Sanofi Pharmaceuticals have opened a joint laboratory, and Bayer Health Care Pharmaceuticals has established an organization at University of California, San Francisco. Academic institutions are also increasingly organizing efforts to identify entrepreneurial scientists and help fund their inventions. For example, The Experiment Fund provides seed funding and/or scientific and corporate expertise at Harvard University.

QUESTIONS TO DISCUSS

1. If you were to repeat Edward Stone's willow bark extract experiment today, how would you design it as a clinical trial?
2. Do you believe that a prodrug, such as Plavix, is likely to be developed by a drug company today? Explain the reasons for your answer.
3. In your opinion, what are the benefits of investing in efforts to support drug development for rare human disorders?
4. Using cystic fibrosis as an example, discuss how highly specific drugs may not be effective in all patients with the disorder.
5. In your opinion, what may be the reasons why the pharmaceutical industry has been unable to increase the rate of drug discovery during the last 25 years?
6. Given the increased efforts to enhance collaboration between the pharmaceutical industry and researchers at academic institutions, what do you believe are the potential benefits and drawbacks of such collaborations?
7. For screening of drug candidates, investigators have developed what is known as "biomimetic tissues on a chip." Discuss the potential and limitations of this technology.
8. Discuss the problems associated with drugs that modify their targets in an irreversible manner.
9. In your opinion, what are the major obstacles to the development of drugs using aquaporin 4 as a target for the treatment of stroke patients?

10. Imagine you are the leader of a drug development group in a major pharmaceutical company. Your group has discovered a compound of low toxicity that inhibits the sprouting of blood vessels in both *in vitro* and *in vivo* assays. However, the compound appears to do this by inhibiting several steps in different signaling pathways.

The company's board of directors informs you that the project will be terminated unless you can find derivatives that are more target-specific. What are the arguments you would use to try to convince the board that the compound you already have is a far better drug candidate than any specific derivatives your group could come up with?

REFERENCES

1. Jeffreys D. *Aspirin: the remarkable story of a wonder drug.* 1st ed. New York, NY: Bloomsbury; 2004.
2. Gerhardt C. Recherches sur les acides organiques anhydrides. *Annales de Chimie et de Physique* 1853;**37**:285–342.
3. Sneader W. The discovery of aspirin: a reappraisal. *BMJ* 2000;**321**:1591–4.
4. Fitzgerald DJ, Fitzgerald GA. Historical lessons in translational medicine: Cyclooxygenase inhibition and P2Y12 antagonism. *Circ Res* 2013;**112**:174–94.
5. Maffrand J-P. The story of clopidogrel and its predecessor, ticlopidine: could these major antiplatelet and antithrombotic drugs be discovered and developed today? *CR Chimie* 2012;**15**:737–43.
6. Mega JL, Close SL, Wiviott SD, Shen L, Hockett RD, Brandt JT, et al. Cytochrome p-450 polymorphisms and response to clopidogrel. *N Engl J Med* 2009;**360**:354–62.
7. Simon T, Verstuyft C, Mary-Krause M, Quteineh L, Drouet E, Meneveau N, et al. Genetic determinants of response to clopidogrel and cardiovascular events. *N Engl J Med* 2009;**360**:363–75.
8. Wadman M. Translational research: medicine man. *Nature* 2013;**494**:24–6.
9. Klein R, Sturm H. Viagra: a success story for rationing? *Health Aff (Millwood)* 2002;**21**:177–87.
10. Field MJ, Boat TF, editors. *Rare diseases and orphan products.* Washington, DC: Institute of Medicine (US) Committee on Accelerating Rare Diseases Research and Orphan Product Development; 2010.
11. Rathbun JC. Hypophosphatasia; a new developmental anomaly. *Am J Dis Child* 1948;**75**:822–31.
12. Mornet E. Hypophosphatasia. *Orphanet J Rare Dis* 2007;**2**:40.
13. Whyte MP, Greenberg CR, Salman NJ, Bober MB, McAlister WH, Wenkert D, et al. Enzyme-replacement therapy in life-threatening hypophosphatasia. *N Engl J Med* 2012;**366**:904–13.
14. Kishnani P, Rockman-Greenberg C, Whyte M, Weber T, Mhanni A, Madson K, et al. Hypophosphatasia: enzyme replacement therapy (ENB-0040) decreases TNSALP substrate accumulation and improves functional outcome in affected adolescents and adults. *Proc Am Coll Med Genet* 2012;303.
15. Marfan A. Un cas de déformation congénitale des quartre membres, plus prononcée aux extrémitiés, caractérisée par l'allongement des os avec un certain degré

d'amincissement. [A case of congenital deformation of the four limbs, more pronounced at the extremities, characterized by elongation of the bones with some degree of thinning]. *Bulletins et Memoires de la Société Medicale des Hôspitaux de Paris* 1896;**13**:220–6.

16. Dietz HC, Cutting GR, Pyeritz RE, Maslen CL, Sakai LY, Corson GM, et al. Marfan syndrome caused by a recurrent de novo missense mutation in the fibrillin gene [see comments]. *Nature* 1991;**352**:337–9.
17. Ramirez F, Sakai LY, Dietz HC, Rifkin DB. Fibrillin microfibrils: multipurpose extracellular networks in organismal physiology. *Physiol Genomics* 2004;**19**:151–4.
18. Neptune ER, Frischmeyer PA, Arking DE, Myers L, Bunton TE, Gayraud B, et al. Dysregulation of TGF-beta activation contributes to pathogenesis in Marfan syndrome. *Nat Genet* 2003;**33**:407–11.
19. Habashi JP, Judge DP, Holm TM, Cohn RD, Loeys BL, Cooper TK, et al. Losartan, an AT1 antagonist, prevents aortic aneurysm in a mouse model of Marfan syndrome. *Science* 2006;**312**:117–21.
20. Dietz H. A healthy tension in translational research. *J Clin Invest* 2014;**124**:1425–9.
21. <http://www.innovation.org/index.cfm/innovationtoday/innovationinrarediseases>.
22. Gong Y, Slee RB, Fukai N, Rawadi G, Roman-Roman S, Reginato AM, et al. LDL receptor-related protein 5 (LRP5) affects bone accrual and eye development. *Cell* 2001;**107**:513–23.
23. Little RD, Carulli JP, Del Mastro RG, Dupuis J, Osborne M, Folz C, et al. A mutation in the LDL receptor-related protein 5 gene results in the autosomal dominant high-bone-mass trait. *Am J Hum Genet* 2002;**70**:11–19.
24. Boyden LM, Mao J, Belsky J, Mitzner L, Farhi A, Mitnick MA, et al. High bone density due to a mutation in LDL-receptor-related protein 5. *N Engl J Med* 2002;**346**:1513–21.
25. Brunkow ME, Gardner JC, Van Ness J, Paeper BW, Kovacevich BR, Proll S, et al. Bone dysplasia sclerosteosis results from loss of the SOST gene product, a novel cystine knot-containing protein. *Am J Hum Genet* 2001;**68**:577–89.
26. Balemans W, Patel N, Ebeling M, Van Hul E, Wuyts W, Lacza C, et al. Identification of a 52 kb deletion downstream of the SOST gene in patients with van Buchem disease. *J Med Genet* 2002;**39**:91–7.
27. Li X, Zhang Y, Kang H, Liu W, Liu P, Zhang J, et al. Sclerostin binds to LRP5/6 and antagonizes canonical Wnt signaling. *J Biol Chem* 2005;**280**:19883–7.
28. Hopkins AL, Groom CR. The druggable genome. *Nat Rev Drug Discov* 2002;**1**:727–30.
29. Owens J. Determining druggability. *Nat Rev Drug Discov* 2007;**6**:187.
30. Liu T, Altman RB. Identifying druggable targets by protein microenvironments matching: application to transcription factors. *CPT Pharmacometrics Syst Pharmacol* 2014;**3**:e93.
31. Cheng AC, Coleman RG, Smyth KT, Cao Q, Soulard P, et al. Structure-based maximal affinity model predicts small-molecule druggability. *Nat Biotechnol* 2007;**25** (1):71–5.
32. Halgren TA. Identifying and characterizing binding sites and assessing druggability. *J Chem Inf Model* 2009;**49**(2):377–89.
33. Preston GM, Carroll TP, Guggino WB, Agre P. Appearance of water channels in *Xenopus* oocytes expressing red cell CHIP28 protein. *Science* 1992;**256**(5055):385–7.
34. <http://www.nobelprize.org/nobel_prizes/chemistry/laureates/2003/agre-lecture.pdf>.
35. Agre P, King LS, Yasui M, Guggino WB, Ottersen OP, Fujiyoshi Y, et al. Aquaporin water channels—from atomic structure to clinical medicine. *J Physiol* 2002;**542**:3–16.
36. Amiry-Moghaddam M, Ottersen OP. The molecular basis of water transport in the brain. *Nat Rev Neurosci* 2003;**4**:991–1001.

37. Manley GT, Fujimura M, Ma T, Noshita N, Filiz F, Bollen AW, et al. Aquaporin-4 deletion in mice reduces brain edema after acute water intoxication and ischemic stroke. *Nat Med* 2000;**6**:159–63.
38. Amiry-Moghaddam M, Otsuka T, Hurn PD, Traystman RJ, Haug FM, Froehner SC, et al. An alpha-syntrophin-dependent pool of AQP4 in astroglial end-feet confers bidirectional water flow between blood and brain. *Proc Natl Acad Sci USA* 2003;**100**:2106–11.
39. Tornroth-Horsefield S, Wang Y, Hedfalk K, Johanson U, Karlsson M, Tajkhorshid E, et al. Structural mechanism of plant aquaporin gating. *Nature* 2006;**439**:688–94.
40. Mola MG, Nicchia GP, Svelto M, Spray DC, Frigeri A. Automated cell-based assay for screening of aquaporin inhibitors. *Anal Chem* 2009;**81**:8219–29.
41. Nielsen S, Nagelhus EA, Amiry-Moghaddam M, Bourque C, Agre P, et al. Specialized membrane domains for water transport in glial cells: high-resolution immunogold cytochemistry of aquaporin-4 in rat brain. *J Neurosci* 1997;**17**(1):171–80.
42. Hiroaki Y, Tani K, Kamegawa A, Gyobu N, Nishikawa K, Suzuki H, et al. Implications of the aquaporin-4 structure on array formation and cell adhesion. *J Mol Biol* 2006;**355**:628–39.
43. Jacobsen O, Klaveness J, Ottersen OP, Amiry-Moghaddam MR, Rongved P. Synthesis of cyclic peptide analogues of the 3(10) helical Pro138-Gly144 segment of human aquaporin-4 by olefin metathesis. *Org Biomol Chem* 2009;**7**:1599–611.
44. Verkman AS, Anderson MO, Papadopoulos MC. Aquaporins: important but elusive drug targets. *Nat Rev Drug Discov* 2014;**13**(4):259–77.
45. Nagelhus EA, Ottersen OP. Physiological roles of aquaporin-4 in brain. *Physiol Rev* 2013;**93**:1543–62.
46. Amiry-Moghaddam M, Xue R, Haug FM, Neely JD, Bhardwaj A, et al. Alpha-syntrophin deletion removes the perivascular but not endothelial pool of aquaporin-4 at the blood–brain barrier and delays the development of brain edema in an experimental model of acute hyponatremia. *FASEB J* 2004;**18**(3):542–4.
47. Haj-Yasein NN, Vindedal GF, Eilert-Olsen M, Gundersen GA, Skare Ø, et al. Glial-conditional deletion of aquaporin-4 (Aqp4) reduces blood–brain water uptake and confers barrier function on perivascular astrocyte endfeet. *Proc Natl Acad Sci USA* 2011;**108**(43):17815–20.
48. Fenton RA, Moeller HB, Nielsen S, de Groot BL, Rützler M. A plate reader-based method for cell water permeability measurement. *Am J Physiol Renal Physiol* 2010;**298**(1):F224–30.
49. Verkman AS. Knock-out models reveal new aquaporin functions. *Handb Exp Pharmacol* 2009;**190**:359–81.
50. Scharfman HE, Binder DK. Aquaporin-4 water channels and synaptic plasticity in the hippocampus. *Neurochem Int* 2013;**63**:702–11.
51. Benfenati V, Caprini M, Dovizio M, Mylonakou MN, Ferroni S, Ottersen OP, et al. An aquaporin-4/transient receptor potential vanilloid 4 (AQP4/TRPV4) complex is essential for cell-volume control in astrocytes. *Proc Natl Acad Sci USA* 2011;**108**:2563–8.
52. Bagley CE, Tvarnø CD. Pharmaceutical public–private partnerships in the United States and Europe: moving from the bench to the bedside. *Lecturer and other affiliate scholarships series. Paper 12.* <http://digitalcommons.law.yale.edu/ylas/12>.
53. Milne C-P, Malins A. Academic–industry partnerships for biopharmaceutical research & development: advancing medical science in the U.S. Tufts Center for the Study of Drug Development; 2012.
54. Rosenblatt M. How academia and the pharmaceutical industry can work together. *Ann Am Thorac Soc* 2013;**10**:31–8.

55. Corillon C, Mahaffy P. *Scientific relations between academia and industry: building on a new era of interactions for the benefit of society. Report from an international workshop on academia–industry relations*, Sigtuna, Sweden; November 2011. p. 22–25.
56. Parson A. Evolving partnerships: academia, pharma, and venture groups adapt to challenging times. *Cell Stem Cell* 2013;**12**:12–14.

FURTHER READING

Meyers MA. *Happy accidents. Serendipity in modern medical breakthroughs.* New York, NY: Arcade Publishing; 2007. pp. 390.

CHAPTER 8

Clinical Research

Eva Skovlund[1] and Kjell Magne Tveit[2]
[1]School of Pharmacy, University of Oslo, Blindern, Oslo, Norway
[2]Department of Oncology, Oslo University Hospital, Nydalen, Oslo, Norway

8.1 INTRODUCTION

Traditionally, medical research has been classified as basic or applied. When the objective is to expand knowledge by exploring ideas and questions, or to develop models and theories to explain certain phenomena, the research is usually defined as basic. Applied research, on the other hand, is conducted specifically to find solutions to practical problems and typically aims to improve the treatment for a specific disease. Examples include clinical trials of new drugs or new indications for existing drugs. The term *translational research* has emerged as a paradigm alternative to the dichotomy between basic and applied research. Basic research is conceptualized and conducted to render findings that are directly applicable to the population under study, for instance by translating the results of laboratory research more rapidly into clinical practice and vice versa—often referred to as "bench to bedside and back." This chapter will focus on applied clinical research, which involves the use of human subjects to test new treatments.

Clinical research, as well as research in general, can be categorized in different ways: as quantitative or qualitative research; observational or experimental research; or as descriptive or analytical research. Quantitative research involves measurements of characteristics and outcomes that can be subjected to formal statistical analyses. Typically a relatively large number of subjects are studied. One example of the aim of such a study would be to draw a general conclusion on the size of the effect of a certain treatment. In contrast, qualitative research gathers information about how a phenomenon is experienced by one or a few individuals based on unstructured interviews and is exploratory, also referred to as hypothesis-generating, by nature.

Qualitative research is typically hypothesis-generating, whereas quantitative research can either be hypothesis-generating or hypothesis-testing. Large hypothesis-testing trials are often referred to as confirmatory.

Another important distinction is whether a study is observational or experimental (see also Sections 4.3 and 4.7). In observational studies the

Research in Medical and Biological Sciences
DOI: http://dx.doi.org/10.1016/B978-0-12-799943-2.00008-2

factors of interest are naturally occurring risk factors or exposures that are not controlled by the investigator (see Chapter 9). In experimental studies the factor of interest can be manipulated by an intervention, such as treating one group of patients with a new treatment and another comparable control group with the same disease with an established treatment, or placebo (i.e., no active treatment). Experimental studies are prospective by nature, whereas observational studies can be either prospective or retrospective.

Finally, research can be subdivided into descriptive or analytical types. Descriptive studies can involve a single case or a large population. They are primarily observational and include surveys, case series, and reviews of medical records. Often such studies are used for hypothesis generation. In analytical studies, comparisons between different groups or within a single group over time (before versus after an intervention) are performed to draw conclusions based on previously stated hypotheses on associations between an outcome and an exposure. Analytical studies can be retrospective or prospective, as well as observational or experimental, but this chapter will deal with experimental studies.

8.2 THE RESEARCH PROCESS—IMPORTANT STEPS

A well-designed research project should begin with a clear statement of an important research problem that needs to be investigated. The research problem should be clear, concise, and lead to testable hypotheses. In general, a clinical research problem is considered important if its solution has the potential to clarify an issue affecting public health, or if it enables the clinician to make a correct decision on which treatment to choose for a specific patient.

Sometimes research problems are too broad to be addressed in a single study. It is generally better to provide valid answers to a small number of questions than to undertake complicated study designs. Clinical research problems can be generated from observations collected in conjunction with medical procedures. Review of published research may also point to gaps in knowledge that could be filled by new investigations.

De Angelis[1] identified a series of steps comprising the research process; a condensed list follows.

1. Identify the problem or question.
2. Restate the question as a hypothesis.
3. Review the published literature to determine whether the hypothesis has already been appropriately evaluated or if it requires further study.

4. Identify all relevant study variables, including an operational definition.
5. Develop a research design and analytical plan to test the hypothesis.
6. Specify or construct data collection instruments and design a data collection plan.
7. Collect and process data.
8. Perform statistical analyses.
9. Draw (careful) conclusions.
10. Draft the research report or scientific publication.

All research on human subjects needs approval from a local ethics committee. For clinical trials including drugs, approval from the relevant national medicines control authority is also necessary, and the trial must be registered in an appropriate database, such as EudraCT or ClinicalTrials.gov. The Good Clinical Practice (GCP) Guideline is an international ethical and scientific quality standard for designing, conducting, recording, and reporting clinical trials. Compliance with this standard assures that clinical trial data are credible, and that the rights, safety, and well-being of the human subjects in a trial are protected, which is consistent with the principles of the Declaration of Helsinki. The GCP Guideline[2] is a unified standard for the European Union (EU), Japan, and the United States, and facilitates the mutual acceptance of clinical data by the regulatory authorities in these jurisdictions.

The guideline on statistical principles for clinical trials[3] provides direction regarding the design, conduct, analysis, and evaluation of clinical trials for a new drug, as well as useful guidance on the statistical aspects of clinical trials in general.

8.3 CONTROLLED CLINICAL TRIALS

The term clinical trial usually refers to any type of planned clinical experiment or intervention involving human subjects. Normally the aim of performing a clinical trial is to evaluate the efficacy of one or more treatments in a defined group of patients. Since a researcher is never able to include all patients with a specific diagnosis in any one experiment, conclusions are based on a trial that includes a sample of patients (or study sample). In order to draw valid conclusions, this sample must be representative of the target population so that results can be generalized. If the patients included in the trial are highly selected and the sample is not representative of the population the trial intends to investigate, then the results may not be valid for this population and interpretations should be made with great caution.

Most clinical trials deal with the efficacy of drugs; thus, most are initiated and conducted by the pharmaceutical industry. Clinical trials can also be used to evaluate the efficacy of other modes of treatment, such as surgery, radiation therapy, or lifestyle interventions. The main focus of this chapter will be on documentation of the efficacy of a new drug. The general principles can of course be applied to any type of clinical trial.

A real challenge when planning and conducting clinical trials, especially large ones, is amassing the necessary human and economic resources. Current and future high-quality clinical studies performed according to GCP guidelines are resource-demanding in all phases: planning, execution, and evaluation. Large, randomized, multi-center trials, involving a high number of institutions in several countries, may be impossible to perform within a sphere of small academic research groups. Thus, such studies may have to be conducted by larger international research organizations or pharmaceutical companies. One should keep in mind that the most important research questions should be raised in the planning phase of a study if it is to render results that will have impact on patient health.

8.4 DRUG DEVELOPMENT

A series of clinical trials must be conducted to document the efficacy of a new treatment.[4] When it comes to new drugs, clinical trials are divided into four phases. Phase I trials study toxicity and side effects, and identify a maximum tolerated dose. As a rule, Phase I trials are conducted in small groups of healthy volunteers. Usually one begins with a small number of test subjects receiving a very low dose. To identify a safe dose for the "first in man" experiment, a so-called no observable adverse effect level from toxicology studies in animals is converted to a human equivalent dose, which is taken from the most appropriate species. Thereafter a safety factor (of at least 10-fold) is used to define the maximum recommended starting dose. If none of the (usually three) first test subjects experience an adverse effect, the dose is increased to a defined level (usually twice the maximum recommended starting dose) and administered to three new test subjects. At the next increment the dose is again increased and administered to three new test subjects, and so on until adverse events or signs of toxicity occur (Box 8.1).

BOX 8.1 Purpose of Different Phases of Clinical Trials in Drug Development

Phase I	Toxicity and side effects
Phase II	Dose-response
Phase III	Comparison with established treatment or placebo
Phase IV	Post-marketing studies

Phase II trials study the treatment efficacy in the form of dose-response, i.e., the effect of different doses of a drug, sometimes also compared to placebo. Relatively small samples are used to study dose-response, and Phase II trials are often referred to as exploratory, as the samples are usually too small to draw firm conclusions on efficacy. Studies on pharmacokinetics (absorption, distribution, metabolism, and excretion of the drug) are usually part of both Phase I and Phase II trials. Phase II trials can also be utilized to make decisions on how to measure efficacy in Phase III trials. Some Phase II trials are designed as randomized trials, in order to ensure that at least two trials are performed with similar patient cohorts. However, these studies should be regarded as hypothesis-generating rather than confirmatory, and the comparisons of clinical benefits in the two groups should be made with great caution. When a drug regimen has shown promising efficacy and acceptable safety in Phase I and II trials, the next step is to test the new treatment against an established treatment.

This is done in Phase III trials, which are larger randomized controlled trials that compare the efficacies of two or more different treatments. A new drug can be compared either to an established treatment, placebo, or both. The term *placebo* is Latin for "I shall please" and refers to an inactive substance with the same appearance, and preferably the same smell and taste, as the drug under investigation. Phase III trials are used to document the efficacy of a new drug before the manufacturer applies for marketing authorization. Some trials combine Phases II and III to investigate the relationship between dose, efficacy, and safety in a randomized, double-blind study.

Phase IV is post-marketing, and covers both long-term studies of safety and pure marketing studies. In the remainder of this chapter, the term "clinical trial" will refer mostly to Phase III trials, unless otherwise stated.

8.5 CLINICAL TRIAL PROTOCOL

A detailed trial protocol must be written before any clinical trial starts, be it a Phase I, II, III, or IV trial. The purpose is to give a detailed description of all aspects of the trial, including the aim, documentation of the trial plan, inclusion and exclusion criteria, description of the treatments and clinical examinations, the evaluation of treatment efficacy, randomization, blinding, number of patients, a plan for statistical analyses, and the reporting of adverse events. Moreover, the trial protocol should include a description of relevant practical and administrative procedures. A trial protocol is required not only to have a written record of how the trial is to be conducted, including planned statistical analyses and reporting, but also to obtain official approval to conduct the trial and to apply for funding, which are both vital aspects for the researcher. The most important parts of a trial protocol are described in Box 8.2 (see also Section 4.10).

For trials that include interventions that go beyond routine clinical treatment, approval must be sought from the relevant ethics committee and, if drugs are involved, the relevant national medicines control authority. Additionally, before including patients in a trial they must be appropriately informed about procedures and potential risks, and an informed consent form must be signed by each subject before he or she takes part in the study.

BOX 8.2 Main Content of a Trial Protocol

- Purpose of the trial
- The research problem, including a testable hypothesis
- Trial design
- Target population (inclusion and exclusion criteria)
- Justification of sample size (number of subjects)
- Description of treatments and justification for the choice of comparator
- Patient follow-up, handling of drop-outs
- Randomization, blinding
- Efficacy variables (endpoints)
- Adverse events
- Planned statistical analysis
- Administrative matters
- Ethics

8.6 TRIAL DESIGNS

The efficacy of a new treatment is assessed by comparing it either with an established treatment or with a placebo. Sometimes it is obvious which treatment should serve as its comparison, or comparator, but often there is no general agreement on what should be considered optimal treatment of the patients in question. A discussion of the choice of control group is given in a guidance document used by regulatory authorities.[5]

In principle there are two kinds of comparative designs to choose from. The most commonly used is called a parallel group trial, which compares two or more independent study samples. Hence, in a parallel group trial the test group receives the experimental treatment and the control group receives established treatment. Treatment efficacy is then compared between the two treatment groups. However, parallel group trials can include more than two treatment groups. In trials of antidepressants, for instance, it is quite common to perform three-armed trials, including a test group that receives the new drug, a control group that receives established treatment, and a placebo group. In this way assay sensitivity can be assessed.

In a factorial trial, combinations of treatments can be studied. The simplest factorial trial contains two factors at two levels. Assume patients are randomized between treatment A or placebo for A, plus treatment B or placebo for B. This leads to four possible combinations: A + placebo, B + placebo, A + B, or placebo only. If there is no interaction between treatment A and B, the number of patients needed is the same as if only one of the interventions were studied.[3] However, it is important to keep in mind that factorial trials are often used specifically to study the interaction between two active treatments to be used in combination, and this strategy could result in a lack of power to study the interaction effect.

The other type of trial design utilizes comparison within subjects. This is often referred to as a crossover trial. Here, all patients receive both treatments, but the sequence of treatments varies between patients. As in parallel group trials, more than two treatments can be compared, but patients act as their own controls, which leads to reduced random variation as observations from a single subject tend to be positively correlated. Accordingly, crossover trials require fewer patients than do parallel group trials, which in some sense make them attractive as time and cost may be reduced. Nonetheless, crossover trials are carried out less frequently than parallel group trials, partly because their applicability is limited to studies of treatments of chronic and stable diseases that cannot be cured, and partly because it may be difficult to generalize results from a small sample of patients. Moreover, results from crossover trials may sometimes be hard to interpret due to so-called interaction or carry-over effects. What should be the conclusion on efficacy if the active treatment is demonstrated to be more effective than placebo, but only in patients who receive the placebo before they received active treatment? Crossover trials are also less well suited to long-term treatment, as the risk of loss to follow-up increases with the duration of a trial.

In longitudinal trials the outcome is measured repeatedly over time, and efficacy may be compared both between treatment groups and within treatment groups over time.

8.7 TARGET POPULATION

In clinical research it is usually impossible to study all individuals in the patient population. Instead a study sample is selected and statistical analyses, results, and conclusions are based on this sample. However, if one wishes to extend conclusions to a larger population of patients, it is important to make sure that the study sample is representative of the patient population to which the conclusion is to be generalized. A smaller number of patients is required to make inferences on efficacy with a homogeneous sample of patients, but that comes at the price of reduced generalizability. Thus, careful thought should be given to the consequences of defining certain inclusion and exclusion criteria. Patient characteristics and the patient's condition at the start of the trial are of major importance to identify the population from which the study sample is collected, and therefore baseline observations must always be recorded.

Clinical trials on new drugs do not usually include women who may become pregnant, the elderly, or patients with comorbidities. While the arguments for not including specific patients may be obvious for safety reasons and valid concerning the conduct of the trial *per se*, such exclusion may reduce the value of the trial results. Indeed, a large proportion of individuals on drug regimens are actually elderly people with comorbidities, but efficacy and safety has often not been appropriately studied in this patient group. Medical treatments for children are especially challenging, but since the EU pediatric regulation came into force, pediatric drug development has become more integrated into overall drug development.

Some studies include a run-in period to identify fluctuations in important variables before initiation of treatment or to assess patient compliance with prescribed drug dosage and frequency of intake. Placebos are most commonly administered during a run-in period, and it is sometimes argued that a run-in period provides a better rationale for deciding whether a patient is eligible for inclusion in the full trial. However, an important drawback is that exclusion of patients with certain characteristics, such as fluctuating baseline values, may reduce the generalizability of the study results.

8.8 HISTORICAL CONTROLS

It is not uncommon for researchers or pharmaceutical companies to conduct studies without a control group. Indeed, Phase I and II trials are often not meant to be comparative; it is only in late Phase II and Phase III (the confirmatory phase) trials that a control group is included. It is difficult to evaluate treatment efficacy in uncontrolled trials, since the result that would have been observed if the patients had not been treated or had received an established treatment is unknown.

One way to try to solve this is to compare the test group with a historical control group. The comparison is then between patients included in the trial who are receiving a new active treatment and a group of patients that received some established treatment in a period of time before the trial was initiated (historical controls). The main difficulty with such a strategy is to ensure a fair comparison of treatment efficacy. If patient characteristics differ between the two groups, any difference in efficacy may be the result of differences between the treatment groups rather than an actual effect of the treatment.

Although the differences between previously treated patients and those included in a trial may not be obvious, it is usually difficult or impossible to ascertain that differences do not exist. As a rule, a historical control group is not subjected to strict inclusion and exclusion criteria, while the inclusion of patients in a treatment group is far more restrictive. Also, the quality of recording of the treatment effect for the historic controls may be substantially lower, as the historical sample was not included in a trial. Retrospective collection of information rarely fully overcomes that problem. Moreover, the criteria for evaluating treatment response, as well as other important factors, may have changed over time. Thus, an unbiased comparison cannot be ensured by using historical controls.

8.9 RANDOMIZATION

The gold standard to demonstrate the efficacy of a new treatment is the randomized controlled trial. Instead of choosing actively the treatment group to which patients should be assigned, patients are randomly allocated to one of the treatment groups. In this way an unbiased comparison of treatments is possible. Correspondingly, bias can be avoided by random allocation of patients between treatment sequences in crossover trials. The purpose of randomization is to prevent systematic differences between treatment groups. Any difference in the distribution of patient characteristics between treatment groups in a randomized trial will be due to chance. If no obvious comparator exists in the form of an established treatment, a placebo can be used (Box 8.3).

In principle, simple randomization is equivalent to tossing a coin to decide to which treatment group or treatment sequence a patient should be allocated. In practice, computer-generated random number tables are used to allocate patients to treatment groups. If a trial is not double-blind (see Section 8.10) potential selection bias can be avoided by obtaining patient consent to participate before randomization.

The term randomized controlled trial should not be confused with the term random sample. In the former, included patients are randomized between treatment groups or treatment sequences. However, this does not necessarily imply that the patients are a random sample from a patient population. In fact this is almost never the case, since it is rarely possible to enumerate all the individuals in such a population. Instead individuals are often selected based on convenience or specific characteristics. This may lead to a sample that is not representative of the population the researcher wishes to study, and thus questionable external validity (see also Section 8.17).

Patients are usually randomized 1:1 between treatment groups, i.e., there is an equal number of patients in each group. However, in comparisons between new and established treatments, or between active treatment and placebo, more patients may be included in the test group in order to

BOX 8.3 Randomization

Randomization between treatment groups or treatment sequences is a tool to enable fair comparison.

gain more information on the new treatment, or possibly for ethical reasons, to keep the number in the placebo group low. A 1:1 randomization ratio is more efficient from a statistical point of view (see Section 8.12).

Simple randomization does not necessarily result in the treatment groups being equally distributed in age, gender, or other characteristics that may be of importance, but any differences will be due to chance and not to systematic allocation of patients to different groups. This means that significance tests and *P*-values will be directly interpretable and that conclusions on causality will be valid, provided appropriate statistical analyses are performed. If it is important to keep a similar distribution of important patient characteristics or prognostic factors, for instance in a small trial, stratified randomization may be applied.

8.9.1 Stratification

There are two main types of stratification: stratification by design, which is described below, and *post hoc* stratification in the analysis phase, which is done to try to overcome imbalances in the distribution of patient characteristics that may affect the result. Such imbalances are not expected in randomized trials, and even if it turns out that prognostic factors are not equally distributed between treatment groups, statistical tests are still valid since any differences must have occurred by chance. Thus, simple statistical methods are usually applied to assess efficacy. In order to preserve the nominal significance level (which is usually set to 5%), stratification factors included in the design phase should in principle also be included in statistical tests performed during the analysis phase. However, the effect of violating this principle is usually limited.

The purpose of stratified randomization is to keep patient groups balanced in terms of important prognostic factors that might otherwise affect estimates of differences in efficacy. Stratification is most often applied in so-called randomized blocks. An example of block randomization in four strata in a breast cancer study is shown in Table 8.1. In this example prognostic factors considered to be of importance for patient survival are estrogen receptor status (positive or negative) and tumor size (T_1 or T_2).

The first step is to identify one or a small number of prognostic factors. If the stratification is based on two prognostic factors, each of them at two levels, there will be four (2×2) strata, or subgroups. Within each stratum, patients are randomized in blocks. A customary choice of block size is four, for which there are six possible permutations of the sequence

Table 8.1 Randomized blocks within strata

Estrogen receptor +	Estrogen receptor +	Estrogen receptor −	Estrogen receptor −
T_1	T_2	T_1	T_2
A	A	A	B
B	A	A	A
B	B	B	A
A	B	B	B
B	B	A	B
B	B	B	A
A	A	B	B
A	A	A	A
A	A	B	A
B	B	B	B
A	B	A	A
B	A	A	B

of patient allocation to treatment A or B: AABB, ABAB, ABBA, BBAA, BABA, and BAAB. This type of randomization scheme means that for every four patients enrolled in a stratum, two will be on treatment A and two will be on B. Hence, a balance between treatments A and B is attained.

Inclusion of a large number of stratification factors should generally be avoided, because the number of strata increases rapidly with an increasing number of factors. For example, if a third prognostic factor is included in the breast cancer study example above, such as number of positive lymph nodes, grouped in three levels (0, 1−3, and 4 +), the number of strata would increase from four to 12 ($2 \times 2 \times 3$). With a large number of strata there is a major risk that some may have too few patients, or none at all, and the entire concept of balance breaks down.

In open trials it is recommended that the investigator be blinded with regard to block size, otherwise it may be possible to foresee which treatment will be given to some of the patients. The allocation is obvious for the last patient enrolled in each block regardless of blinding, and such knowledge may consciously or unconsciously affect the decision to include a particular patient, which may in turn lead to selection bias. Larger blocks or varying block sizes reduce this problem, but there is also a larger risk of imbalance in a stratum in which few patients are enrolled.

Table 8.2 Characteristics of allocation procedures

	Simple randomization	Randomized blocks within strata	Minimization
Implementation	Very easy	Reasonably easy	Requires updated records of all previous treatment assignments
Predictability	None	Some with fixed block size	Potential
Prognostic factors	Imbalance possible by chance	Can balance a small number	Can balance a large number

As a rule, stratification is deemed unnecessary in larger trials, because full (simple) randomization is expected to provide good balance when the number of patients is large. Moreover, stratification should be avoided if the prognostic significance of a particular factor is uncertain.

8.9.2 Minimization

With an increasing number of strata, balance between treatments within each subgroup becomes irrelevant. Instead it is important to ensure that the proportion of patients with a given characteristic is similar in both treatment groups. In statistical terms this is referred to as balancing the marginal totals. Instead of preparing a randomization list in advance, an updated record of treatment assignments by patient characteristics is kept, and the allocation of a new patient to a treatment group will depend on his or her characteristics and the characteristics of patients included in the trial up to that point. A practical description of the implementation of minimization is given by Pocock.[6] True randomization is thus not necessarily applied, but an element of chance is often introduced by assigning the appropriate treatment with probability $p = 3/4$, for instance. The minimization method is frequently used in cancer trials (Table 8.2).

8.10 BLINDING

Randomization of patients into treatment groups does not in itself ensure an unbiased comparison of treatments. Systematic bias may arise whenever a physician knows which patient has received what treatment. Particularly when response is evaluated subjectively, knowledge of the treatment may lead to the physician's (or the patient's) confidence in the treatment consciously or unconsciously affecting the assessment of efficacy.

Ideally, all randomized trials should be double-blind, so that neither the patient nor the treating physician knows which treatment is actually given. Indeed, both patients and treating physicians may be overly optimistic about the efficacy of a new treatment. Consequently, the assessment of patient response, or even the response itself, may be affected. If the comparator is a placebo, it is evident that an open trial cannot be performed.

Blinding in clinical trials is sometimes difficult or even impossible, for example in clinical cancer research, an area in which few trials are actually blinded in practice. This may be due to expected toxicity, which makes it essential to know which treatment is given, or it may be for practical purposes due to differences in posology or dosing regimens. Assume, for instance, that treatment A is oral treatment, two tablets twice daily, and that treatment B is one daily infusion. It is in theory possible to perform a double-blind trial using a double dummy technique in which one patient would be given A in tablet form and placebo infusion for B, while another will be given B in infusion form and placebo tablets for A. Each patient would receive both oral treatment and infusions but only one of these would contain the active drug, depending on whether the patient is randomized to treatment A or B. Most people would question the ethical acceptability of a procedure where patients are exposed to unnecessary infusions. It would probably be easier to accept the use of a double dummy technique with two oral treatments, depending on the patient population.

Trials that are not blinded, or in which the randomization code is broken, can be subject to bias, both in influencing patient expectations and in the evaluation of treatment efficacy. Sometimes treatments have specific adverse effects that clearly reveal which treatment is given even if the trial is double-blind. This problem can be reduced to some extent by having one investigator administer treatment and another evaluate patient response.

Clinical trials comparing other types of treatments, such as psychotherapy or the effect of strength training, are impossible to blind. Again, one might consider using an evaluator that is blinded to treatment allocation, which may reduce the risk of bias.

8.11 CHOICE OF ENDPOINTS

The results of a study with a large number of outcome variables or endpoints may be difficult to interpret unless all or most comparisons point in the same direction. It is expressly recommended not to conclude that there is a difference in efficacy if only one of several outcome variables shows a statistically significant difference between treatments. When multiple significance tests are performed, the significance level becomes inflated, increasing the risk of false-positive findings (see Section 8.13). One or very few outcome variables should ideally be chosen at the planning stage. Typically, a well-designed study will have one primary endpoint and a small number of secondary endpoints.

8.12 SAMPLE SIZE ESTIMATION

8.12.1 Power

Before an experimental or observational study is initiated, it is essential to estimate how many subjects are needed to reach valid conclusions. In order to perform appropriate calculations, one must identify a realistic and clinically relevant difference between two treatments. Of course there is no single correct definition of clinical relevance, and sample size estimation is often performed for a range of possible effect sizes. One needs to compromise between the wish to detect very small differences and the resources and time available to conduct very large trials. However, the major problem is unrealistic expectations, which can lead to trials that are too small.

One of four different results may occur when a hypothesis test is performed. Two events, denoted in Table 8.3 as *A* and *D*, are desirable, while events *B* and *C* lead to erroneous conclusions. The errors that may occur are either rejection of the null hypothesis even though it is true (*B*) or acceptance of the null hypothesis when it is actually false (*C*), often referred to as type I and type II error, respectively. A researcher must always aim to keep the probabilities of these errors small. The upper limit of the probability that *B* will occur, $P(B)$, is termed the significance level of the trial, which as a general rule in medical research is set at $\alpha = 5\%$. Likewise, the probability of erroneously accepting a false null hypothesis, $P(C)$, is often designated β. The power of a test is the probability of

Table 8.3 Possible results of a hypothesis test

	H_0 is true	H_0 is false
H_0 accepted	*A*	*C*
H_0 rejected	*B*	*D*

H_0: null hypothesis.

detecting a certain treatment efficacy, which is designated $P(D) = 1 - \beta$ and is the probability of rejecting the null hypothesis when it is in fact false. This probability should be high, and when planning a clinical trial it is usually set at 80%, 90%, or 95%.

The first step in planning a trial is to set the significance level and power to detect a difference that is considered clinically relevant. Then the number of patients to be enrolled is determined, and the trial is conducted. Finally, statistical analyses are performed.

The smaller the clinically relevant difference, the larger the number of patients that need to be enrolled in the trial. Unless a trial is large, small differences are likely to go undetected due to a high risk of a type II error. Thus, the conclusion that two treatments have equal efficacy cannot automatically be drawn, even when the null hypothesis is not rejected, since the number of patients included may have been too small to detect a difference.

8.12.2 Sample Size Estimation for Categorical Outcomes

Assume that patients receiving established treatment have a 40% probability of surviving 5 years ($p_s = 0.4$) and that one wishes to determine whether a new treatment can increase 5-year survival to 50% ($p_n = 0.5$). A simple and useful formula for calculating the necessary sample size in this case is given in Box 8.4.

BOX 8.4 Sample Size Estimation for Two Binomial Proportions

The number of patients to be included in each treatment group can be estimated as follows:

$$n = \frac{p_n \cdot (1 - p_n) + p_s \cdot (1 - p_s)}{(p_n - p_s)^2} \cdot c \qquad (8.1)$$

The constant c depends on the choice of significance level and power to detect a certain difference between two treatments as statistically significant. Relevant values of c at various significance levels and power are given in Table 8.4.

Table 8.4 Suggested values of the constant c

	Power		
Significance level (two-sided)	**0.80**	**0.90**	**0.95**
0.10	6.2	8.6	10.8
0.05	7.9	10.5	13.0
0.01	11.7	14.9	17.8

Example

Now assume that in the example above a significance level of 5% and a power of 80% are chosen. In this case, it is necessary to enroll $n = (0.5 \cdot 0.5 + 0.4 \cdot 0.6)/(0.1^2) \cdot 7.9 = 387$ patients in each group.

8.12.3 Sample Size Estimation for Continuous Outcomes

The formulas given in this section assume that the outcome is normally distributed. In general the sample size estimate also holds approximately for non-normally distributed data when nonparametric analyses are applied. The exact number of subjects needed to obtain the desired power depends on the true distribution, which is rarely known. If it is suspected that the distribution is far from normal, it may be advisable to add 5% to the estimated number of subjects needed, to avoid a potential loss of power.

8.12.3.1 Pairs of Observations

In a crossover trial all patients receive two (or more) treatments in a randomized sequence. Let Δ be the difference in treatment efficacy considered to be of clinical relevance and σ_d the standard deviation of the difference between the two treatment efficacies. Unfortunately, it may be difficult to make a valid assumption regarding variability before the trial is completed, and therefore σ_d is rarely known. However, if published literature does not provide an estimate of σ_d that is thought to be valid in the setting, a small pilot study can be carried out to estimate the standard deviation. A pilot study can be conducted for other purposes as well, such as to train investigators or to assess the feasibility of planned procedures. The sample size needed in a crossover trial with a continuous outcome variable is given in Box 8.5.

BOX 8.5 Sample Size Estimation for Continuous Outcomes. Pairs of Observations

The number of patients to be included can be estimated by the following formula:

$$n = \left(\frac{\sigma_d}{\Delta}\right)^2 \cdot c \qquad (8.2)$$

where values of c are listed in Table 8.4.

Example

Assume a crossover trial of two asthma drugs is planned and that peak expiratory flow (PEF) rate (L/min) is selected as the primary efficacy variable. A clinically relevant difference might be set at $\Delta = 30$ L/min, and it is assumed here that $\sigma_d = 50$ L/min. The significance level is set at 5% and the power to detect the chosen difference at 80%. According to Table 8.4, the constant is then $c = 7.9$. Thus, the study should enroll $n = (50/30)^2 \cdot 7.9 = 22$ patients.

8.12.3.2 Two Independent Samples (Parallel Group Trials)

Parallel group trials are the most common trials performed. The formula for estimation of sample size in these trials is given in Box 8.6.

Example

As a rule, considerably more patients are needed in a parallel group trial than in a crossover trial. Take the example of a parallel group trial of two asthma drugs, similar to the crossover trial described previously. As before, a clinically relevant difference in PEF is assumed to be 30 L/min, but the standard deviation (SD) is assumed to be $\sigma = 60$ L/min. The SD is expected to be larger than in the crossover trial, because the parallel group trial takes individual data from different patients in the two groups, while the crossover trial used within-patient differences. As it is reasonable to assume that observations from a single patient are in general positively correlated, σ_d is usually less than σ. The estimated number of patients to be enrolled is then $n = 2 \cdot (60/30)^2 \cdot 7.9 = 63$ patients in each group; 126 in all.

8.12.3.3 Unequal Group Size

Most often the two groups in a study with two independent samples are of the same size and randomization is 1:1. However, in some cases it is desirable to have groups of different sizes. One such case may be when a

BOX 8.6 Sample Size Estimation for Continuous Outcome. Parallel Group Trials

The number of patients n to be enrolled in each treatment group is

$$n = 2 \cdot \left(\frac{\sigma}{\Delta}\right)^2 \cdot c \qquad (8.3)$$

where, as before, Δ is the clinically relevant difference to be detected, c is a constant that depends on the significance level and power, and σ is the standard deviation of the observations, which is assumed to be equal for the two treatment groups.

Table 8.5 Loss of power with unequal treatment group size

Ratio r test group: control group	Power
1:1	0.95
2:1	0.92
3:1	0.88
4:1	0.84

trial includes a placebo group that is kept small to minimize the number of patients who receive the placebo. Another case is when more extensive information is sought on a new drug, so a larger number of patients is included in the test group compared to the control group.

Whatever the total number of patients, 1:1 randomization always leads to the highest power, but, as shown in Table 8.5, if randomization is set at 2:1, the loss of power is small. For larger differences in the number of patients between the two groups, it may be wise to slightly increase the total number of patients enrolled in order to maintain power.

A description of how to adjust the sample size follows here:

First, n is calculated as if the treatment groups were of equal size. Then, a modified sample size m is calculated. Assume r is the ratio between the number of patients in each of the two treatment groups. The total number of patients to be enrolled is then

$$m = \frac{n \cdot (1+r)^2}{4r} \tag{8.4}$$

and the sample sizes in each of the two groups respectively are

$$\frac{m}{(1+r)} \quad \text{and} \quad \frac{r \cdot m}{(1+r)}. \tag{8.5}$$

For instance, randomizing in a ratio of 3:1 leads to a modified number m of

$$m = \frac{4}{3} \cdot n. \tag{8.6}$$

8.12.4 Sample Size Estimation Based on Precision of Estimates

Sample size estimates can also be precision-based. The question then becomes, what width of a 95% confidence interval is acceptable, or, in

BOX 8.7 Precision-Based Estimates of a Proportion *p*

Let the width of the confidence interval be 2*d* (i.e., the 95% confidence interval is $\hat{p} \pm d$). The number of observations is

$$n = \left(\frac{1.96}{d}\right)^2 \cdot p \cdot (1 - p). \qquad (8.7)$$

BOX 8.8 Precision of an Estimate of the Population Mean

The number of observations necessary to attain the required precision of an estimate of the population mean is

$$n = \left(\frac{1.96 \cdot \sigma}{d}\right)^2 \qquad (8.8)$$

where, as before, *d* is the half width of the confidence interval and σ the standard deviation.

other words, how imprecise an estimate will be acceptable? Boxes 8.7 and 8.8 show formulas for precision-based estimates of sample size.

8.12.4.1 Categorical Outcome

The 95% confidence interval of a probability p is

$$\hat{p} \pm 1.96 \cdot \mathrm{SE}(\hat{p})$$

where $\hat{p}$ is the estimated proportion, n the total number of observations, and $\mathrm{SE}(\hat{p}) = \sqrt{(\hat{p}(1 - \hat{p}))/n}$.

The true value of the probability p is, of course, unknown. To estimate the number of observations required, we must either substitute the true unknown value with some reasonable value of p, or set $p = 0.5$. The latter procedure is conservative and yields the largest possible value of n for a given confidence interval width, as the product $p(1 - p)$ has its maximum (0.25) when $p = 0.5$. The product is lower for all other choices of p. Therefore if p cannot be "guesstimated" in advance, it is reasonable to use $p = 0.5$ to estimate n. However, if p is presumed to be relatively small (such as less than 10%), $p = 0.1$ would be a more suitable assumption.

This type of calculation is used when assessing the number of subjects to be questioned in a political opinion poll and is also often used in sample size calculations for Phase II trials.

Example

Assume one wishes to estimate the proportion of general practitioners whose first choice in treating sinusitis is to prescribe penicillin. The intention is to send out a questionnaire, and therefore an estimate is needed of how many general practitioners should be queried. First, the desired precision of the estimate must be determined, in other words, how wide a 95% confidence interval will be acceptable. Assume an interval width of 0.1 (that is, $d = 0.05$) is required and set $p = 0.5$ in order not to underestimate the number needed. It is then estimated that $n = (1.96/0.05)^2 \cdot 0.5 \cdot 0.5 = 385$ general practitioners should be included in the study.

8.12.4.2 Continuous Outcome

A 95% confidence interval for the true mean is

$$\overline{x} \pm t_{0.975,n-1} \cdot \frac{s}{\sqrt{n}}$$

where $\overline{x}$ is the estimated mean, s the standard deviation, n the number of observations, and t the relevant fractile in the t-distribution. For practical purposes, $t_{0.975}$ is often replaced with $z_{0.975} = 1.96$, and this approximation holds when $n \geq 30$.

Unfortunately, the true value of σ is unknown and must be estimated, for instance from earlier studies, if they exist.

8.13 STATISTICAL ANALYSIS

8.13.1 Analysis Sets

It is of course desirable to record as much relevant information as possible on all patients enrolled in a clinical trial, but in practice much information may be lacking at the end of the trial. Some patients may be lost to follow-up, some may take excessive or insufficient doses of the drug under investigation, and some may have completely refrained from taking the drug they were randomized to receive. In some cases, adverse events may even lead researchers to stop treatment for some patients.

It is essential to choose which patients will be included in statistical analyses before the data are analyzed. Complete agreement and definite rules concerning which patients should be included in primary statistical analysis do not exist. Plans and justifications for how to handle missing data should therefore be made in advance and described in the trial protocol.

Strategies can be divided into roughly two categories. A per-protocol (PP) analysis includes only compliant patients who took a predefined percentage of the planned dose (such as 80–120%), showed up at all visits, have no missing values for the essential outcomes, and fulfill the inclusion and exclusion criteria of the protocol. In theory, such an analysis might be used to estimate the "true" treatment efficacy. The problem with this strategy is that it often leads to selection bias. If patients who do not respond to treatment are more prone to withdraw from the study and are subsequently not included in the statistical analysis, there is a high risk that treatment efficacy will be overestimated. A different strategy is to perform an intention to treat (ITT) analysis, which includes all randomized patients whether or not they actually adhered to the prescribed treatment. This can be viewed as a pragmatic approach that underestimates rather than overestimates differences in treatment efficacy and probably mimics better real life—i.e., the effect of prescribing a treatment to a patient population. Conclusions on superiority of one drug over another are always primarily based on some kind of ITT analysis. It is nevertheless common practice to also report the results of a PP analysis.

8.13.2 Handling of Missing Values

It is almost inevitable that some data will be missing in a clinical trial. Since it is not recommended to focus on the sample of patients without missing values (PP analysis), the question becomes how can one ensure that an ITT analysis can be performed? An analysis of the full dataset generally requires imputation of values or modeling to estimate the missing values. Unfortunately, there is no universally applicable method that adjusts the analysis, and different approaches may lead to different conclusions. Therefore the robustness of trial results should always be investigated by performing sensitivity analyses based on different assumptions.

If patients are excluded from statistical analyses, this will affect the power of the trial, since the sample size is reduced. More importantly it may lead to bias, and the representativeness of the study sample may also be reduced.

In situations where data collection is interrupted before the predetermined final observation (outcome), the most commonly used imputation method is last observation carried forward. Its statistical properties are not optimal, but will in general lead to conservative efficacy estimates, i.e., underestimate rather than overestimate treatment efficacy. However, it is not a recommended procedure if the patient's condition is expected to deteriorate over time.

Baseline observation carried forward is another simple imputation approach that is sometimes used. It may not be unreasonable to believe that a patient has withdrawn from a trial due to lack of effect on a subjectively measured response (like pain). Yet another simple and conservative approach could be to carry the worst observation forward.

The disadvantage of single imputation methods is that they tend to bias the standard error downwards by ignoring uncertainty in the imputed values. In principle, this may lead to underestimation of the precision of the trial and confidence intervals that are too narrow. This risk can be reduced by more sophisticated implementation techniques like multiple imputation. Such methods generate multiple copies of the original dataset by replacing missing values using a stochastic model, then analyzing them as complete sets, and subsequently combining the different parameter estimates from the multiple copies to produce a single estimate and a corresponding confidence interval, taking into account the uncertainty in the process. Multiple imputation methods are available in most statistical software packages (Box 8.9).

8.13.3 Analysis Strategies

Randomized clinical trials are usually analyzed by simple methods that include only the exposure variable (treatment) and the outcome variable. Sometimes additional analyses that adjust for potential imbalance in prognostic factors are performed even if the trial was randomized. Table 8.6 lists the most commonly used methods for statistical analysis of simple trial designs.

8.13.3.1 Factorial Trials

When analyzing a factorial trial, both main effects and interactions between the exposure variables may be estimated. If the outcome variable is continuous, two-way analysis of variance (ANOVA) is usually applied, whereas logistic regression would be the method of choice for a categorical outcome. If interaction between treatments is present, main effects must be interpreted with this in mind. The statistical test for interaction

BOX 8.9 Primary Analysis Set

The primary analysis to demonstrate differences in efficacy is an ITT analysis with imputation or modeling of missing values.[3]

Table 8.6 Commonly used statistical analyses

	One exposure variable			Several exposure variables or covariates
	Crossover trial	**Two parallel groups**	**Several parallel groups**	
Categorical outcome	McNemar's test	Chi-square test Fisher's exact test	Chi-square test Fisher's exact test	Logistic regression
Continuous outcome				
Normal distribution	One-sample *t*-test	Two-sample *t*-test	One-way ANOVA	Two-way ANOVA Multiple linear regression
Skewed distribution	Wilcoxon signed rank test	Wilcoxon-Mann-Whitney test	Kruskal-Wallis test	
Time to event data		Kaplan-Meier estimates and log rank test	Kaplan-Meier estimates and log rank test	Cox proportional hazards model

may have low power, and true interactions may therefore go undetected if a sufficient number of patients have not been included.

8.13.3.2 Repeated Measurements

In medical research patients are often followed over time (longitudinal trial) and repeated measurements of the outcome variable are made. A commonly encountered way of analyzing such data is to perform separate comparisons between groups at each point in time. Multiple *P*-values are presented and the two groups are reported to be statistically significantly different at certain time points but not others. This analysis strategy is not recommended. Such analyses fail to take into account that measurements at different time points from the same subject are correlated. In addition, there is the problem of multiplicity; as previously mentioned, multiple testing leads to an increased risk of false-positive findings. Also, dividing the results into significant and nonsignificant introduces artificial dichotomy in data that are assumed to develop continuously over time.

There are two different strategies that can be used to analyze repeated measurements. The simplest one is to reduce the problem by calculating a summary measure, such as the mean over all time points.[7] The choice of summary measure depends on the main focus of the study. In bioavailability studies, for instance, the area under the plasma concentration curve and/or the maximum plasma concentration (c_{max}) are usually of major interest. If one deals with growth data, separate linear regression lines may be fitted for each subject. The estimated slopes can then be compared between groups by standard tests for independent samples. If a summary measure will not do, more complex analyses, such as ANOVA for repeated measurements or mixed linear models, can be applied. Such analyses take into account the correlation between measures from the same subject.

8.13.4 Multiple Endpoints

In most clinical trials, efficacy is measured in various ways. For example, the outcome of cancer treatment may be expressed as overall survival, disease-free survival, cancer-specific survival, progression-free survival, tumor response according to specific criteria, toxicity, pain, and quality of life.

Analyses of a large number of outcomes or endpoints can be problematic, because separate analyses at the 5% level will increase the probability of one or more false-positive findings. With a significance level of 5%, 1 out of 20 significance tests will be expected to show a statistically significant difference, even when treatment efficacy does not differ in reality. The larger the number of outcome variables, and thus the number of significance tests performed, the larger the probability that one or more of these variables will by chance result in a low *P*-value.

The Bonferroni correction is one way of reducing the probability of false-positive findings. Each *P*-value is multiplied by the number of tests performed, and the conclusion on whether to reject each null hypothesis is made by comparing each of the adjusted *P*-values with the 5% level. Alternatively, one may divide the total significance level by the number of tests performed and compare each *P*-value with the adjusted level. If, for instance, four tests are planned and the total significance level is 5%, each test is performed at the 1.25% level.

BOX 8.10 Multiple *P*-Values

Analysis of several outcome variables increases the risk of false-positive findings. Adjustments for multiplicity reduce power, and the number of outcome variables should thus be restricted.

The Bonferroni correction is simple, but conservative. It tends to overcorrect, particularly when the various endpoints are correlated. Even more importantly, adjustment will lead to reduced power. A better strategy is thus to restrict the number of significance tests or predefine one or two primary endpoints. Whereas it can be tempting to retrospectively select the outcome variables for which the statistical analyses rendered low *P*-values, such a strategy leads to bias and is scientifically unsound (Box 8.10).

Rather than identify a single primary endpoint, sometimes the solution is to combine several endpoints into a composite endpoint. This is often done when each endpoint is a rare event that would lead to the necessity of unrealistically large trials. One example is trials on cardiovascular disease, where different cardiac events are often combined into a composite primary endpoint (e.g., nonfatal myocardial infarction and stroke), and each separate outcome is regarded as a secondary endpoint. Interpretation of the result is not necessarily straightforward, and ideally composite endpoints should be validated before they are used.

8.13.5 Interim Analyses

For most trials, the number of patients to be included is fixed in advance. If the intended number of patients is large, it may be useful to perform an interim analysis during the course of the trial in order to reach a conclusion at an earlier stage, if possible. Indeed, if conclusions can be drawn early on, the trial can be terminated and additional patients need not be included. When interim analyses are performed, the trial is often termed *group sequential.* If the maximum number of patients to be included is not set in advance and data are continuously analyzed (after every patient outcome recorded), the trial design is termed *sequential.*

Many clinical trials last for several years. Indeed, including the necessary number of patients may take a long time and in some cases patients must be followed for several years before treatment efficacy can be

BOX 8.11 Interim Analyses

Interim analyses inflate the risk of false-positive findings and specific methods for adjustment of the significance level must be applied.

Table 8.7 Probability of a false-positive finding

Maximum number of tests	Total significance level (%)
1	5
2	8
3	11
5	14
10	19

evaluated. One example of this is clinical cancer research, which is primarily concerned with studying the effects of treatments on survival or progression-free survival. In some cases, it is considered to be valuable to conduct one or more interim analyses. If a statistically significant difference in the efficacy of trial treatments can be detected early on, it is ethically as well as financially desirable to stop the trial so that the better treatment can be made available to all patients at an earlier point in time.

Interim analyses require the use of specific statistical methods, and should only be made if they are planned in advance and included in the trial protocol (Box 8.11). As an explanation, consider a fixed-sample study with a recruitment target of 100 patients. Assume that it takes longer to include patients than initially thought, and that after data from 50 patients have been registered an interim analysis is performed. If a difference in efficacy can be demonstrated at this point it would be tempting to conclude that one treatment is superior to the other. If, however, no difference is shown, enrollment of patients would continue. Hence there may be two opportunities to identify differences between the treatments during the course of the trial.

If each of the significance tests is performed at the 5% level, the total probability of a false-positive finding (rejecting the null hypothesis even if it is true) will be considerably greater than 5%, since one would have two opportunities to draw an incorrect conclusion. The probability of false-positive findings increases with each interim analysis, as shown in Table 8.7. Assume that in the above example with a 100-patient target, an interim analysis is conducted after data are available on the first 20 patients,

Table 8.8 Nominal significance level in interim analyses with significance level 5%

		Nominal significance level	
Maximum number of tests	**Analysis number**	**Pocock method**	**O'Brien and Fleming method**
1	1	0.05	0.05
2	1	0.029	0.0051
	2	0.029	0.0475
3	1	0.022	0.0006
	2	0.022	0.0151
	3	0.022	0.0472
4	1	0.018	0.00004
	2	0.018	0.0039
	3	0.018	0.0184
	4	0.018	0.0411
5	1	0.016	0.000005
	2	0.016	0.0013
	3	0.016	0.0085
	4	0.016	0.0228
	5	0.016	0.0417

and the trial is stopped if the treatment difference is statistically significant at the 5% level, but 20 new patients are enrolled if no significant difference is found. In the latter case the above scenario would be repeated until data had been acquired from 100 patients, meaning there may be up to five tests and the total probability of detecting a difference, even though none exists, would be 14%, not 5% as initially planned. The numbers in Table 8.7 are based on responses that follow a normal distribution with known variance, but other distributions will produce similar results.

Inflation of the significance level can be prevented by reducing the nominal level for each interim analysis. The maximum number of analyses is then set before the trial begins. Interim analyses are usually performed at set intervals, either by time elapsed or by number of patients enrolled. A simple method to maintain the overall level has been proposed by Pocock.[6] The level for each analysis is reduced to a constant nominal level, which ensures an overall error probability of 5%, as shown in Table 8.8. If a maximum of five analyses are to be performed, each individual interim analysis must be performed at a 1.6% level.

One disadvantage of the Pocock method is loss of power. In practice, other stopping rules, like the one proposed by O'Brien and Fleming,[8] are more often used. These differ from the Pocock method in that the nominal significance level varies between the interim analyses. The level for the first analysis is very low, while that for the last analysis is nearly 5%. The advantage of the O'Brien and Fleming method is that power is retained and there is thus no need to include more patients than would have been the case with a fixed sample trial. Examples of the O'Brien and Fleming method are also listed in Table 8.8.

8.13.6 Subgroup Analyses

A common way of addressing the question of which patients benefit from treatment is to divide the patients into subgroups based on one or more characteristics and perform separate subgroup analyses. Unfortunately, a number of subgroups may be defined and thus many *P*-values calculated, which leads to multiplicity problems, as described previously. This results in a similar increased risk of false-positive findings seen when there are multiple endpoints. On the other hand, as subgroups are often small, relatively large effects may go undetected due to low power in each separate subgroup.

Separate significance tests do not satisfactorily assess whether a given patient characteristic is a predictive marker, i.e., affects treatment efficacy. Instead, a statistical test for interaction is recommended; that is, a significance test that assesses whether there is a difference in treatment efficacy between different subgroups. This can be done by including an interaction term in a statistical model, such as a multiple regression analysis. A possible result of including an interaction term may be that treatment A shows better efficacy than treatment B overall, and that the difference between A and B is larger for patients with a given characteristic than for other patients.

Only if a statistically significant interaction is demonstrated should separate subgroup results be presented. Unfortunately, as with separate subgroup analyses, interaction tests tend to have low power, so their use needs to be considered during the planning stage. Their main advantage is to reduce the problems linked to multiple *P*-values.

8.14 PERSONALIZED MEDICINE

The question posed when assessing the efficacy of a new treatment is not only whether the treatment "works," but also which patients benefit from this treatment. Treatment efficacy often depends on various patient characteristics. This is of particular interest in oncology. Patient groups with a larger or smaller benefit from a treatment have been identified in several cancer types by defining subgroups of patients based on clinical, histopathological, or molecular characteristics. Analyses of new molecular biomarkers are now widely performed to define smaller patient groups with specific mutations who should have a specific, targeted treatment, or even to identify individual gene profiles to tailor specific drugs to each patient. This approach is commonly termed "personalized medicine," and expectations for current and future breakthroughs in this field are very high.

Biomarkers may be either prognostic or predictive. A prognostic biomarker is a factor that is associated with the clinical outcome in the absence of treatment or with the application of an established treatment that patients are likely to receive. A predictive biomarker is a factor that is associated with response or lack of response to a particular treatment. Many clinical trials in oncology are performed in patient groups with defined prognostic or predictive biomarkers.

Example

Cetuximab is an epidermal growth factor receptor inhibitor used to treat metastatic colorectal cancer. Some studies, including the CRYSTAL study,[9] have shown that the efficacy of cetuximab depends on whether there are mutations in the so-called *KRAS* gene, and suggest that the drug has little or no effect in colorectal tumors harboring a *KRAS* mutation. Other trials, such as the Nordic VII trial,[10] failed to demonstrate any such interaction. Several hypotheses have been generated to explain this discrepancy, one being that the interaction may depend on the type of combination chemotherapy administered. This illustrates the complexity of drawing firm conclusions when trying to identify predictive biomarkers.

Several initiatives in oncology try to provide individualized, targeted treatment based on molecular screening for individual gene profiles. A promising response to an unestablished treatment will lead to hypotheses on the efficacy of new treatment principles but will need to be confirmed in larger randomized controlled trials within the identified subgroups of patients. Such studies have now been started by large international research collaborative groups such as EORTC and MRC.

8.15 NONINFERIORITY TRIALS

Drug trials set up for licensing purposes differ from "traditional" superiority trials (clinical trials that seek to document the superiority of one treatment over another) and are referred to as noninferiority trials, as the main objective is to demonstrate that the new active substance is not inferior to established treatment. In noninferiority trials, it is not sufficient to

BOX 8.12 Superiority and Noninferiority Trials

H_0 is the hypothesis that one seeks to reject.

Superiority Trial

H_0: Treatments A and B are equally efficacious.
H_1: Treatments A and B are not equally efficacious.
Here, H_1 is two-sided: A could be superior to B, or B could be superior to A.

Noninferiority Trial

H_0: Treatment A is inferior to treatment B.
H_1: Treatment A is as good as or better than treatment B.
Here, H_1 is one-sided. The only goal is to determine that treatment A is not inferior to the established treatment B.

demonstrate that the *P*-value for a comparison of the two treatments is not statistically significant, i.e., $p > 0.05$. Indeed, although the two treatments in question may in reality be (approximately) equally efficacious, a high *P*-value may also occur due to low power. In that case, concluding that the treatments do not differ constitutes a type II error, i.e., failure to detect a true treatment difference. The smaller the number of patients in a trial, the lower the power, and hence the larger the risk of type II error. In general, a null hypothesis is defined as the opposite of what one wishes to demonstrate, i.e., it is the hypothesis the researcher seeks to reject (Box 8.12). Therefore, in a superiority trial the null hypothesis (denoted H_0) would be that the efficacy of the two treatments does not differ. The alternative hypothesis (denoted H_1) would be that efficacy differs between treatments. No distinction between positive and negative differences is usually made, and the tests in these trials are almost always two-sided. However, when the aim is to demonstrate noninferiority of a new drug, the null hypothesis is that it is *not* as efficacious as the established treatment.

But what does "not inferior to" actually mean? In fact, a noninferiority trial cannot ascertain that the new drug is not inferior to its comparison. Only a superiority trial can demonstrate that. A noninferiority trial aims to demonstrate that the new drug is not worse than its comparator by more than a prespecified small amount, usually denoted delta (Δ). The choice of Δ is not straightforward and must be justified on both clinical and statistical grounds; a fixed general limit cannot be set. Sometimes good historical data can be of help. The choice of Δ must preclude that the new drug is inferior by such an amount that it would be of clinical importance to treated patients, and of course the new drug must actually work. The best way to do this is by performing a three-armed trial that includes a placebo group.

Even though a noninferiority trial actually tests against a one-sided alternative hypothesis, in practice decisions are usually made based on a two-sided 95% confidence interval but deal only with the lower limit of the interval. This is equivalent to performing a one-sided test at the 2.5% level. Thus, after completion of the study a two-sided 95% (or one-sided 97.5%) confidence interval for the true difference between the two treatments is constructed; to conclude noninferiority the interval should lie entirely on the positive side of the noninferiority margin.

Various possible results from noninferiority trials are illustrated in Figure 8.1.

Trial A shows a statistically significant difference in treatment efficacy, since the lower end of the 95% confidence interval of the efficacy difference is greater than 0. Trial B shows noninferiority. The lower end of the confidence interval is less than 0, so there is no significant difference between the two treatments. Moreover, the lower limit of the confidence interval is greater than $-\Delta$, so the largest plausible difference between the two treatments is less than the difference defined as the minimum clinically important difference. Trial C shows no significant difference between the two treatments, but nonetheless does not demonstrate noninferiority, because the confidence interval overlaps $-\Delta$. The difference between the two treatments may thus be larger than is acceptable. Moreover, the point estimate is on the negative side, which implies that the efficacy of the new drug may truly be inferior to that of the

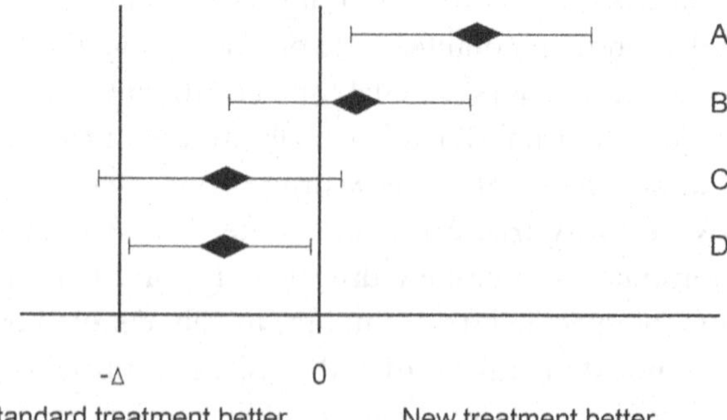

Figure 8.1 Assessment of noninferiority using 95% confidence intervals for the efficacy difference.

established treatment. Trial D shows the same point estimate of the efficacy difference as trial C, but the conclusion is paradoxical. The new treatment is demonstrated to be noninferior because the confidence interval does not include $-\Delta$, but at the same time the new treatment is statistically significantly inferior to the established treatment (upper limit of the confidence interval <0). Should such a result occur in practice, the estimated inferiority must be judged against risks, such as safety problems, before a licensing decision is made. For life-threatening diseases, it is obviously more difficult to accept a treatment that is "almost as good as" established treatment. The difference between the width of the confidence intervals from trials C and D might be that trial D enrolled more patients, and therefore gave a more precise estimate of the efficacy difference, or that the patient sample was more homogeneous.

8.16 SOURCES AND CONTROL OF BIAS

In clinical research, particularly therapeutic trials, the ability to draw valid inferences from data determines whether the results have any practical value. In order to draw valid conclusions from clinical trials, they must clearly demonstrate that the outcome depends on the treatment. Confounding can occur whenever a variable is a common cause for both the exposure (choice of treatment) and the outcome. In order to unambiguously assess the efficacy of an intervention, the potential impact of confounders must be eliminated. This implies that the demonstration of an association between an intervention and an outcome in an observational study is not sufficient. If one or more known or unknown confounders are not included in the statistical model, the result will be biased estimates of the efficacy of an intervention.

The great advantage of randomized controlled clinical trials over epidemiological studies is that randomization and blinding break all possible links between exposure to treatment and external factors that may influence the outcome. It is therefore possible to draw conclusions on direct causal relationships between an intervention and an outcome, and this is why controlled clinical trials are often said to have high internal validity.

Selection bias is a well-recognized threat to internal validity. If study subjects are allocated to treatment groups based on the investigator's choice rather than by randomization, the result will almost always be biased in favor of the new treatment.

In an interventional study the expectations of the investigator may (unintentionally) influence patient outcome. Unless both the treating physician and the patient are blinded to the treatment, bias can occur due either to the expectations of the experimenter or those of the patient, which are likely important "mechanisms" of the placebo effect (see also Section 4.9).

Pre- and post-intervention comparison within one patient sample is also a well-known source of bias. The mere knowledge that one is included in a trial may lead both to a placebo-like effect and true changes in behavior that affect the outcome. Regression to the mean is the tendency for subjects with especially high scores (or low scores) to score closer to the population mean upon subsequent testing. Extreme values may arise by chance, and are less likely to be repeated. Selecting patients based on high blood pressure, for instance, will inevitably yield the result that the mean blood pressure of the whole group is reduced at the subsequent visit, independent of treatment.

Loss to follow-up can lead to what is called attrition bias. Even the comparison of two independent samples from a randomized trial can be biased if loss to follow-up is due to a characteristic of the intervention that is not related to its mechanism of action. This is the main reason that primary efficacy analyses are performed according to the ITT principle rather than restricted to the sample of completers only.

8.17 GENERALIZATION OF TRIAL RESULTS

The main drawback of controlled clinical trials is that they may have low external validity, which implies that the results of a study cannot necessarily be generalized. Clinical trials often employ strict criteria for including patients, both with regard to age, comorbidities, and intake of drugs other than the trial medication. Hence the study sample actually comprises a very select part of the target population, and it is not clear whether efficacy estimates are representative of the patient population that will be treated once the trial is finalized. For example, it has been shown that the characteristics of patients with metastatic colorectal cancer enrolled in clinical trials differ from those of patients receiving chemotherapy outside the clinical trial protocol and have better survival than patients outside the clinical trial protocol, even when given the same treatment.[11] Various levels of patient selection are illustrated in Figure 8.2.

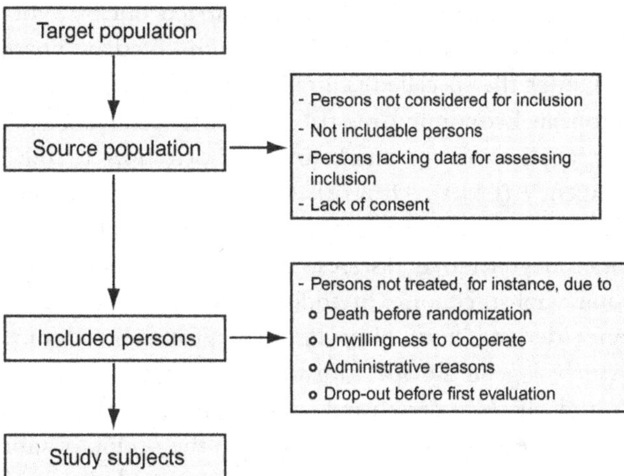

Figure 8.2 Levels of selection in a clinical trial.

Whenever only a small percentage of relevant patients are included, or if a large percentage of patients that fulfill the requirements for inclusion or exclusion are not enrolled in the trial, there may be cause to question the generalizability of the trial results. Some patients do not want to be randomized, or the treating physician may decide that a certain patient be given a specific treatment. All such exclusions can lead to selection bias and consequently weaken the generalizability of the results of a clinical trial.

8.18 REGULATORY ISSUES

By law, a drug can only be marketed once it has received marketing authorization. The scientific evaluation this authorization is based on includes documentation of pharmaceutical quality, preclinical studies in animal models, and safety and efficacy estimates from clinical trials in humans. Upon approval, companies must commit to continue evaluating safety in general use through pharmacovigilance procedures and risk management plans.

Article 26 under the European Directive 2001/82/EC says that "marketing authorization should be refused if benefit risk balance is not considered to be favorable, or therapeutic efficacy not sufficiently substantiated, or qualitative and quantitative composition not as declared." In other words, approval is based on quality, efficacy, and safety only. There is no need clause and no requirement that the benefit-risk balance is not inferior to products already on the market.

The scientific evaluation of new drugs carried out by national medicines agencies is based on detailed dossiers submitted by pharmaceutical companies. Under the so-called centralized procedure in the countries of the EU/European Economic Area (EEA), pharmaceutical companies submit a single application for marketing authorization to the European Medicines Agency (EMA). Drugs that fall under this centralized procedure include human medicines for the treatment of HIV/AIDS, cancer, diabetes, neurodegenerative diseases, auto-immune and other immune disorders, and viral infections. In addition, the procedure is compulsory for medicines derived from biotechnology processes, advanced-therapy medicines (such as gene therapy, somatic cell therapy, or tissue-engineered medicines), and officially designated "orphan medicines" (medicines used for rare human diseases). Companies also have the option of submitting an application for a centralized marketing authorization that is valid throughout the EU and EEA countries, if the product concerned is a significant therapeutic, scientific, or technical innovation, or if its authorization would be in the interest of public health.

In the United States all regulatory work and scientific evaluation of medicinal products is carried out centrally by the Food and Drug Administration (FDA). Regulatory work in the EU/EEA countries is organized differently. National agencies play an important role in the scientific evaluation of new drugs and form a network of national medicines agencies represented in various scientific committees that advise the European Commission on approval of new active substances or potential withdrawal of marketed products due to safety concerns.

Typically, regardless of procedure or agency, the clinical evaluation and discussion on whether or not to authorize a new drug focuses on the following problem areas: insufficient documentation of dose, choice of endpoint, choice of comparator, sample size and duration of trials, inconsistent results from different trials, clinical relevance of estimated efficacy, patient population included in Phase III trials (inclusion and exclusion criteria, as well as handling of loss to follow-up), and potential restriction of indication.

Historically, pharmaceutical regulations varied between countries, which was considered a hindrance to efficient drug development. In 1989 the International Conference of Drug Regulatory Agencies started to plan the harmonization of regulations in Europe, Japan, and the United States, and the steering committee of the International Conference on Harmonization of Technical Requirements for Registration of

Pharmaceuticals for Human Use was established. Since then a number of harmonized guidelines have been developed (http://www.ich.org). Some of them have been referred to previously in this chapter.[2–5] The documents are categorized into four groups: Quality, Safety, Efficacy, and Multidisciplinary. Any pharmaceutical company that applies for a marketing authorization needs to follow this harmonized guidance.

In addition to the harmonized guidelines, the FDA in the United States, the EMA, and the Pharmaceuticals and Medicals Devices Agency in Japan issue a large number of "local" guidelines and discussion papers on less general topics. These include, for instance, statistical guidance, as well as disease-specific guidance on choice of endpoints, duration of trials, and recommended trial design.

QUESTIONS TO DISCUSS

1. What are the main advantages of randomization?
2. Discuss the advantages and disadvantages of stratified randomization.
3. Explain why the number of patients included in a trial is of major importance.
4. Discuss the limitations of crossover trials.
5. List the different phases of drug development and explain the main purpose of each phase.
6. Discuss the differences between superiority trials and noninferiority trials. When is each type of trial applicable?
7. Why is blinding important? What is the difference between a single-blind and a double-blind trial?
8. Discuss the problems related to reporting a large number of *P*-values. How might that affect conclusions?
9. What is the difference between a prognostic and a predictive factor/biomarker? For what purpose is the identification of each of these factors important?
10. What are the main challenges when it comes to demonstrating efficacy in small patient populations, for example, in "personalized medicine"?

ACKNOWLEDGMENTS

We thank Carina V. S. Knudsen, Institute of Basic Medical Sciences, University of Oslo, for producing the figures.

REFERENCES

1. De Angelis C. *An introduction to clinical research.* New York, NY: Oxford University Press; 1990.
2. Note for guidance on good clinical trial practice (CPMP/ICH/135/35). <http://www.ema.europa.eu/docs/en_GB/document_library/Scientific_guideline/2009/09/WC500002874.pdf>.
3. Note for guidance on statistical principles for clinical trials (CPMP/ICH/363/96). <http://www.ema.europa.eu/docs/en_GB/document_library/Scientific_guideline/2009/09/WC500002928.pdf>.
4. Note for guidance on general considerations for clinical trials (CPMP/ICH/291/95). <http://www.ema.europa.eu/docs/en_GB/document_library/Scientific_guideline/2009/09/WC500002877.pdf>.
5. Note for guidance on choice of control group in clinical trials (CPMP/ICH/364/96). <http://www.ema.europa.eu/docs/en_GB/document_library/Scientific_guideline/2009/09/WC500002925.pdf>.
6. Pocock SJ. *Clinical trials. A practical approach.* New York, NY: Wiley; 1983.
7. Matthews JNS, Altman DG, Campbell MJ, Royston P. Analysis of serial measurements in medical research. *BMJ* 1990;**300**:230–5.
8. O'Brien PC, Fleming TR. A multiple testing procedure for clinical trials. *Biometrics* 1979;**35**:549–56.
9. Van Cutsem E, Kohne CH, Lang I, Folprecht G, Nowacki MP, Cascinu S, et al. Cetuximab plus irinotecan, fluorouracil, and leucovorin as first-line treatment for metastatic colorectal cancer: updated analysis of overall survival according to tumor KRAS and BRAF mutation status. *J Clin Oncol* 2011;**29**:2011–19.
10. Tveit KM, Guren T, Glimelius B, Pfeiffer P, Sorbye H, Pyrhonen S, et al. Phase III trial of cetuximab with continuous or intermittent fluorouracil, leucovorin, and oxaliplatin (Nordic FLOX) versus FLOX alone in first-line treatment of metastatic colorectal cancer: the NORDIC-VII study. *J Clin Oncol* 2012;**30**:1755–62.
11. Sørbye H, Pfeiffer P, Cavalli-Björkman N, Qvortrup C, Holsen MH, Wentzel-Larsen T, et al. Clinical enrollment, patient characteristics, and survival differences in prospectively registered metastatic colorectal cancer patients. *Cancer* 2009;4679–87.

CHAPTER 9

Epidemiology

Dag S. Thelle and Petter Laake
Oslo Centre for Biostatistics and Epidemiology, Department of Biostatistics, Institute of Basic Medical Sciences, University of Oslo, Blindern, Oslo, Norway

9.1 INTRODUCTION

There have been dramatic changes in mortality rates in Norway between the 1950s and 2010. These changes are reflected in the increased life expectancy and the demographic distribution of the population in Norway. Indeed, life expectancy at any age has never been higher in Norway than at the end of this observation period.

The mortality curve for cardiovascular disease (CVD) is particularly impressive, showing a rapid increase until 1970, followed by a consistent decline. Cancer deaths and deaths due to accident or violence showed smaller changes, whereas death from other diseases decreased (Figure 9.1). These patterns differ slightly by age and sex, which only underlines the notion that both of these variables are determinant for mortality rates. The role of an epidemiologist is to ask why these changes have occurred. Is the decline in CVD deaths due to a reduced risk, improved treatment, or a change in the fatality of CVD? Why do men and women differ with regard to these changes? And if there is a reduced risk of CVD, what caused this change?

When it comes to disease occurrence, there are two principal questions. For a single patient, one may ask why this individual, and not a neighbor or a friend, developed a disease at a particular time. For populations that differ in their risk of disease, one may ask why there is more disease in one population than another. The answers to these questions may differ depending on whether they concern individuals or populations. An individual may develop a disease at a certain point in time for any number of genetic, lifestyle, or environmental reasons, whereas the differences between populations are more often determined by external factors, such as social or economic inequality. Of course, genetics may also play a role when it comes to population differences, but the principal causes of such differences have to do with living conditions and other external factors.

Research in Medical and Biological Sciences
DOI: http://dx.doi.org/10.1016/B978-0-12-799943-2.00009-4

9.2 THE DEFINITION OF EPIDEMIOLOGY

There are many definitions of epidemiology. Among the more common are as follows:

- Epidemiology is a scientific discipline concerning disease occurrence in the population.
- Epidemiology is a scientific discipline concerning the occurrence and variation of disease in the population regarding factors that determine this variation, and how diseases can be best treated and controlled.

The word epidemic is a composite of *epi* and *demos*, which mean "on or upon" or "over and above" and "population or people." An epidemic is defined as an increase in the occurrence of a disease compared to its normal rate of occurrence in a particular population. There is no set rule as to how many cases of disease must occur before a situation can be

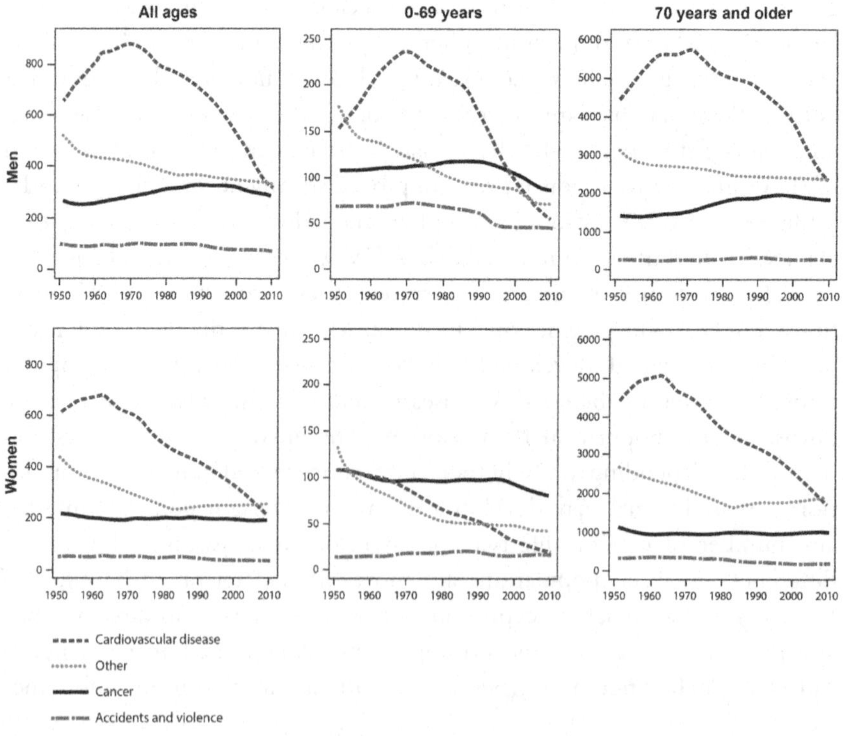

Figure 9.1 Mortality in Norway in 1951–2010 by cause of death.

declared an epidemic. In the past, the word epidemic was applied mostly to contagious diseases, but today any increase in disease relative to a comparable population may be deemed an epidemic.

9.3 THE ROLE OF EPIDEMIOLOGY

Epidemiology describes variations in morbidity and mortality in time and space, and attempts to identify factors, variables, or exposures that explain this variation, including genetic or environmental factors. Epidemiologists also perform risk assessment, for example estimating the risk of leukemia among those living near a nuclear power plant or the risk of cancer among people with radon in their basements. The field of epidemiology consists of both a descriptive and an analytical domain, and is aimed at etiological research, but it is also closely associated with basic biological science, clinical medicine, statistics, and social science. So epidemiology is a quantitative medical discipline that bases its methods on logic and statistics. The concepts of epidemiology are useful in addressing a broad array of problems, and the methods applied are fundamental in assessing knowledge that is essential to clinical medicine and public health. Epidemiologists are less concerned with pathogenesis; their main interest is the etiology of disease, variables, or exposures that either trigger disease or predispose an individual to disease. Etiological research implies that one can determine both the direction and the strength of the association between exposures that explain disease and the disease itself.

Epidemiological studies may be regarded as having three aims, as listed in Box 9.1.

9.3.1 Study Designs

The choice of study design depends upon the research problem, such as how common a disease may be in a population, and what resources are

BOX 9.1 The Three Aims of Epidemiological Studies

1. Describing disease occurrence and distribution in a population, as well as the development over time (trend)
2. Identifying the cause of disease (etiology)
3. Undertaking experiments to assess the effect of treatments or preventive efforts on disease

Table 9.1 Study designs

Study design	**Observation units**
Experimental studies	
Randomized controlled trials	Individuals
Intervention trials	Individuals
Interventions in populations or communities	Groups
Observational studies	
Ecological studies (correlation studies)	Groups
Case-control studies	Individuals
Cross-sectional studies	Individuals
Cohort studies	Individuals

available for the study. The most common study designs and their observation units are listed in Table 9.1 and described further in Section 9.7.

Studies can be classified as experimental or observational studies. The observations themselves may be of either subjects or groups of subjects. Experimental studies are common in basic medical research and clinical research, but may also be used in epidemiology. Observational studies often have an administrative aim, such as describing disease occurrence in a population as precisely as possible, or they may be aimed at identifying etiological associations. Observational studies are so common in epidemiology that the term epidemiological study is used synonymously with observational study, as it will be in this chapter.

9.4 TARGET POPULATION, SOURCE POPULATION, AND STUDY SAMPLE

For epidemiological purposes, some concepts of population need to be defined. The target population is the population to which conclusions are to be generalized. The source population (hereafter referred to as *population*) is defined as a subset of the target population, which is obtained by applying certain selection criteria and will in most cases differ from the target population. The study sample (hereafter referred to as *sample*) is drawn from the source population.

A population may be considered as groups of observation units, all of which have measurable or quantifiable characteristics. Observation units may be an individual human subject, groups of subjects, families, or any other unit that is defined and described. Statistics provide the procedures necessary to draw conclusions and make inferences about the population based on the sample. See also Chapters 4 and 11 for the general principles of making valid statistical inference.

The characteristics to be measured and recorded in a sample are called variables. The variables that are measured determine the subsequent conclusions that can be drawn. A summary of the observations from a sample is a descriptive statistic. Examples of descriptive statistics are mean value, which is the sample analogue of the unknown population mean, and sample proportion, which is an estimate of the population proportion. Indeed, disease prevalence and disease risk, which are both important concepts in epidemiology, are actually proportions, as will be demonstrated later.

In general, effect estimates are calculated based on the sample. It is the sample analogue of the effect measure in the population that describes the population parameter of interest. For more about effect measures and effect estimates, please see Chapters 4 and 11.

It is intuitively obvious that the larger the sample, the more accurately a sample mean or proportion reflects the true values in the population. The magnitude of the sample and the accuracy of the measuring instrument (be it a technical apparatus or a questionnaire) determine the confidence with which conclusions can be drawn about the phenomenon under study. The principles for assessing sample size are discussed in Chapters 4 and 8, and are the same for epidemiological and clinical studies.

Numerous samples from the same population provide a number of mean values ($\overline{X}$), which are likely to differ from sample to sample. It is obvious that all these means represent the mean for the population, but none of them are fully correct. The variation of the mean is called the standard error of the mean (SEM), which is equal to the standard deviation of the sample (s) divided by the square root of the number of observations (n), i.e., $\text{SEM} = s/\sqrt{n}$. Similarly, the standard error of a sample estimate ($\hat{p}$) of the population proportion is given as $\sqrt{\hat{p}(1-\hat{p})/n}$.

A confidence interval gives an upper and lower limit of the population values, calculated from the sample estimates. A 95% confidence interval will contain the population value with a probability of 95%. The estimated 95% confidence interval around an unknown mean value is $(\overline{X} - 1.96\text{SEM}, \overline{X} + 1.96\text{SEM})$, assuming that the measurements follow a normal distribution (see Section 11.6). Similarly the 95% confidence interval for a population proportion is $\left[\hat{p} - 1.96\sqrt{\hat{p}(1-\hat{p})/n}, \hat{p} + 1.96\sqrt{\hat{p}(1-\hat{p})/n}\right]$ (see Section 11.4).

9.5 DISEASE OCCURRENCE, RISK, ASSOCIATION, IMPORTANCE, AND IMPLICATION

Epidemiological measures can be classified into three categories:

1. Measures of disease occurrence and risk
2. Effect measures for the association between a disease and an exposure
3. Measures of importance or implication

Measures of disease occurrence and risk are used to describe causal relationships and in descriptive analyses of the evolution of disease occurrence or mortality over time.

The concept of disease risk is often used to describe the association between an exposure and the probability of having a given disease now or developing that disease later in life. To determine the risk associated with a given exposure, the population studied must be divided into two or more groups, for example an exposed group and an unexposed group, or by level of exposure. The importance of the exposure can be assessed when the number of exposed and unexposed subjects that either have a disease or developed a disease during a defined time period has been determined. With this information, the risk associated with the exposure can be estimated. The association between the occurrence of a disease and a given exposure can be summarized in a contingency table such as Table 9.2, which displays contingencies between the data entered therein.

Table 9.2 is a 2×2 contingency table because it displays two rows and two columns of data. This table has the same form as Table 11.1, but with a, b, c, and d as the four outcomes. The cell counts enable us to estimate the effect measures of interest, provided that a suitable and valid study design has been used. The number of subjects with the disease under study in the exposed group (a) and the unexposed group (c) provides information about disease occurrence according to exposure. These numbers provide the basis for assessing disease risk. The ratio of risk of disease among the

Table 9.2 Cell counts for the association between disease and exposure

	Disease		
Exposure	**Yes**	**No**	**Total**
Yes	a	b	$a+b$
No	c	d	$c+d$
Total	$a+c$	$b+d$	n

exposed to the unexposed groups is a measure of the association, which is the measure of interest. The direction of the association; that is, whether the exposure increases or reduces the risk of disease, is indicated by the ratio, i.e., a ratio with a value greater than 1 shows an increased risk, whereas a ratio with a value less than 1 shows a decreased risk. Thus the ratio of the risks, or the relative risk (RR), as it is usually called, is an effect measure that expresses the association between a disease and an exposure.

The effect measure yields an estimate of the strength of the association between two variables, for example, between a possible causal factor and a disease. Effect measures are based on comparisons of disease occurrence in groups with various levels of exposure to a certain variable. The effect measure may be expressed as the risk or the probability of disease occurrence during a defined time period. One example is a study investigating the association between breast cancer and the use of oral contraceptives. The observations are used to estimate the risk of breast cancer in the groups exposed and unexposed to oral contraceptives. In this case, the effect measure is relative risk, which expresses the strength of the association between a disease and an exposure. An alternative effect measure is odds ratio (OR), which will be covered later in this section.

Measures of importance and implication express the impact of a certain disease or exposure on a population. These measures are of public health interest, meaning that they also help to assess the possible effect of preventive efforts. In the example of breast cancer above, the number of breast cancer cases that may actually be attributed to the use of oral contraceptives would be of interest.

Epidemiological effect measures are listed in Box 9.2 and discussed further in Section 9.7.

BOX 9.2 Examples of Different Effect Measures in Epidemiological Studies

Measures of occurrence and risk	Effect measures for association	Measures of importance or implication
Incidence	Relative risk (RR)	Excess risk
Prevalence	Odds ratio (OR)	Attributable risk (AR)
	Incidence rate ratio (IRR)	Population attributable risk (PAR)
	Correlation	

9.5.1 Denominators, Numerators, and Time

Epidemiology is about denominators, numerators, and time. Risk assessment is based on ratios. A central concept is the assessment of prevalence or proportions and rates, all of which are ratios. A prevalence or proportion is the ratio of the number of subjects with disease in the numerator to the number of subjects at risk in the denominator. A proportion is dimensionless and ranges from 0 to 1. Consequently, it is often expressed as a percentage up to 100%.

A rate expresses a quantitative change in the occurrence of an event with the change of another unit, often time. An example is the speed of a car at an instant in time expressed as the distance covered divided by the time elapsed (km/*t*). A rate always has a derived dimension, and thereby may assume a wide range of values. For the purposes of this chapter, constant or average rates are more practical and will be used. A rate is the ratio of the number of new cases (in the numerator) and the sum of the follow-up times for each subject (total follow-up time) in person-time (in the denominator). For example, annual mortality rate is expressed as the ratio of the number of people who die in 1 year in the numerator, and the total follow-up time, which in this case is the sum of the time all the subjects in the population were alive during that year, in the denominator. For those who die, person-time is calculated from the beginning of the year until date of death. Often the average population or mid-year population is used to estimate the number of person-years in the denominator. The error induced by this is minimal. In order to obtain valid results when comparing populations, the same disease endpoint or cause of death must be used, and age and sex must be taken into account.

For a given exposure, the ratio of the risk of disease in the exposed and unexposed groups renders a measure of the association between an exposure and the disease under study. It can be estimated as the ratio between proportions or rates. For example, when considering the risk of dying from myocardial infarction during a 10-year period for smokers compared to nonsmokers, the ratio between the risks in the two groups is a measure of the relative risk of disease.

The aim of epidemiological studies is often to assess the importance of a particular exposure. However, details on the diagnoses or endpoints must be independent of knowledge of the exposure. This independence between exposure assessment and endpoint ascertainment is necessary to draw valid conclusions concerning possible associations between exposures and disease.

9.5.2 Measures of Disease Occurrence and Risk (Incidence and Prevalence)

There are a number of measures of disease and disease occurrence in the population, but there are three estimates of principal interest:

1. The number of people who have a disease
2. The size of the population
3. The duration of the disease

Incidence rates are used to explore disease variations between populations. Incidence is defined as the number of individuals newly diagnosed with disease in a defined time period, and incidence rate is the incidence divided by the length of this time period.

Each year in Norway, about 100 people become infected with HIV. The incidence rate for HIV infection can be expressed as the number of:

- cases per year
- cases per person-year
- cases per person-day

The numerator of the incidence rate for HIV is the total number of new cases. The total follow-up time for the whole population is the denominator. The dimension is cases per year or per person-day. This scenario assumes that the population remains constant over the time period, save for the few with the disease under study. For all practical purposes, the size of most examined populations will be reasonably stable, and therefore the average population can be used as a denominator.

Incidence rates for disease or death can also be estimated from cohort studies. However, follow-up time in these studies may be difficult to estimate, as the time interval from the beginning of the study until death, emigration, or other endpoint is often not known. This difficulty can be overcome by making the time interval a censored variable, meaning a variable that is observed only under the specific condition of the attained endpoint. As with all rates, the subjects included from beginning to end have the whole study period as their follow-up time.

Incidence assessment assumes that the number of people who developed disease in a population is known. Unfortunately this sort of information is often lacking, and all-cause mortality, or cause-specific mortality (i.e., the number of people who die from a particular disease), may be the only information available. In this case the incidence rate cannot be estimated, and mortality rates should be calculated instead.

Prevalence is defined as the proportion of a population with a given disease at a set point in time. Duration of disease together with the incidence rate will determine the prevalence. Short duration implies decreasing prevalence. Increasing prevalence is seen when the duration of a disease lengthens even when incidence rates are constant. Disease prevalence declines for two obvious reasons: individuals die or they are cured. There are two measures of prevalence: point prevalence, which is the probability an individual will have a given disease at a set point in time (t), and period prevalence, which is an expression of a probability that an individual has been affected by a given disease during a defined time period. This is often used as a substitute for risk when the time at diagnosis of disease is not known. Any measure that reduces the fatality of the disease, but does not allow for complete recuperation, will imply that the prevalence of this disease is increasing, in which case increased pressure on health services is unavoidable. Correspondingly, if fatality increases, the prevalence of disease will decrease.

9.5.3 The Importance of Vital Statistics in an Epidemiological Setting

Registers of vital statistics comprise population-based data on birth, death, cause of death, marriage, divorce, area of residence, and emigration, which are collected and recorded systematically. These data are originally collected for administrative purposes; seldom with epidemiological aspects in mind. Today it is reasonable to include census data, disease and accident registers, and other types of registers when seeking vital statistics (Box 9.3).

Data on vital statistics may be combined with other information about a population to provide the basis for ecological studies. Ecological studies involve comparisons of aggregated data and are not based on individualized information. Therefore, ecological studies can be used to estimate

BOX 9.3 Main Rationales for the Use of Vital Statistics

1. Description of population mortality trends
2. Comparison of age- and sex-specific mortality rates during a defined time period
3. Description and comparison of cause-, age-, and sex-specific mortality rates for different populations

disease occurrence and assess it against aggregated data of known exposure, which may provide insights and enable the development of new hypotheses. Ecological studies can be included in the category of observational studies, but the term is also applied to other types of studies. Ecological studies often are based on vital statistics, which are readily and inexpensively available. An ecological study is almost always conducted before starting specific studies to test a research hypothesis. Epidemiologists often ask whether more than one factor will change an individual's risk of disease. Vital statistics contain information on cause of death and consequently contain some information on disease. Therefore a vital statistics register can be used to track endpoints in an epidemiological study. Linking vital statistics data to data on lifestyle habits, environment and social condition, and possibly genetic properties, forms the basis for the follow-up of cohort studies. This may be the most important use of vital statistics in present-day epidemiological research.

9.5.4 Mortality Rates

We can foresee three categories of mortality statistics:

A. Death due to disease X

B. Death due to other reasons but with disease X

C. Death due to other reasons without disease X

Total mortality or all-cause mortality in a population during a defined time period is the sum of A, B, and C. Mortality can be described in two ways, either as mortality rate or as a mortality risk (the probability of dying during a defined time period). Mortality rates can be looked upon as incidence rates for death. The total number of deaths from all causes is usually expressed as the number of deaths per 1000 people per year. Mortality can be expressed in a cause-specific manner, or can be broken down by age group, sex, and occupation, depending on the associations under investigation.

The expression of mortality will depend on the focus of the study, i.e., mortality under condition A, B, or C. The mortality rate associated with disease X can refer both to those who die as a direct result of disease X (category A), or those who die from other causes but still have disease X (category B). It is often impossible to distinguish between categories A and B in a sample. Another measure of interest is disease fatality rate, the mortality rate in a group of individuals that designates how many die from a certain disease among those who have that disease during a defined time period. Other measures of health status are infant mortality and perinatal mortality, which are often used as indicators of living conditions, as well as quality of the health-care system and mother and child health in a country.

Infant mortality is the number of deaths among children under the age of 1 year per 1000 live births during the same time period. Perinatal mortality is the number of stillborn and deaths within 1 week of birth per 1000 births in the same time period.

The first year of life usually is divided into different periods:

- Under 24 h
- Under 1 week
- Under 4 weeks (neonatal deaths)
- 4 weeks to less than 1 year (post-neonatal deaths)

9.5.5 Cause of Death Statistics

Data entered in death certificates serve as the basis for cause of death statistics. For example, Norwegian death certificates have two parts for the cause of death: Part I for the direct cause or the underlying cause, and Part II for other conditions that contributed to the death, but are not the direct cause. In Part I, all underlying causes of death should be entered, as the intent is to identify the cause that started the disease process. Death certificates are handled in different ways in different countries. However, the classification of diagnoses has been internationally standardized. Present-day classification systems originated in the 1850s, and the first edition—the International List of Causes of Death—was adopted by the International Statistical Institute in 1893. The current International Classification of Diseases (ICD) was adopted by the World Health Organization (WHO) in 1948. The classification system has been revised a number of times and the current system is the ICD Revision 10 (ICD-10).[1]

The number of revisions reflects advances in medicine, and consequently in the classification of diseases and causes of death with regard to etiology, pathogenesis, and morphology. How much of the risk of dying from myocardial infarction should be attributed to the fact that a person had diabetes for many decades? It may be reasonable to record diabetes in a death certificate as a disease leading to myocardial infarction, given the considerable risk conferred by diabetes. However, as the mechanism of this increased risk is not completely clear, in Norway diabetes is recorded in Part II of the death certificate. For a more detailed description of the problems related to the registration of cause of death, coding, and classification, see the ICD-10.[1]

9.5.6 Morbidity Register

A morbidity register includes more than just reports of new cases. It includes permanent documents for all submitted individuals, all of whom are followed up in statistical tabulations of occurrence, frequency, and survival. Various morbidity registers are maintained around the world. The most common are in-patient registers, cancer registers, birth registers, and registers of infectious diseases, some of which focus on chronic diseases such as tuberculosis. A full description of these registers is beyond the scope of this chapter, but is available on the WHO website at www.who.int/classifications/icd/en.[1]

9.5.7 Morbidity Registers or Mortality Registers?

Mortality is a good indicator of disease occurrence provided fatality is high. However, this is not always the case; fatality may change over time and may well vary between populations. Morbidity registers are the obvious choice if the occurrence of different diseases in a population is to be followed. Permanent morbidity registers are large and costly, but their value usually becomes apparent with time, after trends have become evident and can be analyzed. The standard of the registration and the validity of the information collected must be consistent, and consequently such a register must be impeccably managed at all times. This is one of the reasons why there are relatively few permanent morbidity registers that cover the global population. The establishment of a morbidity register to support studies of incidence may not be a sufficient rationale. One can, on the other hand, see the establishment of such registers as useful for particular populations as part of follow-up after surveys or screenings.

9.6 EPIDEMIOLOGICAL STUDY DESIGNS

The possibility of an association between an exposure and a disease usually comes from clinical observations or case studies where a number of patients with the same disease seem to have been exposed to the same variables. Case histories are detailed descriptions of a patients' disease histories, either individual patients or small groups of patients. As case histories often concern rare diseases, they offer insights into the occurrence, prognosis, and treatment of these diseases. However, case histories give only limited insight with regard to etiology. Of course, etiological conclusions can be drawn in certain obvious situations, such as work accidents or other similar events, but for other situations, the causal relationship is far less clear. A sample of healthy controls is needed to assess whether affected people differ from healthy people with regard to more than the disease under study. In principle, there are three main study designs that aim to describe associations. Box 9.4 combines Table 9.1 and Box 9.2, and gives a simplified overview of the different types of studies and the corresponding epidemiological methods. Note that different studies render different effect measures. One way to describe these studies is to say that the investigator is in the present time, studying events that either may have occurred in the past (case-control studies, also called retrospective studies), or are oriented towards the present (cross-sectional studies) or are looking into the future (cohort studies, also called longitudinal or prospective studies). This has given rise to a number of conflicting descriptions and classifications of epidemiological study designs. Hence, there may be cohort studies that are retrospective in the sense that their data have been collected in the past, or there may be case-control studies in which the data of both cases and controls have been collected with a future perspective in mind.

BOX 9.4 Principal Differences Between Case-Control Studies, Cross-Sectional Studies, and Cohort Studies

Category	Case-control	Cross-sectional	Cohort
Sample	Start with ill subjects (cases) and healthy subjects (controls)	All subjects are included	Start with healthy subjects
Measures of occurrence			
Incidence	No	No	Yes
Prevalence	No	Yes	No
Measures of association			
	OR	RR OR	RR OR IRR

9.6.1 Case-Control Studies

9.6.1.1 Design and Structure

The classic approach to identify the cause of disease has been to compare ill subjects (cases) with corresponding healthy subjects (controls) in a case-control study. In these studies, researchers determine how cases and controls differ in regard to a number of possible etiological factors. In order to give valuable results, controls must be sampled among healthy subjects who would have been included as cases if they had the disease under study; ideally controls should belong to the same population as the cases. Case-control studies are often called retrospective, as they attempt to elicit the history of cases and controls, especially that of cases before they were diagnosed with disease. However, as some case-control studies look forward in time, the more generic term case-control studies will be used in the present chapter.

9.6.1.2 Advantages

The main advantage of case-control studies is that they are relatively simple to carry out. Moreover, they are inexpensive, reasonably rapid, and suited to rare diseases. They allow investigators to examine a large number of possible exposures. They can also be organized as multi-center studies, whereby a number of research groups can collaborate and include their respective cases and controls.

9.6.1.3 Disadvantages

The major disadvantage of case-control studies is the inverse direction of the exposure data, as it is collected from cases who were recruited because of their disease and are aware of their disease. Indeed, the fact that someone has been diagnosed with a disease may affect their responses to questionnaires, as well as biological values (biomarkers). In addition, the researcher will be well aware of the problem or the research question, therefore knowing which subjects are cases and which are controls. Thus, the researcher may be more thorough and assess the cases differently from the controls.

There are a myriad of possible sources of systematic error in case-control studies that can affect the study outcome. Systematic error, errors that affect the results in a certain direction, are called bias. Thus a major problem with regard to interpretation of study results from retrospective or case-control studies is the inverse time aspect.

Examples of case-control studies that may induce bias include the relationship between stress and myocardial infarction, or the use of drugs during pregnancy and the risk of congenital malformations. If a subject who has recently survived myocardial infarction is being asked about stressful events that occurred beforehand, he or she may give completely different answers than a subject who had a skiing accident. Likewise, mothers of babies with congenital diseases will scrutinize their past with regard to the use of drugs during pregnancy, whereas mothers with healthy babies will be less prone to do so. This illustrates that cases may tend to search for the causes of their affliction, and may give responses that are colored by these situations, whereas controls will not. In these cases systematic error, or recall bias, can influence the results. To avoid these types of systematic errors, one must establish examination procedures that reduce the possibility that clinical disease will affect the answers. One of the practical problems in case-control studies is the selection of controls. Three options in locating controls are direct contact (by mail or phone) with:

1. Neighbors in the same residential area as the case, or have cases themselves contact friends or relatives in the same age group
2. A random sample of the population based on census registers, including background factors that also are available for the cases
3. Hospital-based controls, i.e., patients with other diagnoses from the same hospital as the cases, also known as hospital-based case-control studies

It is relatively easy to choose hospital-based controls from one's own institution. However, this can raise uncertainty as to whether the controls are representative of the population that cases are recruited from, and whether the exposures examined are related to diseases for which the controls are hospitalized. Assume that a study aims to examine whether cigarette smoking is associated with an increased risk of bladder cancer, and controls are selected among patients in the lung department of a hospital. In this case the smoking habits of the controls likely deviate from those of the source population, and therefore the results of the study would be invalid.

9.6.2 Matched Case-Control Studies

One of the concerns when planning a study is the influence of factors that may contribute to an increased risk of disease, but which are otherwise of no particular interest in the study. One approach to avoid this is matching, by which cases and controls are selected to be as similar as possible with regard to factors such as age, sex, area of residence, and socioeconomic status. However, the drawback of matching is that one may control for factors related to the variables being assessed, thereby reducing the possibility of demonstrating associations. One example is a study of the importance of cigarette smoking on myocardial infarction. Assume that cases and controls are matched with regard to socioeconomic status. This may reduce the possibility of showing the association between smoking and myocardial infarction, as smoking is associated with socioeconomic status.

9.6.3 Cross-Sectional Studies

9.6.3.1 Design and Structure

Cross-sectional studies collect all data at a set point in time. This allows the researcher to assess the prevalence of a disease or of other variables of interest.

9.6.3.2 Advantages

Cross-sectional studies are intuitively clear and allow for the examination of a large number of variables. The methods can be standardized by the researcher, and clear-cut definitions can be applied to the exposure and the endpoints.

9.6.3.3 Disadvantages

By definition, cross-sectional studies have no dimension of time, so they cannot support conclusions on the risk of disease, nor on causal relationships. One way to circumvent this drawback is to formulate questions that assess the subject's past, such as questions regarding previous lifestyle, occupation, or other exposures. This enables the researcher to categorize subjects with regard to previous exposure even if the information is collected at a set point in time. To a large extent the interpretation of cross-sectional studies is similar to that of case-control studies. The lack of the dimension of time implies that cross-sectional studies are poorly suited to the examination of diseases of shorter duration.

Cross-sectional studies may also exhibit recall bias, because disease or assessment of disease may influence subjects' responses to questionnaires, and may even affect biomarkers. In a cross-sectional study in Norway, those who were most overweight in the population were also those who reported consumption of low-fat milk. Although it is unreasonable to infer that low-fat milk increases body weight, the finding does reflect attempts to lose weight. As in this example, the findings of cross-sectional studies should be interpreted carefully. An increasing problem in cross-sectional and cohort studies is the low response rate or attendance rate, an issue that may invalidate the results of the studies with regard to prevalence assessment.

9.6.4 Cohort Studies

9.6.4.1 Design and Structure

Cohort studies follow a defined group of subjects (cohort) over a defined time period. The usual approach is to start with healthy subjects, or subjects without the disease under study. The main purpose is usually to assess the possible effects of different external or internal factors on the risk of disease. Exposure information is collected at baseline for each subject in the cohort. These subjects are followed over time, and the final analysis contains a comparison between those who remained healthy and those who developed the disease under study. In cohort studies, variables at baseline differ between the subjects. Indeed, a homogeneous cohort with regard to the variable under investigation makes it impossible to assess the association of this variable with a given disease. For example, there is no way to find an association between smoking and disease if everyone in the cohort smoked (and at the same dosage) from the age of 15 years until the end of the study period.

9.6.4.2 Advantages

The principal advantage of cohort studies is that they include the dimension of time, which permits the researcher to draw conclusions about causal relationships. Cohort studies are intuitively easy to understand, numerous variables can be assessed at the same time, and one can apply standardized methods with clear definitions with regard to both exposure and endpoints. Ascertainment of exposure and endpoints are independent, as all exposure data are collected while the subjects are still healthy. Prospective cohort studies are assumed to give a more valid, less biased result compared to other study designs.

Both incidence rates and incidence rate ratios can be estimated from cohort studies. Large cohort studies allow for the assessment of more than one endpoint. In this way multiple disease problems can be examined as different exposures.

9.6.4.3 Disadvantages

Cohort studies have several practical and inferential problems. First, cohort studies on diseases with low incidence rates are not amenable to drawing conclusions on causal relationships, as too much time is needed to accrue a sufficient number of events. On the other hand, cohort studies are suitable to the examination of exposures that are stable over time, or of diseases that are either relatively frequent or of a certain duration.

Cohort studies are relatively costly and usually require long follow-up, as well as an infrastructure for follow-up and database updating. In addition, the recording of exposure data can be inflexible. Exposure data are collected only at baseline, and are used to construct different exposure categories. Therefore, any changes in exposures the subjects may undergo during follow-up, such as change in their area of residence or occupation (if occupational exposure is of interest), cannot be taken into account. Without follow-up, these changes are unknown to the researcher and can lead to an underestimation of the risk assigned to a particular exposure. Assume that a sample of subjects is being followed; some are cigarette smokers, whereas others have never smoked. During follow-up, a number of the smokers stop smoking. As the analysis on the associations between disease and smoking will be based on exposure information collected at baseline, no information will be available on who stopped smoking or when. This means that there will be a number of quitters in the smoking category. The result may be a reduced risk of disease, as incidence is lower than that actually associated with smoking. This misclassification usually gives an effect measure that underestimates the true association between smoking and disease. Repeated assessments of the exposure of interest during follow-up, which are performed in longitudinal studies, are one way to avoid this kind of systematic error.

The ascertainment of exposure, including the degree or dose of exposure, is crucial to establishing the baseline cohort. As with all other epidemiological studies, cohort studies contain three principal elements:

1. Choice of target population, which contains the source population
2. Methods for ascertainment of exposure
3. Registration of endpoints

Ascertainment of exposure may involve data collected directly from subjects via questionnaires, biological tests and measurements, or data already included in other registers, such as census data or income data. These data can be collected in a standardized manner to prevent systematic errors. The work involved, however, may be substantial. One example of an area where it is difficult to obtain good, valid data is the area of nutrition and other lifestyle habits. Extensive multi-page questionnaires covering a myriad of topics have been developed for this purpose but may be cumbersome to use if a cohort comprises thousands of subjects. Data collection should therefore be simplified, or else organized at a level that is acceptable for the observation units used. The recording of endpoints consumes a lot of time and resources in cohort studies, as in practice all subjects must be monitored during follow-up to ascertain whether they have developed any of the diseases of interest, or whether they died or emigrated prior to the end of the study (censored data).

9.7 EFFECT MEASURES IN EPIDEMIOLOGICAL STUDIES

In this section three effect measures will be discussed:

1. relative risk (RR)
2. odds ratio (OR)
3. incidence rate ratio (IRR)

and three derived measures including:

1. excess risk
2. attributable risk (AR)
3. population attributable risk (PAR)

All effect measures are based on estimates of disease risk and/or rates. The measure used depends on the study design (see Box 9.4). The conventional term relative risk often is used in the literature to encompass three effect measures—RR, OR, and IRR—independent of the study design underlying the estimates. This is confusing as well as mathematically incorrect, but is acceptable for practical purposes providing certain conditions are met. Accuracy of the effect estimates is given by a confidence interval, usually at a 95% level. For the general principles involved in the calculation of confidence intervals, please see Chapter 11, or chapter 2 of Veierød et al.[2]

A sample of subjects is categorized into exposed and unexposed groups in relation to a possible exposure, and examined with regard to subsequent disease. A fraction of the exposed group and a fraction of the unexposed group develop the same disease during follow-up. The data can be displayed in a 2 × 2 contingency table as is shown in Table 9.2.

Among the $(a + b)$ exposed there are a subjects with the disease under study. This gives a risk (R) of $R_1 = a/(a + b)$. Among the $(c + d)$ unexposed, the risk equals $R_2 = c/(c + d)$. The ratio between these two risks will express the risk of the exposed group relative to the unexposed group. This ratio, called relative risk, is given as:

$$\widehat{\mathrm{RR}} = \frac{a/(a+b)}{c/(c+d)}. \tag{9.1}$$

To calculate the confidence interval the logarithmic transformation must be used, i.e., a confidence interval for ln(RR) is calculated and then transformed back to RR by the exponential function. The 95% confidence interval for RR is then given by

$$\left[\widehat{\mathrm{RR}}\cdot\exp\left(-1.96\cdot\sqrt{\frac{1}{a}-\frac{1}{a+b}+\frac{1}{c}-\frac{1}{c+d}}\right),\right.$$
$$\left.\widehat{\mathrm{RR}}\cdot\exp\left(1.96\cdot\sqrt{\frac{1}{a}-\frac{1}{a+b}+\frac{1}{c}-\frac{1}{c+d}}\right)\right]. \tag{9.2}$$

Here, exp denotes the exponential function.

The OR is another frequently used effect measure, perhaps intuitively understood only by statisticians. Odds are the relationship between two probabilities; namely, the probability of the occurrence of an event divided by the probability of the nonevent provided that there are only two possible outcomes. Let the probability of an event be p and the probability of the nonevent be $(1 - p)$. Odds are then the ratio between p and $(1 - p)$, i.e., $p/(1 - p)$.

In Table 9.2, among the exposed group $(a + b)$, a represents those with disease. Consequently, the odds of having the disease under study are expressed by $a/(a + b)$ divided by $b/(a + b)$, which is a/b. In the unexposed group $(c + d)$, c represents those with disease and d those who are healthy. The odds of having the disease versus being healthy are c/d. The OR, a comparison of the exposed group versus the unexposed group, is obtained by dividing the two odds as follows:

$$\widehat{\mathrm{OR}} = \frac{a/b}{c/d} = \frac{ad}{bc}. \tag{9.3}$$

Going back to Table 9.2, this OR can also be obtained by multiplying the counts, taking the diagonal from top left to bottom right in the numerator and the diagonal top right to bottom left in the denominator, resulting in ad/bc. This product is called the cross-product. The OR is an effect measure which tells us how much larger the odds for developing the disease are among those in the exposed group than those in the unexposed group, and is therefore considered an estimate of relative risk.

The information in Table 9.2 can be used to assess the odds of being exposed based on disease status. There are $(a + c)$ subjects with disease and $(b + d)$ subjects without disease. Among the subjects with disease there are a with the exposure under study, whereas c were unexposed. Therefore the odds for being exposed are $a/(a + c)$ divided by $c/(a + c)$ or a/c. In healthy subjects, the odds for being exposed compared to unexposed are b/d. The two groups can be compared by dividing the odds for being exposed among those with disease by the odds among those without disease. This gives:

$$\widehat{\mathrm{OR}} = \frac{a/c}{b/d} = \frac{ad}{bc}, \tag{9.4}$$

which is the same as Eq. (9.3) for assessing the odds of having a disease for the exposed group relative to the unexposed group.

To construct the confidence interval for OR the same transformations used for RR are applied. The 95% confidence interval for OR is then given by

$$\left[\widehat{\mathrm{OR}}\cdot\exp\left(-1.96\cdot\sqrt{\frac{1}{a}+\frac{1}{b}+\frac{1}{c}+\frac{1}{d}}\right), \widehat{\mathrm{OR}}\cdot\exp\left(1.96\cdot\sqrt{\frac{1}{a}+\frac{1}{b}+\frac{1}{c}+\frac{1}{d}}\right)\right]. \tag{9.5}$$

When information about follow-up time (usually in person-years) is collected, incidence rate for the exposed and unexposed groups can be calculated. The effect measures can then be comparisons of the incidence rates, which by division yield IRRs, or by subtraction yield excess risk.

Let the data for the number of deaths and follow-up time for exposed and unexposed groups be as stated in Table 9.3.

The incidence rate is estimated by the number of deaths divided by total follow-up time. Then the incidence rates among the exposed and

Table 9.3 Cell counts for the association between death and exposure, with follow-up time

Exposure	Number of deaths	Follow-up time
Yes	d_1	t_1
No	d_2	t_2
Total	$d_1 + d_2$	$t_1 + t_2$

unexposed groups are $\widehat{\text{IR}}_1 = d_1/t_1$, and $\widehat{\text{IR}}_2 = d_2/t_2$, respectively. Then the estimated IRR is given by

$$\widehat{\text{IRR}} = \widehat{\text{IR}}_1/\widehat{\text{IR}}_2. \tag{9.6}$$

To estimate the confidence interval for IRR it must be transformed to a logarithmic scale. This is the same transformation used to estimate the confidence interval for RR or OR. The standard error for the logarithm of the incidence rate $\widehat{\text{IR}}_1$ is $1/\sqrt{d_1}$ and for the logarithm of $\widehat{\text{IR}}_2$ it is $1/\sqrt{d_2}$. By transforming back to an ordinary scale, the 95% confidence interval for IRR is:

$$\left[\widehat{\text{IRR}} \cdot \exp\left(-1.96\sqrt{1/d_1 + 1/d_2}\right),\ \widehat{\text{IRR}} \cdot \exp\left(1.96\sqrt{1/d_1 + 1/d_2}\right)\right]. \tag{9.7}$$

Note that the null value is 1 for the relative risk, the OR, and the IRR. Values larger than 1 indicate an increased risk for the exposed group relative to the unexposed group. If the 95% confidence interval does not contain the value 1, the *P*-value for the statistical test of no association between disease and exposure will be less than 0.05 (5%) and the null hypothesis (which says there is no association) is rejected at the 5% level.

9.7.1 Effect Measures in Case-Control Studies (OR)

In case-control studies, cases and controls are studied with regard to past exposures, but disease risk cannot be directly estimated, neither in the exposed nor in the unexposed group. However, the odds for a given exposure in the cases and controls can be estimated. Following Eq. (9.3), the OR can be used as a measure of relative risk of disease:

$$\widehat{\text{OR}} = \frac{a/c}{b/d} = \frac{ad}{bc}. \tag{9.8}$$

Example

A case-control study includes 500 deaths from myocardial infarction (cases), and 500 live controls in the general population. These are compared with respect to their smoking habits in Table 9.4.

Table 9.4 Cell counts for the association between death from myocardial infarction and cigarette smoking

	Deaths		
Smoking status	**Dead**	**Alive**	**Total**
Smoker	140	60	200
Nonsmoker	360	440	800
Total	500	500	1000

When cases and controls are selected, the risk of death due to cigarette smoking cannot be estimated. Thus RR is an inappropriate effect measure; instead, as mentioned previously, OR is most appropriate. The OR is calculated as follows:

$$\widehat{\text{OR}} = \frac{140 \cdot 440}{60 \cdot 360} = 2.85.$$

The 95% confidence interval is (2.04, 3.98). The null value 1 is outside the confidence interval.

9.7.2 Effect Measures in Cross-Sectional Studies (RR, OR)

Since data for cross-sectional studies are collected at a set point in time, they can only support conclusions about associations between an exposure and a disease at that specific point in time. On the other hand, the prevalence of diseases and exposures can be estimated in these studies, and thus RR and OR related to these prevalences can be estimated.

Example

The Low Birth Weight Study, used also in the examples in Chapter 11, is a cross-sectional study, for which the data can be downloaded from http://www.umass.edu/statdata/statdata/stat-logistic.html. The goal of the Low Birth Weight Study was to identify variables associated with having children with low birth weight (weighing less than 2500 g). Data were collected on 189 women, 59 of whom had children with low birth weight. Variables included in these data are the child's birth weight in grams (BWT), low birth weight (LOW: 0 = birth weight ≥2500 g, 1 = birth weight <2500 g) and smoking (SMOKE: 0 = nonsmoking, 1 = smoking).

The results for the association between LOW and SMOKE are given in Table 9.5.

Table 9.5 Cell counts for the association between LOW and SMOKE in the Low Birth Weight Study, $n = 189$

	LOW		
SMOKE	**Yes**	**No**	**Total**
Yes	30	44	74
No	29	86	115
Total	59	130	189

The data in Table 9.5 give a relative risk of

$$\widehat{RR} = \frac{30/74}{29/115} = 1.61.$$

The 95% confidence interval for RR becomes (1.06, 2.44). Thus the risk of having a child with low birth weight increases by 61% for women who smoke, with a corresponding 95% confidence interval of (1.06, 2.44).

Table 9.5 yields an OR of

$$\widehat{OR} = \frac{30 \cdot 86}{44 \cdot 29} = 2.02.$$

The 95% confidence interval for the OR is (1.08, 3.78). Thus, smoking raises the odds of having a child with low birth weight by a factor of 1.02, with a 95% confidence interval of (1.08, 3.78).

9.7.3 Effect Measures in Cohort Studies (RR, OR, and IRR)

Cohort studies yield information on the number of subjects with the disease under study or the number of deaths, and thus risk can be calculated for the exposed and unexposed groups. RR and OR can be calculated from cohort studies. If follow-up time is included, IRR can also be calculated.

Example

The results of a cohort study of 1000 men are given in the 2×2 contingency table presented in Table 9.6. We know that 40 men died from myocardial infarction during a 10-year period; 30 of them were smokers and 10 were nonsmokers.

Table 9.6 permits the estimation of the risk of myocardial infarction among smokers and nonsmokers. The relative risk is

$$\widehat{RR} = \frac{30/500}{10/500} = 3,$$

and by applying Eq. (9.2), the 95% confidence interval is (1.48, 6.07). Since the confidence interval does not contain the value 1, the *P*-value for the test of no association will be

Table 9.6 Cell counts for the association between death from myocardial infarction and cigarette smoking

	Deaths		
Smoking status	**Dead**	**Alive**	**Total**
Smoker	30	470	500
Nonsmoker	10	490	500
Total	40	960	1000

less than 0.05. Therefore the risk of dying from myocardial infarction is three times greater among smokers than among nonsmokers. The excess risk is calculated by subtracting the risk for nonsmokers from that for smokers, or 30/500 − 10/500 = 20/500, indicating that among 100 smokers there will be four more individuals who will die from myocardial infarction than among 100 nonsmokers.

The RR is a better measure of the strength of the association between an exposure and a disease than excess risk, because RR by definition is relative to the underlying risk for the unexposed group.

The OR can also be estimated from cohort studies as the ratio between the odds for having a disease in the exposed group and the odds for developing a disease in the unexposed group. In the 2 × 2 contingency Table 9.6, the cross-product will be:

$$\widehat{OR} = \frac{30 \cdot 490}{470 \cdot 10} = 3.12.$$

Using Eq. (9.5), the 95% confidence interval for the OR is (1.51, 6.47). As for the RR, the confidence interval does not contain the value 1. The OR is slightly higher than the RR, but the difference is small as long as the incidence rate of the disease (or in this case mortality from myocardial infarction) is low. Therefore, in this example 'of myocardial infarction, OR and RR can be used as alternative effect measures.

Table 9.7 Cell counts for the association between death from myocardial infarction and cigarette smoking, with number of person-years

Smoking status	**Number of deaths**	**Pearson-years**
Smoker	30	4850
Nonsmoker	10	4950
Total	40	9800

Assume that Table 9.6 also contains total length of follow-up, as in; Table 9.7.

The incidence rate is the number of deaths divided by total follow-up time, $\widehat{IR} = d/t$. The incidence rate among smokers (the exposed group) is $\widehat{IR}_1 = 30/4850 = 6.2/1000$ and among nonsmokers (the unexposed group) is $\widehat{IR}_2 = 10/4950 = 2.0/1000$. The IRR is defined as $\widehat{IRR} = \widehat{IR}_1/\widehat{IR}_2$, so that IRR = 3.1. The excess risk is 4.2/1000. In the analysis of Table 9.7, $\widehat{IRR} = 3.1$, $d_1 = 30$, and $d_2 = 10$. Using Eq. (9.7) the confidence interval is (1.5, 6.3). Again, the confidence interval does not include 1, and the null hypothesis is rejected.

There are situations in which the excess risks are identical, though the relative risks differ markedly. For example, if the values in Table 9.6 had been 20 and 40 deaths among smokers and nonsmokers, respectively, the excess risk would still have been 0.04 (40/500 − 20/500), but the relative risk would have been 2. The excess risk would be the same as that in Table 9.6, but the relative risk would be two-thirds that of the original value.

The higher the relative risk (the further away from 1), the stronger the assumption that the association is not caused by chance, but is an expression of a direct causal relationship (discussed further in Section 9.13).

9.7.4 Attributable Risk

Attributable risk (AR) is a measure of the proportion of the disease occurrence that can be attributed to a certain exposure. The risks among the exposed and unexposed groups are denoted p_1 and p_2. AR can then be expressed by estimating excess risk as $p_1 - p_2$ divided by the risk for the exposed group, p_1, i.e.,

$$\mathrm{AR} = (p_1 - p_2)/p_1. \qquad (9.9)$$

This gives the proportion of the excess risk of disease that can be attributed to the exposure. As relative risk has been defined as the risk among the exposed group divided by that among the unexposed group, this can be substituted in the equation for AR, giving

$$\mathrm{AR} = (\mathrm{RR} - 1)/\mathrm{RR}. \qquad (9.10)$$

This is a measure of the importance of the exposure in explaining disease occurrence in the exposed group. The association by itself is not sufficient to argue for a causal relationship.

The term prevalence is an epidemiological measure that was described previously as the proportion of the population with a disease at a set point in time (Box 9.2). However, the term can also be used to indicate the proportion of a population with a given exposure at a set point in time. The prevalence of exposure p in the population, which is used to assess the importance of a factor in total disease occurrence, has a value between 0 and 1.

Provided that the prevalence of the exposure is known, the following equation:

$$\mathrm{PAR} = p(\mathrm{RR} - 1)/[1 + p(\mathrm{RR} - 1)] \qquad (9.11)$$

can be used to calculate the AR for the population. PAR is an expression of the proportion of disease in a population that can be attributed a certain exposure given a prevalence p and relative risk RR. The equation follows from the previous definitions; its derivation will not be described here.

Both AR and PAR are mathematical or algebraic assessments of statistical association, but they do not provide any information on causal relationship. PAR is a measure of the magnitude of a given problem from a public health point of view, for instance the proportion of lung cancers that can be attributed to smoking. If the prevalence of smoking is set to 50%, or 0.5 of the population, and the relative risk of lung cancer at 10, this will give us a PAR of 0.82. This implies that 82% of all lung cancer in the population can be attributed to smoking. The equation tells us that if prevalence falls, the importance of the exposure as a public health issue will decline even if the relative risk for the disease remains the same. AR can be used to determine the potential impact of prevention or health promotion if the prevalence of the exposure is reduced.

9.8 EXPERIMENTAL STUDIES AND RANDOMIZED CONTROL TRIALS

Case studies, case-control studies, cross-sectional studies, and cohort studies provide valuable insights into the associations between an exposure and disease risk. However, these studies provide little information on effectiveness and efficiency, or treatment or intervention, where one attempts to determine whether a particular exposure has harmful or beneficial consequences. To tackle these issues, one must turn to the experimental part of epidemiology, the main emphasis of which is on randomized controlled trials. These trials differ from cohort studies in the sense that the investigators actually control the exposure status. Random allocation is used to allocate subjects to the respective exposure groups. These types of studies were discussed in greater depth in Chapter 8.

Interventions, either on individuals or groups, are also of interest in epidemiology. Examples may be studies of changing dietary habits among schoolchildren, or changing smoking habits among adults. Interventions are of interest to public health decision makers.

9.9 TYPES AND SOURCES OF ERROR

All measurement incurs error, or deviation from a variable's true value. The researcher's task is to provide the best possible estimate of this true value. Generally, a measuring instrument is used in the data recording and measurement process. The magnitude of the deviation from a variable's true value depends on the type and accuracy of the measuring instrument used. The deviation from the true value is called measurement error. If the outcome or exposure is categorical, the term misclassification is often used, as when the measurement error causes a subject to be placed in a category other than his or her true one. The magnitude of acceptable measurement error is based on costs and resources, and is a practical question depending upon the nature of the research problem. More accurate instruments and larger samples are more costly. How a researcher copes with this problem depends on the importance attached to measurement error.

9.9.1 Error of the Measuring Instrument

The purpose of a measuring instrument is to get an accurate measurement of a variable. In many situations, only indirect methods of measurement are available. This brings up two questions: To what extent does the measuring instrument used reflect the research problem? How large is the error of the measuring instrument compared to the true value of the variable? One must also consider the repeatability of the measurement, which depends not only on the instrument used, but also on the variable measured. To what extent does this variable vary within a single subject during a relatively short time period? It is difficult to repeat the measurement to a greater degree of accuracy whenever the variation is large. The presence of variation in repeated measurements within single subjects (within-subject variation) may imply that more measurements should be taken from the same subjects to achieve the most valid value possible. If there is considerable variation between the subjects being examined (between-subject variation), more subjects would need to be recruited to the study. The number of measurements taken will depend upon the research problem and the extent of the resources available. The concepts of precision, accuracy, reliability, and validity are discussed in Chapter 4.

9.9.2 Bias, Random Error, and Systematic Error

Bias is traditionally classified as selection bias, information bias, and confounding. Selection bias is a deviation of the results caused by a nonrepresentative selection of subjects. It arises due to a flaw in the study design, when the groups to be compared are selected in different ways. In cross-sectional studies and most case-control studies, the subjects are all survivors. Survivors may differ from those who died from the disease under study, which may lead to a systematic loss of information, and in turn cause survival bias.

Longitudinal studies confer a number of problems with regard to selection: first, the number of subjects at baseline depends on the response rate (selective nonresponse) and second, the extent to which all subjects can be followed up varies. If only people with particular exposures are followed, it can be assumed that there is systematic deviation in the disease, events, or endpoints.

Information bias is a systematic deviation of the results where the disease affects the exposure data. It can arise in case-control studies when the issues under study are known to both the observer and the observed, and in which cases may be able to recall more extensive information than can healthy controls about previous disease, the use of pharmaceutical drugs, and other habits.

Detection bias is a type of systematic error that may occur if exposure data has been used as an argument for the diagnosis.

Confounding implies that an observed association is actually between the endpoint and an exposure other than the one under study. This may occur when there is a strong association between the exposure under study and a confounding variable that is either not measured or is measured with a large random error. Confounding can be corrected by applying regression models that control for covariates (see Chapter 11), provided that the covariates have also been measured. However, adjusting for other forms of bias is far more difficult, if not impossible.

Assume that data on smoking habits are recorded in a cohort, followed by a long follow-up period, during which some subjects develop liver damage. When analyzing the data, it is found that smokers have a higher risk of liver damage than nonsmokers. When analyzing this association, the researcher must investigate other included covariates that are known causal factors for liver damage, such as alcohol consumption. Taking this into account, the present example would imply that smokers had higher alcohol consumption than nonsmokers, which led to the observed result. To clarify this, the researcher must investigate the effect of smoking stratified by alcohol consumption, or adjust for alcohol consumption in a multivariable model. Of course this type of adjustment is only possible if there is information on the variable for which one wishes to control. Age is often a confounding variable, since disease risk usually increases with increasing age. It is therefore appropriate to present age-specific results using age stratification, age-standardized estimates, or using age as a control variable in a multivariable model (see Chapter 11).

If the extent of the error or deviation from the true values are the same, and if it can be assumed that the true value does not affect the magnitude of the error, then the error is classified as random. Systematic error, on the other hand, occurs when the deviations always go in a certain direction or depend on the true value of the variable. Systematic error makes it difficult to interpret results, as it may create artificial associations or bias that consistently deviates from the true values.

9.9.3 Sources of Error

Questionnaires used for charting eating habits and lifestyle habits are crude and inaccurate measuring instruments; genetic and molecular methods are likely to be far more accurate. Measurement error can occur due to the way a measuring instrument is fabricated or the way that questionnaires are formulated, but they can also occur as a result of how the instruments are used and the skills of the individuals performing the measurements. Researchers vary with regard to their own accuracy, and human error often accounts for inaccurate readings and misclassifications.

9.10 TESTS AND VALIDITY

In addition to the validity of the observed association, researchers may also ask about the validity of the tests used to distinguish between individuals with and without disease. The validity of a test must be assessed against a true standard with regard to an observed individual who actually has the disease. Valid tests are necessary in both clinical medicine and preventive efforts.

9.10.1 Sensitivity and Specificity

Tests can be used to determine disease status or to predict disease. The test can be a measuring instrument, such as a questionnaire or an autoanalyzer. The disease is the truth, and the better the tests succeed in identifying individuals with disease, the more valid the test. To assess the validity of a test, those who have been examined are divided into two categories according to disease status and test results. The results are displayed in a 2×2 contingency table, as illustrated in Table 9.8.

Table 9.8 Disease status and diagnostic test results for a binary test

	Test result	
Disease	**Positive**	**Negative**
Yes	True positive	False negative
No	False positive	True negative

Table 9.9 Cell counts for disease status and diagnostic test results

	Test result		
Disease	**Positive**	**Negative**	**Total**
Yes	a	b	$a+b$
No	c	d	$c+d$
Total	$a+c$	$b+d$	n

Table 9.9 displays the number n of the subjects classified by disease status and test result, and shows that a of the subjects with disease had a positive test result. The validity of the test is therefore $a/(a+b)$, since $a+b$ is the total number of subjects with disease. This is an expression of the sensitivity of the test. To judge the ability of the test to identify individuals without disease, the number of subjects with a negative test result d must be estimated and compared to $c+d$, which is the total number of subjects without disease. Thus, $d/(c+d)$ is the expression of the specificity of the test. Sensitivity and specificity both express the validity of a test, or its ability to correctly identify individuals with disease and distinguish them from healthy individuals.

Assume now that the test variable is a continuous variable, which implies that those with a test result above a specific cut-off have positive test results, and those with a test result below the cut-off have negative test results. If the cut-off for positivity is moved, a larger or smaller proportion of healthy individuals will be included. Indeed, the lower the cut-off is set, the more people with disease will be included, the ratio $a/(a+b)$ will approach 1, and sensitivity will increase. At the same time, a larger proportion of the healthy subjects will have a positive test result, whereby the ratio $d/(c+d)$ decreases, and the specificity declines. In other words, if a valid test to identify individuals with disease is needed, some validity in identifying healthy individuals will be sacrificed.

9.10.2 Predictive Power of Tests

Another issue is to estimate how large a proportion of individuals with a positive test result $(a + c)$ actually have the disease, and how many with a negative test result $(b + d)$ are actually healthy. These test characteristics are called the test's predictive power, which is expressed either as the power given a positive test result (positive predictive power), or a negative test result (negative predictive power). Reading Table 9.9 column-wise, one can assess the extent to which subjects with positive test results actually have the disease. This can be calculated by $a/(a + c)$, which is the test's positive predictive power. The corresponding ratio $d/(b + d)$ estimates the proportion of subjects with negative test results who are healthy, which gives the test's negative predictive power.

Example

Consider two examples where the same test (with known specificity and sensitivity) is used for two studies: A and B.

A: A hospital-based study with 2000 subjects where half develop a disease and half are healthy

B: A population-based study with 10,000 subjects where 99% of the population develops a disease and 1% is healthy

Table 9.10 presents the results from the hospital-based sample.

Table 9.10 Cell counts for disease status and diagnostic test result among a hospital-based sample of 2000 subjects

	Test result		
Disease	**Positive**	**Negative**	**Total**
Yes	990	10	1000
No	10	990	1000
Total	1000	1000	2000

In the hospital-based study, 990 of the 1000 subjects with positive test results actually have the disease, so the positive predictive power of the test is 0.99. Only one subject out of 100 is wrongfully declared healthy. The test was valid and distinguished well between subjects with disease and healthy subjects, with a sensitivity of 0.99 and a specificity of 0.99.

Table 9.11 gives the results from the population-based study.

The sensitivity and specificity are both 0.99, as in the hospital-based study, but the positive predictive power is now 0.50 (99/198).

In the hospital-based study, 99 out of 100 have positive test results, and are correctly identified as having the disease, and consequently the test has a very high positive

Table 9.11 Cell counts for disease status and diagnostic test result among a sample of 10,000 subjects in the general population

	Test result		
Disease	**Positive**	**Negative**	**Total**
Yes	99	1	100
No	99	9801	9900
Total	198	9802	10,000

predictive power. When the same test is applied in the general population, only one individual out of two is correctly identified as having the disease.

The predictive power of a test depends on the prevalence of the disease and will vary even if sensitivity and specificity remain the same. This reality must be taken into consideration when deciding whether to test the general population for rare diseases. A prevalence of 0.1% signifies that for each person with disease, about 10 healthy individuals will have positive test results, and thus will be misclassified. The arguments for and against testing must always include an assessment of the consequences for those who will have false-positive results.

9.10.3 Predictive Power and Number Needed to Treat

An old adage advises "when you hear hoofs in the street, don't look for a zebra." Applying this to predictive power, we can change it to: when a rare phenomenon shares characteristics with frequent phenomena, look for the frequent phenomena rather than the rare one. For example, an otherwise healthy person of 60 years of age with a random systolic blood pressure of 200 mmHg is more likely to be alive than dead 5 years later. We may draw similar conclusions on the effects of various treatments. If the underlying risk of death from a certain disease is low and there is a treatment that can reduce related mortality, a large number of treated patients are needed to avoid one death. This phenomenon has led to a concept called number needed to treat, which also expresses the effect measure of a treatment. To what extent has the intent of the treatment been achieved? It may seem reasonable to treat a group of individuals who have a risk of stroke that is 10 times higher than normal, rather than those with a risk that is only three times higher. However, a low absolute risk (which may well correspond to an RR of 10) implies that one must treat a large number of people in order to avoid one case of unwanted disease or death. This is the dilemma when trying to reduce the risk of myocardial infarction, hip fracture, diabetes, or stroke in a population of otherwise healthy subjects. Indeed, among healthy subjects the probability of these events is relatively low, even 5–10 years in the future. Any attempt to further reduce these already low possibilities implies that a large number of people have to be treated or undergo intervention (e.g., quit smoking or change their eating habits) to avoid a single event. In other words, as something can only be said about the probability within a group with regard to disease risk, and nothing can be said about the single individual, the whole group must be approached in order to achieve a reduction in the risk of disease. To achieve such a goal, namely that not one person succumbs to a disease during the defined time period, a large number of people must be treated, but for all those except the one person who actually needs it, the treatment will have no consequences. The problem is that the individuals who will develop the disease cannot be identified.

9.11 CAUSES OF DISEASE

For the following two sections please refer also to Chapter 1 on the philosophy of science.

9.11.1 Causes of Disease—a Definition and a Condition

The cause of disease can be defined as a factor that is needed for that disease to occur. It is evident that all those who are exposed to a certain causal factor do not develop disease. A causal factor may increase the risk of developing disease, but only rarely alone and absolutely. Risk may be defined as the probability of a certain event within a defined time period and varies between 0 and 1. Therefore, a causal factor can be defined as one that increases the probability of disease. A major precondition for being able to identify an exposure as a possible causal factor is the variation of this exposure in a population. If everyone in the population has been exposed to a suspected causal factor, it will be impossible to identify its possible association with disease. Any variation that occurs with regard to the disease has to be attributed to variations in other factors. A theoretical example is to assess the effect of allowing everyone in a population between the ages of 15 and 65 years to smoke 40 cigarettes a day. During this time a number of people will develop a disease, but as they were all smokers, the harmful effect of the tobacco would be impossible to detect.

All the exposures, such as diet and genetic disposition, would emerge as variables that either increased or decreased the risk of disease. This implies that the exposure that triggers a disease in a single individual will differ from the exposures that explain an increase in disease in one population compared to another. This may be the problem when assessing whether the intake of salt has consequences on blood pressure, or if saturated fat increases the risk of breast cancer.

The precondition for being able to identify an association is the presence of between-subject variation that is larger than the within-subject variation. Thus, in overly homogeneous populations or populations with excessive between-subject variations, it is impossible to identify associations. One way to avoid this problem is to examine groups of individuals known to differ with regard to the factor examined. Associations seen in ecological studies but not in studies where individuals are the observation units may lead to the conclusion that there is no association between the exposure examined and the disease. However, this conclusion may be false, because the individual variation may be very large. On the other hand, differences between groups may be due to exposures other than the one examined. The example of salt and blood pressure may, for instance, imply that the populations with low blood pressure are consuming a great deal of fruits and vegetables in addition to a low-salt diet, and thereby they differ in other ways from salt consumers. However, precise estimates of salt intake and blood pressure measurements are difficult to achieve.

9.11.2 Necessary, Contributing, and Sufficient Causal Factors

A necessary causal factor is a factor that must be present for the disease or health problem to occur. Examples are microorganisms and viruses that cause infectious disease or chromosome aberrations leading to genetic defects and disease. For example, there is no tuberculosis without bacillus, and Down syndrome cannot exist without the extra chromosome 21. This last example implies that everyone with Down syndrome must have an extra chromosome 21.

But it must be emphasized that some may be affected without being exposed to a causal factor. The concepts of necessary, contributing, and sufficient factors may help in this discussion. When a light is switched on, the switch, the bulb, the power source, and the connections are all needed in order for the light to work. If one of these components is taken away, there would be no light even if all the other components are intact.

Each of the components in this example has contributed to light and is a necessary factor, but not sufficient by itself.

Contributing factors do not need to be necessary factors, but they may increase the probability of a complex set of necessary factors to result in disease, thereby constituting a sufficient causal complex. This thinking applies to nutrition, genetic disposition, and social or climatic conditions. The causal complex consists of factors that may be considered equally important. However, they will vary across different individuals, for instance, one subject may have a genetic disposition making cigarette smoking more harmful (such as a lack of special repair enzymes), whereas another person may be more susceptible to saturated fatty acids than others. This line of thought has consequences for planning and organizing preventive strategies. Some of the single factors in a causal complex are likely to be easier to handle than others, for instance, environmental and lifestyle factors are simpler to manipulate than genetic dispositions.

Before entering the discussion regarding the distinction between association and causal relationship, the problem of comparing groups or reference groups will be briefly addressed. An extensively studied issue is whether oral contraceptives increase the risk of death, for example, cause an increase in breast cancer, uterine cancer, or other diseases. When assessing this problem, one has to ask what the alternative is. With whom shall the oral contraceptive users be compared? Should they be compared to women with low sexual activity (for instance nuns) who are avoiding the risk induced by pregnancy and childbirth, or to women using alternative methods of contraception, or to those who are not using any contraception? The final conclusion of the study may depend on the choice of reference group that served as the comparison.

9.12 ASSOCIATION VERSUS CAUSAL RELATIONSHIP

The essence of any etiological reasoning leading to a rational intervention or treatment is that a particularly defined exposure, either as a single entity or a causal complex, will have a specific effect on a human body. The word rational implies that the etiology is based upon a recognizable and describable causal complex, logically ending in clinical overt disease.

9.12.1 When Does an Association Express a Causal Relationship?

There are three different types of associations that can be expressed between an exposure and disease.

1. The relationship is causal—variable A leads to disease B, for instance, cigarette smoking leads to bronchial carcinoma.
2. The observed variable is the intermediary part of a causal chain. X, a factor so far unknown to us, has affected the organism, whereby variable A will emerge or be changed and become detectable. This process will precede event B in the causal chain. Examples are cytological changes in the cervix that predict malignant changes, or electrocardiogram changes predicting later myocardial infarction. Such factors will often be called intermediary variables. They can be assessed as exposures as they are associated with disease, but there may also be endpoints in a study where the main interest is to examine why these factors vary. Such variables may also be called surrogate endpoints.
3. The observed variable is associated with an alternative factor, which is the real cause of disease. For instance, alcohol may be associated with bronchial carcinoma, but cigarette smoking is the causal factor. Alcohol has a predictive power, but this depends upon the association between alcohol and cigarettes; alcohol is not the singular causal factor for bronchial carcinoma.

Any of these associations can be used when predicting disease as part of a battery of tests to predict an individual's risk of disease, even if there is no direct causal relationship. Whether an association should be accepted as a causal relationship is always a matter of opinion based upon the available evidence. The questions to ask when making this decision are the following:

- Is this association a random or chance finding?
- Is the association an expression of systematic error—a bias either in selection of the sample because of the measuring instrument being used?
- Can this association be explained by another factor or confounding variables?

If the answer to all these questions is no, it is time to consider whether the relationship really is causal in a biological sense. In his textbook, Sir Bradford Hill[3] stated a number of issues that should be discussed before it is reasonable to reach a conclusion regarding a causal relationship. The list comprises demands with regard to the strength of the

BOX 9.5 Issues to be Discussed When Assessing Causal Relationship

1. Strength of association
2. Consistency
3. Specificity
4. Temporality
5. Biological gradient
6. Biological plausibility
7. Coherence
8. Experiment
9. Analogy

association expressed, for instance, by relative risk; consistent results from different studies (preferably also with different methods); the condition that the causal factor is followed by the effects; and that the association be understandable from a biological viewpoint. The list has been looked upon as criteria that must be fulfilled before determining that an association represents a causal relationship, but it has been criticized by a number of authors. In any case, Hill's original intentions were not that one should fulfill all the demands every time; the list is a guide for how to think about causal relationship within the health sciences (Box 9.5).

Of all these points, there is only one that must be fulfilled with no exceptions: One has to be able to show that the factor in question was followed by disease and is not a consequence of the disease. The time relationship between the exposure and the disease must fit in the causal relationship.

9.13 CAUSAL CALCULUS

The previous discussion on causal relationships and associations focused on logical and verbal arguments for a possible causal effect of a disease-associated exposure, with the notion that causal relationships could not be proven or formally tested. The closest one could come to proof would be a randomized controlled trial, but this is an unrealistic approach to a large number of health issues. Although observational studies are one of the tools available for this purpose, they have inherent problems of controlling for confounding, as well as colliding, regarding the inferences to be drawn.

Epidemiologists have always been aware of these challenges and the problems in drawing causal inferences from different research designs.

During the last 20 years a formal theory of causal inference has been developed, which is intended to address these problems. One of the main contributors to this theory and developer of a new terminology for discussing cause and effect is Judea Pearl. His book, entitled *Causality: Models, Reasoning, and Inference*, lays out the basis for this thinking.[4]

The idea of direct and indirect effects is fundamental in causal thinking. The example of smoking and lung cancer shall be used. A causal model of the association is shown in Figure 9.2. This is a very simple model, in the sense that there is a direct pathway from the exposure (E) to the disease (D).

The association between smoking and lung cancer has been noted in an observational study. An unknown factor cannot be excluded as the real culprit, or as being otherwise involved in this causal complex.

There are three possible explanations:

1. There is no unknown factor playing an additional part in this causal chain; smoking induces lung cancer. This is illustrated in Figure 9.2 with smoking (E) and lung cancer (D).
2. A common factor—a confounder—increases the association between smoking and lung cancer (Figure 9.3). If Figure 9.3 contained no arrow between smoking (E) and lung cancer (D), it would be interpreted as no direct causal effect of smoking on lung cancer, and the exposure-disease association would be explained by the confounder.
3. Lung cancer increases the risk of smoking, or reversed causality.

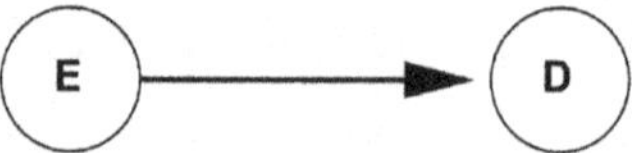

Figure 9.2 A causal model for the association of exposure (E) and disease (D).

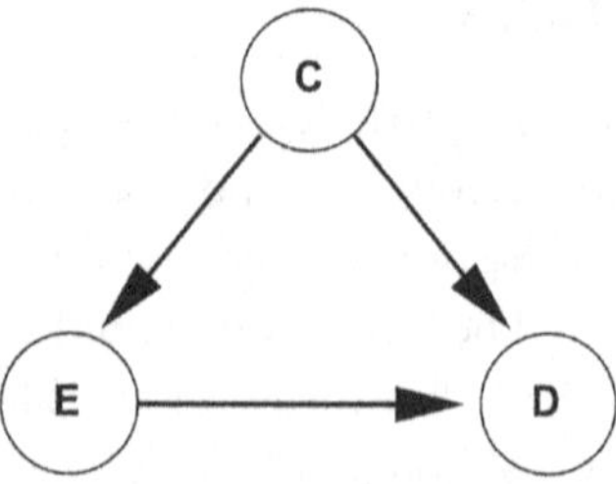

Figure 9.3 A confounder (C) explaining the association between exposure (E) and disease (D).

The analysis indicating that there is an association, for instance by a high OR, is based upon a symmetrical equation. The equation can theoretically go both ways. The OR is symmetric, in the sense that a high OR alone is no indication of which variable represents the cause, and which the effect. It is unlikely that lung cancer increases the risk of smoking, but this interpretation cannot be discarded from the statistical analysis. Indeed, the arrows between the variables smoking and lung cancer imply directions in the theory of causal inference; nevertheless, if it is assumed that smoking leads to lung cancer, not the other way around, then option 3 is discarded. This is done because there is knowledge on cigarette smoking and lung cancer, but reversed causality always has to be considered, based upon additional knowledge.

Colliding is a situation where a variable (the collider) is the consequence of two associated factors. In the example of cigarette smoking and lung cancer there may also be information about body weight. A separate analysis of body weight and lung cancer risk may show that low body weight and lung cancer are associated. This may be the result of colliding as cigarette smoking is associated with low body weight and unrecognized lung cancer can have the same effect. Given that low body weight is a result of both smoking and lung cancer, adjusting for body weight in this situation makes no sense. Figure 9.4 shows the observed association between E (cigarette smoking) and D (lung cancer), and their common effect C (body weight).

The idea of direct and indirect effects is fundamental in causal thinking. If there is a pathway from E to D, we call this a direct effect between cause and effect. But in many cases the exposure will also act through mediating variables. In that case the direct effect is given by the direct pathway from E to D, and the indirect effects are given by the effects of E on D via the mediating variables. In Figure 9.4 there are two mediating variables. In our example of smoking and lung cancer, mediating variables may be coughing (M_1) and weight loss (M_2) (Figure 9.5).

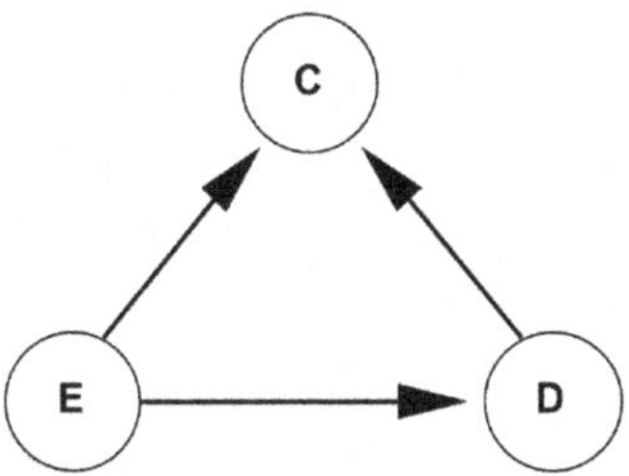

Figure 9.4 Colliding, a common consequence of two variables.

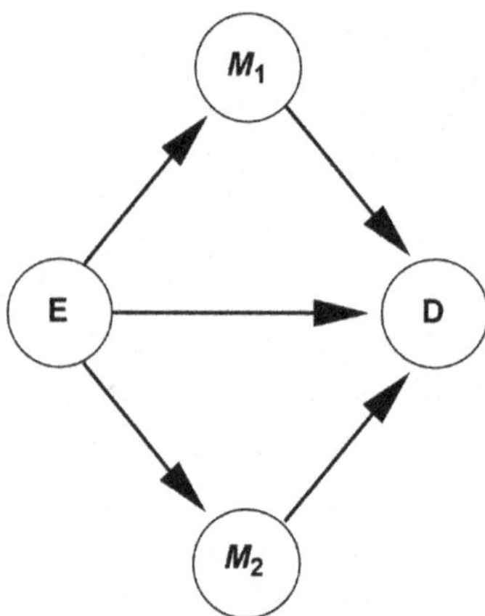

Figure 9.5 A causal model for the association between exposure (E) and disease (D), with two mediating factors (M_1 and M_2).

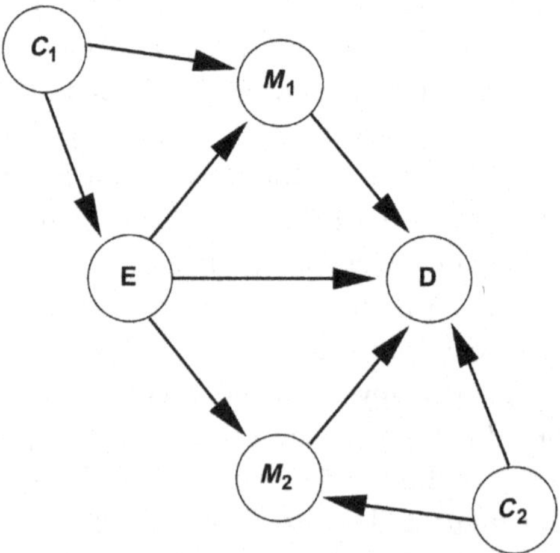

Figure 9.6 A causal model for the association between exposure (E) and disease (D), with two mediating factors (M_1 and M_2) and two confounders (C_1 and C_2).

The directed acyclic graphs (DAGs) discussed above may be extended to more complex situations, such as the one in Figure 9.6.

Here there are two mediators (M_1 and M_2). There are indirect effects of E via M_1 and M_2. The effect of E on M_1 is confounded by C_1 and the effect of M_2 on D is confounded by C_2.

In the terminology of DAGs, the arrows are called edges. These edges can only go one way. The circles are called vertices and can be equal to what was earlier listed as events, or endpoints. One may be familiar with the term vertex from computer graphics, but it is also the general term for the point created by an angle, or where lines and rays meet. The circle represents a breaking point in the causal process.

The smoking–lung cancer model can be looked upon as a general model where multiple variables are involved. In the model above, each vertex represents an event or variable with different outcomes, which for simplicity could be dichotomized to smoking, not smoking, or gene X on or off. The edges point at associations and provide directions.

The graphs in Figures 9.2–9.6 are acyclic, the second word in DAG. To emphasize this point, the cyclic graph in Figure 9.7 would be a nonsense or uninformative graph.

That Y causes Z, which causes X, and X causes Y, does not provide any useful information, but rather is just another example of reversed causation. Thus this cyclic model may well exist, but if so, it should be used as an argument against a causal relationship.

Depending on the sample, the effect sizes in a DAG may be estimated by a multivariable model that mimics the DAG. When the variables are continuous (or normally distributed) the statistical software packages LISREL and SPSS AMOS can be used for calculations.

9.13.1 Conditional Causal Probabilities

The ideal situation for assessing a causal relationship between smoking and lung cancer would be a randomized controlled trial, but this is not possible. The mathematical theory (which is of Bayesian origin) underlying the derivations in this section relies on the distinction between conditional probabilities, which means the probability of finding cancer in a person known to smoke, and interventional probability, or the probability of finding cancer in a person forced to smoke. The first would be the result of an

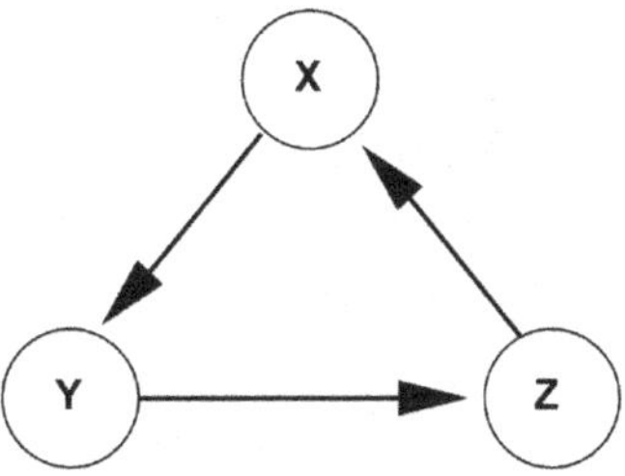

Figure 9.7 A cyclic graph for the association between three causal factors.

observational study, whereas the latter can only be derived from a randomized controlled trial, which in this case is neither possible nor ethical.

Pearl introduced the concept of the conditional causal probability, which is the conditional probability of cancer in an imaginary randomized study, given that someone has been allocated to the smoking group. Pearl also established a set of rules—a causal calculus—that such conditional causal probabilities should obey, and from which probability should be calculated and inferred, but this calculus has no basis in a real randomized study. These rules allow a systematic analysis of the potential paths from an exposure to a disease, while recognizing and avoiding mistakes in the possible causal chain or causal complex.

Whether DAGs and graphical models will contribute substantially to causal analyses beyond what is already possible with current statistical instruments remains to be seen. The main problem with observational studies is the identification of sources of bias, their magnitude and direction; DAGs do not give any further information on these issues.

ACKNOWLEDGMENTS

We thank Carina V. S. Knudsen, Institute of Basic Medical Sciences, University of Oslo, for producing the figures.

QUESTION TO DISCUSS

1. Explain the difference between prevalence and incidence. Give examples of when one would be interested in the prevalence of a disease, and in the incidence of a disease.
2. Outline the advantages and disadvantages of case-control studies, cross-sectional studies, and cohort studies.
3. Which effect estimates can be derived for case-control studies, cross-sectional studies, and cohort studies?
4. Explain why the RR and the OR will be about the same size for rare diseases.
5. A random sample of 1000 individuals from a source population was followed for 10 years.

   At baseline, 500 were nonsmokers. After 10 years, 40 had died due to myocardial infarction. Of these, 10 individuals were nonsmokers (to simplify, assume that all had their myocardial infarction at the end of the study period, and that no one else died).

Assess the impact of being a smoker compared to be a nonsmoker. What is the proportion of deaths that could potentially have been avoided in the study population if nobody had smoked? Which confounders could have influenced the results?

6. A case-control study may reveal an association between an exposure and an outcome that cannot be reproduced in a cohort study. Why is that? Give an example of an exposure and outcome for which this may be the case.
7. You have been asked to design a study to assess whether children of parents with alcohol addiction run an increased risk of developing alcohol-related problems.
   a. Which study design would you choose for this issue?
   b. Provide arguments for and against your choice.
8. The explosion and fire at the nuclear power plant in Chernobyl in Ukraine on April 26, 1986, released large quantities of radioactive particles into the atmosphere. They spread over much of western USSR and Europe. This resulted in an increased number of babies born with congenital defects. You have data from Kiev (close to Chernobyl) and Donetsk (further away) covering the period 1980 until the present.

   Design a study assessing the risk of having babies with congenital defects as a consequence of the Chernobyl catastrophe.
9. Supplementation with folic acid and iron is supposed to reduce the risk of spina bifida and cleft lip/palate. In a study, 498 women who gave birth to children with cleft lip/palate and 2485 women with unaffected children were compared with regard to the use of folic acid. Of the women with affected children, 223 had taken folic acid. The corresponding figure among women with unaffected children was 1944.
   a. Which study design is this?
   b. What is the exposure? What is the endpoint?
   c. Calculate the risk for cleft lip/palate.
   d. What is your conclusion?

REFERENCES

1. World Health Organization (WHO). *International Classification of Diseases (ICD)*, 10th ed. Available at <www.who.int/classifications/icd/en>.
2. Veierød MB, Lydersen S, Laake P. *Medical statistics in clinical and epidemiological research.* Oslo: Gyldendal Akademisk; 2012.
3. Hill AB. *A short textbook of medical statistics.* 10th ed. London: Hodder & Stoughton; 1977.
4. Pearl J. *Causality: models, reasoning, and inference.* 2nd ed. Cambridge: Cambridge University Press; 2009.

FURTHER READING

Kleinbaum DG, Kupper LL, Morgenstern H. *Epidemiologic research: principles and quantitative methods*. New York, NY: John Wiley; 1982.

Kleinbaum DG, Sullivan KM, Barker ND. *ActivEpi*. 2nd ed. New York, NY: Springer; 2013.

Rothman KJ, Greenland S, Lash TJ. *Modern epidemiology*. 3rd ed. Philadelphia, PA: Lippincott, Williams & Wilkins; 2008.

Rothman KJ. *Epidemiology: an introduction*. 2nd ed. Oxford: Oxford University Press; 2012.

Wikipedia. Retrieved from January 3, 2014 <http://en.wikipedia.org/wiki/Causality>.

CHAPTER 10

Qualitative Research Methods

Kåre Moen and Anne-Lise Middelthon
Department of Community Medicine, Institute of Health and Society, University of Oslo, Blindern, Oslo, Norway

10.1 INTRODUCTION

In the quest for new knowledge and insight, a broad variety of research methods are needed, as different methods enable different kinds of knowledge to be developed. Despite this, one can still encounter the idea that the research methods employed in the health sciences can be universally ranked in terms of the "level of evidence" they generate. Inspired by conceptualizations of evidence "hierarchies"[1] and evidence "pyramids,"[2] this idea effectively holds that there is a fixed relationship between different research methods in terms of their quality and utility for health research.

But attempts at ranking research methods do not make much sense unless the objective of a given research project is specified. Ranking a randomized controlled trial over an observational study is rational if the aim is to evaluate the primary treatment effect of a certain drug in a Phase III pharmaceutical trial.[3] However, for a study with the same aim, it would be a waste of time to compare qualitative interviewing (one of the research methods discussed in this chapter) with a randomized controlled trial, as qualitative research has no relevance in the context of that aim; it cannot render the kind of data that are needed to fulfill it.[i] If the aim of the research were to identify and understand taken-for-granted assumptions about health and illness in a certain sociocultural setting or locale, it would be equally meaningless trying to compare case-control studies with participant observation (another research method discussed in this chapter). Again, it would be an irrelevant comparison, because the first research approach does not enable investigators to explore the social domain in ways that would allow them to respond to the question asked. In short, any comparison or ranking of research methods must take into consideration, at its very center, the objective of the inquiry.

[i] Qualitative research can of course play a role in pharmaceutical research, but the research aim would have to be different from the example mentioned here.

Research in Medical and Biological Sciences
DOI: http://dx.doi.org/10.1016/B978-0-12-799943-2.00010-0

Qualitative research methods provide strategies for exploring experiences, practices, and phenomena in sociocultural worlds. That is, they aim to provide researchers with approaches that can be used to discover and examine the ways in which interconnected people encounter, perceive, understand, and bring into being processes, practices, and phenomena. In this chapter we will discuss how qualitative research methods are conducted and how they can contribute to the study of human health, illness, suffering, and healing.

Sections 10.2, 10.3, and 10.8–10.10 discuss issues that pertain to qualitative research in general. Two specific qualitative research methods (qualitative interviewing and participant observation) are discussed in detail in Sections 10.4–10.7. Several other qualitative research methods are also used in the health sciences; a brief overview of them is provided in Box 10.1.

Throughout the chapter, we have chosen to draw upon examples of research we know well rather than referring to studies we have only encountered in text.

BOX 10.1 Overview of Qualitative Research Methods

A detailed discussion of qualitative interviewing and participant observation will be offered in subsequent sections of this chapter. Space does not allow us to cover the full range of qualitative research methods in the same way, but in the following we offer a brief overview of each of the major approaches in qualitative health research.

- *Qualitative interviewing*: Qualitative interviewing is a special form of conversation involving a researcher, a research participant, and a theme. The interview entails interpersonal interaction between epistemologically active subjects and aims at joint construction of knowledge through reflection and articulation (see Sections 10.4 and 10.5).
- *Participant observation*: Participant observation is a research strategy aiming to produce knowledge *on* interaction between people *through* interaction between people. The researcher aims to become part of the social contexts in which the phenomena she or he studies are embedded (see Sections 10.6 and 10.7).
- *Group interviews and/or focus group discussions*: While group interviews/discussions can well be the main—and also the sole—tool for qualitative data production, they are more typically used to complement other qualitative research methods. In group interviews/discussions, insight is generated both through observation of interaction between participants and through analysis of their reflections and discussions. When used as a complementary research method, group interviews/discussions can be valuable early on in

(*Continued*)

BOX 10.1 Overview of Qualitative Research Methods—cont'd

the research process, as forums for initial exploration of a theme (to assist with the identification of topics to be pursued and/or in the development of interview guides). Group discussions may also be used in the later stages of a study as forums where descriptions, analyses, and theorizations generated during the course of other research activities can be further tested, broadened, nuanced, and/or challenged. (Group interviews/discussions can also be used as a complementary method in quantitative research, e.g., to bring out contextual information that can inform the discussion of findings generated through surveys.)

- *Key person interviews*: Key persons have knowledge of particular relevance for a theme under investigation (they could, for example, be experts of various kinds, policymakers, activists, or community elders). However, they are not core study participants in the sense that it is not *their* life, work, communities, or experiences we are inquiring into. While key person interviews are frequently used as a complementary research method (in qualitative as well as quantitative research), they are unfortunately not always visible in study protocols and research publications. In the quest for transparency, one should aim to make the role of key person interviews in knowledge production explicit. Doing so will also more accurately demonstrate the comprehensiveness of the study design.
- *Texts, images, and objects*: Analysis of data other than those produced in interpersonal interaction with study participants can constitute a study on its own or be complementary to other data production methods (quantitative or qualitative). Many things can be analyzed, including:
  - *Texts*, e.g., medical textbooks, policy papers, blogs, online discussion forums, laws, patient journals, nonfictional and fictional literature, fairy tales, sagas, myths, or newspaper articles.
  - *Pictorial material*, e.g., images of illness, disease, suffering, healing, or caring; portraits of people or groups of people (e.g., sick children, people who inject drugs, surgeons, nurses, or hospital directors); and films or videos produced by professional and nonprofessional actors.
  - *Objects*, e.g., medical equipment, food items, compression stockings, drugs, hospital beds, or wheelchairs.
- *Historicizing*: Historicizing is used as a complementary research approach in many qualitative studies. Indeed, most studies can be enriched by a component that elucidates, however briefly, the historical dimensions of the phenomena being scrutinized. To historicize is also to acknowledge the fact that all phenomena dealt with in qualitative research exist in time and space and are historically contingent. (Medical history is also a discipline with long traditions. Medical and health-related phenomena are analyzed in the [historic] context of their time, and/or their development is traced through time.)

10.2 WHAT IS QUALITATIVE RESEARCH?

10.2.1 Qualitative Research Methods Are Exploratory Methods

The most widely known usage of qualitative research methods in the health sciences is probably studies that aim to understand health-related issues from the insider perspective of patients, or of people "at risk" of becoming patients. An example could be a study investigating what it is like to live with a certain type of cancer, or a study aiming to gain insight into the reasons why people do or do not use condoms in a given setting.

Both of these are research questions that typically require open and broad *exploration*. Epidemiological research approaches are of limited utility in this kind of research, as they are not well positioned to discover things the researcher has not already realized were possibilities before the research commences. Indeed, questionnaires used in quantitative studies ask preformulated and identical questions of all study participants and provide them with mostly closed-ended reply options. If a researcher figures there are a dozen significant reasons why a person may not have used a condom during his last sexual encounter, he or she will enter these as reply options in the questionnaire and determine what proportion of study participants say they agree with each of them. However, the provided reply options will likely grasp only a small part of the universe of associations that are relevant to the ways in which condoms play into people's lived sexual lives. In order to obtain a wider and deeper understanding of *that*, the exploration must be much broader than can be offered by a quantitative questionnaire.

Qualitative researchers would typically engage in an exploration of condom use and nonuse, with the aim of mapping out and understanding the wider landscapes in which condoms are perceived, experienced, interpreted, used, and not used. They would characteristically first engage in broad and open discussion with study participants. While data produced through interaction between researcher and study participant would thus stand at the center of the study design, these data would likely be contextualized through a subsequent examination of ongoing discourse in society. This could entail a review of sources, such as media reporting, condom marketing materials, and popular literature, that may reveal something about how condoms are perceived, experienced, and handled in a given community. The findings might be put into an even broader social, cultural, political, and historical context through the identification and analysis of texts that have dealt with topics such as bodies, sex, contraception, and HIV prevention at various points in time.

Whether broad exploration of the type just mentioned, and the development of the finer insight that could follow as a result, is "useful" in a certain context has to be assessed on a case-by-case basis. However, in the context of the condom example above, it is relevant to note that in a review of HIV prevention work globally, Kippax and Race[4] found that the most successful HIV prevention strategies have generally been those that took into account and adapted to the lived experiences of those at most risk. They found that in order to be successful, HIV prevention has to make sense to those at risk, and HIV prevention practices have to fit into their overall lives in a meaningful manner. In order to gain insight into the complex and various ways in which condoms do and do not make sense to people, and into the many webs of meanings within which they are entangled in people's lives, research approaches that facilitate broad exploration are needed.

While qualitative studies are used to explore insider perspectives (among patients and people at risk of becoming patients, but also among health-care providers and decision-makers in the health-care sector), they can be used for much more in the context of human health, illness, suffering, and healing. Box 10.2 gives an overview of some typical questions qualitative research can explore, while Box 10.3 provides an overview of some typical fields of inquiry in qualitative health research.

BOX 10.2 Typical Questions That Can Be Explored in Qualitative Research

- What do people experience, and how do they interpret and articulate their experiences?
- What do people know, believe, and expect, and how does this make sense to them?
- What do people do, and how and why do they form and perform social practices?
- How do people reason, and what is the logic behind their reasoning?
- How do people feel, and how do they make sense of and express their emotions?
- How do people interact, and what are the "codes" and "rules" that underlie interaction?
- What are the sociocultural values and norms?
- What do social phenomena consist of?
- What are the social, ideological, and political processes involved?

BOX 10.3 Typical Fields of Inquiry in Qualitative Health Research

- *Exploring conceptions of health and illness*: Health and illness-related concepts are regularly understood differently in different sociocultural contexts. Such differences, and the diverse experiences and knowledge that shape them, can be identified and explored through qualitative research.
- *Understanding rationales*: Actions and recommendations in preventive and curative medicine build on rationales that make sense to the people that develop and promote them. However, they are almost never the only possible or the only meaningful rationales for a given theme. For example, HIV prevention may at times appear to be built on the assumption that protection against transmission is the paramount issue in a sexual encounter. Of course, this is not (always) the case, and to gain insight into, and an appreciation of, the many alternative rationales associated with sexual practices in the context of HIV is often a prerequisite for meaningful HIV prevention programs. Qualitative research methods offer researchers ways to explore the range of such rationales.
- *Discovering tacit or taken-for-granted knowledge*: Much of the knowledge we have is tacit (silent; implicit) and taken for granted; that is, it is knowledge that seems so obvious to us that we do not even realize that it *is* knowledge. While we are fundamentally dependent on this way of knowing in order to function, that which is taken for granted can frequently cause challenges, especially when people with different tacit and taken-for-granted knowledge meet and interact. Consider the difficulties that sometimes arise when health workers and their clients have backgrounds from different countries, or the tensions that may emerge when healthcare workers from different professional backgrounds work together. The aim of much qualitative research is to identify and understand tacit and taken-for-granted knowledge.
- *Gaining insight into the experiences of health and illness*: Humans are corporeal beings, and issues pertaining to corporeal well-being, suffering, and healing are central to the human experience. Exploration of the range of ways in which health, illness, and disease are experienced, articulated, and made sense of is a common use of qualitative health research.
- *Contextualizing health and illness*: Health, illness, and suffering are shaped by biological factors, but also by a wide range of social, cultural, political, historical, and economic conditions and circumstances. Qualitative research offers the opportunity to explore the contexts in which both good and ill health are shaped and embedded.
- *Exploring patient–health worker relationships*: The relationship between those who experience illness, injury, or disability on the one hand, and persons who engage in practices aiming to heal or mitigate such conditions

(*Continued*)

BOX 10.3 Typical Fields of Inquiry in Qualitative Health Research—cont'd

on the other, is a central issue in all medical systems. What are these relationships really like? What do they consist of? How are they shaped, and how do they play out? These are examples of questions that qualitative research methods may be used to answer.

- *Qualitative exploration ahead of quantitative studies*: Almost all epidemiological research is dependent on some form of previously conducted qualitative exploration. An example is the need for insight that can inform the development of questionnaires. Sometimes such insight is available from previously conducted qualitative research. In other instances, epidemiological studies need to include a qualitative component in their formative stages.
- *Exploring cases*: Case studies offer opportunities to reflect on and describe findings encountered in clinical practice. As an example of a highly influential case study, let us mention the one published by Gottlieb et al.[5] in the United States in 1981. Their short report on the unusual findings of five middle-class men with *Pneumocystis carinii* pneumonia intervened in the world, in the sense that it changed the possibilities for understanding and acting in affected communities, in the health sector, and in politics. It also represented the start of one of the most extensive research undertakings in the history of man: that pertaining to HIV and AIDS.
- *Exploring ways in which medical insight is developed*: Qualitative research can also be used to develop new perspectives on how medical truths are developed in the first place. A seminal example is Latour and Woolgar's[6] ethnographic study in a research laboratory in the United States (see also Blok and Jensen[7]). Their work demonstrated how scientific facts are not merely "discovered" in a laboratory, but *produced* through a mixture of social, literary, and laboratory practices.

10.2.2 Flexibility in Study Design

When one aims to explore in order to discover and understand, a considerable degree of flexibility[ii] is a fundamental requirement. Whenever there is something we have not yet discovered, understood, or been able to articulate, we can hardly know ahead of time what questions we ought to ask ourselves and others, who could help us develop insight, or in what contexts

[ii] That qualitative research needs to be flexible should not be taken to mean that planning of such research is unnecessary or superfluous. On the contrary, a comprehensive and rigorous planning process is a prerequisite for flexibility in qualitative research, and the basis on which changes and amendments can be made as the research process moves forward.[8]

our understanding could evolve. These are decisions that must be made based on the insight that gradually develops as the research project moves forward. A researcher engaged in exploratory research must seek positions from which discovery is possible and broad, and must be flexible enough to allow the insight that evolves during fieldwork to shape the next step in the research process.

That explorative qualitative research needs to be flexible has many practical implications. Among them is that any fixed phrasing of interview questions should normally be avoided (see Section 10.5). Also, the choice of study participants has to be continually assessed and reassessed as the research progresses. The researcher might come to realize a need to involve people with different kinds of experiences than foreseen initially, or the need to interview a larger number of people in order to understand something in sufficient breadth or depth. Furthermore, in many cases the study needs to be flexible with regard to the methods it employs. Group discussions might prove to work better than individual interviews for a certain theme, or interview data may need to be complemented with data produced through participant observation.

Finally, flexibility also means that even the overall research questions articulated at the outset of a study should not be treated as fixed. As Hilden and Middelthon[9] have argued, it is difficult to envision a successful qualitative study in which the research questions do not change to some degree during the course of the inquiry. Indeed, ongoing analysis of the data being generated will almost always suggest ways in which some research questions could and should be refined, and will in some cases demonstrate that some of them need to be changed entirely, or that new research questions need to be integrated into the study. Again, this is something that follows from the exploratory nature of most qualitative research work. It is, of course, difficult to formulate questions about phenomena one does not know exist.

In short, qualitative research designs have to be flexible[iii] and dynamic in the sense that they allow—and facilitate—productive working relationships among methods, empirical fields, practicalities, research themes, and theories (Box 10.4).

[iii] The degree to which flexibility is central to the design of qualitative studies depends on the theme under investigation, the orientation of the researchers involved, and the practical constraints under which a study is implemented (such as the time available or the access granted).

BOX 10.4 Example of a Qualitative Study Using Multiple Data Production Approaches

Olsvold[10] wanted to explore responsibility practices in hospitals. The emphasis on practices meant that the focus was not so much on formally allocated responsibilities, but rather on the way responsibility is "done" as part of everyday work in hospital wards. To undertake the study, Olsvold developed an ethnographic design with relatively short periods (1 month) of fieldwork in multiple sites. The fieldwork included participant observation in the wards as well as in meetings between doctors and nurses. Formal individual interviews with nurses and doctors were also conducted. The interviews proved to be a productive arena for further exploration of fieldwork observations, and for the "trying out" of possible understandings and interpretations that had emerged during participant observation, and the number of interviews were increased. The dynamic relationship between participant observation and qualitative interviewing made it possible not only to check factual information, but also to arrive at nuanced descriptions and analyses. Various texts and material objects were also identified and included in the data. These ranged from handwritten messages on noticeboards to detailed information about complex information systems and advanced medical technology. The exploratory character of the study and its flexible design meant that changes could be made to the original study protocol if and when necessary.

10.2.3 Traveling Along with Others

As we have already mentioned, in exploratory research one must get beyond the point where one seeks to answer only the questions one believes are worth asking. Indeed, the questions we formulate ahead of research will to a large extent reflect our own *a priori* understanding of the theme under study; not necessarily the questions that prevail, or that are most significant, in the lives of people in the context into which we hope to gain insight. Therefore, instead of demanding answers to our own preformulated questions, we need approaches that can help us learn and see the theme under investigation from the point of view of others.

Kvale,[11] and later Kvale and Brinkmann,[12] used the words "miner and traveler" as a metaphor to illustrate this point. A miner is in the business of digging and excavating, much like the researcher who aims to unearth answers to preformulated questions (such as those included in questionnaire studies). The traveler, on the other hand, is on the move, traveling through landscapes together with other travelers, learning to see the world with them. He or she is akin to the researcher who joins people in their engagement with the various situations, tasks, and themes they encounter in their lives:

> *The traveller explores the many domains of the country, as unknown territory or with maps, roaming freely around the territory. The traveller may also deliberately seek specific sites or topics by following a method, with the original Greek meaning of "a route that leads to the goal." The interviewer wanders along with the local inhabitants, asks questions that lead the subjects to tell their own stories of their lived world, and converses with them in the original Latin meaning of conversation as "wandering together with."*
>
> ***(Kvale[11])***

The traveler metaphor highlights two epistemological questions of central significance in many of the theories that inform qualitative research. First, it is untenable to conceive of knowledge as something that is ready-made and can be "dug out" fairly directly by a researcher. It is equally indefensible to assume that researchers can produce knowledge alone. We will return to both of these tenets in more detail shortly.

10.2.4 Required Closeness and Productive Distance

When traveling along with others in qualitative research projects, researchers and study participants are brought into rather close vicinity, either for a short period of time (e.g., in a one-off qualitative interview), at recurrent occasions (e.g., during repeat qualitative interviewing), or for a longer duration (e.g., during participant observation). In contrast to the rather distant relationship that often exists between epidemiological researchers and their study participants, many qualitative researchers do indeed get close to the people with whom they are coexploring one or more research questions.

However, although varying degrees of closeness are essential in qualitative exploration, there is also a concomitant requirement for distance. Indeed, all the research methods presented here must continually negotiate the seemingly opposed positions of required closeness and productive distance between researchers and their fields of inquiry. In his discussion of human relationships, Helm Stierlin succinctly remarks:

> *Seen epistemologically, distance and closeness are always related to each other. As relative measurements, the two concepts require each other in order to be meaningful. The closest closeness which still warrants this term also presupposes some distance between two objects. Where this minimum distance is lacking, closeness fades into fusion and the term is no longer applicable.*
>
> ***(Stierlin[13])***

At some point, a change in distance can radically transform any relationship to either one of alienation (lack of any closeness) or one of fusion (lack of any distance). In both instances, one is no longer positioned to see. Therefore, what we refer to as "productive distance" is a prerequisite for, and forms part of, the closeness qualitative researchers depend upon if they are to succeed in their research endeavors.[iv]

10.2.5 Knowledge Is a Communal Achievement

At the fore in discussions regarding epistemology one usually finds questions about what kind of knowledge research produces (is it true, valid, etc.) and how knowledge is produced (research methods and their use and relevance). The question of *who* produces knowledge is thematized and theorized to a much lesser extent. In many scientific discourses it may seem that the implicit idea at work is that the researcher is the (sole) producer of knowledge. In this chapter, we assert that the people we interact with during research need to be recognized both as epistemologically active subjects and as coproducers of knowledge. To use a term from Lassiter,[14] such people need to be regarded as "epistemic partners" in the research process.

Few will deny that qualitative research is brought into being through subject-to-subject interaction. However, this does not always mean that study participants are acknowledged as subjects—and not objects—of the knowledge production process. Not only is this misleading (and discourteous); how we understand the "who" of knowledge production has implications for the entire research process, including how a study is conceived and designed; how fieldwork is conducted; what issues are noticed, accounted for, and critically examined; and how data are analyzed.

10.2.6 Situating Subjects in Their Life-Worlds and in Their Social, Cultural, Political, and Historical Contexts

> *Regardless of the type of interview, there is always a working model of the subject lurking behind the persons assigned in the roles of interviewer and respondent.*
>
> ***(Holstein and Gubrium[15])***

One cannot take for granted that one recognizes the model of the subject in operation at any given time or in any given context. What one

[iv] The navigation of closeness and distance is of significance not only for researchers, but also for study participants, for whom distance affords possibilities for reflection and self-reflection, among other things.

can take for granted is that such models *are* always in operation, including in research contexts. How researchers conceive of the subject has implications not only for how they relate to the people directly involved in knowledge production, but also for the ways in which they approach the themes and questions at the core of the research process.

The model of the subject at work in this chapter holds that there is no "I" without a "me," i.e., there is no "I" without a capacity for self-reflection and a "me" as the focus of that reflection. Also, there is no "I" without a "we:" all people have a life-world and belong to one or more communities and societies (whether they are life-giving, unsatisfactory, or perhaps even destructive). As subjects, we are situated in place and time, and embedded in social, cultural, political, and historical contexts. Lastly, we are all corporeal beings engaged in diverse sets of intercorporealities.

In sum, subjects are corporeal, reflective, situated, and interconnected beings. Therefore, when designing and conducting research, the involved subjects need to be situated in their immediate life-worlds and in the social, cultural, political, historical, and contemporary contexts within which they live their lives, whatever their explicit awareness of these interwoven contexts might be.[16] It is hardly possible to do full justice to the complexities of individual and collective lives in any single study, but this does not mean that we should not use these facts as the backdrop against which we inquire about—and aim to understand—the people, practices, and phenomena we encounter during the research process. (Needless to say, one should strive for concordance between the way one understands the subject and the designations one uses to refer to those who take part in research. This is discussed in Box 10.5.

10.2.7 Summary

In this section, we have referred to qualitative research methods as exploratory methods. When the aim is to explore, the research approach must be flexible enough to enable the researcher to identify and follow up on clues that can lead to discovery. For this reason, exploratory research can hardly ever rely on preformulated questions. To understand *what* the relevant questions are must to a large degree be a part of the exploratory venture itself. To get into a position from which discovery is possible, qualitative researchers need to get close to the people with and among whom they work, but they also need to maintain a productive distance from their fields of inquiry.

BOX 10.5 Referring to People Taking Part in Research

If we consider the people we interact with during research as coproducers of knowledge, we must refer to them accordingly. In most cases, we prefer the term "participants" in order to emphasize the fact that they have participated in knowledge creation, and to stress the idea that they are subjects (and hence agents). We would normally avoid designations such as "informant" or "respondent," because these terms would seem to mediate rather opposite ideas.

There will be instances when the term "participant" may not coincide with the ways in which a person understands his or her involvement in a study. This could, for example, be the case among persons with whom a researcher interacts briefly during fieldwork (even if they know the researcher is conducting, say, participant observation in the hospital where they work). We would recommend the use of brief descriptions to refer to people like this, for example simply: "people with whom we engaged in informal conversations during the course of the study."

(This is one version of more general advice, i.e., if one does not find an established word or expression that does justice to what one wants to convey, it is usually better to provide a description in one or more full sentences than to use a term that does not accurately or adequately convey the idea one wishes to put across.)

10.3 DOING QUALITATIVE RESEARCH

10.3.1 Reading

One of the aims (and strengths) of qualitative research is to link research findings generated in a certain sociocultural setting to contextual and comparative information available in other ways. In order to be able to contextualize, researchers must spend considerable time on literature review and reading. They must become familiar both with literature on broad aspects of the setting or phenomena to be studied and with thematically relevant literature from other locations and settings. In addition, there is normally a range of theoretical work that needs to be identified and considered for its potential contribution to, and facilitation of, data analysis.

10.3.2 Identifying Whom to Involve

The way study participants are identified and included in qualitative research is considerably different than in quantitative studies. In quantitative research, sampling refers to processes aiming to identify and recruit study participants who are representative of a larger population in a statistical sense. While statistical representativity is a prerequisite for research that produces statistical measures (such as frequencies, distributions, and statistical associations), it is rarely relevant in qualitative studies for at least three reasons.

First, if the objective of qualitative research is to explore in order to discover and understand, representative samples are rarely logical. When exploring a phenomenon it makes more sense to seek out, involve, and pursue people, places, events, situations, positions, and experiences where the phenomenon is the most varied and/or the most pronounced.

Second, although qualitative research takes place with and among people, independent individuals will only rarely be the main or sole focus of the discovery process. Instead the focus is commonly on what goes on *between* people (as in patient–health worker relationships) and/or on what is *shared* between some or many of those who participate in a certain setting or belong to a particular group (i.e., that which is cultural: ideas, norms, patterns of thinking, ways of acting and being in the world that exist between people in a particular social context).[7]

Third, qualitative researchers do not always one-sidedly seek out their settings and study participants, but may also themselves be sought out, in a manner of speaking. This may be particularly true for participant observation. As Geertz[17] described it, participant observers somehow put themselves in a culture's way, and the culture "bodies forth and enmeshes" them.

When the term sampling is used in connection with qualitative research, the sampling rationale is usually significantly different from that of statistics-oriented studies (see Box 10.6). What most qualitative sampling/selection/involvement strategies have in common is a quest to include a wide and diverse range of people, phenomena, positions, and situations that are of potential relevance to the themes under exploration. To strive for broad involvement in order to maximize variation is often essential, in part because most qualitative studies aim to give complex, nuanced, situated, and contextually dense descriptions and analyses, and also because it is important to continually seek opportunities to challenge those very same descriptions, theorizations, and analyses as they are being developed.

Practitioners of participant observation have to make decisions about the setting(s) they wish to join and the people they will aim to engage with. The selection of a suitable site depends on the goal of the research

BOX 10.6 Sampling in Qualitative Studies

The term *theoretical sampling*, which first emerged in grounded theory,[18] refers to the process of selecting study participants according to their theoretical relevance for a study's emerging conceptualizations. In a less restrictive manner, the term is used to mean any theoretically informed selection of study participants.

"Purposive," "purposeful," or "strategic" sampling denote approaches that aim to select people, practices, processes, events, phenomena, places, etc. to be focused on and included in the study based on the study's aims and the various challenges and demands that emerge in its course, be they theoretical, empirical, or practical.

(it could, for example, be the existence of a health problem that prompts a study, a theoretical interest, or a more general quest for new knowledge about a certain setting or locale). Within the chosen site, a main aim in participant observation, as in qualitative research in general, is often to allow exposure to as many different views and perspectives on a phenomenon as possible. One may expect some of the variation in perspectives to be found along axes of social difference, for example income level, gender identity, religious affiliation, work status, sexual behavior, organizational affiliation, ability or disability, health status, or experience with illness, suffering, and healing. But perspectives will invariably also vary within such categories, and variation may also exist along axes one did not expect before the study started. Thus, to identify how, and along what axes, perspectives vary will often form part of the research work itself.

10.3.3 Recording it All

Qualitative researchers seek to preserve the occurrences and utterances that transpire during conversations, interviews, and interaction with study participants. This is typically achieved through the writing of notes and/or through audio recording and transcription.

10.3.3.1 Fieldnotes

Taking scratch notes (i.e., to jot down keywords and short sentences) is a characteristic and constantly ongoing fieldwork activity in participant observation. In order to remember what one has observed and taken part in, and the associations these activities have engendered, one jots down quick notes throughout the day on what one witnesses, experiences, does, and learns.[v] Scratch notes may also be used in qualitative interviewing, either alone, or in addition to audio recording.

At the end of each field day, the researcher characteristically uses the scratch notes as a memory aid to write the fieldnotes: a more coherent and detailed text describing what has transpired. Typically one also incorporates comments to oneself about what one has realized and learned, what one has not understood, issues requiring further exploration, and ideas and proposals about gradually emerging insights and theorizations. These comments form part of the ongoing analytical process that is central in the qualitative research process.

10.3.3.2 Recordings and Interview Transcripts

Anyone who has gone through the process of transcribing a recorded interview knows that (apart from being time-consuming and at times tiresome) such work is useful in at least three ways. First, listening to recordings affords the opportunity to re-experience one's encounters with study participants and their narrations. Second, it allows one to reflect upon one's own way of engaging in conversations, and may serve as an aid to developing and improving interviewing approaches and practices. Finally, listening to oneself provides an opportunity to evaluate the (reflected and unreflected) choices one made about which topics to pursue and which not to pursue. By identifying and thinking through the many potential trajectories of interviews and the paths one did and did not follow, one may become aware of the (analytical) considerations that shaped one's choices and (tacit) evaluations of what was important in the situation.

Interviews are inevitably sensorial encounters. When an interviewee and interviewer meet at a specific point in time at a particular place, the meeting is characterized by corporeal immediacy, and as such the meeting is an intercorporeal event. When converting recorded speech to text, one

[v] We recommend that space be set aside on each notebook page for the researcher's own comments (e.g., in the margins of the page). Ideas, associations, doubts, questions, and reactions that come to mind during interviews or fieldwork can quickly be jotted down here, without the risk of mixing one's own commentary and contributions with those of the study participants. Knowing that one has noted these thoughts may make it easier to concentrate on, and engage in, ongoing fieldwork and interviews, since it may remove distracting thoughts like "I have to remember this."

should keep in mind that it may be difficult later, when the recording is gone, to remember how things were said, and the context in which they were uttered. Was the statement made ironically, with sadness, or in joy? Was it a figure of speech or something intended literally? Questions like these can at times only be answered in the sensorial context of the utterance itself. Although the researcher's memory can compensate to some degree for lack of context, as can listening again to recordings, it is often helpful to integrate notes about the character of utterances or use transcription symbols when creating a transcript.

10.3.3.3 Analytical Writing

The writing of memos and analytical notes throughout the research process is promoted in most qualitative research traditions. The purpose of such writing is to keep track of one's own (and others') reflections as the research progresses, to remember questions that surface, and to record ideas about possible patterns, emerging interpretations, and incipient theorizing.[vi] The relevance and value of this kind of writing may become clear only as the study advances, but we recommend making analytical note writing a regular activity throughout the study. One should not be too self-critical about what is thematized in these notes, at least not initially.

10.3.3.4 Journal Keeping

In addition to the other records produced during the course of a qualitative study, we advise that researchers keep track of the research process, and the notes, recordings, and transcripts produced, in a running journal.

10.3.3.5 Anonymizing and Storing Data

Changes that will ensure the anonymity of study participants must be made when typing up notes and transcribing interviews. Qualitative researchers typically do this by omitting or changing potentially person-identifying information, such as names and revealing circumstantial information.

[vi] It is advantageous to organize memos so they can be easily retrieved and connected to relevant parts of the data material. Many qualitative data management packages can handle memos in this way.

We advocate the use of pseudonyms when referring to study participants, as opposed to numbers or random letter combinations. Numbers can easily introduce a sense of alienation, hiding the fact that real persons took part in the study. Pseudonyms also make it easier for readers to follow individuals, and their stories, across paragraphs (and perhaps even across publications).

Even when identifying information has been removed, it is important to store the data safely. Hard copies should be kept in locked storage when not in use, and digital copies should be stored on computers without an Internet connection. Many research ethics committees set specific storage requirements for data and may demand that data materials be deleted after a specified period of time.

Electronic storage of audio recordings may pose a particular challenge. There is nothing anonymous about a voice, and audio-recorded material therefore demands special precaution. Apart from general data safety measures, options include deleting sensitive sound sections from the original audio files (such as those containing names), and manipulating recorded voices with the assistance of sound filters.

10.3.4 Qualitative Data

The different texts produced during the course of the study (e.g., fieldnotes, transcripts, and analytical notes and memos) make up the bulk of the data in a qualitative study. They are not, however, the only texts that may be included. Texts written by others are also often incorporated, e.g., letters, e-mails, newspaper articles, reports, manuals, or fictional writing identified or acquired during the course of fieldwork.

Apart from texts, qualitative data can also consist of drawings, photographs, audio and video recordings, and/or a collection of objects that are used in the context being studied.

A less tangible part of qualitative data consists of the various memories of fieldwork held in the mind of the researcher. Such memories may be an invaluable tool in the research undertaking. They can assist in the reading of fieldnotes and transcripts, for example by adding information about the tone of utterances, facial/corporeal expressions, and emotional climates. Moreover, they may assist in understanding choices that were made, for example with regard to why a particular topic was pursued or a certain phrasing chosen. (In cases where researchers work on fieldnotes and/or transcripts they did not help create, this is of course not possible.

This is one of the reasons why we would normally promote designs where at least one of the investigators is closely involved in all stages of the study.)

10.3.5 Summary

In this section, we have discussed the reading and writing activities that characterize qualitative studies. Investigators need to read in considerable breadth and depth to be able to contextualize empirical data, and they must write throughout the data production phase in order to preserve, engage with, and reflect on experiences and utterances. We have also drawn attention to how researchers select and include people, phenomena, and settings in qualitative studies. The overriding concern in such selection/inclusion processes is often to maximize variation.

10.4 WHAT IS QUALITATIVE INTERVIEWING?

10.4.1 A Triadic Encounter

[vii]At their core, most qualitative studies involve conversations that take place directly between a researcher and the people whose lives and circumstances are being enquired into. Such conversations are fundamentally interactional, and Kvale[11] has captured the characteristics of this interaction well in his seminal book entitled *InterViews* (see also Kvale and Brinkmann[12]), in which the interview is described as "a professional conversation" and as

> *...an inter-view, where knowledge is constructed in the inter-action between the interviewer and the interviewee. An interview is literally an inter view, an inter-change of views between two persons conversing about a theme of mutual interest.*
>
> ***(Kvale and Brinkmann[12])***

The hyphens in this quote are significant. However guided or free-flowing an interview might be, it will always be an interactional event in which the persons who interact are engaging in a collaborative "meaning-making occasion."[20]

[vii] In her discussion of the history of the qualitative interview, Jennifer Platt[19] pointed out that interviewing has not always been thoroughly theorized. In this section, we offer our take on qualitative interviews.

While we fully agree with the conceptualization of the interview as a collaborative meaning-making occasion involving at least two agents, we are hesitant to accept any implicit or explicit understanding of interviews as essentially dyadic (two-sided) relationships, i.e., involving nothing more than the interviewer and the interviewee. Rather, we propose that an interview should be conceived of as a triadic (three-sided) relationship involving (1) the interviewer, (2) the interviewee(s), and (3) the theme(s) or topic(s) they are engaging with and that engages them. Our suggestion is inspired by the American philosopher Charles S. Peirce and his seminal theorizing of sign-activity (or "semiosis") (Box 10.7).[16] It is also akin to the Norwegian philosopher Hans Skjervheim's[21] understanding of a conversation:

> *The relationship here is triangular, between the other, myself and the subject matter, such that the latter is shared between us.*
>
> ***(Skjervheim[21])***

A conception that incorporates the (unavoidable) presence and role of the theme of the conversation as a third party in the interview relationship captures more accurately what is involved in interviews than conceptions of them as dyadic. The theme is not a passive but a dynamic party to the meaning-making processes of the interview.[viii]

The fact that the third party—the theme of the interview—is shared by the other parties does not mean that it is identical to them. One of the

BOX 10.7 Medicine as a Triadic Relationship

Peirce's theorizing of relationships as irreducibly triadic is at times referred to as "Peirce's theorem." As de Waal[22] points out, this theorem holds "that it is impossible to define triadic relations in terms of simpler ones (e.g., dyads)."

In the context of this book it might be of interest to note that an analogous conceptualization is found in the *Of the epidemics* of the Hippocratic Corpus,[23] where the art of medicine is conceptualized as a practice and relationship of three:

> *The art consists in three things—the disease, the patient, and the physician. The physician is the servant of the art, and the patient must combat the disease along with the physician.*

[viii] Let us quickly add that this does not involve any claim that themes are conscious agents (or persons), but that the topic is nonetheless an inevitable, dynamic, working party in the interview.

obvious differences is that the experiences being inquired into are primarily those of the interviewee (as opposed to those of the researcher).

10.4.2 Coproducing Data

It follows from our reference to interviews as meaning-making encounters that they are occasions for production of data and insight. This view stands in contrast to discourse that portrays research as processes aiming at the "collection" of data or the "gathering" of research information. As Holstein and Gubrium[15] described it,

> *This commonsensical, if somewhat oversimplified, view suggests that those who want to find out about another person's feelings, thoughts, or activities merely have to ask the right questions and the other's "reality" will be revealed.*
>
> ***(Gubrium[15])***

The idea of the subject "lurking behind" the "commonsensical" conception Gubrium referred to in effect portrays people akin to passive reservoirs of preexisting knowledge. Researchers may be able to fish out this knowledge with the right "fishing equipment." While the role of the researcher is conceived of as more active than that of the research participant, the activity he or she is seen as engaging in is limited to that of fishing. Within this conception, the knowledge itself is already there.

We agree with Gubrium and Holstein that conceptions like these are not tenable. Indeed, we think that few people would be prepared to say that their knowledge about the social world is something that has always been there. First, most of us would readily acknowledge that there are many things we do not fully understand or know about the social contexts of which we are part. Second, it would hardly seem plausible that our understandings of, insights into, and beliefs about ourselves and our social contexts could have come about without actively engaging with that world; experiencing it, talking about it, and thinking about it. In short, our insight is ceaselessly produced and reproduced, revised and nuanced, refined or altered, rejected or confirmed.

Seen from this perspective, qualitative interviews are occasions for knowledge *creation*. Knowledge is not extracted in them, nor is it created by one person in isolation. Rather, it is co-constructed through iterative processes of reflection and articulation involving both the interviewee and the researcher.

10.4.3 Reflecting and Articulating

The research interview may be conceptualized as a "space for reflection and a room for articulation" (see Box 10.8).[24] While articulation is of

BOX 10.8 Interviews as Spaces for Reflection

During interviews for a study on being young and gay in the context of the HIV epidemic in Norway in the 1990s, the study participants would often interrupt themselves and say things like, "Oh, *that* may be the reason"; "Maybe *that* was what he meant"; or "I have never thought about *that* before." Utterances like these clearly reflected the ongoing reflection that occurred during these conversations. One study participant emphasized this aspect of the interview experience when, after the last interview, he made the following statement as part of his comments about being part of the study:

> *[When you] sit here and talk; you see yourself [from the] outside in, and [you] become more analytic.*
>
> ***(Middelthon[24])***

course central to the interview process, one would be amiss to conceive of an interview as talk only, or to conceive of narration as nothing more than a person expressing him- or herself verbally. When articulating an experience, narrators inevitably come to form part of their audience; they become a listener of their own stories. A space and possibility for reflection and self-reflection is created as the topic of the inquiry becomes external to the narrator and intrinsic to the endeavor of the inquiry. It is this simultaneous functioning of the topic as both part of and apart from that makes reflection possible. Compare this with our previous discussion of required closeness and productive distance in qualitative research.

10.4.4 Summary

In this section, we have characterized qualitative interviews as interactional, triadic, meaning-making events where insight and data are produced through iterative processes of articulation and reflection.

10.5 DOING QUALITATIVE INTERVIEWING

10.5.1 Engaging in Conversation (Rather Than Asking Questions)

Most qualitative interviews are only partially structured. On the one hand, this means that they lack the strict regimes that are so typical in quantitative survey interviews, where researchers ask preformulated questions in a predetermined order. On the other hand, qualitative interviews are rarely entirely unstructured, since they are usually conducted within the framework of one or more research questions that shape the interaction with study participants.

In most cases, therefore, qualitative interviews can aptly be called semistructured. While they are likely to have some degree of premeditated "direction," they also have considerable degrees of flexibility and openness built into them, making it possible to incorporate and pursue emerging themes and topics as they arise, both during the course of each individual interview and during the course of the study itself. The degree to which interviews are open and flexible varies from study to study, and between different qualitative research traditions.

Semistructured interviews are typically not conducted with the guidance of an interview schedule containing fixed and predetermined phrasing of questions, nor do they have any predetermined order in which to pursue themes. The questions and issues that are dealt with, and how and when they are brought up, depend to a large degree on the stories being told, the perspectives of the study participants, and the way conversations develop *in situ*. Neither the style, content, nor syntax of the interview is fully decided ahead of time; it is established during the course of the interview itself.

10.5.2 Interview Guide

As a memory aid, the researcher often uses a carefully planned interview checklist or guide, typically a list of keywords or short sentences about the issues the study aims to understand. Our recommendation is that the interview guide contain no full sentences, and in particular, no sentences formulated as questions. There are several reasons for this recommendation.

First, in qualitative interviewing the overarching aim is to set the stage for conversations where comments, questions, and probing are grounded in the experiences, stories, and opinions put forward by the study participants. Preformulated questions may shift the focus of the conversation away from participants' perspectives and towards the perspectives of the interviewer and consequently undermine the potential to get at the experiences of the participant.

Second, preformulated questions tend to fix the interviewer in a preconceived logic rather than to promote his or her attentiveness and responsiveness. By responsiveness we mean an orientation where the researcher continues to respond to and ask about topics that have been brought up until some mutual understanding of them has been reached.[25]

Third, as we have already mentioned, we conceive of interviews as rooms for reflection. Reflection requires time and space, and both will often vanish if the interviewer is bound by a set of preformulated questions.

Fourth, the questions that the researcher formulates *in situ* are more likely to fall in line with the style and tone of the individual interview than when they are preformulated.

Fifth, similarly phrased questions do not mean the same thing to everybody. In order to facilitate a situation in which participants understand the intentions behind our statements and questions, there is a need to adapt both to the person with whom we are conversing and to the context of the ongoing conversation.

In most cases, the interview guide should be regarded as a living document. That is, it should be worked upon between interviews and adjusted and refined as the research process moves forward. This allows each new interview to build on those that have preceded them, and the study to deepen and widen its exploration of questions, impressions, insights, and tentative analyses that arise during the course of the inquiry.

An example of how an interview guide can be prepared is offered in Box 10.9. The interview guide can also include (or consist of) vignettes, such as a brief text (e.g., a description of a concrete case or experience) that may serve as an entry point, or as prompts during the course of a conversation or discussion.

BOX 10.9 What an Interview Guide Might Look Like

In an interview guide developed for a study on food, eating, and feeding in the context of the medicalization of everyday life,[26,27] one of the general themes listed was "daily food." This entry was followed by a list of keywords, including:

- "food eaten at breakfast,"
- "food eaten for lunch,"
- "alone or together with someone,"
- "place where food is purchased," etc.

The interview guide did not specify any questions, such as, "What do you normally eat in the morning?" or "What kinds of food do your children prefer?" Although the researcher may have ended up asking these exact questions, hopefully this occurred because they fit well into the situation of the interview; not because this particular phrasing had been predetermined.

10.5.3 Conducting a Qualitative Interview

In order for interviews to be venues for exploration, the perspectives and experiences of the interviewee must be allowed space and time to play out and become the focus of joint consideration. Thus, the interviewer cannot steer the discussion too much. On the other hand, both steering and probing will often be necessary to ensure that the interviewer understands the perspectives presented, and that interviewer and interviewee are actually talking about the same phenomenon.

Our tacit, everyday knowledge of conversations may be a help in the pursuit of these aspects of qualitative interviewing. Even neophyte interviewers are normally seasoned conversationalists who can draw on their vast previous experience of verbal interaction and interpersonal meaning-making. This is a valuable competence that interviewers should and must draw on. At the same time, it should be remembered that many of us have grown up in "interview societies"[28] where some of the interview formats we (and our study participants) are accustomed to are not of a kind that engender open exploration into meaning and understanding. On the contrary, many interviews are conducted with different aims altogether, such as to gather mere factual information (e.g., some news interviews), to evaluate (e.g., job interviews), to adulate (e.g., "star" interviews), to condemn (e.g., some critical journalism), to scandalize (e.g., tabloid media), or to entertain (e.g., gossip media). None of these are aims that are sought after in qualitative research (nor is the elevated position of the interviewer which is sometimes inherent in some other kinds of interviews). Our experience with these kinds of interviews may nonetheless help shape expectations of, and performances in, qualitative interview encounters. So it may be important to be aware of and address this possibility. One can emphasize that the encounter will be rather more like a conversation or a discussion than a media interview, and that the aim is to learn and understand things that the study participant has experience with and insight into.

In most interviews it is wise to strive for conversations that can be grounded in *concrete* events, situations, or relationships to facilitate narration. For example, one may ask the study participant to talk in more detail about the last time she or he experienced something specific, or ask whether it would be okay to talk through an event or experience slowly and step by step. As a narrative develops, the researcher may probe into certain aspects of the story in order to prompt more details.[ix]

Let us finally mention that it may be productive to make deliberate use of metacommunication in interviews; that is, to thematize ongoing communication between the interviewee and the interviewer. For example, metacommunication can be used to explain why one is bringing up a certain issue ("I am asking about this because...") or to request a more detailed explanation ("this is probably trivial to you, but I am not sure I got it completely right"; "it is exactly such small things I think I need to understand properly").

Drawing attention to why one asks a particular question, or why one keeps on probing, may lead to opportunities for inquiry that differ from, and complement, those facilitated by other modes of communication. Metacommunication of the kind just mentioned may also contribute to ensure that interviewees become informed epistemic partners in the study.

10.5.4 Repeat Interviews

Our experience is that qualitative interview studies often benefit from repeat conversations, and our recommendation would be to aim for no less than two encounters with each interviewee.[x]

One of the reasons for this recommendation is that first encounters often provide for somewhat special conversations. Indeed, first encounters bring together people who have to get to know each other from scratch. This invariably takes some attention and time, and has an impact on the ways in which people interact. While a first conversation is no better or worse than other conversations, they are qualitatively different from conversations between people who have become acquainted with each other. Repeat conversations also make more time available, as the duration of the interviewer-interviewee relationship is prolonged, providing for more opportunities to explore. In addition, the time in between conversations provides both the researcher and the study participants with an opportunity to reflect upon issues and questions, and these reflections can be tapped into and discussed in a subsequent meeting. Issues that may

[ix] Ethical considerations must always be integrated into the researcher's ongoing evaluation of how the interview develops. He or she must continually consider whether a certain question could and should be asked in the particular situation.

[x] Many studies will benefit from a higher number of conversations with some or all of the participants.

have been unclear in a previous conversation can also be revisited for clarification, which may prevent misunderstandings and help secure the quality of the data.

On a slightly different note, let us mention that repeat interviews also give the researcher an opportunity to bring up issues raised by one study participant with one or more of the others, and to come back to a discussion in the light of additional perspectives and inputs. By taking advantage of this opportunity, interviewing can be used to cross-check descriptions and test out elements of one's nascent theorizing.

On a practical note, the knowledge that there will be future opportunities to continue a conversation may allow the interviewer to worry less about time constraints, thus contributing to a conversation climate that permits the dialogue to take, or find, its own path. Lastly, repeat interviews provide an opportunity to follow people and phenomena over time.

10.5.5 Summary

In this section, we emphasized the conversationalist role of qualitative interviewers and described how the interview guide can function as a tool for research conversations. We recommended active engagement with, and continuous modification of, this tool throughout fieldwork, so as to grasp, and remind the researcher of, the (currently realized) need for further insight. We also pointed out why qualitative interview studies often benefit from repeat interviews.

10.6 WHAT IS PARTICIPANT OBSERVATION?

10.6.1 Producing Knowledge in and through Interaction

Participant observation is a research strategy that aims to produce knowledge both on and through interactions between people. Social interaction is thus central to participant observation in two significant ways. First, the main objective of participant observation is to explore phenomena as they emerge in interactions between people in a given context. Second, knowledge generation occurs through personal interaction between researchers on the one hand, and the people that make up the social context under study on the other. What this affords are opportunities to engage in a process of socialization and enculturation into the context and become part of the interactions and practices that produce the phenomena being investigated.

10.6.2 Participating

Researchers who practice participant observation join a social context with the aim of gaining insight into and developing understanding of it. They typically spend considerable time together with variously positioned "insiders," witnessing and sharing the events, experiences, joys, worries, and sorrows that occur regularly, and not so regularly, in their individual and collective lives. In this way, the researcher gradually becomes part of, and to some degree a co-insider in, the setting, practice, or phenomenon under study. At the end of fieldwork, he or she might have gained understanding of prevailing points of view, grasped the meaning associated with various phenomena, and developed insight into some of the shared ideas and "rules" that shape social practices.

Learning in and through participation and interaction may happen in a multitude of overlapping ways. It occurs to no small degree because one shares time and space with others. This affords opportunities to notice (consciously, but also subconsciously) how people act, react, and interact in different situations. Other aspects of the learning process take place through talk and discussions. One can learn from what people say, but also from how they express themselves, the emotions and reflections that talk gives rise to, and the actions that accompany, precede, or follow speech events. One may also learn by paying attention to the relationships between verbal and nonverbal communication, and by noticing what people do not deal with verbally in the first place. Learning may be particularly efficient when one engages in the doing of things (socially and practically) together with others. One may learn through observing what others do, through (conscious and subconscious) attempts to imitate others, through verbal feedback on one's attempts to act aptly, and through the complex interactive processes that coperformance of activities and situations entail. In short, participation offers the opportunity to learn through seeing, listening, talking, discussing, feeling, touching, moving, and doing.

The knowledge a researcher may acquire in this way will be of different kinds. On the one hand, there will be the knowledge the researcher can formulate verbally. However, some of what he or she learns will also be "embodied" or "corporeal" knowledge, i.e., knowledge that resides within one's body but that one may not necessarily, entirely, or immediately be able to express in language. The significance of embodied knowledge (or "corporeal knowing") can be substantial.[29,30] Consider how a certain smell can sometimes evoke memories of a past event. In participant observation, a researcher will gain a plethora of such sensorial knowledge that can be drawn on in the research process. Participant observers will also acquire a number of (social and practical) skills, another kind of knowledge that to a large extent resides in the body.

As an example of a skill a researcher might consciously decide to acquire, consider the possibility of learning how to do "intermittent self-catheterization." This is a procedure some patients with urinary bladder problems regularly perform on themselves in order to empty their bladder of urine. How self-catheterization is done can be explained in words, at least to a certain extent. However, to give a full verbal account of all the acts that must be performed in order to carry out the procedure, and the thoughts and emotions that may accompany it, would be exceedingly difficult.[xi] To acquire personal experience with this procedure would make a significant difference for a researcher who aimed to inquire into patient perspectives on a condition that requires self-catheterization to be performed on a regular basis.

Participant-observing researchers may develop a range of social and practical skills during fieldwork. Some of these can be subsequently explicated (following discussion and reflection), whereas others may remain (partially or fully) unarticulated (but will nonetheless contribute to inform the impression the researcher develops of a phenomenon, practice, or process). A study exploring the importance of "knowing bodily" is mentioned in Box 10.10.

As an approach to exploration, participant observation offers opportunities complementary to research methods that rely mainly on verbal acts. There are many reasons why people do not, and cannot, always express

BOX 10.10 Learning about Bodies through Bodies

To explore corporeal ways of knowing was among the reasons why Langaas[31] included participant observation as one research method in her study of the ways in which physiotherapy students learn through and about bodies (their own as well as those of their fellow students and their patients). She participated in the students' learning process in order to gain new insight into how physiotherapists acquire a physiotherapy-specific way of knowing bodily.

[xi] *The tacit dimension* by Michael Polanyi[29] discusses how "we can know more than we can say."

themselves in words. As we have already mentioned, much of the knowledge we have is tacit, and some is too complex to bring into language. Moreover, we never have ready-formed, transparent, complete, or final answers to all the questions others may ask of us, and there are always things we do not say as part of our ongoing process of styling and representing ourselves.

10.6.3 Observing

We have so far highlighted the participatory role of the researcher in participant observation; that is, the opportunities for exploration and knowledge creation that are afforded when one is present and taking part in a certain setting. While participation is essential in participant observation, it is obviously not sufficient. By being an "involved and vulnerable participant,"[32] researchers may learn how a setting or practice is experienced and produced, but to explicate this learning they must also be able to observe, notice, record, and analyze.

As Dewalt et al.[32] have pointed out, combining participation and observation into one concept and practice is in some sense paradoxical. While participation requires closeness and involvement, observation requires a degree of distance and detachment. Thus the two main aspects of participant observation create some tension. Yet, it is this very tension that researchers exploit in participant observation to generate insight and produce knowledge. To have different preconceptions than those that prevail in the study setting—to take other associations and meanings for granted—is a resource that creates a productive contrast. This contrast renders it possible to appreciate and explicate that which one might otherwise have left unnoticed precisely because it was taken for granted.

10.6.4 Summary

In this section, we have seen that in participant observation researchers join people in a few or many of their life contexts. This affords opportunities to learn and understand not only through conversation and discussion, but also through the socialization and enculturation that follows when people experience and perform social and practical acts, tasks, and practices together. While the researcher aims to genuinely join and participate in the setting, he or she is at the same time also always a researcher. We noted that this double role is associated with tension, but that this very tension creates opportunities to explicate patterns that might otherwise have remained tacit and taken for granted.

10.7 DOING PARTICIPANT OBSERVATION

10.7.1 Settings

Participant observation can be employed in a range of different settings. It could be an entire community (e.g., a village, as has been the case in much traditional ethnographic research), a more limited sociocultural unit (e.g., a hospital department or a nongovernmental organization), or a group of people with something in common (e.g., characteristics, experiences, or goals). Participant observation can also take place across countries, communities, and contexts, such as when researchers follow ideas, organs, people, patients, or metaphors across sites and communities (multi-sited ethnography[33]). Box 10.11 presents some examples of settings in which participant observation has been used in qualitative health research.

BOX 10.11 Examples of Settings for Participant Observation

In the three qualitative studies below, researchers engaged in participant observation among health-care workers in a Norwegian hospital,[10] among same-sex attracted men in Tanzania,[34–36] and among young patients attending an obesity clinic in Norway,[37] respectively.

- In Olsvold's[10] study of "responsibility as practice" in medical care, the study setting consisted of several hospital departments. The researcher took part in patient care, attended meetings, and was regularly present to observe daily hospital life. Throughout the fieldwork, she paid particular attention to how responsibility was performed by doctors and nurses.
- In Moen's[34–36] study of male same-sex practices in the context of the HIV epidemic in Tanzania, the research setting was the various situations and contexts that same-sex attracted men are part of and bring into being in the urban environment of Dar es Salaam. Among other things, the researcher spent time with people at home, at work, and when they went out to parties, concerts, and clubs; came along for HIV testing and hospital visits; attended funerals; and visited family members, neighbors, and friends.
- In his study of children with large bodies, Wathne[37] wanted to explore the experience of having a large body in ordinary everyday tasks and activities. He developed an approach where leisure activities (chosen by the individual child), such as bowling or mini golf, became the main setting for participant observation. The bulk of the fieldwork was spent participating in such activities.

10.7.2 Participating in What?

In participant observation, the researcher may take part in a broad range of life settings or join people in a more limited and specified assortment of practices, relationships, and contexts, depending on the thematic scope of the study. The kind of knowledge the study aims to produce must shape the (evolving) decision about the activities in which one should partake. An important question to ask is always whether the range of contexts is sufficiently broad and varied. For example, for a study aiming to investigate patient experiences, it would hardly ever be sufficient to only participate during patient visits to a clinic or hospital.

10.7.3 Types of Research Questions

Participant observation can be used to explore the entire range of themes and topics that qualitative research methods are generally used for, and may be particularly useful whenever the aim is to understand a phenomenon or a setting from the perspective of those who live, experience, and/or are affected by it. It is also a research approach that may be exploited in participatory action research,[38] i.e., in research where the aim is to use knowledge production as part of a process aiming to address and change circumstances that contribute to bring about a certain health problem. An example of such a project could be one in which public health professionals and community representatives jointly engaged in production, evaluation, and application of knowledge with the aim of reducing HIV transmission in a given community.

10.7.4 Degrees of Participation and Participatory Roles

The degree of researcher participation in participant observation varies. The researcher may adopt the role of a "peripheral member researcher" at one end of the spectrum, or a "complete member researcher" at the other.[39] In between these extremes, there is a range of possible roles as "active member researcher," and it is roles within this latter-mentioned range that most researchers strive to emulate. A peripheral member researcher risks remaining too far outside of the social interaction on which the research is focused, whereas a complete member researcher risks becoming fused into the research setting and the loss of productive distance that effective observation depends upon. Active member researchers aim to become actively involved in the setting they study and closely identified with the group they work with and among. However, in order to uphold their researcher role, they must not fuse with it entirely. To guard against this, the researcher must actively strive to maintain awareness of her or his role, practice self-reflexivity, and consider spending time away from the setting from time to time during fieldwork.

Coy[40] has pointed out how finding a "natural role" within the research setting may be an advantage, because insiders will then, explicitly and tacitly, attempt to educate the researcher so that he or she "has as little impact as possible on the social system" they are newcomers to. When the outsider makes mistakes, misunderstands, or causes embarrassment, people will provide implicit and explicit guidance on how to act more suitably. In essence, the researcher becomes an apprentice of the sociocultural setting under study.

Researchers of different ages and with different gender identities, family situations, professional backgrounds, interests, and skills will often find that there are many meaningful roles open to them in most study settings.

While social learning may start as soon as participant observation is under way (and may be particularly intense in the beginning), a considerable amount of time is required to develop familiarity with, and insight into, the complexities of a sociocultural system. Therefore, participant observation has traditionally entailed rather long periods of fieldwork. However, if the issue one aims to explore is narrow and specific, shorter fieldwork periods can also be useful.

10.7.5 Language

In participant observation, most researchers depend on knowledge of the language(s) spoken in the study setting. It could be a general vernacular, a dialect, or a professional or technological vocabulary. To be without (or only have limited) language abilities limits researchers' opportunities to participate and understand, and the use of an interpreter complicates the socialization process, which is central in participant observation.

However, it may not always be necessary to be proficient in a foreign language before fieldwork starts; one can meaningfully integrate the learning of a language, dialect, or vocabulary into the research process itself. As one learns a new language, it is not only the lexicographic understanding of words that has the potential to develop, but also one's insight into the webs of meaning in which words and sentences are entangled.

If interpreters must be used, we recommend that one aims to involve them in roles as coresearchers.

10.7.6 Access

Participant observation presupposes that there are people willing to let a researcher take part in, and inquire into, their setting (be it a clinic, a laboratory, a nongovernmental organization, or a local community). While some people may be unfamiliar with participant observation as a research method, and some may think of it as awkward or intrusive, our experience is that people are often welcoming, and many find it interesting to be part of participant observation.

Dewalt et al.[32] have noted that chances for access are generally better when the researcher has a genuine interest in the topic and setting that he or she intends to study, and when the researcher and study participants discover that they have things in common (which they almost always do). These could be shared interests, characteristics, or experiences. It could also be a shared interest in the study itself.

10.7.7 Observing

Observation is the process of noticing and recording whatever goes on and transpires in the field. While it will never be possible to observe all features of a setting, much of "the devil is in the details." That is, cultural patterns can often most effectively be detected if one pays close attention to the details of mundane, everyday features, occurrences, and experiences. Repeated occurrences are similarly significant.

Different authors have proposed different schemes to help researchers sharpen and improve their observation practices. Wolcott[41] recommended that the researcher systematically alternate among four techniques: observation in broad sweeps, observation of nothing in particular (to facilitate the detection of that which stands out), searches for paradoxes, and searches for problems facing the group under study.

Using a scheme like this may assist a researcher in maintaining an active and conscious awareness of the observation process and help her or him take advantage of the different ways of seeing that may be enabled when different questions are posed.

In another scheme, Glesne[42] recommended that observers alternate between observations of

1. the setting,
2. the actors in that setting (who they are in terms of characteristics and backgrounds, how they dress, what they say, what they do, and how they interact),

3. the events the actors experience and create,
4. the acts these events are made up of, and
5. people's gestures as they carry out those acts.

To notice that which is typical and regular may help the researcher form ideas about patterns, which is a central aim in the research process. However, continuously attempting to contest and challenge one's emerging notions about patterns must also be an integral part of the process.

Significantly, one should not only search for observations that may serve to confirm an observed pattern, but also systematically search for potentially reorienting or disconfirming cases; that is, one should strive to make observations that might not fit with the (imagined) pattern (strive for disconfirmation) and/or observations that might modify it (strive for reorientation). Through these challenging exercises the researcher aims to continually revise and reformulate his or her understanding, descriptions, and explanations, until the range of observations made has been adequately accounted for.[39]

To help avoid misguided impressions and conclusions, one should assume that virtually everything in the field must be learned more or less from scratch. When something seems recognizable, one may at times have grasped it accurately with the help of experiences and insights from another and more familiar cultural setting. However, it is not unlikely that one is projecting understandings and explanations onto the research context. This dilemma demands that researchers systematically and rigorously question the ideas they develop about what is going on around them in the field. "Is this really the way I think it is?" "If I attempt to see it from another perspective; do I see it differently and, if so, in what way(s)?"

10.7.8 Summary

A central idea in participant observation is to join people in their ongoing lives. Health-related participant observation research can take place in a range of settings, e.g., in a community, in an institution, or among people with certain characteristics, challenges, or conditions in common. The participant-observing researcher aims to learn by being present and taking part, while at the same time engaging in systematic observation of what transpires.

10.8 QUALITATIVE DATA ANALYSIS

10.8.1 An Ongoing Analytic Venture

Let us state at the beginning of this section that we consider data analysis to be an inherent (and inevitable) feature of the *entire* research process.

This should not be taken to mean that data analysis takes the same form, or is necessarily equally intense, throughout a qualitative research project. As Coffey and Atkinson have put it:

> *The process of analysis should not be seen as a distinct stage of research; rather it is a reflexive activity that should inform data collection, writing, further data collection, and so forth. Analysis is not, then, the last phase of the research process. It should be seen as part of the research design and of the data collection. The research process, of which analysis is one aspect, is a cyclical one.*
>
> ***(Coffey and Atkinson[43])***

Analysis is an integral part of the qualitative research process from beginning to end, and it is to a significant degree a shaper of that process. Methodological considerations, data production, analytical work, and theorizing are not only mutually dependent elements of qualitative research; they are mutually constitutive components of it.

Some examples of the different kinds and levels of analytical work that take place during the course of a qualitative research project are:

- pondering which themes to pursue,
- choosing one or more research methods,
- making decisions about whom to join and/or talk to,
- making decisions about codes and categories,
- searching for patterns,
- making tentative inferences,
- developing provisional analyses,
- selecting what to write about and how,
- relating an empirical event to a theoretical insight or theoretical framework, and
- developing theories.

10.8.2 Developing an Intimate Acquaintance with the Data

As a study moves forward, qualitative researchers need to develop an intimate acquaintance with the data produced during the course of the research. They must somehow come to "hold" the empirical material mentally if they are going to adequately describe and analyze it (see Box 10.12).

Many experiences, acts, and practices may contribute to becoming intimately acquainted with the data. The most fundamental of these stem from "having been there," i.e., the qualitative researcher's direct engagement with the field. Such engagement generates experiences and insights that are drawn on in the analytical process; they come to form part of the researcher's knowledge horizon. Typing up notes and interview transcripts adds to

BOX 10.12 Widening Epistemic Horizons

Being and engaging with empirical data material over time contributes to the development of an intimate acquaintance with it. It leads to a broad range of (embodied) experiences and memories, which in turn will widen our epistemic horizons. It provides empirical material that can be analyzed, and it sustains, perhaps intensifies, the process in and through which this material becomes accessible (i.e., the process is no less significant than the product in this respect, and the form no less significant than the content). What we refer to here is akin to Polanyi's[29] concept of "indwelling." The force of such *dwelling in* knowledge (explicit and implicit, focal and subliminal) can be experienced when new associations emerge seemingly out of the blue; when we suddenly juxtapose phenomena we had not even envisaged the possibility of relating to each other; or when we unexpectedly become aware of a link between theoretical insight and empirical events.

It should be cautioned that indwelling may make us more prone to take for granted things that might benefit from conscious reflection. Therefore, associations emerging in this way need to be rigorously challenged and systematically tried against the empirical material and against existing knowledge and theories.

the familiarization process, as do a number of other activities, such as reading and rereading notes, pondering the data, engaging in discussions about the data material, and actively considering the data in different perspectives (e.g., through the lenses of empirical research from elsewhere or through the lenses of different relevant theories).

10.8.3 Organizing and Considering the Data

10.8.3.1 Considering the Data Along Axes of Time

A basic, but significant step in the analytical process is to organize the data so that it can be systematically considered in a range of different ways. A first step after qualitative interviewing and/or participant observation is over is often to arrange the data in chronological order.[xii] This makes it possible to consider the chronology of events and experiences that have taken place during the course of the study. Even more important will be to consider the chronologies of experiences and events pertaining to the people and phenomena that were the focus of the research.

[xii] Ordering the data in these ways involves some practical procedures (such as placing chunks of texts into certain locations within a computer program), and such ordering of the data is sometimes referred to as "data management." While the data certainly need to be "managed," procedures like these are essentially tools for analysis.

10.8.3.2 Considering People

Another strategy is to pull together portions of the data material that pertain to individual participants. This can be done manually or with the assistance of qualitative data management software. The aim of the process is to engage with individual participant's experiences and narratives as holistically as possible.

Equally important may be to pull together and consider groups of people who have something in common. Examples include participants with the same occupation, those living under similar conditions, patients with similar treatment experiences, people in the same age range, or people with the same sexual preferences.

10.8.3.3 Considering Themes

A next step in the analytical process will often be to find ways to juxtapose portions of the data material that are thematically related (e.g., all utterances relevant to a certain topic; all accounts mediating a similar type of experience; or a number of narrations that represent a diverse range of views on, or interpretations of, a phenomenon). The researcher will do some of this work mentally, but there are a number of techniques that can aid in the process, both manual procedures (e.g., color-coding to index written material) and computer-aided procedures where codes are inserted into the written material. Whatever the method, the aim is to place text portions that are thematically related next to each other so that they can be considered in conjunction.

How researchers identify and select themes varies. At times, themes used during coding are predetermined, either by a research question and/or by a theoretical framework. For example, if a researcher wants to let the empirical data enter into dialogue with Appadurai's[44] theorizing about the "landscapes" that may contribute to shape people's imagined realities, the data could be coded so that experiences with each of these landscapes[xiii] can be identified.

While this kind of coding is useful for some types of analysis, a different strategy would be to identify concepts, practices, meanings, and themes that emerge from the empirical material itself, and code the data

[xiii] Appadurai[44] has termed these landscapes the "ethnoscape" (the landscape of people), the "ideoscape" (the landscape of ideas), the "financescape" (the landscape of material resources), the "mediascape" (the landscape of media reporting), and the "technoscape" (the landscape of technological inventions).

with the help of these. One would then select keywords that reflect emic (insider) constructs or experiences as coding labels (as opposed to categories generated by the researcher or on the basis of a theory developed in a different sociocultural setting).

A third possibility could be to use both of the approaches mentioned in combination. One might identify emic categories in a first step, and then relate these to "external" categories in a second step. One would then have two different strategies for making sense of the same data. While categorizing and coding is important in the analytical process, it inevitably leads to fracturing of the data, and researchers must guard against the danger that fragments acquire a life of their own. See Box 10.13.

10.8.4 Comparing

Making comparisons within a data material, and between the data and other sources, is another way of engaging with it (and a practice that is

BOX 10.13 Coding Leads to Fragmentation

When data are organized, categorized, and coded, they are unavoidably fractured.[43,45]

While no researcher would probably ever think of categorizing texts by the letters they consist of, qualitative data are at times broken down by word frequencies. Whereas the resulting word tallies (sometimes illustrated as "word clouds") give an overview of the words most frequently used, they offer little to no assistance in making sense of what people meant by them.

Preparing word frequencies is an extreme way of disintegrating a text, but all categorizing and coding of data unavoidably leads to fragmentation and decontextualization. Portions of text and meaning are pulled loose from the totality of an interview or encounter, from the more complete and complex situation of the person it derives from or pertains to, and from the wider social, cultural, political, and historical context in which that person lives his or her life. It is also removed from the totality of the overall data produced during the course of the study, and from the sensorial dimensions of the encounters that gave rise to it.

Whereas one may indeed need to isolate, focus on, and link together fragments of the data, there is always the danger that fragments may acquire a life of their own. Researchers must guard against this, and any interpretation of a fragment has to be made in light of the meaning afforded in the whole. The analytical process should therefore entail cyclical shifts between engagement with fragments and engagement with the notes and transcripts they came from. One advantage of qualitative data management software is that it provides a simple way of moving back and forth between fragments to the context to which they belong.

unavoidably present in all kinds of coding and categorization—and indeed in any meaning-making process). While horizontal comparisons compare between instances (e.g., people, phenomena, practices, etc.), vertical comparisons compare across different levels of generality (e.g., an instance or event is compared with a model, concept, or theory). The aim of these comparisons may be to identify patterns, the lack of patterns, uniformities, diversities, and nuances.

10.8.5 Exploring Concepts

Exploring a concept can be a powerful way of detecting meaning. For example, what do nurses in a given setting really have in mind when they use terms such as "patient," "care," "overwork," "difficult patient," or "remission?" Questions like these may offer insight into the ways in which nurses experience their work and see the people they care for. Depending on the aim of the study, one could explore whether and how the way nurses understand these concepts differs from people from a different professional cultural context (e.g., medical doctors). What do they make of the terms just mentioned? What about patients or hospital administrators?

One can never take for granted that a given term embodies the same concept in different settings, and exploring conceptual differences across cultural boundaries may add to our understanding of a setting or phenomenon. Box 10.14 provides an example of how a conceptual discrepancy opened up a new way of seeing and understanding.

10.8.6 Analyzing Metaphors

Figures of speech are abundant in and integral to human interaction and meaning-making.

A metaphor is "a figure of speech in which a word or expression is transferred from its customary domain to an unusual one."[46] For example, when the term "bridge" is used to represent people, as in the statement "women who sell sex are a bridge for HIV transmission."

A metonym is a figure of speech in which a part of something is used to represent the whole, for instance, when hospital workers refer to a patient as "the appendix in room 269" (where appendix represents a patient occupying that room), or when we say "he is a drug user."

Exploring the connotations and associations metaphors and metonyms give rise to, and their roles in meaning-making, will often enrich the analysis of empirical data. This may reveal complex webs of meaning that are at work even in seemingly simple processes of sense-making.

BOX 10.14 Gay and *Gei*

In the study of same-sex attracted men in in Dar es Salaam, one of the research objectives was to gain insight into the ways in which same-sex relationships were understood locally. The researcher, who is from northern Europe, had noted that the English word gay had been adopted in Swahili, and that it was one of several terms in use to refer to same-sex attracted men.[35] He initially assumed that the term could be taken to mean more or less the same in Dar es Salaam as in, say, London or New York. Although some men did use the term in this way, this was not typical in Dar es Salaam. Rather, gay (or *gei*) was normally used to refer to only a subset of men who engage in same-sex relations, i.e., those expected to take an inserted position in sexual intercourse and express (some degree of) womanliness. The partner of someone *gei* could not meaningfully be labeled with the same term, and would typically be referred to simply as a "man" (*bwana*). The ways in which these words were used drew attention to a central feature of same-sex practices in Dar es Salaam, and prompted additional inquiry into the ways in which same-sex relationships are sexually and socially structured around gendered and corporeal positionality. This, in turn, was used to reflect on how the sexual and social differences between gays (*magei*) and their men would have to inform, and be taken into account in, HIV prevention programs.

10.8.7 Unpacking and Experimenting

Unpacking is an analytic strategy that aims to break down acts and situations into the bits and pieces they consist of, and may help identify what we do not know about those acts or situations. If one wanted to unpack, say, nursing care for hospitalized patients, one would aim to describe and analyze the care-related acts and situations patients encounter and are involved in during a hospital stay. At a minimum, such an analysis would have to consider some corporeal agents, their acts, a number of locations, some objects, a time frame, and some procedures.

The following list is an example of some of the issues one might need to consider when unpacking an occasion where a nurse provides care for a hospitalized patient.

- Who are the people involved?
- What are the positions they take or hold (formally and/or informally; habitually or after consideration)?
- What are the corporealities of acts (the bodies involved in enacting them)?

- What are the procedures (formal and informal, compulsory and optional, habitual and deliberate, mechanically applied or individually/situationally adapted)?
- What language(s) are used and how (the use of dialects, vernaculars, jargons, tropes; the tone of voice; various speaking styles—e.g., speaking as if to a friend, a colleague or a child)?
- How objects are handled in this context?
- What documents are used (e.g., forms, manuals)?
- What is the locale, including its layout, equipment, and furniture, like?
- What are people's perceptions of what is going on?
- How are acts carried out (with compassion, interest, indifference, or out of duty)?
- What are the temporalities, including schedules, working hours, shifts, time spent in bed, prognoses, experiential aspect of time (e.g., is it experienced as slow or fast by various actors), and hierarchies of time (such as whose time counts, what is done first and last, etc.)?
- What other forms of meaning-making exist in the context?

Unpacking often involves, or leads to, experimentation and/or thought experimentation.[8] A thought experiment that might be productive in this example is trying to envisage how the situation would have been if existing power structures in the hospital had been reversed, or if the care had been given at another place or at another time. Thought experiments like these may be a way of exposing that which we take for granted.

10.8.8 Using Models

To go beyond thematic data analysis, we may let the empirical data enter into dialogue with external sources and resources. One example of this kind of analytical process is the use of models. Models are devices that highlight something typical or characteristic about a phenomenon in a simplified way—they are "ideal types" in Weber's sense. Because a model is simplified, it can never do justice to the full scope, diversity, or complexity of the phenomenon it models, but it may nonetheless act as a mediator in a researcher's efforts to think, explore, and compare.

Figure 10.1 is a graphical representation of a model of how the sociocultural world might be conceptualized. It places people in interaction at the center and envelops them in concentric circles representing their

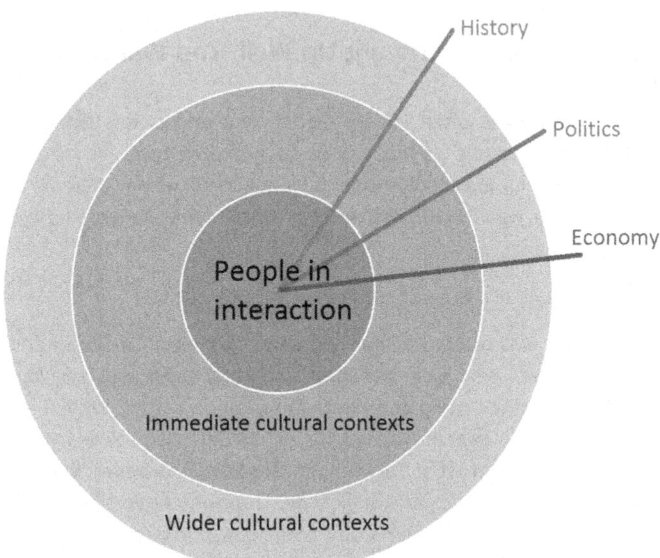

Figure 10.1 A model summarizing theoretical insights that may be used to inform data analysis.

(more immediate as well as wider) cultural contexts. These circles are in turn traversed by three lines representing the political, historical, and economic contexts in which the people at the center of the model live their lives.

This model is an example of a heuristic device. It proposes dimensions that could or should be considered in qualitative research. When considering empirical data in the light of this model, one's reasoning is steered in particular directions. Given this model, it is hardly possible, for example, to consider individuals in isolation, as atoms that can be studied apart from their context.

Box 10.15 gives an example of how this model can inform and guide analyses in a qualitative study.

10.8.9 Using Theory

Using models to guide analysis is one way in which theory may inform analytical work. A model summarizes a theoretical proposition or insight and may be used to shape the way we look at and make sense of data.

Theories of various kinds always inform human thinking; thinking is by necessity theory-laden. When a researcher recommends that an informational campaign be implemented in order to reduce the occurrence of

BOX 10.15 If You Are Brought Up Well, You Will Give in Return

A new line of inquiry is often initiated when one encounters something that disrupts one's expectations (as many of us have experienced, we may not even know that we have an expectation until it is breached). In the example below, a new line of inquiry was made possible by a single sentence uttered by Lars, a 20-year-old gay man:

If you are brought up well, you are supposed to reciprocate.[47]

This sentence was uttered at the end of a story about a sexual encounter that involved unsafe sex. Lars had had a partner who had swallowed his semen, and Lars had felt obligated to do the same in return. While he uttered the sentence as if its meaning would be self-evident, it took a while for the researcher to understand what Lars meant. Gradually, however, that very sentence proved to be of particular significance. Through conversation and reflection together with Lars and other study participants, it became clear that Lars was referring to how the general code of reciprocity may play a significant role in sexual lives.

Since it cannot be taken for granted that reciprocity is universally practiced and understood, the researcher decided there was a need to examine what reciprocity means in a Norwegian context. Because a study of reciprocity was beyond the scope of the study, she consulted existing knowledge. Previous anthropological work suggested that reciprocity in the Norwegian context was typically of a kind the researcher came to term "isomorphic reciprocity," i.e., a form of reciprocity where one is expected to return the same (in form and content) as one receives (and preferably as soon as possible). This is a general cultural code for reciprocity in the Norwegian context; i.e., it is a code for how to treat other people decently and right. Indeed, after having discussed Lars' experiences with other study participants, it became clear that he was far from alone in feeling the force of this code in sexual encounters (although this would not always lead to unsafe sex).

If we apply the model in Figure 10.1 here, we have two loving men in the inner circle and a particular cultural context (i.e., isomorphic reciprocity) in the outer. In order to continue exploring whether other phenomena might have played a role when Lars made the reciprocity code universally valid and did not even consider adapting or modifying it, we may let this particular situation enter into dialogue with historical and political aspects.

The sexuality Lars practices was criminal by law until 1972 in Norway (and is still not fully accepted by all). In another conversation where Lars talked about a different instance of unsafe sex, he gave a clue as to how history and politics may have had an impact on his sexual encounters:

If it is not nice what you do—then it is very, very difficult to protect yourself.

(Continued)

BOX 10.15 If You Are Brought Up Well, You Will Give in Return—cont'd

To protect something that not only society, but even you as a member of that society, has internalized as shameful may be difficult and add to the shame. The question may then arise: could being in accordance with general rules for interpersonal conduct (here reciprocity) be particularly imperative among people who are engaging in practices that are perceived as shameful (by others and perhaps even by oneself)?

Even if we might think of (and experience) a sexual encounter as being beyond time and place, like other human encounters it is not. The model lends no support to thinking that treats an action as a result of purely rational calculation, nor does it see the subject as atomized and insulated from (sub) culture and society. In addition, it recognizes the influence of political and historical forces on our lives.

unsafe sex discovered through his or her research, the recommendation rests on the assumptions that informational campaigns might modify sexual practice and that modified sexual practice may reduce HIV transmission. These are examples of theoretical propositions, derived from empirical research into human and viral behavior.

One of the aims of many qualitative studies is to be explicit about the theoretical assumptions that inform the data analysis. It is helpful to become aware of the assumptions we draw on for several reasons, not the least because it may help us realize what our theories are not able to cover, include, or understand. For example, in the aforementioned example of HIV transmission, the researcher concluded that an informational campaign was needed. If he or she had applied a different theoretical assumption, e.g., that conversation between people is more likely to change behavior than informational campaigns, his or her recommendation would probably have been different. The researcher might have determined that there was a need for health workers to engage in more interaction, conversation, and collaboration with the communities that shoulder the largest burden of HIV infection.

A range of theories are drawn upon in the analysis of qualitative research data. A few such theories have already been mentioned in this chapter, e.g., Kvale's[12] theorizing of interviews, Stierlin's[13] theorizing about closeness and distance, Peirce's theorizing of signs and sign-activity,[16,22] and Polanyi's[29,30] theorizing about tacit knowing, to mention a couple. There are of course many more theories that inform contemporary qualitative health research. Some of them are large theoretical frameworks, whereas others are more limited or specific theoretical schemes or propositions. The theories one can productively draw upon vary substantially depending on one's research question(s) and theme(s). For example, if sense- or meaning-making is at the fore of the inquiry, semiotics may prove valuable; if gender is being explored, feminist theories might be consulted; if exploration of life-worlds or lived experiences is central, one might benefit from phenomenological theoretical approaches.

It is beyond the scope of this chapter to discuss any one theory in detail, but let us emphasize that familiarity with theoretical work is essential in qualitative research. Indeed, having knowledge on a wide array of theoretical insights is one way of developing one's skills as a qualitative researcher. Here we would like to echo Coffey and Atkinson:

> *As we have suggested, reading should be pursued actively. An active engagement with the published literature means that ideas are available for use in research. The work of others is inspected not only for its empirical findings, but also for how its ideas can inform the interpretations of one's own research setting. Theories are not added only as a final gloss or justification; they are not thrown over the work as a final garnish. They are drawn on repeatedly as ideas are formulated, tried out, modified, rejected or polished.*
>
> ***(Coffey and Atkinson[43])***

When making choices about theoretical approaches and resources, one must ask whether a particular theory is capable of operating in continuation with the experiences of those whose lives are being studied. In our view, this enhances the prospect of arriving at an adequate understanding of the empirical field under investigation. Moreover, if one uses a theory that cannot operate in continuation with the phenomena under study, one risks violating the very experience one is inquiring into.

10.8.10 Serendipitous and Revelatory Moments

> *Revelatory moments have been "strikingly instructive for understanding the settings in which we have worked."*
>
> ***(Trigger et al.[48])***

In exploratory research, a line of inquiry or an analysis is not always the result of systematic probing or planned exploration; it can also occur in serendipitous moments. One may stumble onto something beneficiary, or fall into unexpected opportunities for inquiry and analysis. Such moments may occur both during fieldwork and after the researcher has left the field.

The moments we discover phenomena or aspects of phenomena we were not previously aware of may be very significant. In addition to the insight they give rise to, they also present opportunities to analyze the taken-for-granted assumptions that rendered us unaware of them in the first place.

In such analyses, it may not be sufficient to conceive of our ignorance simply as missing knowledge. An entirely new line of inquiry can be opened if we instead try to explore *why* the ignorance existed in the first place. How was it produced? Indeed, just as we need to inquire into epistemological questions about what we know, we should also inquire into those about what we do not know. Among writers who have developed our insight into epistemic practices that produce ignorance are Mills[49] and Tuana.[50,51] Mills explores how ignorance is linked to the production and maintenance of racism, whereas Tuana furthers our understanding of female bodies through her exploration of the ways in which our ignorance about them has been historically and culturally shaped.[xiv]

10.8.11 Summary

In this section, we first pointed out how data analysis is a continuous process in qualitative studies, and how qualitative researchers are dependent on becoming intimately acquainted with their data. We then discussed some strategies that are used to organize empirical data so that they can be considered in a range of ways. While the aim of thematic content analysis is to present the central elements of the accounts of participants, qualitative researchers often want to "ask rather more of their data, and other techniques are needed if we want to move beyond the 'emic' summaries and typologies of participants that a thematic analysis provides."[52] In this section, we therefore also discussed ways in which data analysis may be brought into dialogue with resources from outside the research context via models and theories.

10.9 MAKING INSIGHTS PORTABLE AND APPLICABLE

In this section, we discuss *some* of the ways in which researchers seek to give knowledge produced during the course of qualitative inquiry portability and applicability beyond the context of its origin. In other words, this is a discussion of how qualitative insight can "travel" to, and be useful in, new settings and contexts. In qualitative research, generalizing[xv] occurs

[xiv] See also the volume edited by Sullivan and Tuana.[51]

[xv] Note that here we do not mean generalization in a statistical sense.

to a large extent through theorizing, i.e., through development, on the basis of empirical inquiry, of smaller and larger theoretical constructs or frameworks. It could be the generation of a new concept or the modification of an existing one, the construction of a model, or the building of a whole new theoretical framework.

10.9.1 Widening or Nuancing Repertoires and Horizons

A basic way qualitative research may be made applicable outside of the context in which a study was conducted is through the sharing of knowledge that may widen, or nuance, existing repertoires for understanding something (e.g., structures, processes, or relationships). Let us consider as an example a study carried out among a group of drug users in substitution therapy who had been convicted of a violent crime.[53,54] A number of practices and phenomena were described in this group, e.g., its dominant moral values, the illegal distribution of prescription drugs, violent practices, and a phenomenon the researchers termed "pharmacological agency." Given that the study was qualitative, the researchers could not establish the prevalence and distribution of these phenomena in the larger population, nor did they seek to. What they could establish was that these phenomena existed among drug users in substitution therapy at a certain place and time. By providing descriptions and analyses of these phenomena, knowledge about their existence and characteristics could travel to other contexts where drug use and substitution therapy is on the agenda, including other drug treatment centers around the world. If nothing else, these insights should form part of the repertoire of phenomena clinicians consider in encounters with their patients.

10.9.2 Generating New Concepts

Another way insights from a qualitative study can become applicable elsewhere is through the development of new concepts. Concepts are resources that assist us in thinking and articulation, and a new concept can contribute to our abilities to identify, make sense of, and express experiences, occurrences, and practices. A new concept could help us think about something in a way that is more nuanced, more specific, or more adequate than was possible using previously existing concepts. A concept is a "general" construct in the sense that, by its very nature, it applies to more instances than itself.

Goffman's[55] "stigma" is an example of a concept that has had great impact and acquired a vast field of operation. As a more recent example, consider how Kebede et al.[56] introduced the concept of "negotiated silence" to portray a phenomenon identified in a study among unmarried women who had had abortions in Ethiopia. Abortion (and premarital sex) was morally condemned by many in the society where these women lived, and the study found that abortion practices were associated with several types of silence and modes of silencing. The researchers coined the term "negotiated silence" about one kind of silence they identified and described. This was a silence that was produced jointly and tacitly by two parties, in this case by a mother and a daughter. While the daughter obviously knew that she had a sexual life and had undergone abortion, and the mother suspected this to be the case, both would refrain from talking about these topics. This prevented the mother from becoming someone who had heard (and thus someone who knew). To have heard would have placed the mother in a situation where she was expected to act in a particular way (she would be expected to ostracize her daughter).

When a new concept has been developed, it may function as a resource that can be drawn on in other settings. "Negotiated silence" could potentially be used to pose questions far removed from the context of abortion in Ethiopia. One could, for example, look for negotiated silences in encounters between doctors and patients, or in collaboration between qualitative and quantitative researchers in the health sciences, and explore the effect of such silences in these encounters.

10.9.3 Questioning and Challenging Existing Concepts and Understandings

Just as development of *new* concepts can give a qualitative study general applicability, qualitative research can also be used to question and challenge existing concepts and understandings. As an example of a qualitative study that generated knowledge and critique applicable well beyond the group of people who were involved in the research that produced it, let us consider the study of same-sex attracted men in the context of the HIV epidemic in Tanzania.[34] While homosexual men in Africa had been neglected by HIV programming and research for decades, there had at the same time existed a discourse in public health that portrayed such men as immensely "hard to reach." However, it became clear already early on in the Dar es Salaam study that the hard-to-reach idea did not fit well with observed reality there. Indeed, it was neither complicated nor time-consuming for a researcher to identify, get to know, and interact with a large and diverse group of such men. Based on this empirical finding, the idea that same-sex attracted men in Africa were generally "hard to reach" had been destabilized. In light of this, the researcher returned to the literature and found that while the hard-to-reach idea was often referred to, it was never accompanied by reference to empirical findings that could support it. Notwithstanding this, it circulated widely, as a taken-for-granted belief that seemed likely to have discouraged both research and programming among homosexual men in Africa, and may indeed have contributed to excluding such men from HIV programming and research. Attention could in turn be drawn to the consequences of unexamined ideas and claims more generally, and to the question: Are hard-to-reach claims about other groups of people, about which the term is in use across the world, as misleading as it was when used about same-sex attracted men in Africa?[xvi]

10.9.4 Building Theory

To develop new concepts and critically revise existing ones based on insight generated by empirical work are examples of theorizing—the building of theory. Other examples of theorizing could be to develop a typology or categorization, or to flesh out an explanation for a phenomenon.

At its core, theorizing is about developing new ideas to assist in our understanding of what goes on in social worlds. While theorizing must build on empirical work, the aim is not to summarize that work, but to abstract ideas embedded within it in order to develop new ways of seeing and understanding.

In his discussion of how theories may be developed, Swedberg[57] outlined four conceptual steps in the process. First, the researcher considers the data with the aim to identify and zoom in on something interesting. Second, he tries to give a name to what he has observed and

[xvi] The study also critiqued the hard-to-reach concept itself. It was pointed out how hard-to-reach claims are not positioned (i.e., they do not answer the question "for whom are homosexual men claimed to be hard to reach?") and how they work to construct "us" as actors that must have attempted to reach "them" (how would "we" otherwise know that they cannot be reached?). In effect, the hard-to-reach concept frees "us" from responsibility for "their" exclusion, and portrays "them" as fundamentally responsible for this exclusion.

"formulate a central concept based on it." Third, the theory is built out in accordance with such instructions as:

> *Give body to the central concept by outlining the structure, pattern or organization of the phenomenon. Use analogies, metaphors and comparisons—and all in a heuristic way to get a better grip on the phenomenon under study.*[57]

In a final step, the theorizer aims to explain and justify a coherent set of theoretical propositions. The entire process entails repeated challenging of the emerging theory in light of the data that inspires it. Through such challenging, the theoretical proposition is continually revised until it can adequately account for the observations made during the course of empirical work.

To be meaningful, theorizing must indeed be firmly grounded in empirical findings. However, theorizing also requires creativity and the ability to *abduct*. Abduction, a concept developed by Peirce, begins with "some surprising phenomenon, some experience which either disappoints an expectation, or breaks in upon some habit of expectation of the inquisiturus [inquirer]."[58] It is followed by a pondering of these phenomena, in all their salient aspects, before provisionally arriving at a plausible hypothesis, which must be tested empirically and against other theories.

As Coffey and Atkinson[43] see it, Peirce's conceptualization of "abductive reasoning" captures productively how researchers "actually think and work" and thus "allows for a more central role for empirical research in the generation of ideas as well as a more dynamic interaction between data and theory" than traditional accounts have claimed.

10.9.5 Summary

In this section, we have discussed some ways in which qualitative researchers work to generalize knowledge and insight in order to make it portable and applicable in new settings. We have described how identification of phenomena in one context can widen and nuance the repertoire of tools for understanding in others, how qualitative research can be used to generate new concepts or modify existing ones, and how qualitative research projects often aim to build or modify theoretical constructs and frameworks.

10.10 THE QUALITY OF QUALITATIVE RESEARCH

10.10.1 Epistemic Values

> *[E]pistemology is the study of knowledge and justified belief. As the study of knowledge, epistemology is concerned with the following questions: What are*

> *the necessary and sufficient conditions of knowledge? What are its sources? What is its structure, and what are its limits? As the study of justified belief, epistemology aims to answer questions such as: How we are to understand the concept of justification? What makes justified beliefs justified? Is justification internal or external to one's own mind?*[59]

Epistemological issues have been addressed, either explicitly or implicitly, throughout this chapter. We return to these matters here in order to discuss the key epistemic values of qualitative research, as well as the requirements these values set for qualitative research projects.

We have organized this discussion around four questions,[xvii] which we believe should be asked in all stages of qualitative studies when they are planned, implemented read, and/or evaluated.

10.10.2 Is the Study Transparent?

The first question that needs to be asked is to what extent a qualitative study is transparent. Transparency is quintessential if others (e.g., colleagues, readers, and study participants) are to be able to assess the quality of a qualitative study. To achieve transparency requires ceaseless efforts throughout the research project. One must strive to document and make explicit choices, steps, and turns during the knowledge production process, and document them well so that they can be tracked and examined critically by others.

10.10.3 Is the Researcher Positioned to Understand?

The second question that should be asked is to what extent the researcher is positioned to capture and understand that which he or she endeavors to explore (e.g., that which occurs in a sociocultural setting, the qualities of an experience, the complexities of a practice, or the intricacies of a

[xvii] These questions may come across as unfamiliar to researchers who are more familiar with quantitative research, where the central epistemic values are articulated through the concepts of validity, reliability, and generalizability. In the manner that these constructs are conventionally defined, they are not directly applicable to qualitative research. As Stewart[39] has pointed out, it is nonetheless possible to see a degree of parallelism between the epistemic values underpinning quantitative and qualitative research. More specifically, the question of whether a researcher is "positioned to understand" has something in common with the question of validity in statistics-oriented research, the question of whether a study "is positioned to transcend perspectives" is in some sense parallel to questions about "objectivity," and the questions pertaining to a study's portability ask to what degree insights are generalizable.

meaning-making process). The degree to which researchers are able to position themselves adequately for observing, interacting, and learning is of critical importance in regard to the degree of insight and understanding they may potentially attain.

Several challenges are involved in such positioning. Some of these are related to the researchers themselves (e.g., their limited abilities to observe, grasp, and remember), others to the design of their studies (e.g., studies that provide for fieldwork that is too short), and others to the contexts in which research is undertaken (e.g., the variability of phenomena and challenges relating to accessing them in breadth).

Among research strategies that may be applied to promote the development of insight and understanding that adequately captures what is going on are

- to allow ample time for data production,
- to have knowledge of locally used languages and vocabularies,
- to involve insiders throughout the research process,
- to participate in insider activities,
- to strive to gain access to the broadest possible representation of perspectives and experiences,
- to observe speech and interaction in multiple contexts,
- to consciously seek out cases and accounts that could disconfirm emerging patterns and understandings,
- to pay attention to context, and
- to generate data in several ways.

10.10.4 Is the Study Positioned to Transcend Perspectives?

A third question that should be asked of qualitative studies is the extent to which they are positioned to transcend singular perspectives.

In quantitative research, the concept of reliability asks whether study results are free from bias and whether they are replicable. The kind of objectivity that is conceptualized by this construct requires that study results be as independent of the researcher as possible. This requirement is not tenable in the context of qualitative research. As we have repeatedly pointed out in this chapter, qualitative researchers need to be involved in order for knowledge to be produced. However, as we have also pointed out, researchers can never be alone in this venture. On the contrary, researchers and participants are coproducers of knowledge. They must form—to use a term informed by Peirce—small "epistemic communities."

These, in turn, must relate to other epistemic communities (such as those made up of researcher colleagues and relevant others).

Together, members of epistemic communities can strive to arrive at ways of observing, describing, and analyzing the world that are understandable from—and transcend—the perspectives of both the researcher and the study participants. As Stewart[39] has expressed it, researchers who engage in such processes "deal in one of the ways in which people can overcome the limited purchase of any particular perspective on the world. Their inquiry is at root an effort at intersubjective, often intercultural, communication."

To strive for transcendence of perspectives, qualitative researchers should:

- record information about the participants, how they were encountered and included, and what the interaction between researcher and participants consisted of,
- strive to represent a multitude of perspectives,
- give insiders opportunities to provide comments on descriptions and analyses ("respondent validation"),
- reflect on the researchers' own position(s) in the research process, and
- engage in discussions about the study with relevant "outsiders."

10.10.5 Is the Study Positioned to Produce Knowledge and Insight That Is Portable?

In quantitative research, "generalizability" is a question of whether findings in a study sample are applicable to the population from which the sample was drawn. As we have discussed previously, data and insight generated through qualitative inquiry cannot be generalized in this particular way. However, qualitative and statistical research do share the aim of making knowledge and insight generated in one setting, context, or group valuable and applicable in other settings, contexts, or groups.

The question of generalizability in qualitative research must first be a question of whether the research offers insight that is sound. This has been the main theme throughout this chapter. It is secondly a question of the degree to which any sound insight produced may be portable to and applicable in contexts outside the research setting. As we pointed out in Section 10.8, this may, for example, be achieved through the provision of descriptions and analyses, through generation (or modification) of concepts, through development of models, and/or through the building of theories.

QUESTIONS TO DISCUSS

1. How would you explain what it means to explore?
2. Discuss the claim that qualitative research is dependent on flexible study designs.
3. What are some main differences between participant observation and qualitative interviewing, with respect to the kinds of data they generate?
4. Participant observation and qualitative interviewing are commonly used in combination. In what ways could these two research approaches complement each other in a study of life and health after heart surgery?
5. Discuss the idea that a qualitative interview should strive to be more like a conversation than a Q&A session.
6. What characterizes an "active member researcher" role in participant observation?
7. Think of a thematic field with which you are familiar. Could you suggest some research questions that could be productively explored using qualitative research methods?
8. How might the model presented in Figure 10.1 help shape and inform the development of a study of patient–health worker relations in a hospital setting?
9. Why is maximizing the variation in perspectives often regarded as an overarching aim in qualitative research?
10. Discuss what contextualization may mean and entail during the planning, implementation, and analysis of a qualitative study.
11. Try to think of an example of taken-for-granted knowledge and discuss ways in which a qualitative study could help explicate such knowledge.
12. What is the difference between using and building theory in qualitative research?
13. Discuss some ways in which knowledge generated through qualitative research in a specific place and at a specific time may be made applicable to other times and places.
14. This chapter has highlighted a number of ways in which study participants may play crucial roles in qualitative knowledge production. How would you summarize these?

REFERENCES

1. Melnyk BM, Fineout-Overholt E, editors. *Evidence-based practice in nursing and healthcare: a guide to best practice*. 2nd ed. Philadelphia, PA: Wolters Kluwer Health/ Lippincott Williams & Wilkins; 2010.
2. Parker C. *Evidence-based nursing research guide: strengths & levels of evidence*. DePaul University Library. Available from: <http://libguides.depaul.edu/content.php?pid=448090&sid=3712020>.
3. Grossman AH, Kerner MS. Self-esteem and supportiveness as predictors of emotional distress in gay male and lesbian youth. *J Homosex* 1998;**35**(2):25–37.
4. Kippax S, Race K. Sustaining safe practice: twenty years on. *Soc Sci Med* 2003;**57**(1): 1–12.
5. Gottlieb M, Schanker H, Fan P, Saxon A, Weisman JD, et al. Pneumocystis pneumonia—Los Angeles. *CDC Morb Mortal Wkly Rep* 1981;**30**:250–2.
6. Latour B, Woolgar S. *Laboratory life: the construction of scientific facts*. Princeton, NJ: Princeton University Press; 1986.
7. Blok A, Jensen TE. *Bruno Latour: hybrid thoughts in a hybrid world*. New York, NY: Routledge; 2011.
8. Maxwell JA. *Qualitative research design: an interactive approach*. Los Angeles, CA: Sage; 2013.
9. Hilden PK, Middelthon AL. Kvalitative metoder i medisinsk forskning—et etnografisk perspektiv. *Tidsskr Nor Laegeforen* 2002;**122**(25):2473–6.
10. Olsvold N. *Ansvar og yrkesrolle: om den sosiale organiseringen av ansvar i sykehus*. Oslo: Unipub; 2010.
11. Kvale S. *InterViews: an introduction to qualitative research interviewing*. Thousand Oaks, CA: Sage; 1996.
12. Kvale S, Brinkmann S. *InterViews: learning the craft of qualitative research interviewing*. Los Angeles, CA: Sage; 2009.
13. Stierlin H. *Conflict and reconciliation: a study in human relations and schizophrenia*. Garden City, NY: Anchor Books; 1969.
14. Lassiter LE. *The Chicago guide to collaborative ethnography*. Chicago, IL: University of Chicago Press; 2005.
15. Holstein JA, Gubrium JF. *Inside interviewing: new lenses, new concerns*. Thousand Oaks, CA: Sage; 2003.
16. Colapietro VM. *Peirce's approach to the self: a semiotic perspective on human subjectivity*. New York, NY: State University of New York; 1989.
17. Geertz C. *After the fact: two countries, four decades, one anthropologist*. Cambridge: Harvard University Press; 1995.
18. Glaser BG, Strauss AL. *The discovery of grounded theory: strategies for qualitative research*. Chicago, IL: Aldine; 1967.
19. Platt J. The history of the interview. In: Gubrium JF, Holstein JA, editors. *Handbook of interview research context and method*. Thousand Oaks, CA: Sage; 2001. p. 33–54.
20. Gubrium JF, Holstein JA. *The new language of qualitative method*. New York, NY: Oxford University Press; 1997.
21. Skjervheim H. *Selected essays: in honour of Hans Skjervheim's 70th birthday*. Bergen: Filosofisk Institutt; 1996.
22. de Waal C. *Peirce: a guide for the perplexed*. London: Bloomsbury; 2013.
23. Hippocrates. *Of the epidemics*. Whitefish, MT: Kessinger Publishing; 2004.
24. Middelthon A-L. *Being young and gay in the context of HIV: a qualitative study among young Norwegian gay men*. Oslo: University of Oslo; 2001. 1b.
25. Rubin HJ, Rubin IS. *Qualitative interviewing: the art of hearing data*. London: Sage; 2011.

26. Middelthon A-L. Når maden blir frelser eller bøddel. In: Glasdam S, editor. *Folkesundhed—i et kritisk perspektiv*. København: Nyt Nordisk Forlag Arnold Busck; 2009. p. 223–39.
27. Middelthon A-L. The duty to eat right. *AM Rivista della societa italiana di antropologia medica* 2009;**27–28**:209–25.
28. Gubrium JF, Holstein JA. From the individual interview to the interview society. In: Gubrium JF, Holstein JA, editors. *Handbook of interview research context and method*. Thousand Oaks, CA: Sage; 2001. p. 3–32.
29. Polanyi M. *The tacit dimension*. Garden City, NY: Doubleday; 1967.
30. Polanyi M. *Personal knowledge: towards a post-critical philosophy*. New York, NY: Harper & Row; 1962.
31. Langaas AG. *Å berøre og bli berørt—å bevege og bli beveget: om fysioterapeutstudenters læring om og gjennom kroppen*. Oslo: Unipub; 2013.
32. Dewalt KM, Dewalt BR, Wayland CB. Participant observation. In: Russel HB, editor. *Handbook of methods in cultural anthropology*. New York, NY: AltaMira Press; 2000. p. 259–99.
33. Marcus GE. *Ethnography through thick and thin*. Princeton, NJ: Princeton University Press; 1998.
34. Moen K, Aggleton P, Leshabari MT, Middelthon A-L. Not at all so hard-to-reach: same-sex attracted men in Dar es Salaam. *Cult Health Sex* 2012;**14**(2):195–208.
35. Moen K, Moen P, Aggleton M, Leshabari A-L, Middelthon AL. Gays, guys, and mchicha mwiba: same-sex relations and subjectivities in Dar es Salaam. *J Homosex* 2014;**61** (4):511–39.
36. Moen K, Moen P, Aggleton M, Leshabari A-L, Middelthon AL. Situating condoms in sexual lives: experiences of same-sex-attracted men in Tanzania. *Int J Sex Health* 2013;**25**(3):185–97.
37. Wathne K. Movement of large bodies impaired: the double burden of obesity: somatic and semiotic issues. *Sport Educ Soc* 2011;**16**(4):415–29.
38. Minkler M, Wallerstein N, editors. *Community-based participatory research for health. From process to outcomes*. San Francisco, CA: Jossey-Bass; 2008.
39. Stewart A. *The ethnographer's method*. London: Sage; 1998.
40. Coy MW. Being what we pretend to be. The usefulness of apprenticeship as a field method. In: Coy MW, editor. *Apprenticeship*. Albany, NY: State University of New York Press; 1989. p. 115–35.
41. Wolcott HF. Confessions of a trained observer. In: Popkewitz TS, Tabachnik BR, editors. *Field-based methodologies in educational research and evaluation*. New York, NY: Praeger; 1981. p. 247–63.
42. Glesne C. *Becoming qualitative researchers: an introduction*. Boston, MA: Pearson Education; 2011.
43. Coffey A, Atkinson P. *Making sense of qualitative data: complementary research strategies*. Thousand Oaks, CA: Sage; 1996.
44. Appadurai A. *Modernity at large: cultural dimensions of globalization*. Minneapolis, MN: University of Minnesota Press; 1996.
45. Hollway W, Jefferson T. *Doing qualitative research differently*, vol. 6. London: Sage; 2000. p. 166.
46. Colapietro VM. *Glossary of semiotics*, vol. 16. New York, NY: Paragon House; 1993. p. 212.
47. Middelthon AL, Aggleton P. Reflection and dialogue for HIV prevention among young gay men. *AIDS Care* 2001;**13**(4):515–26.
48. Trigger D, Forsey M, Meurk C. Revelatory moments in fieldwork. *Qual Res* 2012; **12**(5):513–27.

49. Mills CW. *The racial contract*. Ithaca, NY: Cornell University Press; 1997.
50. Tuana N. The speculum of ignorance: the women's health movement and epistemologies of ignorance. *Hypatia* 2006;**21**(3):1–19.
51. Sullivan S, Tuana N, editors. *Race and epistemologies of ignorance*. Albany, NY: State University of New York Press; 2007.
52. Green J, Thorogood N. *Qualitative methods for health research*, vol. 15. London: Sage; 2009. p. 304.
53. Havnes IA, Clausen T, Middelthon A-L. "Diversion" of methadone or buprenorphine: "harm" versus "helping". *Harm Reduct J* 2013;**10**(1):24.
54. Havnes IA, Clausen T, Middelthon A-L. Execution of control among "non-compliant", imprisoned individuals in opioid maintenance treatment. *Int J Drug Policy*. 2014;**25** (3):480–5.
55. Goffman E. *Stigma: notes on the management of spoiled identity*. New York, NY: Simon and Schuster; 1963.
56. Kebede MT, Hilden PK, Middelthon A-L. Negotiated silence: the management of the self as a moral subject in young Ethiopian women's discourse about sexuality. *Sex Educ* 2014;1–13.
57. Swedberg R. Theorizing in sociology and social science: turning to the context of discovery. *Theory Soc* 2012;**41**(1):1–40.
58. Peirce CS. In: Hartshorne C, Weiss P, editors. *Collected papers of Charles Sanders Peirce*, vol. 6. Cambridge, MA: Harvard University Press; 1974.
59. Stanford Encyclopedia of Philosophy. Available from: <http://plato.stanford.edu/archives/spr2014/entries/epistemology/> [accessed 07.07.14].

FURTHER READING

Bloor M, Frank J, Thomas M, Robson K. *Focus groups in social research*. London: Sage; 2002.

Burnham JC. *What is medical history?* Cambridge: Polity; 2005.

Coffey A, Atkinson P. *Making sense of qualitative data*. London: Sage; 1996.

Green J, Thorogood N. *Qualitative methods for health research*. London: Sage; 2009.

Hammersley M, Atkinson P. *Ethnography. Principles in practice*. 3th ed. London: Routledge; 2007.

Helman CG. *Culture, health and illness: an introduction for health professionals*. 5th ed. New York, NY: Hodder Arnold; 2007.

Holstein J, Gubrium JF, editors. *Inside interviewing. New lenses, new concerns*. London: Sage; 2003.

Kvale S, Brinkmann S. *Interviews. Learning the craft of qualitative interviewing*. London: Sage; 2009.

Mason J. *Qualitative researching*. London: Sage; 2010.

Maxwell JA. *Qualitative research design. An interactive approach*. London: Sage; 2013.

Silvermann D. *Doing qualitative research*. London: Sage; 2013.

CHAPTER 11

Statistical Inference

Petter Laake[1] and Morten Wang Fagerland[2]
[1]Oslo Centre for Biostatistics and Epidemiology, Department of Biostatistics, Institute of Basic Medical Sciences, University of Oslo, Blindern, Oslo, Norway
[2]Oslo Centre for Biostatistics and Epidemiology, Research Support Services, Oslo University Hospital, Oslo, Norway

11.1 INTRODUCTION

Statistics play an important role in all research, but especially in medicine and the biological sciences. Statistical inference provides the link between the study sample and the source population (hereafter referred to as sample and population for the sake of brevity), and statistics must be used to provide valid inferences about the relationship between an outcome variable (also known as the effect variable, the response variable, or the dependent variable) and an exposure variable (also known as the causal variable, the explanatory variable, the independent variable, or the covariate).

Chapter 4 discusses the basis for research and places the role of statistics in perspective. The different types of data are reviewed in Section 4.6, and the statistical methods to use with different types of data are given in Section 4.4. In this chapter, the most relevant statistical methods will be covered in more detail, and recommendations given as to which methods to use with different types of data. To analyze categorical data, contingency tables will be used. For continuous data, normally and nonnormally distributed data will be analyzed. The analysis of count data (where the number of events has been counted) will be briefly covered, and finally, the analysis of data on time to event, also known as survival analysis, will be covered. Regression models play an important role in all these analyses. This chapter will demonstrate how to perform these analyses and how to interpret the results using examples from data sets that are readily available on websites, which will be referred to when needed. Statistical software packages, and when they can (and should) be used for these analyses, will also be briefly discussed.

This chapter will focus on effect estimates, confidence intervals, and *P*-values. Veierød et al.[1] will serve as a standard reference work for the methods described herein. In that textbook, the reader will find details

Research in Medical and Biological Sciences
DOI: http://dx.doi.org/10.1016/B978-0-12-799943-2.00011-2

that are not given here. The QuickCalcs[2] application will be used to compute P-values; it can be downloaded from http://graphpad.com/quickcalcs/PValue1.cfm.

11.2 EFFECT ESTIMATE, CONFIDENCE INTERVAL, AND *P*-VALUE

The main concepts involved in all statistical analyses are summarized in Box 11.1.

11.2.1 Concepts

Many studies in medical research entail the assessment and measurement of a health outcome. The outcome may be the result of an animal experiment, a treatment in a randomized controlled trial, or an epidemiological study, such as a study of the effect of smoking during pregnancy on a child's birth weight. Regardless of the study type, an effect measure must be chosen. In an epidemiological study, the effect of smoking during pregnancy could be assessed by the difference in the average birth weight of children born to nonsmoking and smoking mothers. The effect measure is a quantity in the population and is thus unknown. Based on the sample, the effect measure must be estimated, and the manner in which it is estimated depends on the study design and type of data collected for

BOX 11.1 Main Concepts in Statistical Analysis

- The aim is to estimate the relationship between an outcome variable and an exposure variable in the population. The outcome variable is the variable of interest, whereas exposure variables are variables that influence the outcome.
- The effect measure expresses the relationship between an outcome variable and an exposure variable in the population and is thus unknown. The effect estimate is the sample analogue of the effect measure.
- A 95% confidence interval includes the effect measure in the population with 95% probability. The confidence interval quantifies the uncertainty of the effect estimate.
- The null hypothesis is the hypothesis that the study seeks to reject. A null hypothesis is tested by performing a statistical test and computing the associated *P*-value. The *P*-value is the calculated probability of the observed result, or results that agree even less with the null hypothesis than the observed result, when the null hypothesis is correct.

the outcome and the exposure. A confidence interval is used, most often with a 95% confidence level, to quantify the uncertainty of the effect estimate. A 95% confidence interval includes the unknown effect measure with 95% probability.

The aim of a study is based on its research hypothesis, for example, that the outcomes of two treatments differ, or that there is an association between smoking during pregnancy and a child's birth weight. Then, a testable statistical hypothesis is formulated, which is denoted the null hypothesis, i.e., the hypothesis that a study seeks to reject. The null hypothesis states that there is no association between the outcome variable and the exposure variable and is denoted H_0; the alternative hypothesis is denoted H_1. An example of a research hypothesis is that the mean of two normal distributions differ, in which case the null hypothesis is "no difference." The *P*-value provides the basis for rejecting the null hypothesis and, consequently, for accepting an alternative hypothesis. Computation of the *P*-value will be discussed later.

11.2.2 Null Hypothesis, Alternative Hypothesis, and *P*-Value

Statistical testing always calls for a null hypothesis. Assume, for instance, that one wishes to determine whether the effects of two drugs differ. The null hypothesis always represents the *status quo*, which in this case is that the two drugs have equal effects. The effect measure under the null hypothesis therefore equals the null value, which is usually zero or one, depending on the type of effect measure. The alternative hypothesis in this case is that the effects of the two drugs differ, for which the effect measure is smaller or greater than the null value. The alternative hypothesis may be one-sided or two-sided. It is one-sided if changes are bound to be in one direction. It is two-sided if changes to either side are of interest, such as either positive or negative effects. Two-sided alternative hypotheses are the most common in medical and biological research.

A null hypothesis is tested by performing a statistical test, which is a method used to determine whether the outcomes of a study allow for the rejection of the null hypothesis. A statistical test is based on a test statistic, which can be used to determine the validity of the null and the alternative hypotheses. The test statistic is used to calculate the *P*-value, which indicates the level of evidence against the null hypothesis. The computation of *P*-values is inextricably linked to the null and alternative hypotheses.

In all statistical analyses, the goal is to learn if an effect is larger than what would be expected from random variation. The uncertainty (or precision) of the effect estimate is expressed by the standard error. In most cases, the statistical test is performed by taking the size of the effect estimate divided by its standard error. This is the standard form for all parametric tests and will be denoted the Wald statistic. The *P*-value is the calculated probability of the observed result, or results that agree even less with the null hypothesis than the observed result, when the null hypothesis is correct. Computation of the *P*-value for a one-sided alternative hypothesis differs from that of a two-sided alternative hypothesis. For a two-sided alternative hypothesis, the *P*-value is normally twice that of a one-sided alternative hypothesis. Most often, a researcher decides that a *P*-value less than 5% indicates a statistically significant result, and such a result means that the null hypothesis can be rejected. A statistically significant result may also be interpreted as a strong indicator of a real difference. This does not, however, mean that a high *P*-value, such as one greater than 5%, is synonymous with the null hypothesis being correct.

11.2.3 Confidence Interval

An effect measure is estimated in an experiment. Its form depends on the design and on the statistical model employed. The result is the effect estimate, which in turn is associated with uncertainty. The size of the uncertainty is given by the standard error. Uncertainty is conveyed by the confidence interval, which expresses in a percentage, usually 95%, the probability that the interval includes the unknown value of the effect measure. If the effect measure is a mean difference, the value 0, which corresponds to no difference between the effects of two treatments, is the null value. If a 95% confidence interval does not include the value 0, the *P*-value of a corresponding statistical test will be less than 0.05. However, if a ratio is considered, such as a relative risk (RR) (see Section 11.3), odds ratio (OR) (see Section 11.3), or hazard ratio (HR) (see Section 11.13), the value 1 is the null value, and the conclusions will be based on whether or not the 95% confidence interval contains the value 1.

To sum up: Box 11.2 presents the four concepts that are used to present the results of statistical analyses. Please see also the book by Lang and Secic,[3] which gives a nice review of reporting statistics in medicine.

BOX 11.2 Concepts Used to Present the Results of Statistical Analyses

- The null hypothesis to be tested
- Effect estimate
- Uncertainty of the effect estimate, reflected by the 95% confidence interval
- Results of one or more statistical tests, expressed by *P*-values

Table 11.1 Outcomes for two binomial samples: the cell counts of a 2 × 2 contingency table

	Outcome		
Exposure	**Success**	**Failure**	**Total**
Yes	n_{11}	n_{12}	n_{1+}
No	n_{21}	n_{22}	n_{2+}
Total	n_{+1}	n_{+2}	n

11.3 TWO BINOMIAL SAMPLES

If a sequence of n independent trials with one of two possible outcomes is considered, such as childbirth or coin tossing, the outcome of interest is called "success." If the probability of success is the same in each trial, and herein Y will denote the number of successes, then the number of successes in the n trials is binomially distributed. If success in childbirth indicates the birth of a girl, the number of girls born in a maternity ward is binomially distributed. But most frequently, it is the comparison of the probabilities of success in two binomial samples that is of interest. Examples include the probabilities of success in a randomized clinical trial, or a dichotomous health outcome comparing the two sexes. The outcome of two binomial samples is presented here in a 2 × 2 contingency table (Table 11.1). The table summarizes the associations between an outcome variable and an exposure variable, with the exposure variable in the rows and the outcome variable in the columns. There is no generally accepted practice for presenting such tables.

The probabilities of the four outcomes are given as p_{11}, p_{12}, p_{21}, and p_{22}. The subscripts of the n's and the p's relate to the cell numbers, with the first index indicating the row number and the second index indicating the column number.

When the margins n_{1+} and n_{2+} are fixed, which is the case when interest is focused on the effect of an exposure variable on an outcome variable, it is assumed that $p_{11} + p_{12} = 1$, $p_{21} + p_{22} = 1$, $p_1 = p_{11}$, and $p_2 = p_{21}$. The effect measure is then given by relating p_1 to p_2. The estimates of p_1 and p_2 are given as $\hat{p}_1 = n_{11}/n_{1+}$ and $\hat{p}_2 = n_{21}/n_{2+}$. The "hats" indicate that these are estimates.

The term risk will be used synonymously with probability of a disease in this chapter. There are many ways an effect measure can be defined, but medical statistics most often employ risk difference (RD), RR, and OR as measures. The absolute risk reduction (ARR), which is equivalent to RD, is less often used, though it might be just as natural as the other measures. Regardless of the effect measure chosen, the attention is generally focused on its associated confidence interval, as well as the presence of a significant association, determined by performing statistical tests.

Equations for calculating confidence intervals and *P*-values are based on an approximation between the binomial distribution and the normal distribution. The approximation is quite good when all of the expected numbers of observations in the cells (the *m*'s), defined by:

$$\begin{gathered} m_{11} = n_{1+} \cdot n_{+1}/n, \quad m_{12} = n_{1+} \cdot n_{+2}/n, \\ m_{21} = n_{2+} \cdot n_{+1}/n, \text{ and } m_{22} = n_{2+} \cdot n_{+2}/n, \end{gathered} \tag{11.1}$$

exceed 5. The expected number of observations is simply the number that can be expected in each cell when there is no association between an exposure and an outcome. As per the equation above, the expected number of observations is the product of the margins for the rows and columns, divided by the total number of observations. Table 11.2 is an example of a table where all of the expected numbers of observations in the cells are greater than 5. When at least one of the expected numbers of observations is less than 5 (such as in Table 11.3), a small-sample method is used to compute confidence intervals and *P*-values, rather than a normal distribution approach.

In general, if there are r rows and c columns, they will render an $r \times c$ contingency table. Here, only the analysis of 2×2 contingency tables will be summarized. For a discussion of $r \times c$ contingency tables, please see chapter 2 in Veierød et al.[1]

11.4 MEASURES OF ASSOCIATION IN 2 × 2 CONTINGENCY TABLES

The various measures of association between exposure variables and outcome variables are based on the occurrence of a given outcome among exposed and unexposed groups. An effect measure can be defined either by examining the (absolute) difference between them (RD) or the (relative) relationship between risks (RR or OR). As noted earlier, RD is identical to ARR. Let $\widehat{\text{RD}}$, $\widehat{\text{RR}}$, and $\widehat{\text{OR}}$ denote the effect estimates calculated from the cell numbers in Table 11.1.

11.4.1 Risk Difference and its Confidence Interval

The RD is defined as:

$$\text{RD} = p_1 - p_2. \tag{11.2}$$

Note that RD expresses the difference in risks between the exposed and the unexposed groups. Box 11.3 gives two versions of the confidence interval for the risk difference. RD is estimated by

$$\widehat{\text{RD}} = \hat{p}_1 - \hat{p}_2 = n_{11}/n_{1+} - n_{21}/n_{2+}. \tag{11.3}$$

BOX 11.3 Confidence Interval for RD

The standard 95% Wald confidence interval is given as:

$$\left[\hat{p}_1 - \hat{p}_2 - 1.96\sqrt{\frac{\hat{p}_1(1-\hat{p}_1)}{n_{1+}} + \frac{\hat{p}_2(1-\hat{p}_2)}{n_{2+}}}, \hat{p}_1 - \hat{p}_2 + 1.96\sqrt{\frac{\hat{p}_1(1-\hat{p}_1)}{n_{1+}} + \frac{\hat{p}_2(1-\hat{p}_2)}{n_{2+}}}\right]. \tag{11.4}$$

Here, 1.96 is the 97.5-percentile in the normal distribution.

The Wald confidence interval is the recommended method when the sample size is large. When the sample size is small, the Agresti–Caffo confidence interval[4] is recommended:

$$\left[\tilde{p}_1 - \tilde{p}_2 - 1.96\sqrt{\frac{\tilde{p}_1(1-\tilde{p}_1)}{n_{1+}+2} + \frac{\tilde{p}_2(1-\tilde{p}_2)}{n_{2+}+2}}, \tilde{p}_1 - \tilde{p}_2 + 1.96\sqrt{\frac{\tilde{p}_1(1-\tilde{p}_1)}{n_{1+}+2} + \frac{\tilde{p}_2(1-\tilde{p}_2)}{n_{2+}+2}}\right], \tag{11.5}$$

where

$$\tilde{p}_1 = \frac{n_{11}+1}{n_{1+}+2} \text{ and } \tilde{p}_2 = \frac{n_{21}+1}{n_{2+}+2}. \tag{11.6}$$

The Wald confidence interval can be calculated in any statistical software package, such as SPSS. The Agresti–Caffo confidence interval can be calculated in Stata. Note that both these intervals are in closed form and can also be calculated easily by hand.

Many of the examples in this chapter use data from the Low Birth Weight Study. These data were also used by Hosmer et al.[5] and can be downloaded from http://www.umass.edu/statdata/statdata/stat-logistic.html. Variables included in these data are child's birth weight in grams (BWT), low birth weight (LOW: $0 =$ birth weight ≥ 2500 g, $1 =$ birth weight <2500 g), smoking (SMOKE: $0 =$ nonsmoking, $1 =$ smoking), hypertension (HT: $0 =$ normotensive, $1 =$ hypertensive), history of premature labor (PTLD: $0 =$ no, $1 =$ yes), and whether the mother saw a physician during the first trimester (FTVD: $0 =$ no, $1 =$ yes). Note that PTLD and FTVD are recoded from PTL and FTV, respectively, by collapsing the codes greater than 1 into code 1.

Example

With data from the Low Birth Weight Study, the associations between LOW and SMOKE and between LOW and HT are investigated. The results of the first relationship are shown in Table 11.2.

Here, the RD between smokers and nonsmokers for having a child with low birth weight is $\widehat{\text{RD}} = 30/74 - 29/115 = 0.405 - 0.252 = 0.153$. This shows that smokers have a 15-percentage-point increase in risk of having a child with low birth weight compared with nonsmokers. Since all of the expected numbers of observations in the cells in Table 11.2 are greater than 5, the Wald confidence interval can be used. That renders a 95% confidence interval for the RD of (0.016, 0.290), and so it can be said with 95% confidence that the population value of RD is between 1.6 and 29 percentage points.

In a second example, the association between LOW and HT is investigated, with the results presented in Table 11.3.

The RD between hypertensive mothers and normotensive mothers for having a child with low birth weight is $\widehat{\text{RD}} = 7/12 - 52/177 = 0.583 - 0.294 = 0.289$. This shows that hypertensive mothers have a 29-percentage-point increase in risk of having a child with low birth weight compared with normotensive mothers. Because one of the expected numbers of observations in the cells of Table 11.3 is less than 5 (3.7), the Agresti–Caffo

Table 11.2 Cell counts for the association between LOW and SMOKE in the Low Birth Weight Study, $n = 189$

	LOW		
SMOKE	**Yes**	**No**	**Total**
Yes	30 (40.5%)	44 (59.5%)	74 (100%)
No	29 (25.2%)	86 (74.8%)	115 (100%)
Total	59 (31.2%)	130 (68.8%)	189 (100%)

Table 11.3 Cell counts for the association between LOW and HT in the Low Birth Weight Study, $n = 189$

	LOW		
HT	**Yes**	**No**	**Total**
Yes	7 (58.3%)	5 (41.7%)	12 (100%)
No	52 (29.4%)	125 (70.6%)	177 (100%)
Total	59 (31.2%)	130 (68.8%)	189 (100%)

confidence interval is applied instead of the Wald confidence interval, which yields a 95% confidence interval of (0.008, 0.543). It can therefore be stated with 95% confidence that the population value of RD might be as low as 0.8 percentage points or as high as 54 percentage points. The confidence interval for this example is considerably wider than that for Table 11.2, even though the total sample size is the same ($n = 189$). This difference reflects the increased uncertainty introduced by the small number ($n = 12$) of hypertensive mothers in Table 11.3 and illustrates an important general principle: the precision of any statistical analysis depends not only on the total sample size, but also on how the observations are distributed into the categories.

In randomized controlled trials, it can sometimes be helpful to report the results with an effect measure called the number needed to treat (NNT). This is the reciprocal of the absolute RD, 1/RD. The NNT is the number of patients that need to be treated to prevent a single negative outcome. A large treatment effect leads to a small NNT.

11.4.2 RR and its Confidence Interval

The RR is defined as the ratio of the probabilities of disease in the exposed group versus the unexposed group:

$$\mathrm{RR} = p_1/p_2. \tag{11.7}$$

Note that the unexposed group composes the reference group, i.e., the group relative to which risk is calculated. Based on the observations in Table 11.1, RR can now be estimated as:

$$\widehat{\mathrm{RR}} = \frac{\hat{p}_1}{\hat{p}_2} = \frac{n_{11}/n_{1+}}{n_{21}/n_{2+}}. \tag{11.8}$$

A transformation is required to calculate the confidence interval for RR. This is because the estimated RR does not follow a normal

BOX 11.4 Confidence Interval for RR

The standard error of ln $\widehat{RR}$ is estimated by:

$$\widehat{SE}(\ln \widehat{RR}) = \sqrt{\frac{1}{n_{11}} - \frac{1}{n_{1+}} + \frac{1}{n_{21}} - \frac{1}{n_{2+}}}. \qquad (11.9)$$

When the sample size is large, the 95% confidence interval of ln RR is provided by the normal distribution:

$$[\ln \widehat{RR} - 1.96 \cdot \widehat{SE}(\ln \widehat{RR}), \ln \widehat{RR} + 1.96 \cdot \widehat{SE}(\ln \widehat{RR})]. \qquad (11.10)$$

The 95% Katz log confidence interval, which is recommended in this chapter, is given by the inverse transformation back to RR via the exponential function (exp):

$$\{\exp[\ln \widehat{RR} - 1.96 \cdot \widehat{SE}(\ln \widehat{RR})], \exp[\ln \widehat{RR} + 1.96 \cdot \widehat{SE}(\ln \widehat{RR})]\}. \qquad (11.11)$$

When the sample size is small, the Koopman confidence interval is recommended. Unfortunately, this is not expressed in a closed form. For details, please see chapter 2 in Veierød et al.[1] The Katz confidence interval can be calculated in any statistical software package. The Koopman interval can be calculated in Stata.

distribution, but rather a lognormal distribution, which is skewed to the right. This skewness can be explained by the fact that RR can take any value between zero and infinity, with one as its null value. A logarithmic transformation is therefore used, namely ln RR instead of RR. Box 11.4 provides confidence intervals for the RR for large and small sample sizes.

Example

With the data from the Low Birth Weight Study, the association between LOW and SMOKE (see Table 11.2) is estimated by an RR of

$$\widehat{RR} = \frac{30/74}{29/115} = 1.61.$$

Since the sample size is large, the Katz log confidence interval is used. For this, the standard error must be calculated:

$$\widehat{SE}(\ln \widehat{RR}) = \sqrt{1/30 - 1/74 + 1/29 - 1/115} = 0.214,$$

and the 95% confidence interval for RR becomes:

$$[\exp(1.61 - 1.96 \cdot 0.214),\ \exp(1.61 + 1.96 \cdot 0.214)] = (1.06, 2.44).$$

The RR of 1.61 implies that the risk of having a child with low birth weight increases by 61% for women who smoke. The corresponding confidence interval says that one can be 95% confident that the increase in risk might be as low as 6% and as high as 144%; however, the best estimate is that it is 61%.

The estimated association between LOW and HT (see Table 11.3) is

$$\widehat{RR} = \frac{7/12}{52/177} = 1.99.$$

Thus, hypertensive mothers have a 99% increase in risk of having a child with low birth weight compared with normotensive mothers. Since the sample size is considered to be small because of the few hypertensive mothers ($n = 12$), a 95% Koopman confidence interval of (1.05, 3.03) (calculated in Stata) is used. As was observed for RD, the confidence interval for RR for the data in Table 11.3 is noticeably wider than the confidence interval for RR for the data in Table 11.2.

11.4.3 OR and its Confidence Interval

As its name implies, OR is the ratio between two odds. The odds of an event are defined as the ratio of two probabilities. When disease is the outcome, the odds represent the probability of having the disease relative to the probability of being healthy. Odds are calculated both for the exposed group and for the unexposed group. Based on the data in Table 11.1, the odds for the exposed group are $p_1/(1-p_1)$, whereas the odds for the unexposed group are $p_2/(1-p_2)$.

OR is the ratio of the odds for the exposed group to that of the unexposed group:

$$\text{OR} = \frac{p_1/(1-p_1)}{p_2/(1-p_2)} = \frac{p_1(1-p_2)}{p_2(1-p_1)}. \tag{11.12}$$

OR is calculated by inserting the estimates for the probabilities into Eq. (11.12), which yields the following:

$$\widehat{\text{OR}} = \frac{n_{11}/n_{12}}{n_{21}/n_{22}} = \frac{n_{11} \cdot n_{22}}{n_{12} \cdot n_{21}}. \tag{11.13}$$

As for RR, a logarithmic transformation is used to compute the confidence interval for OR. In Box 11.5 confidence intervals for ORs for large and small sample sizes are given.

BOX 11.5 Confidence Interval for ORs

For large sample sizes, the normal distribution is used. The standard error of $\ln \widehat{OR}$ is estimated by:

$$\widehat{SE}(\ln \widehat{OR}) = \sqrt{\frac{1}{n_{11}} + \frac{1}{n_{12}} + \frac{1}{n_{21}} + \frac{1}{n_{22}}}. \tag{11.14}$$

The 95% confidence interval for ln OR is then given in the usual manner, as:

$$[\ln \widehat{OR} - 1.96 \cdot \widehat{SE}(\ln \widehat{OR}),\ \ln \widehat{OR} + 1.96 \cdot \widehat{SE}(\ln \widehat{OR})]. \tag{11.15}$$

The inverse transformation back to $\widehat{OR}$ via the exponential function results in the 95% Woolf logit confidence interval:

$$\{\exp[\ln \widehat{OR} - 1.96 \cdot \widehat{SE}(\ln \widehat{OR})],\ \exp[\ln \widehat{OR} + 1.96 \cdot \widehat{SE}(\ln \widehat{OR})]\}. \tag{11.16}$$

For small sample sizes, the Gart adjusted logit confidence interval is recommended. It is a simple modification of the Woolf logit confidence interval obtained by adding 0.5 to every cell count.

The 95% Gart adjusted logit confidence interval is given by:

$$\left\{\exp\left[\ln\frac{(n_{11}+0.5)(n_{22}+0.5)}{(n_{12}+0.5)(n_{21}+0.5)} - 1.96 \cdot \sqrt{\frac{1}{n_{11}+0.5} + \frac{1}{n_{12}+0.5} + \frac{1}{n_{21}+0.5} + \frac{1}{n_{22}+0.5}}\right],\right.$$
$$\left.\exp\left[\ln\frac{(n_{11}+0.5)(n_{22}+0.5)}{(n_{12}+0.5)(n_{21}+0.5)} + 1.96 \cdot \sqrt{\frac{1}{n_{11}+0.5} + \frac{1}{n_{12}+0.5} + \frac{1}{n_{21}+0.5} + \frac{1}{n_{22}+0.5}}\right]\right\}. \tag{11.17}$$

Both the Woolf logit confidence interval and the Gart adjusted logit confidence interval can be calculated readily by hand. The Woolf logit confidence interval can be calculated in SPSS and Stata, whereas the Gart adjusted logit confidence interval can only be calculated in Stata.

Example

Consider once again the association between LOW and SMOKE found in the Low Birth Weight Study (see Table 11.2). The estimated OR is

$$\widehat{OR} = \frac{30/44}{29/86} = 2.02.$$

Since the sample size is large, the Woolf logit confidence interval is used. The standard error must be calculated as follows:

$$\widehat{SE}(\ln \widehat{OR}) = \sqrt{1/30 + 1/29 + 1/44 + 1/86} = 0.320,$$

and the 95% confidence interval for OR becomes:

$$[\exp(\ln 2.02 - 1.96 \cdot 0.320),\ \exp(\ln 2.02 + 1.96 \cdot 0.320)] = (1.08, 3.78).$$

The OR estimate of 2.02 implies that smoking raises the odds of having a child with low birth weight by about 100%. The confidence interval shows that one can be 95% confident that the population OR is larger than 1.08 and smaller than 3.78.

The association between LOW and HT (see Table 11.3), estimated by the OR, is

$$\widehat{\mathrm{OR}} = \frac{7/5}{52/125} = 3.37.$$

In this case, a 95% Gart adjusted logit confidence interval is used:

$$[\exp(\ln 3.26 - 1.96 \cdot 0.585),\ \exp(\ln 3.26 + 1.96 \cdot 0.585)] = (1.04, 10.3).$$

Thus, hypertension raises the odds of having a child with low birth weight by 237%. The confidence interval, however, is wide, which suggests that the OR was estimated with low precision and that a wide range of ORs are compatible with the data.

Thus far, the focus has been on the computational aspects of the three effect measures RD, RR, and OR and their confidence intervals. RD stands apart from RR and OR because it expresses an absolute difference. Although much has been written about absolute and relative effect measures, there is still confusion surrounding these concepts. When investigating a rare event, it might be particularly confusing to report a relative effect instead of an absolute effect. Suppose that an event occurs among 2% of untreated patients and among 1% of treated patients. The RD is 1%, which is read as a one-percentage-point change. The RR is 0.5. That is, treatment halves the risk of the event. RR has a low value (i.e., a value <1), despite the fact that it represents a small absolute change of just 1%. Apparent disagreements on the magnitudes of effects can be ascribed to differences in effect measures. Regardless of the effect measure, to be germane the changes must be assessed relative to initial values, and the changes must be of scientific interest.

Likewise, opinions differ concerning the choice between RR and OR. Often, the effect measure is referred to as RR, though OR has actually been estimated. This is unfortunate, particularly as it can create unnecessary confusion. The two measures yield approximately the same results for rare events; however, they may differ markedly for events that happen frequently.

The probability of disease can be estimated in both cohort and cross-sectional study designs. Hence, RR is normally chosen for such studies. In case-control studies of disease, the outcome variable is characterized by ill/healthy. These in turn become cases (ill) and controls (healthy). As the ratio between the numbers of cases and controls does not reflect the ratio in the population, RR cannot be used as an effect measure. Instead, OR should be used. However, it would be an oversimplification to say that RR is used for cohort and cross-sectional studies, and OR is used for case-control studies, because OR is often used regardless of the study design. Indeed, OR permits simpler statistical analyses in more complex models, such as logistic regression (see Section 11.10). Hence, there are computational reasons for the extensive use of OR. However, interpretation of RR may be simpler, and so many prefer to use RR whenever possible.

11.5 STATISTICAL TESTS FOR COMPARING TWO PROPORTIONS

As explained in Section 11.3, when the column margins are fixed, there are two proportions, p_1 and p_2. When these two proportions are compared, the null hypothesis to be tested is usually

$$\mathrm{H}_0\colon p_1 = p_2$$

versus

$$\mathrm{H}_1\colon p_1 \neq p_2.$$

Under the null hypothesis, the proportions for the exposure categories are the same, and there is consequently no association between exposure and outcome. Note that the null hypothesis is written differently depending on the effect measure employed: H_0: RD = 0, H_0: RR = 1, and H_0: OR = 1.

The following text describes two recommended tests: the Pearson chi-squared test for large sample sizes and the Fisher mid P test for small sample sizes. A large sample size, as outlined in Section 11.3, is defined as one where all the expected number of observations in the cells $(m_{11} = n_{1+} \cdot n_{+1}/n, m_{12} = n_{1+} \cdot n_{+2}/n, m_{21} = n_{2+} \cdot n_{+1}/n,$ and $m_{22} = n_{2+} \cdot n_{+2}/n)$ exceed 5.

11.5.1 The Pearson Chi-Squared Test

The Pearson chi-squared test compares the observed with the expected number of observations in each cell. The details of the Pearson chi-squared test are outlined in Box 11.6. The number of observed observations in a cell (i,j) is n_{ij}, and the number of expected observations is m_{ij}.

BOX 11.6 The Pearson Chi-Squared Test

The Pearson chi-squared test is a statistical test for the null hypothesis of equal proportions. It is based on the sum of the differences between the observed number of observations and the expected number of observations in each cell, as follows:

$$\chi^2 = \sum_{\text{cells}} \frac{(\text{observed} - \text{expected})^2}{\text{expected}} = \sum_{i,j} \frac{(m_{ij} - n_{ij})^2}{m_{ij}}. \tag{11.18}$$

For large sample sizes, χ^2 is chi-squared distributed with one degree of freedom. Let χ^2_{obs} denote the observed χ^2-value. The null hypothesis can be rejected if $P(\chi^2 \geq \chi^2_{\text{obs}}) \leq 0.05$. The P-value can be found in QuickCalcs. The Pearson chi-squared test and its P-value can be calculated in SPSS and Stata.

Example

The observed number of observations and the expected number of observations (expected in parentheses) in Table 11.2 are 30 (23.1), 29 (35.9), 44 (50.9), and 86 (79.1). All expected values are well above 5, in which case $\chi^2 = \sum_{i,j} (m_{ij} - n_{ij})^2 / m_{ij} = 4.92$. The corresponding P-value for the null hypothesis of no association between LOW and SMOKE in the Low Birth Weight Study is $P = 0.026$. Thus, there is a statistically significant association between smoking and having a child with low birth weight.

11.5.2 The Fisher Mid *P* Test

The Pearson chi-squared test is inappropriate when the sample size is small. There are several alternatives, including exact tests based on summing probabilities. The P-value of the Fisher exact test is computed by fixing the margins of the table, which results in probabilities that can be calculated. Fisher's exact test can be calculated in most statistical software packages, like SPSS and Stata; however, this test is overly conservative in the sense that is produces P-values that tend to be larger than necessary. Better tests are available, and a modified Fisher exact test, known as the Fisher mid P test (Box 11.7), is recommended.

BOX 11.7 The Fisher Mid *P* Test

The one-sided Fisher mid *P*-value is calculated as:

$$\text{One-sided Fisher mid } P\text{-value} = \text{One-sided Fisher exact } P\text{-value} - 1/2 \cdot \text{the probability of the observed outcome.} \quad (11.19)$$

Furthermore,

$$\text{Two-sided Fisher mid } P\text{-value} = 2 \cdot \text{One-sided Fisher mid } P\text{-value.} \quad (11.20)$$

The one-sided Fisher exact *P*-value and the probability of the observed outcome can both be calculated in SPSS and Stata.

Example

The association between LOW and HT (see Table 11.3) can be investigated using the Fisher mid *P* test. A statistical software package such as SPSS or Stata provides the one-sided Fisher exact *P*-value = 0.042 and the probability of the observed outcome, which is 0.032. Then, the two-sided Fisher mid *P*-value is $2 \cdot 0.042 - 0.032 = 0.052$.

Because this *P*-value is slightly greater than 0.05, the null hypothesis cannot be rejected at the 5% level. Note that for the data in Table 11.3, none of the 95% confidence intervals for RD, RR, or OR include the null value. This example illustrates that the results of statistical tests are not always consistent with the results of confidence intervals when, as here, the calculations of the confidence interval and the statistical test are based on different statistical principles. This is particularly noticeable in cases where associations are of borderline statistical significance, as is the case in this example.

11.6 NORMAL DISTRIBUTION

The normal distribution plays a key role in the application of statistical methods. The normal distribution is symmetrical, and its form is described by its location (expectation) and its variation (standard deviation). The location is denoted by the symbol μ and the standard deviation (SD) by the symbol σ. The variance is σ^2. A variable X that is normally distributed with a location μ and a variance σ^2 is written $X \sim N(\mu, \sigma^2)$. A normal distribution is bell-shaped, with its center on the x-axis determined by μ, and its width determined by σ.

When a variable is normally distributed, the probability of a variable lying within an interval or being less than a given value can be calculated.

For example, for variable X and a particular measured value a, $P(X \leq a)$ can be calculated. However, when using this quantity, the concept of probability density must be considered. Probability density is represented by a curve, and the area under the density function corresponds to the probabilities. $P(X \leq a)$ equals the area between a normal distribution curve and the x-axis to the left of a. The entire area under the curve is unity. The 90th percentile is denoted a if $P(X \leq a) = 0.90$ (90%). The 25th percentile is often called the lower quartile, the 50th percentile the median, and the 75th percentile the upper quartile. The interquartile range comprises values between the lower and upper quartiles and covers 50% of the distribution.

For a normal distribution with a location μ and an SD σ, $P(\mu - 1.96\sigma \leq X \leq \mu + 1.96\sigma) = 0.95(95\%)$. This gives a simple interpretation of σ: 95% of values from a normal distribution lies within 1.96 times the SD from the location. Equivalently, if an interval within which the variable will fall can be specified with a probability of 95%, the width of that interval is about four times the SD.

Many statistical models are based on the assumption that the data follow a normal distribution. This assumption can be assessed with a histogram or a normality plot (Q–Q plot). In general, methods that assume normally distributed data are fairly robust to small and moderate deviations from normality, at least when the sample size is not small. The purpose of assessing the distribution of the data is not to confirm that the data are normally distributed, but rather to investigate whether the deviations from normality are too large for normality to be assumed. Unfortunately, there are no distinct rules to determine when normal distribution can be assumed; however, experienced statisticians often develop a "sixth sense" for when the assumption of normal distribution is incorrect. Histograms and normality plots can be created in any standard statistical software package.

And now a closer look at how to estimate the location of a distribution with a confidence interval when the data follow a normal distribution. In a statistical experiment comprising n observations, such as of n persons, the observations $X_1, X_2, \ldots, X_n$ follow a normal distribution, with a location μ and a variance σ^2. The average of the observations is calculated as $\overline{X}$. If the experiment is to be repeated, the distribution of $\overline{X}$ also will be normal, with a location μ, but a variance of σ^2/n, and accordingly $\overline{X} \sim N(\mu, \sigma^2/n)$. This means that the distribution of the average of the observations has the same location, but with a variance equal to $1/n$ times that of a single observation.

The empirical variance is given by:

$$s^2 = \frac{1}{n-1}\sum_{i=1}^{n}(X_i - \overline{X})^2, \tag{11.21}$$

and the square root of the empirical variance is used as the estimate of the SD. Accordingly, the location of the normal distribution is estimated as the mean of the observations, and its variation is estimated by the standard error of the mean (SEM), which may be written

$$\mathrm{SE}(\overline{X}) = \mathrm{SEM} = s/\sqrt{n}. \tag{11.22}$$

This gives the simple relationship between SEM and the SD.

If the distribution of the individual data points is of interest, as might be the case in assessing the uncertainty of measuring instruments or in describing the variation in populations, the magnitude of s must be determined. Thus, the SD is a measure of the variation per observation. However, if the focus of interest is the location of the distribution, it is the SEM that must be determined, because it estimates the uncertainty of the average. Increasing the number of observations will not necessarily decrease the SD. On the other hand, increasing the number of observations will decrease the SEM by a factor of $\sqrt{n}$.

11.7 COMPARISON OF MEANS

t-tests are often used to compare means. They were first proposed in 1908 by W. S. Gosset, who wrote under the pseudonym "Student," which is why they are often referred to as "Student's *t*-tests." One type of *t*-test is used to draw conclusions based on the mean of one sample of observations (one-sample *t*-test), and another type is used to compare the means of two samples of observations (two-sample *t*-test). Pairs of observation, such as two observations from the same subject, may be examined by treating them as a single sample of the differences between the two values. Here, results for one-sample and two-sample situations are outlined. For details on the *t*-distribution, *t*-tests, and the related confidence intervals, please see chapter 7 in Kirkwood and Sterne.[6]

11.7.1 *t*-Distribution

Since t-tests play such an important role in the comparison of means, it is important to review some of the basics of the t-distribution. Take a data set comprising n observations from a normal distribution with a location μ and an SD σ. Let $\overline{X}$ be the mean, and let SEM $= s/\sqrt{n}$ be the estimated standard error of $\overline{X}$. The statistical test used is the Wald test, which in this case uses the test statistic $T = \overline{X}/\text{SEM}$, with observed value denoted by t_{obs}. To calculate the P-value, one should use $P(T \geq t_{\text{obs}})$, and to find that probability, the distribution of T must be calculated. The test statistic T follows a t-distribution with $n - 1$ degrees of freedom. The t-distribution depends on the number of observations, or more precisely, the number of degrees of freedom.

$\overline{X}$ follows a normal distribution. The only difference between the normal distribution and the t-distribution is that the SD (σ) is now regarded as unknown, and is estimated by s. The t-distribution is symmetrical and bell-shaped, as is the normal distribution, but the t-distribution has heavier tails. As the number of observations increases, the t-distribution more closely resembles the normal distribution. The difference between the two distributions is almost negligible for 100 or more degrees of freedom.

The QuickCalcs application can be used to find P-values for a given t_{obs}, as well as to find percentiles in the t-distribution for various degrees of freedom. The 97.5-percentile in a t-distribution with 50 degrees of freedom is 2.009. With 100 degrees of freedom, the result is a percentile of 1.98. As that is nearly 1.96, the 97.5-percentile of the normal distribution, the t-distribution with 100 degrees of freedom can safely be approximated by a normal distribution.

11.7.2 One-Sample *t*-Test for Pairs of Observations

The one-sample t-test is principally applied to pairs of observations, such as the results of crossover trials (see Chapter 8), or for longitudinal data where the comparison is between baseline and follow-up values. The difference between the two observations is calculated in each pair, and the data set becomes a single sample. One-sample effect estimates and confidence intervals for paired data are defined in Box 11.8.

BOX 11.8 One-Sample Effect Estimate and Confidence Interval for Pairs of Observations

Let D denote the difference between the two observations in each pair. The effect estimate is the mean of these differences, denoted by $\overline{D}$. The recommended 95% confidence interval for the difference is given by the equation:

$$[\overline{D} - t_{0.975,n-1} s_d/n,\ \overline{D} + t_{0.975,n-1} s_d/n], \tag{11.23}$$

or the equivalent form

$$[\overline{D} - t_{0.975,n-1}\text{SEM},\ \overline{D} + t_{0.975,n-1}\text{SEM}]. \tag{11.24}$$

Here s_d is the SD of the differences, given by $s_d^2 = (1/n-1)\sum_{i=1}^{n}(D_i - \overline{D})^2$, and SEM $= s_d/n$. As before, n is the number of pairs. The constant $t_{0.975,n-1}$ is the 97.5-percentile in the t-distribution having $n-1$ degrees of freedom. The 97.5-percentile for any value n can be found in QuickCalcs. The confidence interval can be calculated readily in the statistical software packages SPSS and Stata.

Example

Here, data from the Medical Birth Registry in Norway on the birth weight of children in 1967–1996 are used. Data are available for 267 mothers who gave birth to at least two children. For the purposes of this analysis, data from the first and the second birth are included. These data were analyzed for another purpose in chapter 7 of Veierød et al.[1] Data for a subsample of 20 are presented in Table 11.4.

Table 11.4 Ranked differences in birth weight in the Medical Birth Registry of Norway, 1967–1996, $n = 20$

Absolute difference	**100**	**100**	**100**	**100**	**200**	**200**	**200**	**200**	**300**	**300**
Sign	+	+	−	−	+	+	+	+	+	+
Rank value	2.5	2.5	2.5	2.5	6.5	6.5	6.5	6.5	10	10
Absolute difference	**300**	**500**	**500**	**500**	**500**	**600**	**600**	**900**	**900**	
Sign	+	+	−	−	−	+	+	+	−	
Rank value	10	13.5	13.5	13.5	13.5	16.5	16.5	18.5	18.5	

One pair of observations with a birth weight of 4000 g for both first and second child (tied observations) has been deleted.

The purpose of the present analysis is to determine whether there are any differences in birth weight between the first and second child. For this, let D = birth weight of the second child − birth weight of the first child (measured in grams), which renders $\overline{D} = 103.0$, $s_d = 538.6$, and SEM $= 38.1$. Since $n = 267$, it follows that $t_{0.975,266} = 1.97$.

Inserting these results into the formula for the confidence interval gives a result of (27.9, 178.1). The null value lies well outside this confidence interval, and there is sufficient evidence to suggest that the average birth weight of the second child is larger than the average birth weight of the first child. The estimate shows that the second child is 103 g heavier than the first child; however, the confidence interval indicates that this difference may be as low as 28 g or as high as 178 g.

The following text will explore how to test a null hypothesis of equal means. Let the pair of observations consist of one observation before and one observation after an event, such as a treatment or the birth of a second child. The null hypothesis to be tested is

H_0: the mean before the event equals the mean after the event.

The alternative hypothesis is then

H_1: the means before and after the event differ.

By calculating the difference (D) between a pair of observations for each subject, one may express the null hypothesis as:

H_0: the mean of the distribution of D is 0.

For this hypothesis setup, the recommended t-test is given in Box 11.9.

BOX 11.9 One-Sample *t*-Test for Paired Data

The one-sample t-test for paired data is based on the mean of the differences between the two observations for each pair, denoted by $\overline{D}$. Under the null hypothesis, the mean value of the difference is 0, and the test statistic is written as follows:

$$T = \frac{\overline{D}}{\text{SEM}} = \frac{\overline{D}}{s_d/\sqrt{n}}. \tag{11.25}$$

Now T follows a t-distribution with $n-1$ degree of freedom. For a particular value of t_{obs} for the test statistic, the null hypothesis can be rejected if the P-value is $2P(T \geq t_{obs}) \leq 0.05$. Note that here and in the following the P-value is calculated as $2P(T \geq t_{obs})$, since the testing is always against two-sided alternatives.

QuickCalcs can be used to calculate the P-value. Statistical software packages, such as SPSS or Stata, can also be used to perform the statistical test and calculate the P-value.

BOX 11.10 The Two-Sample *t*-Test for Samples with Equal SDs

Assume that the two samples comprise n_1 and n_2 independent observations. The means of the samples are $\overline{X}_1$ and $\overline{X}_2$. The two-sample *t*-test statistic is the ratio between the effect estimate and its standard error,

$$T = \frac{\overline{X}_1 - \overline{X}_2}{s\sqrt{(1/n_1) + (1/n_2)}}, \tag{11.26}$$

where s is the estimate of the common SDs of the two samples, expressed by:

$$s = \sqrt{\frac{s_1^2(n_1 - 1) + s_2^2(n_2 - 1)}{n_1 + n_2 - 2}}. \tag{11.27}$$

Here,

$$s_1^2 = \frac{1}{n_1 - 1}\sum_{i=1}^{n_1}(X_{1i} - \overline{X}_1)^2 \text{ and } s_2^2 = \frac{1}{n_2 - 1}\sum_{i=1}^{n_2}(X_{2i} - \overline{X}_2)^2.$$

In this case, T is *t*-distributed with $n_1 + n_2 - 2$ degrees of freedom. Let the observed *T*-value be t_{obs}. H_0 can be rejected if $2P(T \geq t_{obs}) \leq 0.05$. The *P*-value for an observed *T*-value corresponding to $n_1 + n_2 - 2$ degrees of freedom can be found in QuickCalcs. The statistical test can be performed and its *P*-value calculated in any statistical software package, such as SPSS or Stata.

Example

The data on birth weight of children from the Medical Birth Registry in Norway showed that $\overline{D} = 103.0$ and SEM = 38.1. Then $t_{obs} = 103.0/38.1 = 2.70$. In QuickCalcs, a *t*-distribution with 266 degrees of freedom gives a *P*-value of $2P(T \geq 2.70) = 0.007$. The null hypothesis of an equal birth weight for the first and second child can thus be rejected.

11.7.3 Two Independent Samples

One of the most commonplace statistical analyses is the comparison of two independent samples of observations. One example is the parallel group design in clinical research (see Chapter 8). In these cases, the goal is to determine if there is a difference in the location of the distributions of the two samples. Let the locations of the two distributions be μ_1 and μ_2. There are two sets of observations, $X_{11}, X_{12}, \ldots, X_{1n_1}$ from sample 1 and $X_{21}, X_{22}, \ldots, X_{2n_2}$ from sample 2, and it is assumed that the n_1 and n_2 observations of the two samples are independent and follow normal distributions with means μ_1 and μ_2. Additionally, it is assumed that the SDs in the two samples are the same and, as usual, equal to σ (Box 11.10). Later in this section, recommendations will be given for situations in which the two samples have different SDs.

Now, the following null hypothesis can be tested:

H_0: there is no difference in location between the groups, that is, $\mu_1 = \mu_2$,

as opposed to the two-sided alternative hypothesis:

H_1: the locations of the groups are different, that is, $\mu_1 \neq \mu_2$.

Example

Data from the Low Birth Weight Study are used to determine if there is a difference in the birth weight of children born to smoking and nonsmoking mothers. The null hypothesis of no difference in the mean birth weight of children born to smokers and those born to nonsmokers is tested.

In this example, $n_1 = 115$ and $n_2 = 74$, and the mean birth weight is $\overline{X}_1 = 3055$ for children born to nonsmokers and $\overline{X}_2 = 2773$ for children born to smokers. The SDs are estimated as $s_1 = 752.4$ and $s_2 = 660.1$, respectively. An estimate of the joint SD is

$$s = \sqrt{\frac{s_1^2(n_1 - 1) + s_2^2(n_2 - 1)}{n_1 + n_2 - 2}} = \sqrt{\frac{752.4^2 \cdot 114 + 660.1^2 \cdot 73}{115 + 74 - 2}} = 719.7,$$

and

$$t_{\text{obs}} = \frac{3055 - 2773}{719.7\sqrt{(1/115) + (1/74)}} = 2.63.$$

There are $155 + 74 - 2 = 187$ degrees of freedom, and QuickCalcs is used to find that $P = 0.009$.

In the example above, the two samples are assumed to have equal SDs, which is a credible assumption given the minor differences between the SDs of the two samples; however, this may not always be the case. We may test the assumption of equal SDs by Levene's test for equality of variances (details omitted), which is available in statistical software like SPSS and Stata. In our example, the *P*-value of Levene's test is 0.22, and we do not reject the null hypothesis of equal SDs. We proceed with the analysis of testing differences in the birth weights under the assumption of equal SDs in the two samples. The problem with Levene's test is that it is likely to reject the null hypothesis of equal SDs when the sample sizes are large (due to large power), even though the difference in SDs may be small and unproblematic. Furthermore, it is also likely that we fail to reject the null hypothesis of equal SDs for small sample sizes, even for cases with considerable differences between the SDs. Levene's test must be used with caution to determine whether the assumption of equal SDs is satisfied. We recommend a pragmatic approach, as outlined in Box 11.11.

BOX 11.11 The Two-Sample *t*-Test for Samples with Unequal SDs

First, the size of the two SDs must be calculated. If the largest is more than 150% of the smallest, the Welch *U* test (also called the two-sample *t*-test with adjustment for unequal variances) should be used. Indeed, the Welch *U* test is recommended whenever two sample sizes differ greatly, even if the largest SD is less than 150% of the smallest SD.

Suppose that the two samples comprise n_1 and n_2 observations and that the means of the samples are $\overline{X}_1$ and $\overline{X}_2$, respectively. The Welch *U* test statistic is the ratio between the effect measure and its standard error under the assumption of unequal SDs:

$$U = \frac{\overline{X}_1 - \overline{X}_2}{\sqrt{(s_1^2/n_1) + (s_2^2/n_2)}}, \tag{11.28}$$

where s_1^2 and s_2^2 are as in Box 11.10. *U* is then *t*-distributed with degrees of freedom given as:

$$\mathrm{df} = \frac{s_1^2/n_1 + s_2^2/n_2}{s_1^4/(n_1^3 - n_1^2) + s_2^4/(n_2^3 - n_2^2)}. \tag{11.29}$$

Let the observed *U*-value be u_{obs}. The null hypothesis can be rejected if $2P(U \geq u_{obs}) \leq 0.05$. The *P*-value can be found in QuickCalcs. The Welch *U* test, and the corresponding *P*-value, can also be calculated in statistical software packages such as SPSS or Stata.

Example

Data from the Low Birth Weight Study are used to study the association between BWT and HT. The null hypothesis states that there is no difference in BWT. Here $n_1 = 177$ and $n_2 = 12$, and the mean BWT is $\overline{X}_1 = 2972$ for normotensive mothers and $\overline{X}_2 = 2537$ for hypertensive mothers. The SDs are estimated respectively as $s_1 = 709.2$ and $s_2 = 917.3$. The relative SD is 1.2; however, there are almost 15 times as many normotensive as hypertensive mothers. The recommended test is therefore the Welch *U* test. In this case,

$$U = \frac{2972 - 2537}{\sqrt{(709.2^2/177) + (917.3^2/12)}} = 1.61,$$

with

$$\mathrm{df} = \frac{(709.2/177 + 917.3/12)^2}{709.2^4/(177^3 - 177^2) + 917.3^4/(12^3 - 12^2)} = 11.9.$$

QuickCalcs renders $P = 0.13$. There is no evidence to reject the null hypothesis of equal birth weight for children of hypertensive and normotensive mothers.

BOX 11.12 Effect Estimate and Confidence Interval for Comparing Two Sample Means

The effect estimate for the difference in the means of two samples is

$$\overline{X}_1 - \overline{X}_2. \tag{11.30}$$

When it is assumed that the SDs are the same in both samples, a 95% two-sample confidence interval for the difference is

$$\left(\overline{X}_1 - \overline{X}_2 - t_{0.975,n_1+n_2-2} \cdot s\sqrt{\frac{1}{n_1} + \frac{1}{n_2}},\ \overline{X}_1 - \overline{X}_2 + t_{0.975,n_1+n_2-2} \cdot s\sqrt{\frac{1}{n_1} + \frac{1}{n_2}}\right). \tag{11.31}$$

The constant $t_{0.975,n_1+n_2-2}$ is the 97.5-percentile for a *t*-distribution with $n_1 + n_2 - 2$ degrees of freedom.

When it is assumed that the SDs are different in the two samples, a 95% Welch *U* confidence interval is given by:

$$\left(\overline{X}_1 - \overline{X}_2 - t_{0.975,\mathrm{df}} \cdot \sqrt{\frac{s_1^2}{n_1} + \frac{s_2^2}{n_2}},\ \overline{X}_1 - \overline{X}_2 + t_{0.975,\mathrm{df}} \cdot \sqrt{\frac{s_1^2}{n_1} + \frac{s_2^2}{n_2}}\right). \tag{11.32}$$

Here, df is given in Box 11.11, and $t_{0.975,\mathrm{df}}$ is the 97.5-percentile for a *t*-distribution with df degrees of freedom.

Both these intervals can be calculated readily in SPSS or Stata.

11.7.4 Effect Estimate and Confidence Interval

In Box 11.12, an appropriate effect estimate and confidence interval for comparing two sample means is defined.

Example

For the association between BWT and SMOKE in the Low Birth Weight Study, the effect estimate is 3055 − 2773 = 282. Thus, those born to nonsmokers are 282 g heavier than those born to smokers. The standard two-sample *t*-confidence interval is used (see Box 11.12). A 95% confidence interval then is (70.7, 492.7).

For birth weight of children born to normotensive and hypertensive mothers, the effect estimate is 2972 − 2537 = 435, i.e., children born to normotensive mothers are 435 g heavier than those born to hypertensive mothers. Since the sample size is greatly unbalanced, the Welch *U* confidence interval is used. A 95% confidence interval then is (−153.5, 1024.6).

11.8 NONPARAMETRIC METHODS

The approaches described in Section 11.7 are based on the assumption that the data follow a normal distribution. The veracity of this assumption can be investigated using normality plots (Q–Q plots) and histograms.

Two alternative methods exist for cases in which data do not follow a normal distribution. A data set may either be transformed so that it (approximately) follows a normal distribution, or nonparametric methods may be used. Transformation is necessary whenever the tails of a distribution are heavier than they would be in a normal distribution. A logarithmic transformation is most often used in such cases, although in principle many mathematical functions may be used. The drawback of transformation is that the resulting analyses and interpretations refer to the transformed scale instead of the original scale. If the original scale is important for a proper interpretation of the results, nonparametric methods are recommended and explained in more detail below. Note, however, that even though nonparametric methods are distribution-free, they are not free of assumptions and must be interpreted within the scope of their definitions. As will be discussed later, for some methods these assumptions may be quite restrictive, which limits their usefulness in many cases. For details on nonparametric methods, please see also chapter 30 in Kirkwood and Sterne.[6]

11.8.1 Effect Estimate for Nonparametric Situations

The mean of observations is a sensible estimate of the location of a distribution whenever the data completely or approximately follow a normal distribution. But the mean is excessively influenced by extreme observations. The median is a more appropriate effect measure when a distribution is skewed, because the median is not influenced by extreme observations. When a distribution is symmetrical, the mean and the median are the same.

11.8.2 The Wilcoxon Signed Rank Test for Paired Nonparametric Data

The Wilcoxon signed rank test is the most commonly used nonparametric test for paired data. When the observations D_i are assumed to follow a symmetric distribution, the null hypothesis of interest is

$$H_0\text{: the median (and mean) of the } D_i\text{s is equal to zero,}$$

and the alternative hypothesis is

$$H_1\text{: the median (and mean) of the } D_i\text{s is not zero.}$$

BOX 11.13 The Wilcoxon Signed Rank Test for Paired Nonparametric Data

Tied pairs of observations occur when the differences equal zero. Ties add no information to whether the differences are shifted towards positive or negative values, and they are generally deleted from the analysis. Let the number of pairs of observations be n. For n less than or equal to 15, the Wilcoxon signed rank test, presented here in four steps, is recommended.

First, calculate the absolute difference for each pair of observations, and next arrange them in ascending rank order. Third, sum the rank values for the positive (or the negative) absolute differences. Let T^+ be the summed rank value, which now constitutes the test statistic.

Finally, find the P-values from http://faculty.washington.edu/heagerty/Books/Biostatistics/TABLES/Wilcoxon/.

For n greater than 15, the summed rank value can be approximated by a normal distribution. The location and standard error of the normally distributed T^+ are $E(T^+) = n(n+1)/4$, and $\mathrm{SE}(T^+) = \sqrt{n(n+1)(2n+1)/24}$. This test statistic is equal to the difference between the observed value of T^+ and the location divided by the standard error, i.e.,

$$Z = \frac{T^+ - E(T^+)}{\mathrm{SE}(T^+)}. \tag{11.33}$$

Let the observed Z-value be z_{obs}. The null hypothesis can be rejected if $2P(Z \geq z_{\mathrm{obs}}) \leq 0.05$, where the P-value is found from the normal distribution, for instance, in QuickCalcs. The Wilcoxon signed rank test is available in SPSS and Stata.

Under the alternative hypothesis, the D_is will tend to be either positive or negative. The method is described in Box 11.13.

Example

The first 20 mothers are selected from the data on birth weight of children collected from the Medical Birth Registry in Norway in 1967–1996. The absolute differences for each pair of observations of birth weight are arranged in ascending rank order (see Table 11.4). Thereafter, the summed rank value is calculated to determine the positive (or negative) absolute differences.

The sum of the rank values for positive absolute differences is $t^+ = 2.5 + 2.5 + 6.5 + 6.5 + 6.5 + 6.5 + 10 + 10 + 10 + 13.5 + 16.5 + 16.5 + 18.5 = 126$.

The normal approximation is applied, and $z_{\mathrm{obs}} = (126 - 95)/(24.85) = 1.25$. The P-value is $2P(Z \geq 1.25) = 0.21$. The null hypothesis cannot be rejected, meaning that the data from these 20 mothers do not provide evidence that the birth weight of the second child is different from that of the first child.

BOX 11.14 Effect Estimate and Confidence Interval for Paired Nonparametric Data

The effect estimate and confidence interval are calculated in three steps.

First, calculate the difference for each pair of observations. Second, run a median regression model, with the constant term as the only covariate. The resulting constant term $\hat{\beta}_0$ is the estimated median difference, and its estimated standard error is $\widehat{SE}(\hat{\beta}_0)$. The confidence interval is given as:

$$[\hat{\beta}_0 - t_{0.975,n-1}\widehat{SE}(\hat{\beta}_0),\ \hat{\beta}_0 + t_{0.975,n-1}\widehat{SE}(\hat{\beta}_0)]. \qquad (11.34)$$

Here $t_{0.975,n-1}$ is the 97.5-percentile in a t-distribution with $n-1$ degrees of freedom. The estimated median difference, its standard error, and the confidence interval can be calculated in Stata.

11.8.3 Effect Estimate and Confidence Interval for Paired Nonparametric Data

In this case, as for paired normally distributed data, the information in the data lies in the differences between the pairs of observations, D_i. Then the confidence interval must be found for the median of the D_is. There are several ways to compute the median and the confidence interval. Box 11.14 outlines how to estimate the effect estimate and the confidence interval for paired nonparametric data. Here, as later in the two-sample case, median regression is recommended. More details on median regression will be given in Section 11.11.

Example

As in the previous example, the 20 observations from the Medical Birth Registry are used. The results of running a median regression in Stata are shown in Table 11.5.

Table 11.5 Results of fitting a median regression model of the difference in birth weight with a constant term in the Medical Birth Registry of Norway, 1967–1996, $n = 20$

Covariate	Estimate	SE	t	P-value	95% CI
Constant	200	171.8	1.16	0.259	(−159.6, 559.6)

SE, standard error; CI, confidence interval.

The median difference is 200. The confidence interval is calculated as $(200 - 2.093 \cdot 171.8, 200 + 2.093 \cdot 171.8) = (-159.6, 559.6)$. The *P*-value for the present *t*-test ($P = 0.26$) is slightly different from that of the Wilcoxon signed rank test ($P = 0.21$). This is to be expected, since the two test statistics are quite different. One advantage of using median regression (over the Wilcoxon signed rank test) is that both an effect estimate with a confidence interval for the median and a *P*-value for testing the null hypothesis of zero median are obtained in one analysis. A disadvantage is that median regression requires a sophisticated software package, such as Stata, to calculate the standard error.

11.8.4 The Wilcoxon-Mann-Whitney Test for Two Independent Samples

This test has many names, including Wilcoxon's two-sample test, Wilcoxon's rank sum test, and the Mann-Whitney *U* test. The null hypothesis to be tested is

H_0: the observations of the two samples are from the same distribution,

and the alternative hypothesis is

H_1: the observations of the two samples are from two distributions that have the same shape, but there is a shift in location.

The Wilcoxon-Mann-Whitney test is a test for equality of location that is applicable only when the two underlying distributions have an equal shape. This is a fairly strict assumption that is difficult to verify in practice. If the assumption cannot be accepted, the test can still be used; however, the null and alternative hypotheses are then more general, as is the interpretation of the results. The details of the latter situation are not discussed, and in general, median regression (see Section 11.11) is preferred for the analysis of non-normally distributed data.

Let n_1 be the number of subjects in sample 1, and let n_2 be the number of subjects in sample 2. The total number of subjects is $n = n_1 + n_2$. The observations can be plotted in ranked order so the sample to which they belong is obvious. Thereafter, rank values are assigned to the observations. A sum of ranks for one of the samples that is considerably higher or lower than anticipated supports the alternative hypothesis. The critical values depend on the choice of significance level and the probability distribution of W under the null hypothesis.

The Wilcoxon-Mann-Whitney test is summed up in Box 11.15.

BOX 11.15 Wilcoxon-Mann-Whitney Test for Two Samples

Let the number of observations in the two samples be n_1 and n_2. For n_1 and n_2 less than or equal to 10, the following statistical test, presented in three steps, is recommended.

First, rank observations for both samples together. Second, calculate the sum of ranks (W) for one of the samples. Finally, find the P-values from http://faculty.washington.edu/heagerty/Books/Biostatistics/TABLES/Wilcoxon/.

For n_1 and n_2 greater than 10, the distribution of the summed rank value (W) can be approximated by a normal distribution. The location and standard error of W are $E(W) = (n_1(n_1 + n_2 + 1))/2$ and $SE(W) = \sqrt{(n_1 \cdot n_2(n_1 + n_2 + 1))/12}$. The test statistic is equal to the difference between the observed value of W and the location divided by the standard error, i.e.,

$$Z = \frac{W - E(W)}{SE(W)}. \tag{11.35}$$

Let the observed Z-value be z_{obs}. H_0 can be rejected if $2P(Z \geq z_{obs}) \leq 0.05$, where the P-value is found from the normal distribution, for instance, in QuickCalcs. This test can be performed in statistical software packages like SPSS and Stata.

Example

Data from the Low Birth Weight Study and the Wilcoxon-Mann-Whitney test are used to determine whether there is a difference in BWT by HT. Since $n_1 = 177$ and $n_2 = 12$, it is a good idea to let a statistical software package do the calculations of W. Let W be the sum of ranks for the 177 normotensive mothers. Then, $W = 17103$. Since both sample sizes are above 10, the normal approximation is used. Then, $E(W) = 177 \cdot 190/2 = 16815.0$ and $SE(W) = \sqrt{177 \cdot 12 \cdot 190/12} = 183.4$. These values give $z_{obs} = (17103.0 - 16815.0)/183.4 = 1.57$, and the P-value is $2P(Z \geq 1.57) = 0.12$. There is no conclusive evidence that can lead to the rejection of the null hypothesis of equal birth weight for children born to hypertensive and normotensive mothers.

11.8.5 Effect Estimate and Confidence Interval for Two Samples

The effect estimate for observations from two samples is the median difference between the two samples (Box 11.16). Again, median regression is recommended, but details will only be discussed in Section 11.11.

BOX 11.16 Effect Estimate and Confidence Interval for Data from Two Samples

The effect estimate and confidence interval are calculated using the following procedure: Keep the observations in two samples. Let the sample variable be an indicator variable, for instance, by letting the indicator values be 0 and 1 for the two samples.

Run a median regression model with a constant term and the indicator variable (for the samples) as the only predictor variable. In the median regression model, let the constant term and the sample effect be β_0 and β_1, respectively. The estimated sample effect, $\hat{\beta}_1$, is the estimated median difference, and its estimated standard error is $\widehat{SE}(\hat{\beta}_1)$.

The confidence interval is given as:

$$[\hat{\beta}_1 - t_{0.975,n-2}\widehat{SE}(\hat{\beta}_1),\ \hat{\beta}_1 + t_{0.975,n-2}\widehat{SE}(\hat{\beta}_1)]. \quad (11.36)$$

Here $t_{0.975,n-2}$ is the 97.5-percentile in a t-distribution with $n-2$ degrees of freedom. The estimates, standard error, and confidence interval can be calculated in Stata.

Example

Data from the Low Birth Weight Study are used to determine the median difference in BWT by HT. There are 177 normotensive and 12 hypertensive mothers. A median regression of BWT on HT run in Stata renders the results shown in Table 11.6.

Table 11.6 Results of fitting a median regression model of BWT on HT in the Low Birth Weight Study, $n = 189$

Covariate	Estimate	SE	t	P-value	95% CI
HT	−482	317.2	−1.52	0.13	(−1108, 144)
Constant	2977	79.9	37.3	<0.001	(2819, 3135)

SE, standard error; CI, confidence interval.

The median difference is −482 g, which means that children born to hypertensive mothers have a median birth weight that is 482 g lower than the median birth weight of children born to normotensive mothers. The effect is not significant, with a P-value of 0.13 and a confidence interval of (−1108, 144). The confidence interval includes a wide range of possible differences in median birth weights, including a large negative difference (−1108 g) and a small positive difference (144 g).

11.9 LINEAR REGRESSION ANALYSIS

Regression analyses describe the relationship between one outcome variable and one or more exposure variables. The purpose of regression analysis is to adjust the association between an outcome and an exposure for the effect of other variables, such as age and sex, or to investigate the simultaneous effects of several exposure variables on the outcome. In regression analysis, it is common not to refer specifically to exposure variables or other variables, but rather call them all covariates, which will be used hereafter. This and the following sections will explore linear and median regression for continuous data, logistic regression for categorical data, Poisson regression for count data, as well as Cox regression for data on time to event. The topic of this section is linear regression. The simple linear regression model will be discussed and then extended to the multiple linear regression model, but details will not be covered. For a thorough treatment of linear regression, please see chapter 4 in Veierød et al.[1] or the book by Kleinbaum et al.[7]

11.9.1 Simple Linear Regression

Simple linear regression analysis comprises the study of the association between a continuous outcome variable and a continuous covariate. The relationship is assumed to be linear, i.e., a straight line in the slope-intercept form, where x is the covariate and y the outcome: $y = \beta_0 + \beta_1 x$. Here β_0 is the intercept of the straight line and the y-axis, and β_1 is the slope of the line, i.e., the change in y for a 1-unit change in x. Since β_0 is the y-value when x is zero, it is of very little practical importance, and its estimated value is rarely discussed.

Consider n pairs of observations of (x_i, Y_i). Observed data will not fit the slope-intercept equation exactly; the observations have an additional randomness or "noise." This is expressed by adding a term to the slope-intercept equation:

$$Y = \beta_0 + \beta_1 x + \varepsilon, \tag{11.37}$$

where β_0 is the intercept, β_1 the slope, and ε the noise. In this interpretation, the observations are not linear due to individual variations. For a given value of x, there is a $\hat{Y}$ on the regression line, given by the equation $\hat{Y} = \hat{\beta}_0 + \hat{\beta}_1 x$. This $\hat{Y}$-value is called the predicted value, and $\hat{\beta}_0$ and $\hat{\beta}_1$ are the estimated values of the intercept and the slope (defined in

BOX 11.17 Simple Linear Regression

Suppose that there are n pairs of observations of (x_i, Y_i), and let $\bar{x}$ and $\bar{Y}$ denote the respective means of the x's and the Y's. The intercept β_0 and slope β_1 are estimated by:

$$\hat{\beta}_0 = \bar{Y} - \hat{\beta}_1 \bar{x}, \tag{11.38}$$

and

$$\hat{\beta}_1 = \frac{\sum_{i=1}^{n}(x_i - \bar{x})(Y_i - \bar{Y})}{\sum_{i=1}^{n}(x_i - \bar{x})^2}. \tag{11.39}$$

$\hat{\beta}_0$ is called the constant, and $\hat{\beta}_1$ is called the regression coefficient.

The residual sum of squares is

$$s^2 = \frac{1}{n-2}\sum_{i=1}^{n}(Y_i - \hat{\beta}_0 - \hat{\beta}_1 x_i)^2. \tag{11.40}$$

The standard error of the estimated regression coefficient is then

$$\widehat{\mathrm{SE}}(\hat{\beta}_1) = \frac{s}{\sqrt{\sum_{i=1}^{n}(x_i - \bar{x})^2}}. \tag{11.41}$$

The 95% confidence interval of the regression coefficient β_1 is given in the usual manner, as:

$$[\hat{\beta}_1 - t_{0.975,n-2}\widehat{\mathrm{SE}}(\hat{\beta}_1),\ \hat{\beta}_1 + t_{0.975,\,n-2}\widehat{\mathrm{SE}}(\hat{\beta}_1)]. \tag{11.42}$$

The value of $t_{0.975,\,n-2}$ is the 97.5-percentile in a t-distribution with $n-2$ degrees of freedom. It can be found in QuickCalcs for any given n.

The hypothesis that the slope is zero is tested using the test statistic:

$$T = \frac{\hat{\beta}_1}{\widehat{\mathrm{SE}}(\hat{\beta}_1)}. \tag{11.43}$$

The observed value of T is t_{obs}. H_0 can be rejected if $2P(T \geq t_{obs}) \leq 0.05$. The P-value can be found in QuickCalcs. Regression analyses can be performed in statistical software packages such as SPSS and Stata.

Box 11.17). The difference between the predicted and the corresponding observed Y-value is called the residual. The size of the noise will be estimated by the sum of squares.

The values of β_0 and β_1 are estimated by the least squares method, i.e., by minimizing the squared distance from the observations to the regression line. This means that the residual sum of squares is also minimized.

The goal is to estimate β_1 with a confidence interval and to test a null hypothesis about β_1. When $\beta_1 = 0$, there is no association between x and y. Therefore, the null hypothesis is defined as:

$$H_0: \beta_1 = 0,$$

versus the alternative hypothesis:

$$H_1: \beta_1 \neq 0.$$

Simple linear regression is summarized in Box 11.17.

Example

Data from the Low Birth Weight Study are used to study the association between BWT and LWT. The results of fitting a simple linear regression model of BWT on LWT are shown in Table 11.7. The estimated slope is $\hat{\beta}_1 = 4.43$ and the intercept is $\hat{\beta}_0 = 2370$. The interpretation of the estimated slope is that BWT increases by 4.4 g for each 1-unit (pound) increase in LWT. To determine the 95% confidence interval for the slope, the 97.5-percentile of the t-distribution with $189 - 2 = 187$ degrees of freedom is needed, which is 1.97. From the formula in Box 11.17, a confidence interval of (1.05, 7.81) is obtained. The confidence interval is quite wide, which indicates a low precision for the estimate; however, the interval does not contain zero. The test statistic for the hypothesis of a zero slope (i.e., no association between BWT and LWT) is $t_{obs} = 2.59$, which gives $P = 0.010$.

Table 11.7 Results of fitting a simple linear regression model of BWT on LWT in the Low Birth Weight Study, $n = 189$

Covariate	Estimate	SE	*t*	*P*-value	95% CI
LWT	4.43	1.71	2.59	0.010	(1.05, 7.81)
Constant	2370	228	10.4	<0.001	(1919, 2820)

SE, standard error; CI, confidence interval.

The example above illustrates the relationship between an outcome variable and one covariate. But most often, the focus of the investigation is on the simultaneous associations between several covariates and the outcome variable. An observed association between the outcome variable and a covariate might be due to, or confounded by, associations with other covariates. Even in cases focusing on the effect of a specific covariate, a regression model must also include other covariates. These other covariates may be regarded as control or confounding variables and are used to verify the effect of the covariate of interest.

BOX 11.18 Multiple Linear Regression

Suppose that there are n pairs of observations of $(x_{1i}, x_{2i}, \ldots, x_{pi}, Y_i)$. The regression parameters $\beta_0, \beta_1, \ldots, \beta_p$ are estimated by the least squares method. The standard error of $\hat{\beta}_j$ is denoted by $\widehat{\text{SE}}(\hat{\beta}_j)$. A 95% confidence interval for the regression coefficient β_j is given by:

$$[\hat{\beta}_j - t_{0.975,n-p-1}\widehat{\text{SE}}(\hat{\beta}_j),\ \hat{\beta}_j + t_{0.975,\ n-p-1}\widehat{\text{SE}}(\hat{\beta}_j)], \tag{11.45}$$

where $t_{0.975,n-p-1}$ denotes the 97.5-percentile of the t-distribution with $n-p-1$ degrees of freedom.

The test statistic for the null hypothesis:

$$\text{H}_0\colon \beta_j = 0$$

versus the alternative hypothesis:

$$\text{H}_1\colon \beta_j \neq 0$$

is

$$T = \frac{\hat{\beta}_j}{\widehat{\text{SE}}(\hat{\beta}_j)}. \tag{11.46}$$

A P-value for the null hypothesis is obtained by comparing the observed value of T with the 97.5-percentiles in a t-distribution with $n-p-1$ degrees of freedom. The P-value can be calculated by QuickCalcs. The calculation of the estimates, confidence intervals, and P-values can be performed in any statistical software package, including SPSS and Stata.

11.9.2 Multiple Linear Regression

Extending a simple linear regression model to several covariates is known as multiple linear regression. The estimated regression coefficients are calculated with the least squares method in the same manner as for simple linear regression. Box 11.18 gives the main results for multiple linear regression. The computations, however, are more complicated and will not be discussed further here. This presentation of the multiple linear regression model is restricted to describing the relationship between the outcome variable Y and the p covariates as:

$$Y = \beta_0 + \beta_1 x_1 + \beta_2 x_2 + \cdots + \beta_p x_p + \varepsilon, \tag{11.44}$$

and the estimated regression coefficients as $\hat{\beta}_1, \hat{\beta}_2, \ldots, \hat{\beta}_p$. Here, β_1 is interpreted as the change in the outcome variable caused by a 1-unit change in x_1 when $x_2, x_3, \ldots, x_p$ are held constant. This might be an appropriate interpretation if one is interested in specific values of $x_2, x_3, \ldots, x_p$, but most often, the interpretation of β_1 is rephrased as the effect of x_1 after controlling for $x_2, x_3, \ldots, x_p$. The other regression coefficients are interpreted in the same manner.

A multiple linear regression model is not restricted to continuous covariates. Binary and nominal categorical variables may also be included. Binary variables are straightforward to include and to interpret. The estimated regression coefficient of a binary variable—coded 0 and 1—is the mean difference in the outcome between the two subgroups of observations defined by the binary variable. In some cases, it might be useful to recode a nominal categorical variable into a binary variable, for example, when some categories are infrequent. The variable can then be recoded into a binary variable by collapsing the infrequent categories.

Categorical variables with three or more categories require the selection of a reference category and the creation of dummy variables to include in the regression model. Dummy variables (design variables) are two or more binary variables that are used to compare each category with the reference category. A variable with k categories needs $k-1$ dummy variables. Note that when dummy variables are included in a regression model, they must all be included at the same time, and one cannot be removed without removing them all. For more details on dummy variables for linear regression, please see section 4.3.5 of Veierød et al.[1] Dummy variables are created automatically in most statistical software packages.

To build a multiple regression model, the following strategy is recommended. Fit simple linear regression models containing each covariate that is a candidate for inclusion in the multiple model (candidate covariates). Include all covariates at the 20–25% significance level in a multiple model. Use the P-values from the Wald tests for the individual coefficients in the multiple model to identify covariates that are no longer significant. Then delete the covariate with the largest P-value and refit the model. Repeat the last two steps until all covariates are either significant or important for explaining the outcome variable.

Example

Data from the Low Birth Weight Study is used to consider the covariates LWT, AGE, RACE (RACE2 and RACE3), SMOKE, PTLD, and FTVD in a multiple regression model with BWT as the outcome variable. Two dummy variables (RACE2 and RACE3) are defined based on the three-level nominal categorical variable RACE, with RACE = white as the reference category. All the candidate covariates are significant at the 25% level in simple linear regression models.

A multivariable model that includes all the candidate covariates is then fitted (Table 11.8).

Table 11.8 Results of fitting a multiple linear regression model of BWT on LWT, AGE, RACE, SMOKE, PTLD, and FTVD in the Low Birth Weight Study, $n = 189$

Covariate	Estimate	SE	*t*	*P*-value	95% CI
LWT	3.60	1.73	2.07	.040	(0.17, 7.03)
AGE	−0.08	10.1	−0.01	0.99	(−20.0, 19.8)
RACE[a]				0.001	
RACE2	−482	156	−3.08	0.002	(−791, −174)
RACE3	−357	122	−2.93	0.004	(−598, −116)
SMOKE	−337	113	−2.98	0.003	(−560, −114)
PTLD	−301	140	−2.14	0.033	(−578, −24.0)
FTVD	60.7	105	0.58	0.56	(−147, 268)
Constant	2823	320	8.83	<0001	(2192, 3454)

[a]Wald test for the overall *P*-value for the two coefficients for RACE.
SE, standard error; CI, confidence interval.

Table 11.9 Results of fitting a multiple linear regression model of BWT on LWT, AGE, RACE, SMOKE, and PTLD in the Low Birth Weight Study, $n = 189$

Covariate	Estimate	SE	*t*	*P*-value	95% CI
LWT	3.57	1.73	2.06	0.041	(0.15, 6.99)
AGE	1.17	9.84	0.12	0.91	(−18.2, 20.6)
RACE[a]				<0.001	
RACE2	−486	156	−3.11	0.002	(−794, −178)
RACE3	−371	119	−3.11	0.002	(−607, −136)
SMOKE	−349	111	−3.15	0.002	(−568, −131)
PTLD	−297	140	−2.12	0.035	(−573, −21.0)
Constant	2837	318	8.91	<0.001	(2209, 3465)

[a]Wald test for the overall *P*-value for the two coefficients for RACE.
SE, standard error; CI, confidence interval.

AGE and FTVD are not significant, therefore FTVD is excluded but AGE is kept in the model, as it is of biological interest regardless of its significance. Table 11.9 shows the results of the refitted model.

All covariates, apart from AGE, are statistically significant. As expected, SMOKE is important, but PTLD and RACE also show interesting effects.

In Tables 11.8 and 11.9, the overall *P*-value for a Wald test of RACE is given in addition to *P*-values for each dummy variable of RACE. Details on the Wald test are given in section 4.3.4 in Veierød et al.[1]

11.10 LOGISTIC REGRESSION

In the previous section, the relationship between a continuous outcome variable and one or more covariates was studied. The relationship between an outcome (Y) that has only two values, 0 or 1, and one or more covariates (x's) can be described using logistic regression, rather than linear regression. Usually the value 1 will indicate the event of interest, for instance, the presence of a disease. The probability of the outcome having the value 1 is expressed as a nonlinear function of the covariate(s). The function usually used to express this relationship is the logistic, hence the term logistic regression analysis. Here, the results will be reviewed only briefly. For details, please see chapter 3 in Veierød et al.[1] or the book by Hosmer et al.[5]

11.10.1 Simple Logistic Regression

A simple logistic regression model is given by the function:

$$P(Y=1|x) = p(x) = \frac{\exp(\beta_0 + \beta_1 x)}{1 + \exp(\beta_0 + \beta_1 x)}. \tag{11.47}$$

To interpret the values of β_0 and β_1, the equation for $p(x)$, the probability of disease, is transformed, and the odds of disease are calculated by $p(x)/(1 - p(x))$. Hence,

$$\ln \frac{p(x)}{1 - p(x)} = \beta_0 + \beta_1 x, \tag{11.48}$$

which is called the log-odds, or logit, of $Y = 1$. Because the log-odds is linear in β_0 and β_1, the constant and the regression coefficient are readily interpreted. β_0 is interpreted as the log-odds for $x = 0$; however, as for linear regression, this interpretation is of little practical importance. Because

$$\ln \frac{p(x)/[1 - p(x)]}{p(x-1)/[1 - p(x-1)]} = \beta_1, \tag{11.49}$$

the value of β_1 shows how much the log-odds for $Y = 1$ changes for a 1-unit change in x. It is more amenable to interpret β_1 via a change in odds. From the above equation, it follows that:

$$\frac{p(x)/[1 - p(x)]}{p(x-1)/[1 - p(x-1)]} = \exp(\beta_1), \tag{11.50}$$

which indicates that the odds of $Y = 1$ changes by $\exp(\beta_1)$ units for a 1-unit change in x. Thus, $\exp(\beta_1)$ is the effect measure in logistic regression, and it assesses the degree of association (interpreted as the OR) between the covariate and the outcome variable.

BOX 11.19 Simple Logistic Regression

There are n pairs of observations of (x_i, Y_i). The values of β_0 and β_1 can be estimated by the maximum likelihood method so the data best fit the logistic function. A statistical software package that supports logistic regression, such as SPSS or Stata, provides maximum likelihood estimates for $\hat{\beta}_0$ and $\hat{\beta}_1$, as well as their estimated standard errors $\widehat{\mathrm{SE}}(\hat{\beta}_0)$ and $\widehat{\mathrm{SE}}(\hat{\beta}_1)$.

The confidence interval for β_1 is found in the customary manner, by associating the estimated $\hat{\beta}_1$ with $\widehat{\mathrm{SE}}(\hat{\beta}_1)$. The estimated $\hat{\beta}_1$ has a distribution that is very close to a normal distribution, provided there are sufficient data, so in the usual manner, an approximate 95% confidence interval for β_1 is

$$[\hat{\beta}_1 - 1.96\cdot\widehat{\mathrm{SE}}(\hat{\beta}_1), \hat{\beta}_1 + 1.96\cdot\widehat{\mathrm{SE}}(\hat{\beta}_1)]. \tag{11.51}$$

Here, 1.96 is used as the percentile, since it is the 97.5-percentile of the normal distribution. Through transformation via an exponential function, a confidence interval for $\exp(\beta_1)$ is found as follows:

$$\{\exp[\hat{\beta}_1 - 1.96\cdot\widehat{\mathrm{SE}}(\hat{\beta}_1)], \exp[\hat{\beta}_1 + 1.96\cdot\widehat{\mathrm{SE}}(\hat{\beta}_1)]\}. \tag{11.52}$$

The test of the null hypothesis of no association between Y and x is conducted in the usual manner, by dividing the effect estimate with its standard error (the Wald test). The test statistic used is

$$Z = \frac{\hat{\beta}_1}{\widehat{\mathrm{SE}}(\hat{\beta}_1)}. \tag{11.53}$$

Denote the observed value of Z by z_{obs}. The null hypothesis is rejected if $2P(Z \geq z_{obs}) \leq 0.05$. The P-value can be found in QuickCalcs, or it can be calculated by a statistical software package.

As for the linear case, there is no association between a covariate and an outcome variable when $\beta_1 = 0$, since then the odds of disease are independent of the covariate. Therefore, the null hypothesis of interest in logistic regression is

$$\mathrm{H}_0\colon \beta_1 = 0,$$

and the alternative hypothesis is

$$\mathrm{H}_1\colon \beta_1 \neq 0.$$

Simple logistic regression is summarized in Box 11.19.

Example

In the Low Birth Weight Study, LOW is fitted on LWT in a simple logistic regression model. This yields the result in Table 11.10.

Table 11.10 Results of fitting a simple logistic regression model of LOW on LWT in the Low Birth Weight Study, $n = 189$

Covariate	Estimate	SE	z	P-value	OR	95% CI for OR
LWT	−0.014	0.006	−2.28	0.023	0.99	(0.97, 1.00)
Constant	0.998	0.785	1.27	0.20		

SE, standard error; CI, confidence interval.

Here, the regression coefficient is $\hat{\beta}_1 = -0.014$, $\exp(\hat{\beta}_1) = 0.99$, and the 95% confidence interval for $\exp(\beta_1)$ is (0.974, 0.998). This means that the odds of having a child with birth weight less than 2500 g is decreased by a factor of 0.99 for each 1-unit (pound) increase in LWT, or, equivalently, a 1.0% decrease. The null hypothesis of no effect of LWT on LOW can be rejected ($P = 0.023$).

11.10.2 Multiple Logistic Regression

As discussed above, an observed association between the outcome variable and a covariate might be due to or confounded by associations with other covariates. The effect of one covariate, controlled for other covariates, or the simultaneous associations between the outcome variable and several covariates can be studied with multiple logistic regression. Assume in general that there are p covariates and that the logistic regression model is expressed as:

$$P(Y = 1 | x_1, x_2, \ldots, x_p) = p(x_1, x_2, \ldots, x_p) = \frac{\exp(\beta_0 + \beta_1 x_1 + \beta_2 x_2 + \cdots + \beta_p x_p)}{1 + \exp(\beta_0 + \beta_1 x_1 + \beta_2 x_2 + \cdots + \beta_p x_p)}. \quad (11.54)$$

This expression is equivalent to

$$\ln \frac{p(x_1, x_2, \ldots, x_p)}{1 - p(x_1, x_2, \ldots, x_p)} = \beta_0 + \beta_1 x_1 + \beta_2 x_2 + \cdots + \beta_p x_p. \quad (11.55)$$

Here, β_1 is interpreted as the change in log-odds caused by a 1-unit change in x_1 when $x_2, x_3, \ldots, x_p$ are held constant. It is often said that β_1 is the effect of x_1 when $x_2, x_3, \ldots, x_p$ have been controlled for. The other regression coefficients are interpreted in the same manner. Again, an interpretation of the effects via changes in odds rather than log-odds is often more informative. When there is a 1-unit change in x_1 and the other covariates are held constant, the odds of $Y = 1$ change by $\exp(\beta_1)$ units. The interpretations for the other covariates are similar.

Example

With data from the Low Birth Weight Study, LWT, AGE, RACE via the dummy variables RACE2 and RACE3, SMOKE, PTLD, FTVD are selected as candidate covariates for a multiple logistic regression model with LOW as the outcome variable. All the candidate covariates are significant at the 25% level in simple logistic regression models. A multiple logistic regression model that includes all the candidate covariates is then fitted. The results are given in Table 11.11.

AGE, LWT, and FTVD are not statistically significant. However, because AGE and LWT are both of biological interest, they are retained in the model. The model is then refitted without FTVD, and the results are shown in Table 11.12.

In Tables 11.11 and 11.12, the overall *P*-value for a likelihood ratio test of RACE is given in addition to *P*-values for each dummy variable of RACE. The likelihood ratio test for logistic regression is similar to the Wald test for linear regression (see Section 11.9). It is important to include that *P*-value, because it gives the overall *P*-value of RACE, and not only for each dummy variable of RACE. Details on that test are given in section 3.2 in Veierød et al.[1]

The results in Table 11.12 are quite similar to those shown in Table 11.11. Except for LWT and AGE, all the covariates are statistically significant.

Table 11.11 Results of fitting a multiple logistic regression model of LOW on LWT, AGE, RACE, SMOKE, PTLD, and FTVD in the Low Birth Weight Study, $n = 189$

Covariate	Estimate	SE	*z*	*P*-value	OR	95% CI for OR
LWT	−0.010	0.007	−1.60	0.111	0.99	(0.98, 1.00)
AGE	−0.039	0.038	−1.05	0.295	0.96	(0.89, 1.04)
RACE[a]			6.14	0.047		
RACE2	1.154	0.526	2.19	0.028	3.17	(1.13, 8.90)
RACE3	0.798	0.441	1.81	0.070	2.22	(0.94, 5.27)
SMOKE	0.811	0.406	2.00	0.046	2.25	(1.02, 4.99)
PTLD	1.347	0.452	2.98	0.003	3.85	(1.59, 9.32)
FTVD	−0.216	0.363	−0.59	0.552	0.81	(0.40, 1.64)
Constant	0.474	1.157	0.41	0.682		

[a]Likelihood ratio test for the overall *P*-value for the two coefficients for RACE.
SE, standard error; CI, confidence interval.

Table 11.12 Results of fitting a multiple logistic regression model of LOW on LWT, AGE, RACE, SMOKE, and PTLD in the Low Birth Weight Study, $n = 189$

Covariate	Estimate	SE	*z*	*P*-value	OR	95% CI for OR
LWT	−0.011	0.007	−1.61	0.108	0.99	(0.98, 1.00)
AGE	−0.043	0.037	−1.18	0.238	0.96	(0.89, 1.03)
RACE[a]			6.62	0.037		
RACE2	1.163	0.527	2.21	0.027	3.20	(1.14, 9.00)
RACE3	0.847	0.431	1.96	0.050	2.33	(0.94, 5.27)
SMOKE	0.860	0.396	2.17	0.030	2.36	(1.00, 5.43)
PTLD	1.327	0.449	2.96	0.003	3.77	(1.56, 9.08)
Constant	0.445	1.152	0.39	0.699		

[a]Likelihood ratio test for the overall *P*-value for the two coefficients for RACE.
SE, standard error; CI, confidence interval.

11.11 MEDIAN REGRESSION

Median regression is similar to linear regression; both models describe the relationship between a continuous outcome variable and one or more covariates. The difference between the models is that linear regression studies the mean of the outcome variable, whereas median regression studies the median of the outcome variable. For a readable introduction to quantile regression, which also includes median regression, please see the book by Hao and Naiman.[8]

11.11.1 Simple Median Regression

Suppose that there are n pairs of observations of (x_i, Y_i). The relationship between Y and x is given as

$$Y_{\text{median}} = \beta_0 + \beta_1 x + \varepsilon, \tag{11.56}$$

where, as for linear regression, ε is the residual (noise). The covariate of interest is β_1, i.e., the slope of the line, or the change in the median of Y for a 1-unit change in x. Again, the goal is to test the null hypothesis:

$$\text{H}_0\text{: } \beta_1 = 0$$

against the alternative hypothesis:

$$\text{H}_1\text{: } \beta_1 \neq 0.$$

Based on the n pairs of observations, the parameters of the median regression model are estimated by minimizing the sum of the absolute residuals, rather than the sum of squared residuals as is done in linear regression. The standard error of the regression coefficient is estimated by a complicated procedure, the details of which are not discussed here. The use of a statistical software package is recommended to calculate the estimates. Stata is used in the examples below.

Example

In the Low Birth Weight Study, a simple median regression model of BWT was fitted on LWT. The results are given in Table 11.13. The estimated slope is $\hat{\beta}_1 = 6.37$ and the intercept is $\hat{\beta}_0 = 2132$. The interpretation of the slope is that the median BWT increases by 6.4 g for each 1-unit (pound) change in LWT. The confidence interval indicates the level of precision for the estimated slope and is somewhat wide in this case but does not contain zero. The hypothesis of a zero slope (no effect of LWT on median BWT) is rejected at the 0.4% level ($P = 0.004$).

Table 11.13 Results of fitting a simple median regression model of BWT on LWT in the Low Birth Weight Study, $n = 189$

Covariate	Estimate	SE	*t*	*P*-value	95% CI
LWT	6.37	2.19	2.91	0.004	(2.05, 10.7)
Constant	2132	292	7.30	<0.001	(1556, 2708)

SE, standard error; CI, confidence interval.

11.11.2 Multiple Median Regression

As for simple linear and logistic regression, models with several covariates are needed. The multiple median regression model is given as:

$$Y_{\text{median}} = \beta_0 + \beta_1 x_1 + \beta_2 x_2 + \cdots + \beta_p x_p + \varepsilon. \tag{11.57}$$

Again, β_1 is interpreted as the change in the median of the dependent variable caused by a 1-unit change in x_1 when $x_2, x_3, \ldots, x_p$ are held constant. The other regression coefficients are interpreted in the same manner. The estimation of the regression coefficients, standard errors, P-values, and confidence intervals in multiple median regression models are straightforward extensions from the simple model.

Example

In the Low Birth Weight Study, LWT, AGE, RACE (via RACE2 and RACE3), SMOKE, PTLD, FTVD, and HT are considered candidate covariates for a multiple median regression

Table 11.14 Results of fitting a multiple median regression model of BWT on LWT, AGE, RACE, SMOKE, PTLD, and HT in the Low Birth Weight Study, $n = 189$

Covariate	Estimate	SE	*t*	*P*-value	95% CI
LWT	5.62	2.44	2.30	0.022	(0.81, 10.4)
AGE	−3.39	13.5	−0.25	0.80	(−30.0, 23.2)
RACE[a]				0.092	
RACE2	−396	214	−1.85	0.066	(−817, 25.7)
RACE3	−276	163	−1.69	0.093	(−598, 46.6)
SMOKE	−470	152	−3.10	0.002	(−770, −171)
PTLD	−366	192	−1.91	0.058	(−743, 12.4)
HT	−433	285	−1.52	0.13	(−995, 130)
Constant	2738	438	6.25	<0.001	(1873, 3603)

[a]Wald test for the overall *P*-value for the two coefficients for RACE.
SE, standard error; CI, confidence interval.

Table 11.15 Results of fitting a multiple median regression model of BWT on LWT, AGE, RACE, SMOKE, and PTLD in the Low Birth Weight Study, $n = 189$

Covariate	Estimate	SE	*t*	*P*-value	95% CI
LWT	5.99	2.34	2.56	0.011	(1.38, 10.6)
AGE	−2.88	13.3	−0.22	0.83	(−29.1, 23.3)
RACE[a]				0.048	
RACE2	−401	210	−1.91	0.058	(−816, 14.3)
RACE3	−336	161	−2.09	0.038	(−653, −19.0)
SMOKE	−433	149	−2.90	0.004	(−728, −139)
PTLD	−391	189	−2.07	0.040	(−763, −18.3)
Constant	2655	429	6.19	<0.001	(1808, 3502)

[a]Wald test for the overall *P*-value for the two coefficients for RACE.
SE, standard error; CI, confidence interval.

model, with BWT as the outcome variable. All the candidate variables, except for AGE and FTVD, are significant at the 25% level in simple median regression models. A multiple model that includes all the candidate covariates, except for FTVD, is then fitted. AGE is retained in the model, because it is of biological interest regardless of its significance. The results are given in Table 11.14.

AGE, RACE, PTLD, and HT are not significant. As before, AGE is retained in the model, and the variable with the largest nonsignificant *P*-value, which in this case is HT, is excluded. The refitted model is shown in Table 11.15.

All the variables, except for AGE, are now statistically significant, including RACE and PTLD, which were nonsignificant in Table 11.14. The model in Table 11.15 is therefore denoted the final model. As expected, SMOKE is important; however, the estimated effect of PTLD is almost as large as that of SMOKE, as are the estimated effects of RACE2 and RACE3.

11.12 POISSON REGRESSION

In clinical and epidemiological studies, the effect measure is often based on the incidence rate (IR) (see Chapter 9). A useful measure is the ratio of the IRs among the exposed and the unexposed groups, denoted by incidence rate ratio (IRR). A feasible way to study an IR is to assume that the total number of events follows a Poisson distribution. Let Y be the number of cases, IR the incidence rate, and t the total length of follow-up time or observation time. $Y \sim \text{Poisson}(\text{IR} \cdot t)$ is written to signify that Y follows a Poisson distribution with mean $\text{IR} \cdot t$. The IR is estimated by $\widehat{\text{IR}} = Y/t$ and is covered in Chapter 9. As for RR and OR, IR must be transformed into a logarithmic scale to estimate the standard error. The standard error of $\ln \widehat{\text{IR}}$ is $1/\sqrt{Y}$.

If there are two groups, one exposed and one unexposed, the IRs are $\widehat{\text{IR}}_1 = Y_1/t_1$ and $\widehat{\text{IR}}_2 = Y_2/t_2$, respectively. Then, the confidence interval for $\text{IRR} = \text{IR}_1/\text{IR}_2$ is as stated in Section 9.7.

As for linear and logistic regression, several covariates may be controlled for in Poisson regression models. If there are p covariates, $x_1, x_2, \ldots, x_p$, the Poisson regression model may be expressed by:

$$Y \sim \text{Poisson}[\text{IR}(x_1, x_2, \ldots, x_p) \cdot t], \tag{11.58}$$

where $IR(x_1, x_2, \ldots, x_p) = \exp(\beta_0 + \beta_1 x_1 + \beta_2 x_2 + \cdots + \beta_p x_p)$ is the incidence rate, and t is the total length of follow-up time. As usual, β_0 is the constant term and $\beta_1, \beta_2, \ldots, \beta_p$ are the regression coefficients. In Poisson regression, the coefficients are interpreted via the IRR. When there is a 1-unit change in x_1 and the other covariates are held constant, the change in IR is $\exp(\beta_1)$.

A statistical software package for general linear models, such as SPSS or Stata, can be used to calculate estimates and confidence intervals for the β's.

For more details on Poisson regression, please see chapter 6 of Veierød et al.[1] or chapter 24 of Kleinbaum et al.[7]

Example

For the data in Table 9.7, Poisson regression can be used to estimate the association between death from myocardial infarction and cigarette smoking. Let smoking be the covariate (with two categories). A generalized linear model with a Poisson loglinear link in SPSS gives the results shown in Table 11.16.

The estimates of the regression coefficients are on a logarithmic scale, so the estimated IRR is $\widehat{\text{IRR}} = \exp(1.12) = 3.06$, and the 95% confidence interval is given by (1.50, 6.27). These results agree with those calculated by hand in Section 9.8.

Table 11.16 Results of fitting a Poisson regression model of death from myocardial infarction on smoking, $n = 9800$

Covariate	Estimate	SE	z	P-value	IRR	95% CI for IRR
Smoking	1.12	0.365	3.07	0.002	3.06	(1.50, 6.27)
Constant	−6.21	0.316	−19.6	<0.001		

SE, standard error; CI, confidence interval.

11.13 SURVIVAL ANALYSIS AND COX REGRESSION

Survival analysis plays an important role in medicine and the biological sciences. Time until an event, like time until death, is the outcome variable. Because not every patient is expected to experience the event during the observation period, and because some patients might leave the study before the observation period is ended (loss to follow-up), the time-to-event variable is seldom without missing values. This is called censoring.

The survival function, hazard rate, and hazard function are important concepts in survival analysis. The survival function, which is given such a name regardless of what the event might be, is defined as $S(t) = P(\text{event does not occur until time } t)$. When the event is developing a disease, the hazard rate is the rate of being affected in a short interval around time t, provided that the subject does not experience disease before time t. The hazard function, which is the hazard rate at time t, is defined as:

$$h(t) = \frac{1}{\mathrm{d}t} P(\text{event occurs in } (t, t+\mathrm{d}t), \text{given that no event occurs before time } t), \tag{11.59}$$

where $\mathrm{d}t$ is a (small) unit of time. The survival function $S(t)$ and the hazard function $h(t)$ are related by:

$$S(t) = \exp\left[-\int_0^t h(u)\mathrm{d}u\right], \tag{11.60}$$

where $\int_0^t h(u)\mathrm{d}u$ is the integral of $h(t)$ from time zero to time t.

11.13.1 Kaplan-Meier Survival Function Estimate

The survival function is estimated by the Kaplan-Meier survival function estimate. Let the n time to events be $t_1, t_2, \ldots, t_n$ in increasing order. Let r_j

be number of patients at risk for the event of interest (for instance, dying) at time t_j. Then the Kaplan-Meier (KM) estimated survival function is

$$\hat{S}_{\mathrm{KM}} = \prod_{t_j < t} \left(1 - \frac{1}{r_j}\right), \tag{11.61}$$

where Π is the product sign.

11.13.2 The Logrank Test

If there are two samples, for example one for treatment and one for placebo, the goal is to test whether the hazard functions in the two samples are equal. Let $h_1(t)$ and $h_2(t)$ be the two hazard functions, and define the null hypothesis as:

$$\mathrm{H}_0\colon h_1(t) = h_2(t), \tag{11.62}$$

and the alternative hypothesis as:

$$\mathrm{H}_1\colon h_1(t) \neq h_2(t). \tag{11.63}$$

To construct a test statistic, let r_{1j} and r_{2j} be the numbers of observations just before time t_j in samples 1 and 2, with $r_j = r_{1j} + r_{2j}$. Finally, let i_j be a variable that takes the value 1 if the event at time t_j occurs in sample 1, and 0 otherwise. Then, the logrank (LR) test is given as

$$Z_{\mathrm{LR}} = \frac{\sum_{j=1}^{n} i_j - \sum_{j=1}^{n} \frac{r_{1j}}{r_j}}{\sqrt{\sum_{j=1}^{n} \frac{r_{1j} \cdot r_{j2}}{r_j^2}}}. \tag{11.64}$$

Z_{LR} can be approximated by a normal distribution. When z_{obs} is the observed value of the test statistic, the null hypothesis can be rejected if $2P(Z_{\mathrm{LR}} \geq z_{\mathrm{obs}}) \leq 0.05$.

Example

Here, the cardiovascular disease risk data set (CVD Risk Study) from the website http://www.umass.edu/statdata/statdata/stat-survival.html is used, which has also been used by Hosmer et al.[9] (see section 9.6 in that book). Follow-up time or time until death after hospital admission is recorded for 65 patients. There are 32 deaths from CVD, and the other 33 patients contribute censored observations. In addition to data on time until

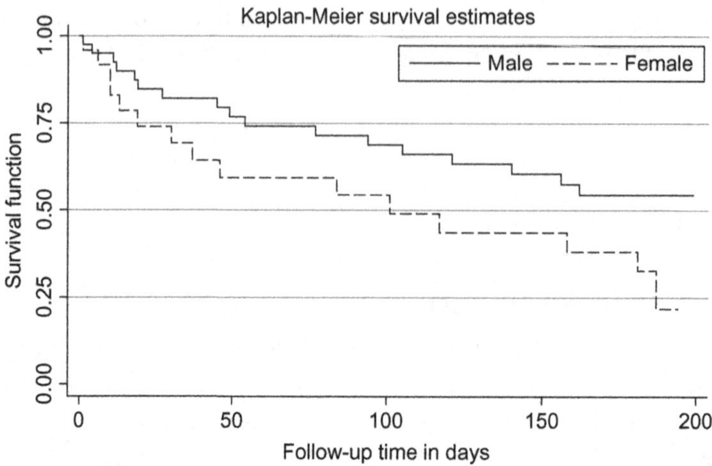

Figure 11.1 Kaplan-Meier survival estimates of CVD death, for males and females, CVD Risk Study, $n = 65$.

death or censoring, the covariates AGE (in years), body mass index (BMI in kg/m^2), and SEX (0 = male, 1 = female) are included.

In this example, differences in survival time by SEX are studied. Figure 11.1 shows the Kaplan-Meier plot for the two sexes.

Further, $Z_{LR} = 1.956$, and the *P*-value for the logrank test is $2P(Z_{LR} \geq 1.956) = 0.051$. Males seem to have better CVD survival than females, although one must be careful about the conclusion in this case; the difference is not statistically significant, and the confounding effects of covariates such as AGE and BMI were not considered.

11.13.3 Cox Regression

Studying the survival curves and performing a logrank test is usually the first step in a survival analysis, especially in clinical studies, where the samples are defined by whether a treatment is received or not. A second step is often to control for other covariates, which is done with Cox regression, where the hazard function can be expressed as:

$$h(t) = h_0(t) \cdot \exp(\beta_1 x_1 + \beta_2 x_2 + \cdots + \beta_p x_p). \tag{11.65}$$

Here, $h_0(t)$ is the baseline hazard, $x_1, x_2, \ldots, x_p$ are the covariates, and $\beta_1, \beta_2, \ldots, \beta_p$ are regression coefficients that express the relationship between the covariates and the time to event. For the Cox regression model, the regression coefficients are interpreted via the HR. A 1-unit change in, say, x_1 will induce a change of $\exp(\beta_1)$ for the hazard rate, which means that the HR $= \exp(\beta_1)$.

To estimate the parameters $\beta_1, \beta_2, \ldots, \beta_p$, the partial maximum likelihood method is used. This produces estimates of the regression coefficients, standard errors, the statistical test with *P*-value, and confidence interval. The procedure is very similar to that of logistic regression in Box 11.19, and the output is also very similar, with OR estimated by logistic regression and HR by Cox regression. Note in particular that the confidence interval for HR in Cox regression is constructed similarly to that of OR in logistic regression.

A statistical software package like SPSS or Stata is required to run Cox regression, but it is technically as simple as running logistic or linear regression.

For a more thorough introduction to survival analysis or Cox regression, please see chapter 5 of Veierød et al.[1] or Hosmer et al.[9]

Example

Data from the CVD Risk Study were used to run a Cox regression analysis with AGE, BMI, and SEX as covariates. The results are shown in Table 11.17.

Both AGE and BMI are statistically significant, whereas SEX is not. The association between SEX and time to CVD death, which was of borderline significance when *P*-values were calculated by the logrank test, is not independent of AGE and BMI.

Table 11.17 Results of fitting a Cox regression model of time to CVD death on AGE, BMI, and SEX in the CVD Risk Study, $n = 65$

Covariate	Estimate	SE	*z*	*P*-value	HR	95% CI for HR
AGE	0.059	0.013	4.64	<0.001	1.06	(1.04, 1.09)
BMI	0.094	0.037	2.54	0.011	1.10	(1.02, 1.18)
SEX	0.307	0.380	0.81	0.419	1.36	(0.65, 2.86)

SE, standard error; CI, confidence interval.

QUESTIONS TO DISCUSS

1. Formulate a hypothesis based on your own research problem. What is the null hypothesis, and what is the alternative hypothesis?
2. Explain the following concepts: effect measure, effect estimate, confidence interval, and *P*-value.
3. The Framingham Heart Study is a large cohort study aimed at studying risk factors for coronary heart disease. A random selection of 500 persons from a total of 1615 persons aged 31–65 can be downloaded from http://www.stat.tamu.edu/~carroll/data.php. Here, the association

Table 11.18 Cell counts for the association between FIRSTCHD and HYPERTENSION in the Framingham Heart Study, $n = 500$

	Evidence of CHD		
HYPERTENSION	**Yes**	**No**	**Total**
Yes	32 (9.3%)	312 (90.7%)	344 (100%)
No	5 (3.2%)	151 (96.8%)	156 (100%)

between coronary heart disease (FIRSTCHD) and elevated systolic blood pressure (HYPERTENSION), defined as a blood pressure over 120 mmHg, will be studied. Table 11.18 gives the results.

Use the data in Table 11.18 to calculate the RD, the RR, and the OR, with their corresponding confidence intervals.

4. Formulate a testable null hypothesis for the association in question 3. Calculate the *P*-value for the statistical test you use. What is your conclusion?
5. The values of RR and OR for the data in Table 11.18 are quite similar. Why is that? Give an example in which the values of RR and OR differ.
6. Table 11.19 shows the results of a subsample of $n = 100$ from the Framingham Heart Study described in question 3.

Table 11.19 Cell counts for the association between FIRSTCHD and HYPERTENSION in the Framingham Heart Study, $n = 100$

	Evidence of CHD		
HYPERTENSION	**Yes**	**No**	**Total**
Yes	6 (8.8%)	62 (91.2%)	68 (100%)
No	1 (3.1%)	31 (96.9%)	32 (100%)
Total	7 (7.0%)	93 (93.0%)	100 (100%)

Use the data in Table 11.19 to calculate RD, RR, and OR, with corresponding confidence intervals.

7. For the data in Table 11.19, calculate the *P*-value for the association between coronary heart disease and elevated systolic blood pressure.
8. Studies have indicated that some calcium antagonists may interact with constituents of grapefruit juice. Apparently, intake of grapefruit juice leads to increased calcium antagonist concentration in plasma. If that is the case, the result may be a more pronounced antihypertensive effect and a greater risk of adverse effects.

Table 11.20 Plasma concentration (c_{max} in nmol/l) of calcium antagonist taken with and without grapefruit juice

Subject	Calcium antagonist + water	Calcium antagonist + grapefruit juice
1	22.5	31.8
2	8.2	12.4
3	11.3	15.7
4	5.9	4.2
5	20.7	23.1
6	19.3	18.5
7	10.4	12.0
8	12.1	14.9
9	18.3	21.2

Assume that a trial is conducted with $n = 9$ volunteers who on one occasion take a calcium antagonist with water and on another occasion take it with a measured quantity of grapefruit juice. The maximum plasma concentration of the calcium antagonist is c_{max}, measured in nmol/L. The results shown in Table 11.20 are from a hypothetical trial that is nonetheless plausible as it reflects previously published results. Assume that the observations are normally distributed.

Calculate the appropriate effect estimate and confidence interval.

Set up a relevant testable null hypothesis. What is the *P*-value? Sum up your findings.

9. Lung capacity is measured by peak expository flow (PEF), which is the maximal flow achieved during a maximally forced expiration. PEF is measured in L/min. Increased PEF values are an indication of decreased airway resistance. The aim of the study is to find out whether training decreases airway resistance among patients with cystic fibrosis. The study sample was divided into two samples, one that underwent training and one that did not. There were $n = 20$ patients in each sample. In the sample with training, there was an increase in PEF of 100 L/min, with an SD of 50 L/min. In the sample without training, there was an increase of 60 L/min, with an SD of 30 L/min. It is assumed that the observations are normally distributed. Calculate a 95% confidence interval for the between-sample difference in increase in PEF. Formulate a null hypothesis and test it. What is your conclusion?
10. Use the data in Table 11.20 to test the hypothesis with a nonparametric method. What is your conclusion?

REFERENCES

1. Veierød MB, Lydersen S, Laake P. *Medical statistics in clinical and epidemiological research.* Oslo: Gyldendal Akademisk; 2012.
2. QuickCalcs. *Free online application for P-value calculation for the most frequently used statistical distributions.* Available at: <http://graphpad.com/quickcalcs/PValue1.cfm>.
3. Lang TA, Secic M. *How to report statistics in medicine.* 2nd ed. Philadelphia, PA: American College of Physicians; 2006.
4. Agresti A, Caffo B. Simple and effective confidence intervals for proportions and differences of proportions result from adding two successes and two failures. *Am Stat* 2000;**54**:280–8.
5. Hosmer DW, Lemeshow S, Sturdivant RX. *Applied logistic regression.* 3rd ed. Hoboken, NJ: John Wiley & Sons; 2013.
6. Kirkwood BR, Sterne AC. *Essential medical statistics.* 2nd ed. Malden, MA: Blackwell Science; 2003.
7. Kleinbaum DG, Kupper LL, Nizam A, Muller KE. *Applied regression analysis and other multivariable methods.* 4th ed. Belmont, CA: Brooks/Cole; 2008.
8. Hao L, Naiman DQ. *Quantile regression.* Thousand Oaks, CA: Sage Publications; 2007.
9. Hosmer DW, Lemeshow S, May S. *Applied survival analysis.* 2nd ed. Hoboken, NJ: John Wiley & Sons; 2008.

CHAPTER 12

Evidence-Based Medicine and Systematic Reviews

Morten Wang Fagerland
Oslo Centre for Biostatistics and Epidemiology, Research Support Services, Oslo University Hospital, Oslo, Norway

12.1 INTRODUCTION

Evidence-based medicine (EBM) is an approach to answering clinical questions. A widely quoted definition of EBM is given in Box 12.1.

EBM marks a "democratization of medical decision-making" and a shift from "authoritarian medicine" to "authoritative medicine."[2] Expert opinion, which once governed medical decision-making, now carries less weight than scientific evidence. Nowadays, every medical doctor is expected to be up to date on the latest research results, and thus able to form medical decisions based on the best available evidence. This was brought about by a change not only in the social and cultural environment in which clinical decisions are made, but also in the standards for collecting evidence and the tools available to analyze it.

The roots of EBM can be traced back to eighteenth-century France and the publication of the first volume of the *Encyclopédie*.[3] Although this was not the first encyclopedia to be published, it was fundamentally different from previous endeavors because it combined theory and practice. The *Encyclopédie* contained articles on the arts and sciences, as well as on trades and crafts. It embraced the scientific method, replacing tradition, authority, and superstition with rational explanations and first-hand observations of the world. The *Encyclopédie* was an effort to systematically assemble all human knowledge—and to change the world.

During the last century, the standards for collecting evidence have evolved from single case reports and cohort and case-control studies to randomized controlled trials (RCTs), which are now considered the gold standard for evaluating the efficacy of interventions. One early supporter of the RCT, the Scottish doctor named Archibald (Archie) Cochrane, went one step further. He recognized that even though a well-designed and properly carried out RCT may provide valuable evidence regarding

Research in Medical and Biological Sciences
DOI: http://dx.doi.org/10.1016/B978-0-12-799943-2.00012-4

BOX 12.1 Definition of EBM

Evidence based medicine is the conscientious, explicit, and judicious use of current best evidence in making decisions about the care of individual patients. The practice of evidence based medicine means integrating individual clinical expertise with the best available external clinical evidence from systematic research

(Sackett et al.[1])

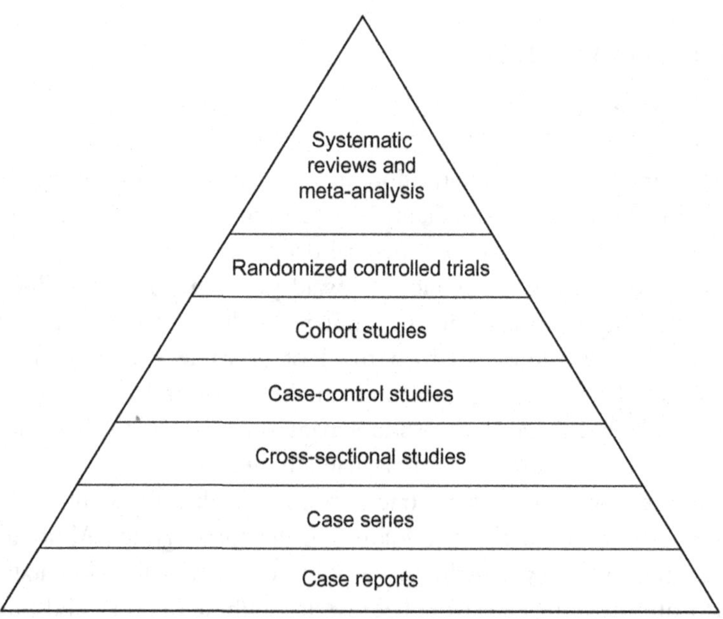

Figure 12.1 Pyramid representing the hierarchy of scientific evidence by study design (based on Greenhalgh[4]). Study designs at higher positions carry greater authority than those at lower positions.

the effects of a treatment, no single trial could provide the whole truth; therefore, he put forth the idea that clinical decision-making should be based on systematic reviews of all available evidence. The systematic review and its statistical component, meta-analysis, are the tools used to synthesize a large body of evidence on a particular scientific question. Systematic reviews represent the top of the hierarchy of scientific evidence (Figure 12.1) and the best estimate of the factual situation. The leading international institution for the systematic collection of health-related evidence is the Cochrane Collaboration (http://www.cochrane.org/), which is named after Archie Cochrane. The Cochrane Collaboration maintains an extensive and comprehensive compilation of systematic reviews (>5000 by December 2013) and the world's largest collection of RCT records.

Today, the amount of available scientific evidence is overwhelming. Davidoff[3] states: "our medical knowledge system is highly chaotic [...] with the result that we are starving for information in the midst of plenty." EBM strives to bring order to this information chaos by providing a systematic approach to the scrutiny of scientific literature.

12.2 THE FIVE ELEMENTS OF EBM

The EBM approach can be broken down into five elements (Box 12.2): asking, accessing, appraising, applying, and assessing. Element 1 (asking) refers to the process of transforming a research idea or a clinical question into a tangible form. This form is called a research question, which is a phrase characterized by a high degree of concreteness. In a clinical setting, a research question should include information about patients, interventions, comparators, and outcomes. This approach to formulating a research question is considered in more detail in Chapters 4 and 5, and also makes an appearance in Section 12.4. Element 2 (accessing) refers to the process of finding the relevant evidence. Most of Chapter 5 (on literature search) is devoted to this task. The focus in this chapter is on element 3 (appraising): how to critically appraise the best available evidence, in the form of systematic reviews. Integrating scientific evidence into clinical practice (elements 4 and 5, applying and assessing) requires detailed and specific knowledge about day-to-day clinical routines. It involves comparing patient characteristics, expected treatment effects, and event rates between study participants and actual patients, as well as appropriateness

BOX 12.2 The Five Elements of EBM

1. *Ask*: Refine the clinical question into an answerable research question.
2. *Access*: Search the literature to find relevant evidence.
3. *Appraise*: Critically evaluate the evidence.
4. *Apply*: Turn the findings into clinical action—if possible.
5. *Assess*: Evaluate the approach, results, and practical implications.

Del Mar et al.[8]

(in terms of patients' values and preferences), alternatives, feasibility, and economic considerations. These broad topics will not be covered in detail here; for more information please refer to the books by Heneghan and Badenoch[5] and Straus et al.,[6] and the book chapter by McAlister.[7]

The remaining sections of this chapter discuss how to conduct and report systematic reviews; particular attention is paid to meta-analysis. A critical appraisal of systematic reviews requires a considerable understanding of the statistical concepts and methods used therein, which may be many and complex. Although the ability to perform meta-analyses is not necessary to evaluate a systematic review, the way in which these analyses are employed (or omitted) may highlight (or hide) important aspects of the research question.

12.3 INTRODUCTION TO SYSTEMATIC REVIEWS

A systematic review is a study of studies. Its purpose is to address a precisely defined research question using all available studies in a specific field—which will be referred to in this chapter as the primary studies—that comply with certain standards. Systematic reviews use formal, structured, and unbiased methods for synthesizing scientific evidence. Unlike narrative reviews, which often contain major flaws such as selective use of evidence and subjective criteria for drawing conclusions, systematic reviews are characterized by transparency, reproducibility, and the capability of being repeatedly updated (Box 12.3).

The design of the primary studies finally included in a systematic review (included studies) is usually determined by the research question (Box 12.4). When the purpose of a systematic review is to summarize treatment effects, the primary studies are often confined to RCTs. Cohort and case-control studies may exist for the topic, but the evidence from cohort and case-control studies is considered inferior to that of RCTs. Systematic reviews that study causes of disease or the accuracy of diagnostic tests (diagnosis) are also quite common, whereas systematic

BOX 12.3 Fundamental Characteristics of a Systematic Review

- Transparency
- Reproducibility
- Capability of being updated

BOX 12.4 Research Question and Corresponding Appropriate Study Design for Systematic Reviews

- *Effect of therapy/intervention*: RCTs
- *Causes of disease (etiology)*: Cohort or case-control studies
- *Diagnosis*: Cross-sectional studies
- *Prognosis*: Cohort studies
- *Prevalence*: Cross-sectional studies (surveys)
- *Experience of disease*: Qualitative studies

reviews that investigate the progress of disease (prognosis), the prevalence of disease, or individual experience of disease are rare. This chapter is focused on systematic reviews of RCTs; however, the principles discussed are general, and therefore apply to systematic reviews of other study designs as well. Some challenges associated with systematic reviews of observational studies are noted in Section 12.14.

As with other study designs, the planning of a systematic review begins with the formulation of a protocol detailing the plan of the review. The Preferred Reporting Items for Systematic Reviews and Meta-Analyses (PRISMA) statement,[9] which is the prevailing guideline for conducting and reporting systematic reviews, recommends that every systematic review protocol be registered in a repository (or published in a suitable journal) prior to performing the review. The systematic review protocol should include the items in Box 12.5, and the published systematic review report itself should indicate where the systematic review protocol was registered and how to access it, e.g., by providing a website.

For the actual publication of a systematic review report, several additional items need to be included. The PRISMA statement provides a checklist of 27 items (see table 1 in Moher et al.[9]) that cover every step in the production of a systematic review report. It is recommended that this list be consulted to aid in the critical appraisal of a systematic review; however, not every item on the list requires further comments and explanations (e.g., item 1 on the checklist: Title). In Sections 12.4–12.12, the important components that should be included in every well-designed and properly reported systematic review are presented and discussed step by step. Section 12.13 presents discretionary elements that often surface in systematic reviews. Please refer to Hirji et al.[10] for a more detailed and statistically oriented exposition of these topics, with particular attention to meta-analysis.

BOX 12.5 Items That Should Be Described in a Systematic Review Protocol

- A precisely defined research question
- A strategy for searching the literature for primary studies
- Inclusion and exclusion criteria
- Primary and secondary outcomes
- Plan for data extraction
- Methods for assessing risk of bias (internal validity) of the included studies
- Statistical issues for the meta-analysis:
  - A summary effect measure (e.g., risk difference, mean difference)
  - A statistical model (e.g., fixed or random effects) or a strategy for deciding an appropriate model based on the included studies
  - Methods for investigating between-study heterogeneity (e.g., I^2, meta-regression)
  - Methods for investigating publication bias
  - Subgroup and sensitivity analyses
  - Additional statistical analyses

Throughout this chapter, the systematic review by Haase et al.[11] on the use of hydroxyethyl starch versus crystalloid or albumin for the treatment of patients with sepsis will be used as an example. This article was published under *BMJ*'s Open Access model and is freely available without subscription or payment. Two supplementary documents (referred to as "data supplements") published alongside the main article give detailed information about the systematic review, including a summary of the included trials and additional figures. The protocol for this systematic review was published in the PROSPERO register at www.crd.york.ac.uk/PROSPERO, with the registration number CRD42011001762. The authors stated that their systematic review complies with the recommendations of the Cochrane Collaboration.[12]

12.4 THE RESEARCH QUESTION

The purpose of a systematic review is to address a well-defined research question. The formulation of a research question for a systematic review is the same as that for a clinical trial or an epidemiological study, and thus should include information on the Population/patient/problem (P), Intervention (I), Comparison (C), and Outcome (O) in one sentence. This is known as the PICO approach (see Chapter 5). PICO is not always used in study reports—whether the study is a clinical trial, an epidemiological study, or a systematic review—but its four elements should be easily identifiable in both the Abstract and the Introduction section.

Example

In the review by Haase et al.,[11] PICO is used to define the research question. The first part of the abstract reads:

> ***Objective***: *To assess the effects of fluid therapy with hydroxyethyl starch 130/0.38–0.45 versus crystalloid or albumin on mortality, kidney injury, bleeding, and serious adverse events in patients with sepsis.*

This sentence contains information about the patients ("patients with sepsis"), the intervention ("fluid therapy with hydroxyethyl starch 130/0.38–0.45"), the comparison ("hydroxyethyl starch 130/0.38–0.45 versus crystalloid or albumin"), and the outcome ("mortality, kidney injury, bleeding, and serious adverse events"). The Introduction section of the article ends with a similar sentence.

To limit the probability of false-positive findings—and other problems associated with multiple statistical analyses—the CONSORT statement[13] recommends that clinical trials explicitly indicate one primary outcome, and that all secondary outcomes be identified and precisely defined. This is also true for systematic reviews. Also note that the number of secondary outcomes should be limited to avoid analyses of subgroups of the included studies. This is because not every included study on the topic under investigation will have reported every outcome of interest. Although missing outcome data can sometimes be obtained by contacting the investigators of the included studies directly, this is not a fail-safe method. A systematic review that looks at a large number of outcomes runs the risk of analyzing some of the outcomes within only a subgroup of the included studies.

Example

Haase et al.[11] defined two (as opposed to the recommended one) primary outcomes: "overall mortality and number of patients still receiving renal replacement therapy at maximum length of follow-up." There were seven secondary outcomes related to kidney injury, amount of blood cell transfusion, and adverse events. Both the primary and secondary outcomes were defined in the systematic review protocol.

12.5 SEARCH STRATEGY

The search for relevant primary studies is the foundation of a systematic review. The search strategy should be as unrestrictive as possible and encompass both published and unpublished studies from several sources (e.g., electronic databases, reference lists, study protocol registries, study authors, relevant manufacturers). A comprehensive literature search should also include studies published in languages other than English. In the systematic review report, the search strategy must be presented in sufficient detail, including the date limitations placed on the search, so that others can assess the comprehensiveness and completeness of the search. This also allows the interested reader to reproduce the search and compare the results with those in the systematic review.

Example

Haase et al.[11] reported that the following databases were searched: the Cochrane Central Register of Controlled Trials, Medline, Embase, Biosis Previews, Science Citation Index Expanded, and the Cumulative Index to Nursing and Allied Health Literature. The reference lists of included trials and the reference lists of other systematic reviews on similar topics were also searched for additional relevant trials. Three trial registries, www.controlled-trials.com, www.clinicaltrials.gov, and www.centerwatch.com, were searched for unpublished trials, and relevant pharmaceutical companies were contacted for unpublished data. The date of the last updated search was reported as September 10, 2012. The actual search terms, the filters used, and the number of hits for each database were presented in a supplementary document.

Chapter 5 explains the process of literature searches in more detail, including how to combine search terms using Boolean operators and setting up methodology filters.

12.6 INCLUSION AND EXCLUSION CRITERIA AND THE SELECTION OF PRIMARY STUDIES

The literature search may return a large number of potentially eligible studies, perhaps several thousand. Each study must then be assessed for eligibility using prespecified (i.e., defined in the systematic review protocol) inclusion and exclusion criteria. The criteria are defined in relation to the research question and specify details about admissible study designs, patients, treatments or exposures, dosages, number of treatment or exposure groups, and other relevant study information.

Example

The inclusion criteria in the systematic review by Haase et al.[11] were as follows:

- prospective and randomized study design or quasi-randomized observational studies with more than 500 patients,
- patients with sepsis,
- one intervention group that received hydroxyethyl starch 130 (with substitution ratios between 0.38 and 0.45),
- at least one other intervention group that received either crystalloid or human albumin, and
- no restrictions on language, publication status, patient's age, indication for fluid therapy, or predefined outcomes.

The exclusion criteria were as follows:

- animal studies,
- patients without sepsis,
- hydroxyethyl starch products of other molecular weights or substitution ratios,
- crossover studies, and
- studies comparing hydroxyethyl starch with other synthetic colloid solutions.

The initial screening process and how the remaining studies were evaluated after the inclusion and exclusion criteria are applied should be stated, including information about which authors did what. It is also recommended that the flow of primary studies through the study selection process be summarized in a flow chart, where the number of studies included and excluded in each phase of the systematic review can be displayed. A standardized flow diagram is provided in the PRISMA statement (see figure 1 in Moher et al.[9]).

Example

Under the heading "Study selection," Haase et al.[11] stated:

> *Two authors (NH, LIH, BL, or MW) independently reviewed all titles and abstracts identified in the literature search and excluded trials that were obviously not relevant. The remaining trials were evaluated in full text. Disagreements were resolved with JW.*

The initials refer to the authors of the systematic review. Figure 12.2 is a reproduction of a figure published in Haase et al.,[11] showing the flow of primary studies through the study selection process. Note that there are two boxes that describe the sources of the trials: the top box with the text "Records identified by database searching..." and the box with text "Additional trials identified from other sources...," which is somewhat confusingly positioned between the other boxes.

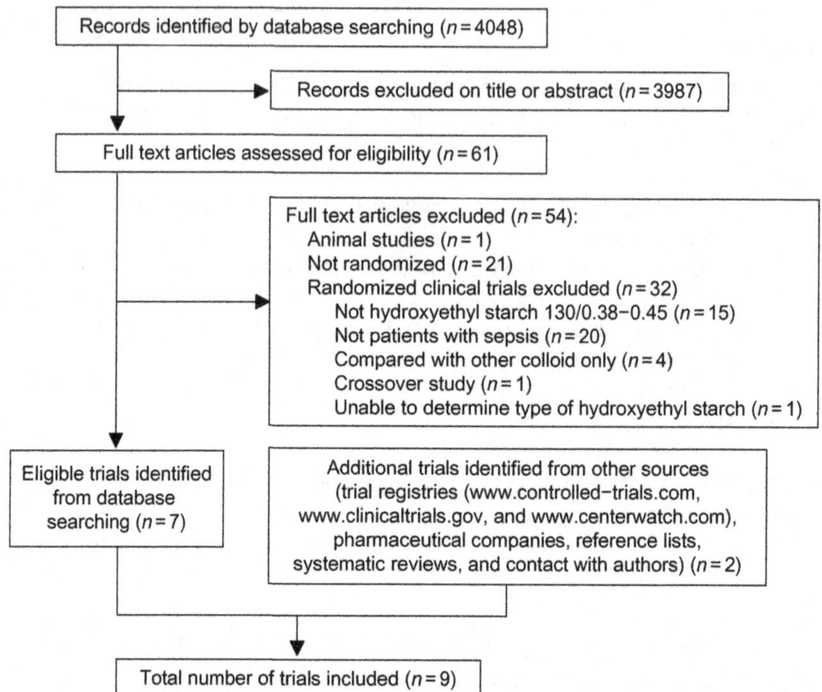

Figure 12.2 The study selection process for the systematic review by Haase et al.[11] Reproduced with permission.

12.7 DATA EXTRACTION FROM INCLUDED STUDIES

The authors of a systematic review must describe how the data collection process was carried out. All relevant information concerning study design, patients, interventions, and outcomes must be extracted from the included studies. A prewritten data extraction form, perhaps piloted on a sample of studies, greatly facilitates the process. One common method is for two reviewers to independently extract information and to resolve disagreements by discussion or by consultation with a third reviewer. A plan for obtaining data not reported in the primary studies and how to deal with non-English language articles should also be included in the systematic review report.

Example

In the systematic review by Haase et al.,[11] a data extraction form was developed. Two of the authors extracted information independently from each included trial (see Section 12.8 for a list of items). It was not reported whether any disagreements occurred or how they were resolved. The authors contacted the corresponding authors of six trials to obtain unreported data. Translators were used to extract data from one article in Spanish, one in Japanese, four in Russian, and four in Chinese.

The outcome data required to perform a standard meta-analysis need only be collected at the study level. If the plan is to perform a meta-analysis at the subject level (see Section 12.13)—which is seldom done—outcome data need to be extracted at the subject level for each study. When the outcome of interest is binary, the number of events and the total number of participants in each group are needed for each study. If the outcome is a continuous measurement, the mean, standard deviation (SD), and total number of participants in each group are needed for each study. A meta-analysis may also be performed for rate and time to event outcomes, for which data extraction is slightly more complicated. Because the use of these outcomes is not that common, details will not be presented here. In Section 12.10, examples of meta-analyses with binary and continuous outcomes are given, for which the relevant outcome data at the study level are shown in Figures 12.3 and 12.4.

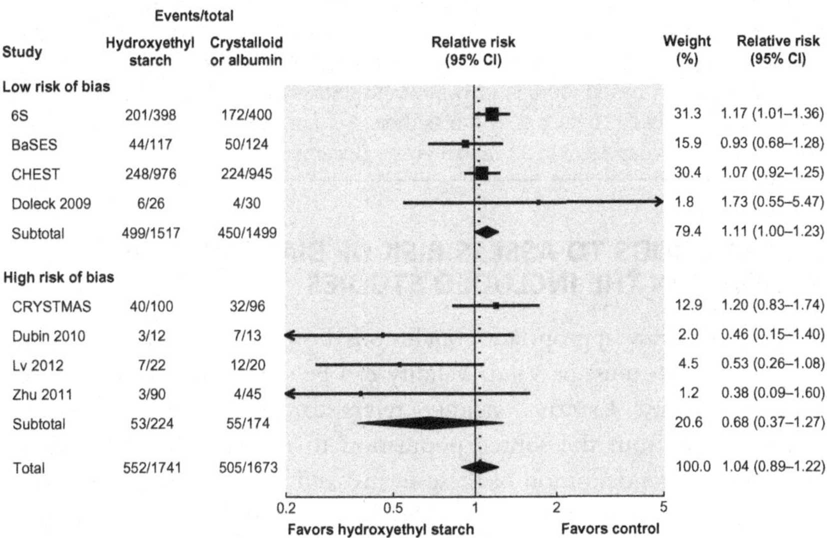

Figure 12.3 Forest plot for the primary outcome all-cause mortality from the systematic review by Haase et al.[11] Reproduced with permission.

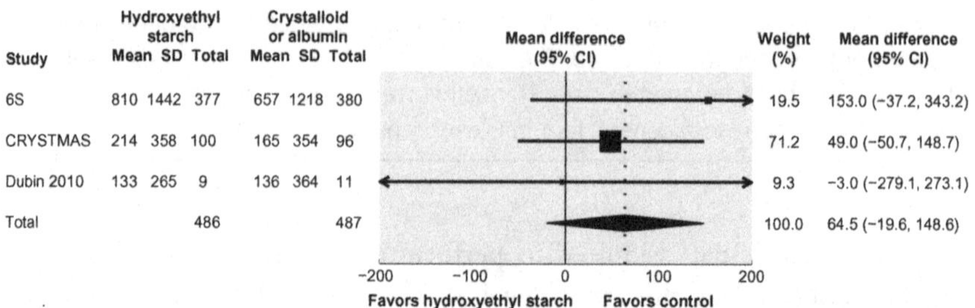

Figure 12.4 Forest plot for the secondary outcome of volume of red blood cells transfused from the systematic review by Haase et al.[11] Reproduced with permission.

12.8 STUDY CHARACTERISTICS

The main characteristics of the included studies should be presented, preferably in a table. For each study, the name and a citation for the source of the study, the study size, follow-up period, and other information relevant for the research question should be given. This information allows the reader to assess the relevance of the included studies.

Example

Table 1 in Haase et al.[11] presents 12 essential characteristics of the nine included trials: the name and where in the reference list to find the trial, the number of patients, whether the trial data were collected at one or more intensive care units, whether the trial used blinding, the number of intervention groups, diagnostic group information, indication for intervention, particulars of the hydroxyethyl starch, the comparator treatment, intervention period, dosage of hydroxyethyl starch, and whether the reviewers managed to establish contact with the authors. Further details on each study were presented (1–2 pages each) in a supplementary document.

12.9 METHODS TO ASSESS RISK OF BIAS (INTERNAL VALIDITY) IN THE INCLUDED STUDIES

If one is to draw appropriate conclusions from a systematic review, the included studies must be valid. Validity can be separated into internal and external validity. External validity refers to the generalizability of a study's results from the source population to the target population (see Section 4.9 for a definition of these terms) and is a measure of the practical utility of the results. External validity is often difficult to assess, and a complex interplay of study design, target and source population, study sample, and statistical analyses must be considered. Internal validity is the extent to which the study answers its research question. In other words, is the study free from bias?

Internal validity has traditionally been understood to mean methodological quality. In recent years, however, the difference between bias and quality has been given increased attention.[14] Bias refers to the extent to which the results of a primary study should be believed. Study quality, on the other hand, refers to the extent to which a study adheres to the highest possible standards. Consider a trial that investigates the effects of surgical versus nonsurgical treatment. It is not likely that patients and study personnel are blinded to the treatment, therefore methodological quality could be of the highest possible standard (for answering the research question), but the risk of bias may still be high. As the most important assessment for a systematic review is bias, not quality,[14] it is no longer recommended that systematic reviews include traditional tools to assess study quality, such as scales or checklists. Instead, an assessment of the risk of bias should be performed.

One way to assess bias is the Cochrane Risk of Bias Tool, which consists of six general domains (random sequence generation, allocation concealment, blinding of participants and personnel, blinding of outcome assessment, incomplete outcome data, and selective outcome reporting) in addition to "other potential threats to validity." The latter element is topic- or study-specific, and it is up to the authors of the systematic review to decide which items to include. Box 12.6 describes how each domain is to be judged. Each included study is assessed for each domain and given a judgment of either "low risk," "high risk," or "unclear risk." "Unclear risk" may be used if information is lacking or if the potential for bias is uncertain.

BOX 12.6 The Six General Domains of the Cochrane Risk of Bias Tool

- *Random sequence generation*: Is the method used to generate the allocation sequence described in sufficient detail to allow an assessment of whether it should produce comparable groups?
- *Allocation concealment*: Is the method used to conceal the allocation sequence described in sufficient detail to determine whether intervention allocations could have been foreseen in advance of or during enrolment?
- *Blinding of participants and personnel*: Are the measures used, if any, to blind study participants and personnel from knowledge of which

intervention a participant received described in sufficient detail to determine whether the intended blinding was effective?

- *Blinding of outcome assessment*: Are the measures used, if any, to blind outcome assessors from knowledge of which intervention a participant received described in sufficient detail to determine whether the intended blinding was effective?
- *Incomplete outcome data*: Are the participants included in the analysis exactly those who were randomized into the trial? Is the completeness of outcome data, including attrition and exclusions from the analysis, for each main outcome described? Are the reasons for attrition/exclusions reported?
- *Selective outcome reporting*: Are there indications that the study authors have failed to report outcome data that seem sure to have been recorded?
- *Other potential threats to validity*: Are there important concerns about bias not addressed in the other domains?

Higgins et al.[14]

Example

Table 3 in the systematic review by Haase et al.[11] presents an assessment of the risk of bias in the nine included trials. The authors used the Cochrane Risk of Bias Tool. Only one domain for blinding was used, and it is unclear whether this assessment included blinding of participants and personnel, or blinding of outcome assessment, or both. The table also includes the general domains random sequence generation, allocation concealment, incomplete outcome data, and selective outcome reporting, in addition to three specific domains: baseline imbalance, vested financial interests, and academic bias. Academic bias refers to bias as a result of the authors' previous research on the same topic, such as if one of the authors has published a preference for one of the treatments under study. The authors defined a trial to have a "low risk of bias" if all domains for that trial were judged to be of low risk. Four of the nine trials were deemed to have a low risk of bias. One trial had potential academic bias, and one trial did not use blinding. The risk of bias could not be fully judged in the remaining three trials, each of which also had at least one domain that was assessed as high risk. To support their judgments, the authors provided explanations in a supplementary document.

12.10 METHODS TO ESTIMATE THE COMMON EFFECT

Two aims of a systematic review are to investigate to what extent (and why) the results of the included studies differ and, if appropriate, to estimate a common effect across the studies. This section considers the latter effort, while Section 12.11 deals with study heterogeneity.

Before any attempts are made to estimate a common effect across the included studies, the authors of a systematic review must first assess whether or not the data from the studies can be combined. This is first and foremost a clinical decision, although statistical considerations (see Section 12.11) are important for the interpretation of the results and for the conclusions of the review. The clinical considerations may involve an assessment of the individual research questions of the included studies, the results from the risk of bias assessment, characteristics at the study level, and other, more subjective criteria. For instance, should studies with a low risk of bias be combined with those with a high risk of bias? No consensus exists concerning general rules or specific guidelines here; however, the decision to combine the results or not—and the criteria used—should be described in the systematic review report. In the following, it is assumed that a decision to combine the results of the included studies has been made.

12.10.1 Effect Measure

An effect measure is a statistical measure used to compare the outcomes of two treatment or exposure groups. It is used in a meta-analysis to summarize the results at the study level and to calculate the common effect across the studies. For binary outcomes, the most common effect measures are risk difference (RD), risk ratio or relative risk (RR), and odds ratio (OR). For continuous outcomes, mean difference (MD), standardized mean difference (SMD), and mean ratio (MR) are most often used. Meta-analyses on survival (time-to-event) outcomes use hazard ratio (HR) or median survival time (MST) as effect measures. Studies that investigate incidence rates may also be the object of a meta-analysis, in which case incidence rate ratio (IRR) or incidence rate difference (IRD) are appropriate effect measures (Box 12.7).

It is useful to distinguish between relative (ratio-based) and absolute (difference-based) effect measures. Relative effect measures, such as RR

BOX 12.7 Common Effect Measures for Meta-Analyses

- *Binary outcomes*: RD, RR, OR
- *Continuous outcomes*: MD, SMD, MR
- *Survival outcomes*: HR, MST
- *Incidence rates*: IRR, IRD

and OR, often suggest more optimistic treatment effects than do absolute effect measures, such as RD. If 2 of 1000 patients in the control group and 1 of 1000 patients in the treatment group experience an unwanted event, the RR = 0.5, a risk reduction of 50%, whereas the absolute value of the RD = 0.001, an absolute risk reduction of 0.1%. The former arguably sounds more impressive than the latter. Whether or not this is a clinically interesting effect depends on the specific research problem, and factors such as patient preference, other outcomes, costs, and practical considerations all play a part.

For binary outcomes, the study design has a bearing on the choice of effect measure. Case-control studies can only estimate ORs (not RDs or RRs), because the incidence of disease, which is needed to calculate risk, is unknown in case-control studies. The natural effect measures for RCTs and cohort studies are RD and RR, respectively, although ORs are commonly used for both designs. The popularity of OR is in part due to the widespread use of logistic regression, for which OR is the inherent effect measure. Chapter 11 shows how to estimate these effect measures with confidence intervals.

There are several other important issues related to the choice of effect measure in meta-analyses. Box 12.8 briefly considers some of these issues, which can be statistical, interpretational, or both. Please refer to Hirji et al.[10] for more details. Note that it is possible to use one measure for statistical analysis and another measure for presenting the results.

BOX 12.8 Some Issues Regarding the Choice of an Effect Measure for Meta-Analyses

- Ratio-based effect measures (e.g., RR, OR, IRR) often indicate a more optimistic treatment effect than do difference-based effect measures (e.g., RD, IRD).
- Ratio-based effect measures have greater stability across different risk groups than do difference-based effect measures.
- Difference-based effect measures reflect the baseline risk of individuals, whereas ratio-based effect measures do not.
- If some of the included studies have zero events, difference-based measures are better than ratio-based measures.
- For binary outcomes, OR has better statistical properties than RD and RR; however, OR is not as easy to interpret or communicate.
- For continuous outcomes, SMD has the best statistical properties, MD is most easily interpreted, and MR is preferable with skewed data.

12.10.2 Fixed Effects Models and Random Effects Models

To estimate the common effect, a statistical model must be chosen. If the simple average of the individual study results were taken as the common effect, small studies would have an impact equal to that of large studies, which is clearly inappropriate; small studies are more subject to the role of chance, and should have less influence than large studies.

There are two main categories of statistical models for meta-analyses: fixed effects models and random effects models. The difference between the two lies in how they treat variability between study results. The fixed effects model assumes the existence of one true effect, and that all variation between studies is due to random variation. Let $f_i(x)$ denote a function for the effect measure for study i ($i = 1,2,\ldots,n$), and let x be an indicator of treatment or exposure ($x = 0$: control/unexposed; $x = 1$: treated/exposed). The fixed effects model is

$$f_i(x) = \alpha_i + \beta_i x,$$

where α_i represents the control/unexposed effect, and β_i is the fixed effects measure for study i, measured as the additive effect of treated/exposed relative to control/unexposed. The fixed effects model makes the assumption that the β_is are equal (i.e., study homogeneity), which can be tested by the null hypothesis:

$$\beta = \beta_1 = \beta_2 = \cdots = \beta_n.$$

The random effects model rejects the idea of one common study effect. Instead, each study is viewed as having its own underlying effect, and the included studies represent a sample from a population of studies with a (normal) distribution of effects. The common effect is taken as the mean of this distribution. The random effects model is

$$f_i(x) = \alpha_i + \beta x + z_i x,$$

where β is the average common effect, and $z_i \sim N(0, \sigma^2)$ is a normal random variable with a mean of zero and SD σ. The level of heterogeneity is given by σ, for which higher values indicate more heterogeneity.

When there is low study heterogeneity, the fixed effects model and the random effects models produce similar estimates of the common effect; however, the random effects model tends to provide wider confidence intervals than the fixed effects model. The choice between the two models is not obvious, and neither model is entirely satisfactory in all cases.[15] The random effects model naturally incorporates study heterogeneity, whereas the fixed effects model does not. Thus, one common strategy is to use the fixed effects model if the study results are similar, and the random effects model if the study results show apparent inconsistency. If the purpose is to investigate heterogeneity and its sources, the random effects model may hide important heterogeneity; moreover it is not helpful in explaining the existence of heterogeneity.

Whichever model is used, the central idea for computing the common effect is to assign a weight to each study, which is usually determined as the inverse of the study's estimated variance. The common effect is then the weighted average across the studies. The common effect should be accompanied by a confidence interval. The calculations of the weights and the standard errors used for computing the confidence intervals depend on the choice of effect measure, type of model, and other choices, such as whether to use exact methods. The particulars of the different techniques used for estimation and significance testing are not covered here; instead please see Hirji et al.[10] for details and technical references. In all but the simplest of cases, software resources are needed to perform the analyses. Many options exist, including general purpose software packages such as Stata (StataCorp LP, College Station, TX), and specific software packages such as RevMan (Review Manager), which is developed and maintained by the Cochrane Collaboration.

12.10.3 Forest Plot

A forest plot is an essential tool to summarize information on individual studies, give a visual suggestion of the amount of study heterogeneity, and show the estimated common effect, all in one figure.

Example

Figure 12.3 shows a forest plot of one of the primary outcomes (all-cause mortality), reproduced from the systematic review by Haase et al.[11] Each of the eight included trials (mortality data was unavailable for one of the nine trials) is represented by one row in the plot, and the trials are grouped according to their assessed risk of bias. The number of events (deaths) and the total number of participants are shown for each treatment arm. The effect measure used was RR, where values lower than 1.0 indicate that hydroxyethyl starch has lower risk of mortality than the control, and vice versa for values greater than 1.0. The solid vertical line at RR = 1.0 indicates no treatment effect. For each trial, the estimated RR with a 95% confidence interval is shown both in numerical and graphical form. The size of the square for each estimated RR in the plot is proportional to the weight of the trial, which indicates its relative impact on the calculations of the common effect.

The two rows titled "Subtotal" display the common effect in each of the two subgroups defined by low risk of bias or high risk of bias. The bottom row titled "Total" represents the common effect across all eight trials. The common effects in the graphical part of the forest plot are drawn in the shape of diamonds; the widths of the diamonds indicate the confidence interval. The dotted vertical line is drawn at the value of the overall common effect. The authors stated that they used the fixed effects model when the I^2 statistic (see Section 12.11) was zero, and both the fixed effects and the random effects models when $I^2 > 0$. The common effect for the subgroup of low risk of bias trials was thus calculated using the fixed effects model, whereas the common effects for the subgroup of high risk of bias trials, and across all eight trials shown in Figure 12.3, was calculated using the random effects model. From a visual inspection of the figure, one gets the impression that the variability of the high risk of bias trials is greater than that of the low risk of bias trials. The meta-analysis showed no significant difference in mortality between patients treated with hydroxyethyl starch and crystalloid or albumin (RR = 1.04, 95% confidence interval [0.89, 1.22]). A subgroup analysis of the low risk of bias trials showed a borderline significant treatment effect (RR = 1.11, 95% confidence interval [1.00, 1.23]); however, a test for differences between the low risk of bias trials and the high risk of bias trials was not significant ($P = 0.13$).

Forest plots are not confined to binary outcomes. Figure 12.4 shows a forest plot of one of the secondary outcomes, volume of red blood cells transfused, where MD is the effect measure. Instead of listing the number of events, the forest plot now displays the mean and SD for each treatment arm in each trial. Only three trials provided data on transfusions: one small, one medium, and one large trial. The medium-size trial is given more weight than the large trial due to much smaller SDs in both treatment groups. Figure 12.4 shows that the confidence interval for the treatment effect is much narrower for the medium-size trial than for the large trial. The overall estimated effect across the three trials favors the control group; however, the effect is not statistically significant.

Meta-analyses of other secondary outcomes (results not displayed here) showed that patients treated with hydroxyethyl starch had a significantly increased risk of renal replacement therapy and serious adverse events.[11]

12.11 METHODS TO ASSESS HETEROGENEITY

Study heterogeneity among primary studies in systematic reviews is to be expected. To quote Higgins[16]: "it would be surprising if multiple studies, performed by different teams in different places with different methods, all ended up estimating the same underlying parameter."

Heterogeneity can be viewed as having two distinct sources: statistical and clinical. Statistical heterogeneity refers to random variation, which is an inevitable byproduct of performing measurements within a study sample. Clinical heterogeneity refers to variation in the characteristics of the included studies. This may include key patient characteristics, such as age and disease severity, methods for diagnosis and evaluation, follow-up, treatment doses and duration, and study design features, such as level of blinding. It is important that authors describe and quantify the heterogeneity in systematic reviews.

Example

A first indication of the extent of heterogeneity may be obtained through a forest plot. Figure 12.3 shows a forest plot for the eight trials with mortality data from Haase et al.[11] For the low risk of bias trials, the estimated RRs agree fairly well. The widths of the confidence intervals vary; however, that is to be expected due to the widely different sample sizes of the trials. The high risk of bias trials consisted of three small trials (with very wide confidence intervals) and one medium-size trial. The three small trials were the only ones that suggested the existence of a treatment effect in favor of hydroxyethyl starch. The plot indicates a greater amount of heterogeneity for the high risk of bias trials than for the low risk of bias trials.

The forest plot can only provide a hint as to the amount of study heterogeneity; formal statistical methods are needed to investigate this matter further. For fixed effects models, the level of heterogeneity is represented by the difference between the β_is (see Section 12.10), and a test for homogeneity is a test for the null hypothesis $\beta = \beta_1 = \beta_2 = \cdots = \beta_n$. In random effects models, the between-study variance, σ^2, measures the extent of heterogeneity.

For binary outcomes, there are many relevant statistical tests available, including the chi-squared (Cochran's) Q test, the Breslow-Day test, and exact tests of homogeneity. A common disadvantage with these tests is related to their power. Systematic reviews commonly include a small number of studies. The resultant power of the tests to detect heterogeneity is poor, and a nonsignificant *P*-value might lead to an incorrect conclusion of homogeneity. On the other hand, if a systematic review includes a large number of studies, the power to detect heterogeneity is also large, and small amounts of heterogeneity—which may be expected and do not affect the results—may result in a small, significant *P*-value. Moreover, the *P*-value for a test of homogeneity does not describe the amount of heterogeneity, and is a poor guide for identifying clinically important subgroups.

A better option than using a formal test is the I^2 statistic.[17] The I^2 statistic, also called the index of heterogeneity, is based on the Q statistic, which is obtained as the sum of the squared deviations of each study's estimate from the overall estimate, weighted by the study's impact on the calculation of the overall estimate. The Q statistic is calculated from a fixed effects model, and its expression depends on the type of effect

BOX 12.9 A Rough Guide to Interpreting I^2 Values

- $I^2 = 0\%$: no heterogeneity
- $I^2 = 25\%$: low heterogeneity
- $I^2 = 50\%$: moderate heterogeneity
- $I^2 = 75\%$: high heterogeneity

Further details are given in Higgins et al.[17]

measure used.[10] It can be calculated in any software package capable of performing meta-analyses. The I^2 statistic is calculated as:

$$I^2 = 100\%(Q - \text{df})/Q,$$

where df are the degrees of freedom (for the chi-squared Q test), which are equal to the number of studies minus one. The I^2 statistic measures the percentage of total variation in study results that is due to heterogeneity (as opposed to random variation). A low I^2 value thus indicates low heterogeneity, except that which can be explained by chance, and larger I^2 values indicate greater amounts of study inconsistency. Box 12.9 gives a rough guide to interpreting different I^2 values. The I^2 value should be accompanied by an uncertainty interval to indicate its level of precision.[18]

Example

Haase et al.[11] calculated I^2 statistics (without uncertainty intervals) and tests of heterogeneity (of unspecified type) for several outcomes and subgroups. For the outcome of all-cause mortality (Figure 12.3), $I^2 = 0\%$ for the low risk of bias trials and $I^2 = 59\%$ for the high risk of bias trials. The *P*-values for the test of heterogeneity were $P = 0.48$ (low risk of bias trials) and $P = 0.06$ (high risk of bias trials). When all nine trials were combined, $I^2 = 37\%$ and the test for heterogeneity gave $P = 0.13$.

Another visual tool to examine heterogeneity for binary outcomes is the L'Abbe plot. It is obtained by plotting, for each study, the observed proportion of events for the control group (x-axis) versus the observed proportions of events for the treatment group (y-axis). If the effect measure is the OR, a logit scale is used on both axes; that is, $\ln[\hat{p}_i/(1 - \hat{p}_i)]$ is used instead of $\hat{p}_i$, where $\hat{p}_i$ denotes an observed proportion and i indexes the treatment groups. If there is study homogeneity, the points (studies) form a straight line. A strong violation of this condition indicates study heterogeneity. Figure 12.5 shows L'Abbe plots for 12 hypothetical studies using RD (top left panel), RR (top right panel), and OR (bottom panel) as the effect measure. The points (studies) show excellent adherence to a straight line in all three cases, indicating study homogeneity.

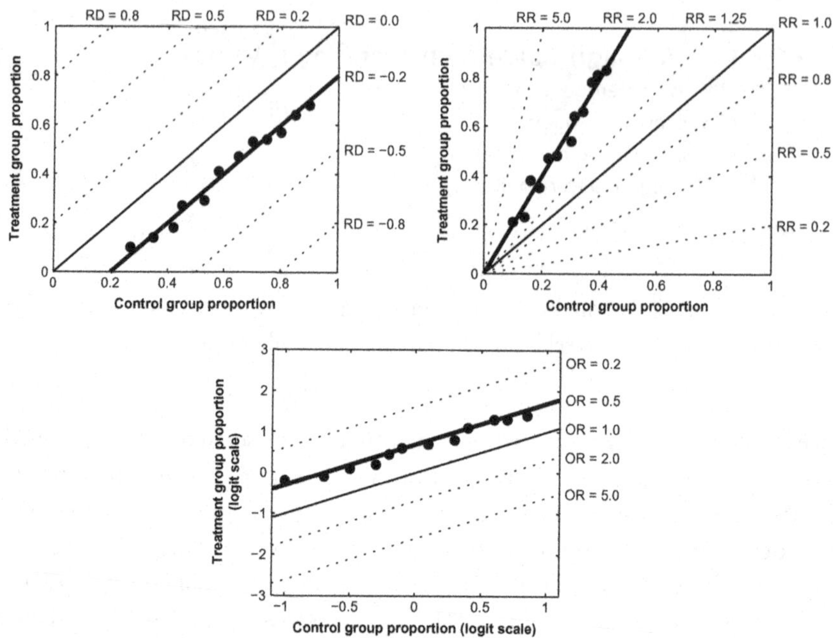

Figure 12.5 L'Abbe plots on hypothetical data showing no signs of heterogeneity with three different effect measures.

Example

Based on the number of events and the total sample size given in Figure 12.3, a L'Abbe plot can be created for the included studies in Haase et al.[11] (Figure 12.6). The solid line

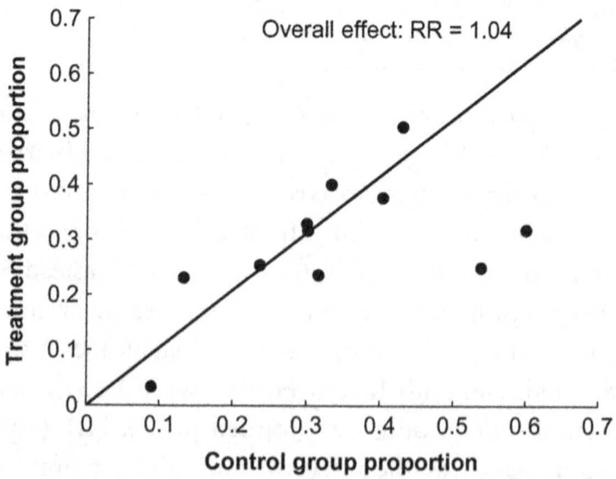

Figure 12.6 L'Abbe plot for the studies in the systematic review by Haase et al.[11]

represents the estimated common effect of RR = 1.04. The points do not adhere well to a straight line, and therefore we can conclude that study homogeneity is in doubt. The random effects model is thus the recommended statistical model, and was used by Haase et al. to estimate overall mortality across the eight trials that included this information.

12.12 METHODS TO ASSESS PUBLICATION BIAS

Publication bias (also called reporting bias) refers to absence of information caused by either nonpublication of entire studies (missing studies), or selective outcome reporting in published studies based on their results (missing outcomes). The latter problem is also called outcome reporting bias. Studies that report a statistically significant result ($P < 0.05$) are more likely to be published, and published sooner, than studies that do not show a statistically significant result ($P > 0.05$). Similarly, selective outcome reporting frequently occurs, which is biased and usually inconsistent with study protocols. This and other types of scientific misconduct are discussed in Chapter 2. Publication bias is particularly problematic in RCTs, as it leads to inflated and unreliable results regarding the benefits of different treatments. Identification and control of publication bias is essential to preserve the validity of a systematic review.

Missing studies may be found in protocol registries; however, the time period between publication of the protocol and publication of the study report varies, and in some cases can be quite long. Missing outcomes may be detected by comparing the published report with the protocol. Although publication of a protocol is now required for most RCTs before study recruitment starts (a prerequisite for publication in many journals), this practice is not equally common for observational studies. For this reason, it is difficult to verify the existence of missing outcomes in observational studies.

The main graphical tool used to identify publication bias in the form of missing studies is the funnel plot,[19] which specifically targets small study bias, in which small studies tend to show larger estimates of effects and greater variability than larger studies. The funnel plot is a scatter plot with effect estimates on the x-axis and some measure of study precision (or study size) on the y-axis. The standard errors of the effect estimates are commonly used. The scale of the y-axis is reversed such that studies with low precision are placed at the bottom and studies with greater precision at the top of the plot. If the effect measure is a ratio measure

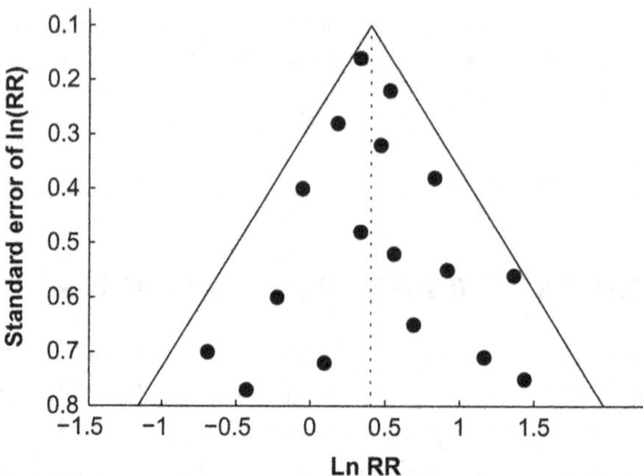

Figure 12.7 Ideal funnel plot for hypothetical data with no signs of asymmetry or holes.

(such as RR), the *x*-axis is log transformed. In the absence of missing studies, the shape of the scatter plot should resemble a symmetrical inverted funnel with a wide base (consisting of small studies with large effect estimate variability) and a narrow top (consisting of large studies with small effect estimate variability). Presence of large "holes"—most often seen close to the bottom—or asymmetry in the plot indicates publication bias, though these holes may have other causes, such as study heterogeneity.[19] An ideal funnel plot based on 17 hypothetical studies with no signs of asymmetry or holes is shown in Figure 12.7. The dotted vertical line represents the estimated common effect.

Several tests for funnel plot asymmetry exist. One of the most common methods, the Egger test for asymmetry,[20] is obtained by regressing the standardized effect size on the inverse of the standard error (the precision):

$$\text{Effect}/\text{SE} = \alpha + \beta(1/\text{SE}) + \varepsilon,$$

where Effect is the estimated effect, SE is the standard error of Effect, and ε is random noise. When there is no asymmetry in the plot, the regression line is expected to go through the origin ($\alpha = 0$), with the slope (β) indicating the size and direction of the common effect. A test of $\alpha = 0$ is a test for funnel plot asymmetry, and the size of α indicates the extent of asymmetry. The test has low power, especially when the number of studies is smaller than 10 and the asymmetry is less than considerable.

Example

Haase et al.[11] did not include an assessment of publication bias. Based on the data for mortality in Figure 12.3, the standard error of the logarithm of the RR can be estimated for each study by rearranging the terms in the expression for the standard 95% confidence interval of the RR. The 95% confidence limits of ln(RR) are given by:

$$\{\ln(\mathrm{RR}) - 1.96 \cdot \mathrm{SE}[\ln(\mathrm{RR})], \quad \ln(\mathrm{RR}) + 1.96 \cdot \mathrm{SE}[\ln(\mathrm{RR})]\}.$$

Next, one arbitrary confidence limit is chosen (*U*; the upper):

$$\ln(U) = \ln(\mathrm{RR}) + 1.96 \cdot \mathrm{SE}[\ln(\mathrm{RR})]$$

and rearranged to obtain

$$\mathrm{SE}[\ln(\mathrm{RR})] = [\ln(U/\mathrm{RR})]/1.96.$$

Figure 12.8 shows the corresponding funnel plot, in which the low risk of bias trials and high risk of bias trials are represented by different symbols. Also shown in the figure is the estimated common effect at RR = 1.04 (dotted vertical line) and a triangular region defined by the solid lines and the *x*-axis, within which 95% of trials are expected to lie in the absence of publication bias and heterogeneity. Interpreting this funnel plot is not easy. The number of trials is small, so there is a high probability that departures from the ideal funnel shape may occur due to chance. Furthermore, given the assessment of heterogeneity (see Section 12.11), one can conclude that study heterogeneity may be a factor. For example, the presence of subgroup effects may distort the shape of the scatter plot. A difference in the estimated common effect between the low and high risk of bias trials was indicated (see Section 12.10), although not strongly established, and this subgroup effect may be the cause of some of the asymmetry observed in Figure 12.8.

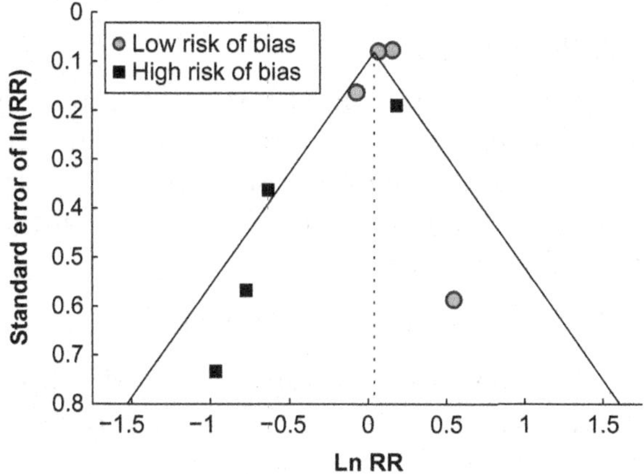

Figure 12.8 Funnel plot for mortality using the studies in the systematic review by Haase et al.[11]

Funnel plot asymmetry can also be assessed by the Egger's test using the log RR and standard errors in Figure 12.8. Any statistical software package capable of running linear regression will render the estimated intercept of −1.15 with 95% confidence interval (−2.63, 0.33). The P-value for a test of $\alpha = 0$ is $P = 0.11$. There is no conclusive evidence of publication bias, but with only eight trials the power of the test is poor.

12.13 ADDITIONAL ANALYSES

12.13.1 Sensitivity Analysis

A sensitivity analysis examines to what extent the inclusion or exclusion of certain types of studies changes the results of a meta-analysis.[21] For instance, high risk of bias studies, studies below a certain size, or studies published before a particular year may be included or excluded. Sensitivity analysis may thus be used to explore study heterogeneity. The criteria that dictate inclusion or exclusion in a sensitivity analysis need to be specified in the systematic review protocol.

12.13.2 Subgroup Analysis

Study heterogeneity may be due to subgroup effects, and meta-analyses performed in subgroups of the study sample are commonly used to detect (or account for) heterogeneity. Subgroups are often defined by factors such as risk of bias assessment, study size, or some other characteristic at the study level. If not pre-specified in the systematic review protocol and few in numbers, subgroup analyses have a high chance of false-positive or inconsistent results.[22] Therefore, a better option is meta-regression.

12.13.3 Meta-Regression

Meta-regression refers to a fixed effects model or random effects model that includes one or more study features as covariates. Let y denote a covariate, for instance, $y = 0$ for low risk of bias studies and $y = 1$ for high risk of bias studies. A fixed effects meta-regression model that investigates the effects of y is written as:

$$f_i(x, y) = \alpha_i + \beta_i x + \gamma y,$$

where γ is the common effect of covariate y, and all other terms are as defined in Section 12.10. A random effects meta-regression model is given by:

$$f_i(x, y) = \alpha_i + \beta x + z_i x + \gamma y.$$

A test for the null hypothesis $\gamma = 0$ is a test to determine whether the covariate y contributes to the study heterogeneity. Meta-regression reduces the number of tests and estimations (as compared with subgroup analysis) and uses all included studies. The power of the analysis is thus greater and the probability of false-positive findings is reduced. Nevertheless, the covariates included in a meta-regression should be few and specified in the systematic review protocol. A review of statistical methods for meta-regression is given by Thompson and Sharp.[23]

12.13.4 Cumulative Meta-Analysis

Cumulative meta-analyses consider study data as they accumulate.[24] The studies are sorted according to publication date, and a meta-analysis may be performed on the studies available at any chosen point in time. One of the purposes of a cumulative meta-analysis is to identify the point in time at which the evidence for a treatment or exposure effect first reached statistical significance. A cumulative forest plot, where the first row represents the first published study and subsequent rows represent the common effect after adding the next published study, can be particularly revealing.

12.13.5 Trial Sequential Analysis

Trial sequential analysis[25] is the meta-analytic version of the group sequential analysis for a single RCT. In a standard cumulative meta-analysis, the number of performed meta-analyses may be large, thus increasing the chance of spurious results. The aim of trial sequential analysis is to perform a cumulative meta-analysis with adjustment for multiple (interim) analyses. Trial sequential analysis has the potential to detect when firm evidence has been established and to prevent initiation of unnecessary trials.[25] An example of trial sequential analysis is given in the Haase et al.[11] systematic review.

12.13.6 Individual Subject Meta-Analysis

When the data from all included studies in a systematic review are available at the subject level, an individual subject meta-analysis is possible. This form of meta-analysis greatly increases power and facilitates the investigation of heterogeneity. Multilevel or hierarchical models are required to perform an individual subject meta-analysis.[26,27]

12.14 SYSTEMATIC REVIEWS OF OBSERVATIONAL STUDIES

Systematic reviews of RCTs investigate the effects of interventions. If the research question involves causes of disease (etiology), the relevant study designs are cohort and case-control studies. Systematic reviews (and meta-analysis) of observational studies are as common as systematic reviews of RCTs.[28]

The main principles and statistical methods for systematic reviews of observational studies are the same as those for systematic reviews of RCTs. However, whereas precision is the main challenge for systematic reviews of RCTs, confounding and selection bias are the main concerns for systematic reviews of observational studies. Lack of precision can be amended by increasing the sample size. By synthesizing the results of several studies on the same topic, systematic reviews are naturally adept at solving issues of precision. Bias in observational studies due to lack of random allocation of exposure cannot be overcome by increasing the sample size. Observational studies can adjust for confounding factors; however, this only applies to observed data. Confounding by unmeasured factors (residual confounding) is an inherent problem in observational studies and one that cannot be solved by statistical combination of studies. Observational studies are usually much larger than RCTs, and meta-analyses of observational studies may therefore provide very precise but biased results.[28]

The benefit of systematic reviews of observational studies lies in the opportunity to examine possible sources of heterogeneity, not in the statistical calculation of a common effect across studies. Sensitivity analysis, meta-regression, and individual subject meta-analyses are the key statistical tools to explore heterogeneity, and they should be routinely used in systematic reviews of observational studies. Such analyses may also be used to examine the dose-response relationship between an exposure and a disease, which is one of the criteria for assessing causation. One recent example is a review that uses meta-regression to estimate the dose-response relationship between circulating folate concentrations and colorectal cancer.[29]

The PRISMA statement[9] mentioned in Section 12.3 is the prevailing guideline for reporting systematic reviews. PRISMA is focused on RCTs but can also be used for systematic reviews featuring other study designs, such as observational studies. The Meta-analysis Of Observational Studies in Epidemiology (MOOSE) recommendations—though they are now somewhat outdated—can be used as a supplementary guide.[30]

QUESTIONS TO DISCUSS

1. Which parts of a systematic review promote transparency, and how are those parts different from a narrative review?
2. Which parts of a systematic review promote reproducibility, and how are those parts different from a narrative review?
3. How often should a systematic review be updated, and how can authors of a systematic review facilitate future updates?
4. Occasionally, the results of a systematic review and those of a subsequent large randomized trial are in disagreement.[31,32] What may be the reasons for such discrepancies? Does this invalidate meta-analysis as a tool to calculate summary effect measures?
5. Why is it important to search for unpublished studies and to obtain data from relevant manufacturers?
6. Apply the Cochrane Risk of Bias Tool to a selection of recently published studies in your field. Discuss the methodological quality of the studies and to what extent the risk of bias and the quality of the studies agree.
7. For the mortality data in the systematic review by Haase et al.,[11] the authors used the RR as the effect measure. Use the data in Figure 12.3 to compute the RD instead. Do the RR and RD communicate the results similarly?
8. Use Box 12.8 as a starting point. Which effect measure is indicated for the mortality data in the Haase et al.[11] review?
9. Figure 4 in Haase et al.[11] shows a forest plot for one of the secondary outcomes: renal replacement therapy (at any time during follow-up). Only five trials have data on this outcome and there are no events in one or both treatment arms for two of the trials. Which effect measure would you recommend in this case?
10. Summarize the use of additional analyses (Section 12.13) in the Haase et al.[11] review. What new information do the additional analyses supply?

REFERENCES

1. Sackett DL, Rosenberg WMC, Gray JAM, Haynes RB, Richardson WS. Evidence based medicine: what it is and what it isn't. *BMJ* 1996;**312**:71–2.
2. Davidoff F. Evidence-based medicine: why all the fuss? [Editorial]. *Ann Intern Med* 1995;**122**:727.

3. Davidoff F. In the teeth of evidence: the curious case of evidence-based medicine. *Mt Sinai J Med* 1999;**66**:75–83.
4. Greenhalgh T. How to read a paper: getting your bearings (deciding what the paper is about). *BMJ* 1997;**315**:243–6.
5. Heneghan C, Badenoch D. *Evidence-based medicine toolkit*. 2nd ed. Malden, MA: Blackwell Publishing; 2006.
6. Straus SE, Glasziou P, Richardson WS, Haynes RB. *Evidence-based medicine: how to practice and teach it*. 4th ed. London: Churchill Livingstone; 2011.
7. McAlister FA. Applying the results of systematic reviews at the bedside. In: Egger M, Smith DG, Altman DG, editors. *Systematic reviews in health care: meta-analysis in context*. London: BMJ Publishing Group; 2001.
8. Del Mar C, Glasziou P, Mayer D. Teaching evidence based medicine. *BMJ* 2004;**329**:989–90.
9. Moher D, Liberati A, Tetzlaff J, Altman DG, The PRISMA Group. Preferred reporting items for systematic reviews and meta-analysis: the PRISMA statement. *PLoS Med* 2009;**6**:e1000097.
10. Hirji KF, Fagerland MW, Veierød MB. Meta-analysis. In: Veierød MB, Lydersen S, Laake P, editors. *Medical statistics in clinical and epidemiologic research*. Oslo: Gyldendal Akademisk; 2012.
11. Haase N, Perner A, Hennings LI, Siegemund M, Lauridsen BL, Wetterslev M, et al. Hydroxyethyl starch 130/0.38–0.45 versus crystalloid or albumin in patients with sepsis: systematic review with meta-analysis and trial sequential analysis. *BMJ* 2013;**346**:f839.
12. Higgins JPT, Greeen S, editors. *Cochrane handbook for systematic reviews of interventions*. Version 5.1.0 [Updated March 2011]. The Cochrane Collaboration, 2011. Available from: <www.cochrane-handbook.org>.
13. Moher D, Hopewell S, Schulz KF, Montori V, Gøtzsche PC, Deveraux PJ, et al. CONSORT 2010 explanation and elaboration: updated guidelines for reporting parallel group randomized trials. *BMJ* 2010;**340**:c869.
14. Higgins JPT, Altman DG, Sterne JAC, editors. Chapter 8: assessing risk of bias in included studies. In: Higgins JPT, Green S, editors. *Cochrane handbook for systematic reviews of interventions*. Version 5.1.0 [Updated March 2011]. The Cochrane Collaboration, 2011. Available from: <www.cochrane-handbook.org>.
15. Thompson SG, Pocock SJ. Can meta-analysis be trusted? *Lancet* 1991;**338**:1127–30.
16. Higgins JPT. Heterogeneity in meta-analysis should be expected and appropriately quantified. *Int J Epidemiol* 2008;**37**:1158–60.
17. Higgins JPT, Thompson SG, Deeks JJ, Altman DG. Measuring inconsistency in meta-analyses. *BMJ* 2003;**327**:557–60.
18. Higgins JPT, Thompson SG. Quantifying heterogeneity in a meta-analysis. *Stat Med* 2002;**21**:1539–58.
19. Sterne JAC, Sutton AJ, Ioannidis JPA, Terrin N, Jones DR, Lau J, et al. Recommendations for examining and interpreting funnel plot asymmetry in meta-analyses of randomized controlled trials. *BMJ* 2011;**342**:d4002.
20. Egger M, Smith GD, Schneider M, Minder C. Bias in meta-analysis detected by a simple, graphical test. *BMJ* 1997;**315**:629–34.
21. Egger M, Smith GD, Phillips AN. Meta-analysis: principles and procedures. *BMJ* 1997;**315**:1533–7.
22. Yusuf S, Wittes J, Probstfield J, Tyroler HA. Analysis and interpretation of treatment effects in subgroups of patients in randomized clinical trials. *JAMA* 1991;**266**:93–8.
23. Thompson SG, Sharp SJ. Explaining heterogeneity in meta-analysis: a comparison of methods. *Stat Med* 1999;**18**:2693–708.

24. Lau J, Antman EM, Jiminez-Silva J, Kupelnick B, Mosteller F, Chalmers TC. Cumulative meta-analysis of therapeutic trials for myocardial infarction. *New Engl J Med* 1992;**327**:248–54.
25. Wetterslev J, Thorlund K, Brok J, Gluud C. Trial sequential analysis may establish when firm evidence is reached in cumulative meta-analysis. *J Clin Epidemiol* 2008;**61**:64–75.
26. Turner RM, Omar RZ, Yang M, Goldstein H, Thompson SG. A multilevel model framework for meta-analysis of clinical trials with binary outcomes. *Stat Med* 2000;**19**:3417–32.
27. Higgins JPT, Whitehead A, Turner RM, Omar RZ, Thompson SG. Meta-analysis of continuous outcome data from individual patients. *Stat Med* 2001;**20**:2219–41.
28. Egger M, Schneider M, Smith GD. Spurious precision? Meta-analysis of observational studies. *BMJ* 1998;**316**:140–4.
29. Chuang SC, Rota M, Gunter MJ, Zeleniuch-Jacquotte A, Eussen SJPM, Vollset SE, et al. Quantifying the dose-response relationship between circulating folate concentrations and colorectal cancer in cohort studies: a meta-analysis based on a flexible meta-regression model. *Am J Epidemiol* 2013;**178**:1028–37.
30. Stroup DF, Berlin JA, Morton SC, Olkin I, Williamson GD, Rennie D, et al. Meta-analysis of observational studies in epidemiology: a proposal for reporting. *JAMA* 2000;**283**:2008–12.
31. Cappelleri JC, Ioannidis JPA, Schmid CH, de Ferranti SD, Aubert M, Chalmers TC, et al. Large trials vs meta-analysis of smaller trials. How do their results compare? *J Am Med Assoc* 1996;**276**:1332–8.
32. Lau J, Ioannidis JPA, Schmid CH. Summing up evidence: one answer is not always enough. *Lancet* 1998;**351**:123–7.

FURTHER READING

The book by Straus et al.[6] is a classic introduction on how to practice and teach EBM. Heneghan and Badenoch[5] is a short pocket guide with many useful lists, tables, and figures. The book edited by Egger et al.[7] is a broad overview of the rationale and principles of systematic reviews. It also includes a comprehensive description of statistical methods for meta-analyses.

CHAPTER 13

Scientific Communication

Haakon Breien Benestad
Department of Physiology, Institute of Basic Medical Sciences, University of Oslo, Blindern, Oslo, Norway

13.1 INTRODUCTION

Communication, which comes from the Latin word *communis*, means common, i.e., to share a message, not just send it out and hope that it will be understood. When writing a scientific paper, the authors are trying to send a message, and as such they must know what the people who will receive that message want, as well as their level of knowledge and ability to understand the subject matter. Indeed, authors of scientific articles and textbooks, newspaper journalists and actors all have different ways of sending messages. Although they are smaller, similar differences exist within disciplines; in science, for example, messages can be sent via scientific papers, poster presentations, lectures, or communication with the public.

13.2 SCIENTIFIC PAPERS

13.2.1 First Things

There is only one way to learn how to write good scientific prose: practice, practice, and more practice. Other helpful tools include reading good professional literature, learning from others, and studying texts on scientific writing and the improvement of writing skills, some of which are listed under "Further Reading" at the end of this chapter. However, even with all these tools, there is no substitute for practice. With each draft of a scientific paper, authors should seek and heed the critiques of colleagues, as these "private referees" can be valuable advisers. The input of colleagues working in other specialties may also prove valuable, as it may be easier for them to spot flaws in logic or statements to which an author may be blind. It is useful to create and nurture a network of friends among colleagues, within which mutual assistance in the writing process can be offered.

Once an audience has been chosen for a scientific paper, the author can choose a suitable journal to which to submit. Of course, that is easier said

Research in Medical and Biological Sciences
DOI: http://dx.doi.org/10.1016/B978-0-12-799943-2.00013-6

than done. On the one hand, the authors may hope to publish in a major international journal, but on the other hand they may prefer to avoid the delays of successive rejections by major journals like *Science*, *Nature*, and *Cell*. Nevertheless, if the work is not worthy of publication in an international journal, it is likely not good enough to be part of a PhD thesis either.

When preparing a scientific paper for submission, it is important to read the journal's instructions or guidelines to authors carefully, and to consult an article from a recent issue of that journal, both of which are usually available from the journal's website (paper versions may also be accessed through the local academic library). Thereafter, authors should preferably summarize their results and the basic data for the article in figures and tables according to the instructions for authors. It is helpful to draft several versions. Indeed, a researcher may have spent years on experimental work, so it is worth spending a few days to ensure that data are presented well and clearly.

After the authors have completed their tables and figures, it is time to write. It is best to start by thinking the paper through thoroughly, and then compile an outline that logically connects its different sections. One option used by many authors is to start with the Material and Methods section, then move on to the Results section, followed by the Discussion and Introduction sections, and finally the Abstract and Title Page. Others prefer to start with the Results section, which can make it easier to gauge the extent of description needed in the Materials and Methods section. Finally, Keywords, Acknowledgments, References, and Figure Legends must be prepared. All these sections are discussed below in their customary order of appearance in a published paper. Although some journals have dispensed with the Introduction, Methods, Results and Discussion (IMRAD) recipe, in one way or another all scientific papers must cover these aspects.

Ideally, authors should write when they have the time, peace, and a desire to write. Some authors have no desire to write and therefore must create the conditions that will make them do it and then force themselves to start, for example by setting aside a couple of hours in the morning for writing and never using these hours for other pursuits. An experienced writer maintains: "Do not think that you can do any other demanding work in a productive writing period. For most of us, writing is so tiring that we work at the limit of our abilities" and "Force yourself to stop writing when you are well on the way with a point or a paragraph." That may be torturous, he admits, but it is best to quit before the author is running on empty, so that there is something to start on the next day. The points do not vanish.

When it comes to sentence length, in general it is a good idea to avoid very long sentences. Indeed, there is nothing wrong with adding full stops. A reader-friendly sentence is seldom more than 22–25 words long. If the author has several things in mind, they should be presented one at a time. It is wise to read through the manuscript to check sentence length, in addition to reading the manuscript aloud to check sentence flow and rhythm. The writer should try to avoid "noun ailment." For example, "This chapter is a summary of current concepts..." is better written: "This chapter summarizes current concepts...." Moreover, to make a sentence easy to understand, the author should ensure that the (main) verb is mentioned early in the sentence.

Writers should not be afraid to express themselves concretely. Although it may seem daring to change "It is clear that much additional work will be required before a complete understanding..." to "Presently, we do not understand...," the example is illustrative. It is important that a writer not waste words and letters; phrases like "effectuate an alteration" and "produce an inhibitory effect" should be discarded in favor of simple, straightforward words like "change," and "inhibit." Empty phrases such as "it is interesting (important) to note that" or "it should be remembered that" are often best deleted. Finally, authors should not be afraid of the personal tone. It is natural to write in the passive voice in Material and Methods, but otherwise it is increasingly common to write "I/we found that...." "It is the opinion of the present author..." can be "I think...."

A summary of common grammatical issues is given in Boxes 13.1 and 13.2, and may be used as a checklist. The IMRAD manuscript structure is summarized in Box 13.3.

13.2.2 Title

The title should be short and specific. It may be divided into two parts, the second of which narrows the first. If possible, state the principal findings or conclusion in the title. It also helps to indicate the experimental animals used in an animal study, or the human subjects used in clinical or other trials.

Example: "Dendritic cells as effector cells: gamma interferon activation of murine dendritic cells triggers oxygen-dependent inhibition of *Toxoplasma gondii* replication."

BOX 13.1 Things to Avoid in Manuscript Revision

- Use of nouns when verbs would be better
- Overuse of prepositions

For example, "There had been marked *changes in* the presentation related *to* the data accumulated as a consequence *of study of* the results *of treatment in* cancer *of* the breast" can be rewritten, "*The surgeon markedly* changed *her presentation after she* studied *the results of* treated *breast cancers.*"

- Overuse of the verb "to be"

"This short chapter *is* an excellent summary of..." could be written, "This short chapter *excellently summarizes...*"

- Empty phrasing

"*The fact is that (It is clear that)* our findings falsify generally held hypotheses..." might be simplified as "*Indeed*, our findings falsify..."

- Impersonal "it"

"*It is* possible that ..." can be substituted with "*Perhaps...*"

- Present participle with no subject ("dangling participles")

"Assaying the cytokine concentrations, the results showed that..." should be rephrased, as the results did not measure the cytokine: "Assaying the cytokine concentrations, we found that ..."

[But such incorrect use of "using" is seemingly becoming correct!]

- Convoluted constructions

"The experimental cell cultures had a very rapid growth rate, *compared to* the controls" could often be changed to "...grew more rapidly *than* the controls"

"The results for the test group were comparable to those for the control group" for variety could be "...*similar to those...*"

- Long sentences

Another clear title is "Minocycline inhibits cytochrome c release and delays progression of amyotrophic lateral sclerosis in mice."

"Pathogenesis of human cytomegalovirus infections. The role of bone marrow stem cells and vascular endothelial cells" is an informative title for a doctoral dissertation. However, "Vasoactive intestinal peptide as a modulator in the neuro-immune axis; the influence of stress" is not very good, particularly because it does not state that the work was performed on human monocytes.

The author should try to avoid phrases such as "The effect of...," "Results from..." in the title. If possible, abbreviations and colloquialisms should be avoided. The title should not include strings of modifiers, such as "tobacco mosaic virus transformed long-passaged cell lines," or "...growth from peripheral blood human megakaryocyte progenitor cells."

BOX 13.2 A Participial Phrase at the Beginning of a Sentence Must Refer to the Grammatical Subject

Walking slowly down the road, he saw a woman accompanied by two children.

The word *walking* refers to the subject of the sentence, not to the woman. If the writer wishes to make it refer to the woman, he must recast the sentence:

He saw a woman, accompanied by two children, walking slowly down the road.

Participial phrases preceded by a conjunction or by a preposition, nouns in apposition, adjectives, and adjective phrases come under the same rule if they begin the sentence. The examples in the left-hand column, below, are wrong; they should be rewritten as shown in the right-hand column.

On arriving in Chicago, his friends met him at the station.	When he arrived (or, On his arrival) in Chicago, his friends met him at the station.
A soldier of proved valor, they entrusted him with the defense of the city.	A soldier of proved valor, he was entrusted with the defense of the city.
Young and inexperienced, the task seemed easy to me.	Young and inexperienced, I thought the task easy.
Without a friend to counsel him, the temptation proved irresistible.	Without a friend to counsel him, he found the temptation irresistible.

Adapted from Strunk and White[1]

BOX 13.3 IMRAD Structure

Title

Include the *keywords* of the paper. Avoid superfluities and ponderous, verbose titles. State the type of work, the experimental animals handled, and the main method used.

Abstract

The Abstract should present the following elements:

1. Why, how and what was done
2. Results
3. Conclusions, or inferences of the results.

Include a maximum of 200–300 words (consult the journal's instructions to authors). Avoid doubt in the abstract; include only what is certain. The abstract should be informative, not indicative, i.e., avoid statements such as: "The findings will be discussed in relation to the. . .hypothesis."

Add up to 10 *keywords* or *index words*, after the abstract, preferably medical subject heading (MeSH) terms.

(*Continued*)

BOX 13.3 IMRAD Structure—cont'd

Introduction

1. Start with two or three sentences that put the topic in perspective.
2. Add a background, perhaps a historical overview, including only what is needed to lead to the next point. Use the past tense, but note that some general truths must be placed in present tense. Refer readily to review articles and meta-analyses. The text should make clear the object of reference citations!
3. Mention the approach, purpose, and perhaps the hypotheses to be tested. Be precise. Explain why the work is important.
4. One may want to clarify the progression of the presented project and results.

   Summary: The Introduction section should answer the question: "*Why?*"

Material and Methods

Order this section according to importance, or put in logical order, starting with a description of the study sample, be it experimental animals or human subjects (and approval by relevant ethical committees): species, number, gender, weights/ages, food, and drink. If applicable, designate the time of day of the various interventions or operations. Italicized headings may be used to help organize this section (consult the instructions to authors). Include *manufacturer's name, city, state* (when applicable) and *country*, and sometimes *batch number* for biological reagents. Donor's name: "provided by Dr. Hansen...." Cite methods that have been described previously! If these methods have been modified, describe these modifications. If applicable, include a subsection: "Methodological considerations" with methodological discussions; then these can be avoided in the Results or Discussion sections. The experimental setup must be described in a manner that allows competent researchers in the field to replicate the trials. Indeed, the Material and Methods section is like a recipe book! Finish by giving an account of the statistical methods under the heading *Statistics*.

Summary: The Material and Methods section should answer the question: "*How?*"

Results

1. Give a *brief* description of the findings, in logical order.
2. *Do not repeat numbers listed in tables or appearing in figures, but describe trends and courses*, citing "Table x" or "Fig y" in parentheses. If possible, state how large the changes are, such as: "... about 25% larger than the control." Real discussion should not be included in the Results section, but some interpretations and explanations are often necessary to facilitate reading.
3. The Results section must sometimes start with a documentation of presumptions (validation of methods or study setup).

(Continued)

BOX 13.3 IMRAD Structure—cont'd

4. State the *uncertainty of the localization parameter* using the mean ± standard error of the mean (SEM), or the mean or median with a 95% confidence interval. If relevant, give an *effect estimate* (difference between means or medians, relative risk, regression coefficients, etc.) with a 95% confidence interval. Preferably, results are presented in tables or graphs. Now and then the *variation* of the measurements—standard deviation (SD), coefficient of variation (CV), or quartile interval—is stated, but this often is better placed in the Material and Methods section. Present the *P*-values, number of replicated measurements, and number of trials, preferably in connection with tables/figures.
5. Data normalization—conversion of data to percentages of median or mean in a control group, such as to permit combining the results of different trials without including the interexperimental variation—can be taken too far!
6. If the results are the same for five trials, the author can occasionally describe the course of one experiment and say that the other four gave "in principle the same result." Another possibility would be: "A representative result from three trials. . ." and the reason for it.

Summary: The Results section should answer the question: "*What was found?*"

Discussion

1. *Briefly point out the principal findings, in relation to the approach* described in the Introduction section.
2. *Interpret* the results—if possible present several possibilities (references), by diminishing probability. How reliable are the results? Do they agree with the findings of others? What is the transfer value or generalizability?
3. State why others have not found this. Give accurate references!
4. Consider adding something on the *conditions* for a successful result.
5. Include unanswered questions and future tasks.
6. End with *Conclusions* that summarize the most important aspects of the paper. The authors should state their opinions and preferably include item 5.

Summary: The Discussion section should answer the question: "*What do the findings mean?*"

13.2.3 Abstract

The Abstract should be comprehensible without reference to the body of the paper. Most readers read only the title and sometimes the abstract of a paper. Only those especially interested in the topic will continue to the figures and tables, and perhaps more. The Abstract should start with the background, purpose or principal problem, continue with the principal methods and the most reliable and certain results, and end with a conclusion that can be easily understood. Authors should study published abstracts from the journal to which the paper will be submitted, and follow their style. It may be best not to include data or *P*-values in the Abstract. It should be obvious that the statements in the Abstract are backed up by significant data. If unusual abbreviations are used, they should be defined upon their first mention. For example, "The plasma concentration of interleukin 1 (IL-1) increased after...."

After the Abstract, or alternatively on the Title Page of the manuscript, in compliance with the instructions to authors, include the keywords or index words that are not already present in the title, preferably medical subject headings (MeSH) terms from the Medline (PubMed) database.

13.2.4 Introduction

The Introduction section should state the scientific problem and why it should be addressed. Authors should avoid stating the obvious in the first sentences. While writing, it is important to keep the target audience in mind. Introduce the theme of the paper and put it in perspective in one or two sentences.

A short historical review may be suitable, preferably with references to review articles or meta-analyses. Throughout the paper, the findings of referenced works should be stated meticulously, clearly, and accurately. It is all too easy to make several statements followed by a reference that appears to support all the statements, but only underpins the last of them. Likewise, the writer should know the field sufficiently well to give credit where credit is due.

Authors should pretend they are storytellers, spinning a thread that can be easily followed, and that leads to the *what* and the *how* of what they did. Usually, something should be mentioned about the relevance of the research project and why the scientific problem in question has not been addressed before. Some writers end an introduction with a summary of the principal findings, but it may be argued that such a summary in the Introduction section is overly repetitious, and thus unnecessary.

A suitable length for an Introduction section is one or two double-spaced, 12-point font manuscript pages.

13.2.5 Material and Methods

Here again, it is wise to study the style of the journal to which the authors will submit their paper. The Material and Methods section of a scientific paper is the "handbook" or recipe; it will be read only by those especially interested in the topic. It is prudent to use subsections to guide the reader to particular points of interest.

The study sample, be it human subjects or experimental animals, should be described in the first paragraph. Characteristics of human subjects (gender, age, weight, etc.) may be presented in a table or flowchart that shows those enrolled, drop-outs, etc. Inclusion and exclusion criteria, among other things, should also be described (see Chapters 1, 4, 7, and 9). Moreover, ethical approval, including the name of the body granting this approval, must be stated, along with information on written informed consent. Experimental animals should be described in terms of species, gender, weight or age, food and drink, and supplier's name and address. It may also be necessary to mention whether experimental animals are inbred or outbred, specific pathogen free, or transgenic, and if so, by what means. Here also, ethical approval must be stated, for example, "All animal experiments were approved by the Experimental Animal Board/Committee under the Ministry of ... and conducted in conformity with The European/... Convention for the Protection of Animals used for Experimental and other Scientific Purposes"—or a shorter version of that statement. Randomization procedures must be stated for both human subjects and experimental animals. If the variables measured are subject to diurnal variations, observation times should be given.

The next subsection might have a heading such as chemicals, reagents, or biological materials, and would comprise a listing, preferably alphabetical, of preparations used, with the trade names, manufacturers, and their locations (city, state if applicable, and country) in parentheses: "cyclophosphamide (Sendoxan™ [alternatively: Sendoxan®], Orion, Espoo, Finland)." Whenever the quality of a biological material varies, the *batch*, *lot*, or *code number* must be stated. If the preparation used was a gift, name the donor in parentheses (*kind gift from Dr. NN, the XX institute, Cambridge, UK*). The donor must also be thanked in the Acknowledgments.

When a procedure has been published previously, it should be mentioned with a reference to its original description: "In short, leucocytes were isolated with a density gradient method... (ref.)." Modifications to previously published procedures, as well as new methods, should be described in sufficient detail to enable other scientists in the field to replicate the experiments. Authors should remember to state all essential physiochemical parameters, such as temperature (18–20°C is preferable to "room temperature"), pH, osmolarity, concentrations (preferably in SI units, such as mmol/L), gas pressure, solution volume, storage conditions, centrifuging data (G- or g- is preferable to rotations per minute or rpm).

It may be useful to include a subsection called "Experimental design," in which the author describes the course of the trial or experiment. The author may also add a subsection entitled "Methodological considerations," in which the results of methodological research germane to the methods are outlined, so as not to disturb the flow of the Results section. Clinical studies, observational studies, and qualitative studies often have more extensive methodological considerations than do laboratory studies. A report on the study design, the outcome and exposure variables, as well as a discussion of the choice of effect measure may be appropriate in this subsection.

Usually, the last subsection is "Statistics" or "Statistical methods," which describes data presentation and processing. The writer may state that the data are presented in means or medians, with their confidence intervals. The relevant effect estimates are also presented, with their 95% confidence intervals. Further, the method(s) used for statistical testing (including the name and supplier of any statistical software package used) should be given. The *P*-values for testing the effect estimates must also be stated. In multiple comparisons, a modified version of the significance level is often used. The most common version for multiple testing is the Bonferroni multiple comparison correction, in which the significance level is equal to 5% divided by the number of statistical tests. This means that if one compares one group of control data with each of five groups of test data, the significance level should be set to 1%, and only effects with *P*-values less than 1% should be reported. The author should also consider stating whether the testing is one-sided or two-sided, although two-sided alternatives are the most common in medicine and the biological sciences. Authors should report the type of statistical model used for the analysis, for instance, whether parametric or nonparametric methods are used, as well as choice of univariable and multivariable model. For more information please see Chapters 4 and 11. Finally, the author should report whether—or better, how—a power analysis has been conducted.

The Material and Methods section should be written in the passive voice, as what has been done is more important than who did it.

There are different guidelines that describe the standard of reporting for specific fields of research: Animal Research: Reporting *In Vivo* Experiments (ARRIVE, www.nc3rs.org.uk/ARRIVE), Consolidated Standards of Reporting Trials (CONSORT, www.consort-statement.org), Preferred Reporting Items for Systematic Reviews and Meta-Analyses guidelines (PRISMA, formerly QUOROM, www.prisma-statement.org) and for epidemiological studies MOOSE (Meta-analyses of Observational Studies in Epidemiology, www.consort-statement.org/Initiatives/MOOSE), MIAME (Minimum Information About a Microarray Experiment, www.fged.org/projects/miame) and MIAPE (Minimum Information About a Proteomics Experiment, www.psidev.info/node/91).

13.2.6 Results

The Results section should present what was found in a logical order. The author can choose to tell one story or many stories, adding subsections if allowed in the instructions to authors. Results should be presented briefly and in the past tense. The most important results should be mentioned first, and pivotal words should be placed early in sentences. Negative or trivial findings are best mentioned toward the end of this section. Journals often recommend that the Results section of a paper reporting clinical trials begin with an overview of the test subjects. Often such overviews are so unexciting that they are better included in the Material and Methods section, as mentioned above, but the journal's instructions always take precedence in the layout of a scientific paper.

Occasionally the author may start the Results section by referring to an experiment that documents the premises of the main scientific problem. Although the Results section is not the place to discuss findings, occasionally authors might have to clarify why the experiment was conducted, what the results mean, and how they lead to the next experiment.

There are a number of opinions on the best way to write a Results section. One viewpoint is that this section can be telegraphic. Indeed, concise language often is best, as some journals have considerable page charges. Hence, the concise "Blood pressure remained constant

(Figure 1)" may be preferable to the more verbose "It can be seen from Figure 1 that blood pressure remained constant...."

Authors should not repeat data in the text that is already included in cross-referenced figures and tables. However, the writer should present data on the description of the course of the experiment and the differences observed, e.g., between test and control groups, in the text. Biological significance is just as important as statistical significance, so if the outcome variable is doubled or halved, or if it declines or increases linearly or exponentially with time, a description is in order. If there is space, it may be best to state *P*-values together with figures or tables, instead of in the text of the Results section.

A figure or table that summarizes the raw data of all replicated experiments is more convincing than the presentation of a single experiment with the ancillary explanation that "three replicate experiments gave similar results." When the interexperimental variation is large, authors may be obliged to present the data this way. In any case, authors should state that representative, not typical, experiments have been selected, as after all, most authors do present the best experiments as examples. Alternatively, a summary of all experiments can be presented, provided that the data are normalized, for example by setting the median or the mean for the control group at 100%. Of course, in that case the author must also state the absolute value of the 100% figure, such as in a table's footnote: "The 100% values ranged from 32 to 56 mmol/L."

Murphy's law states that anything that can go wrong will go wrong. But an author should have a good reason for excluding atypical results, and exclusion criteria should be set before the research project starts. Therefore, the author may need to describe the common characteristics of any atypical results at the end of the Results section.

13.2.7 Discussion

The Discussion section presents the interpretation of the results against the backdrop of the consensus of a scientific field. It also puts forth the relevance of the findings. It should be phrased in the past tense, but general conclusions should be in the present tense. A good rule of thumb is to start the Discussion section with a very short summary of the principal findings. This way the reader can interpret the results in light of the main problem that was presented in the Introduction section of the paper. What do the findings mean?

In the Discussion section the authors should answer the questions: How do the results fit in with the common body of scientific knowledge? Which previous scientific papers support the present findings, and which deviate from these results? As mentioned above, it is essential that a writer be familiar with the scientific literature in the field. In this section the authors should cite primary publications, not review articles. This may present a dilemma if the initial publication was qualitatively weak, but the subsequent publication from the same group of authors was more fundamental, as authors must economize on the number of references cited in the paper.

Occasionally, the meaning of the findings may not be obvious, but there are several solutions to that. The most plausible interpretation should be stated first. Hans Selye allegedly said, "Our facts must be correct. Our theories need not be, if they help us to discover important new facts."

It may be relevant to point to the necessary and sufficient conditions for the findings, and to explain why others have not obtained the same findings. Perhaps there were differences in the experimental setup, the animal model, or another factor.

Often authors must say what is new and important about their paper in an unassuming way to get it published. Indeed, the findings reported should be both new and important to avoid the response once given by a referee who quoted Dr. Samuel Johnson's famed remark: "Your manuscript is both good and original; but the part that is good is not original, and the part that is original is not good."

Occasionally a writer will need to discuss groups of observations concerning dissimilar phenomena or parts of problems, in which case it might be best to start with the most important. In order to decide what is most important, the author can refer to the order of presentation in the Results section. If the instructions to authors so permit, subsections can be used to encapsulate these issues in the Discussion section.

The final section should contain a few summarizing or concluding sentences, perhaps with a rough sketch of further research. If allowed by the instructions to authors, these can be preceded by the heading "Conclusions."

What to avoid in the Discussion section:

- *Repetition of points made in the Introduction section*, which is intended to introduce the background of the project.
- *An excessively long account of the Results section.*

- *An overwritten discussion.* The author should not seek to justify a place in history or show a command of the literature—that may be speculative. The discussion should be of the actual findings and should not expand beyond them. Avoid excessive generalization.
- *Faulty gradation of reality.* Indeed, the authors must point to what is new and important in their study, and to how the results may be generalized and used, but they must not exaggerate.
- *Discussion of trivialities.* Still, it may be advantageous to discuss negative results.
- *Too few explanations.* Disagreements with the findings of others should not be omitted, nor is it inappropriate to criticize colleagues in scientific papers. Instead the authors should discuss any incongruous findings with an understated tone.

As displayed in this section, writing the Discussion section is like sailing between Scylla and Charybdis. Here, as elsewhere in science, common sense is the best guide. A suitable length for the Discussion section is three double-spaced, 12-point font manuscript pages (Box 13.4).

13.3 GENERAL GUIDELINES FOR TABLES AND FIGURES (GRAPHS)

After the authors have collected their raw data and chosen the audience for the paper, they must ask themselves which data *must* be presented, and which data may be replaced by qualitative mention in the text. Would a summary of these data in the main text (means—or medians—with their 95% confidence interval, or SEM, and *n* data points) be sufficient? If the data are vital or copious, authors may choose to present them in tables or figures (graphs). Some data cannot be summarized by numbers alone, such as the results of gel-electrophoresis or DNA microarrays.

A general rule for choosing between tables and figures is: use tables whenever exact values or comparison with the data of others is essential; use figures when the relationship between an independent variable (x) and a dependent variable (y) is the principal point, such as an event over time. In other words, figures ease understanding of main points, and tables provide more comprehensive documentation.

Authors know that many readers will only read the title and abstract of their paper, and perhaps scan the figures. This can of course influence the choice of figure presentation—curves or bar charts, photographs of gels or recordings of action potentials, etc. The authors should remember

BOX 13.4 Can a Speech Make a Good Journal Article?

A speech and a journal article, even if they deal with the same problem and present the same data, are different means of communication, each with its own qualities and rules. One does not succeed readily in the other's form.

. . ., the transcript of a well delivered speech, based on a few notes, supported by 11 white-on-blue slides and kept as short as possible to allow for a lively discussion, probably would produce a very unpromising manuscript. . .

The talk can be enlivened by humorous digression, personal observations and local references; these devices are distractions in a written presentation. The format of the talk can be flexible and the speaker can allow time for immediate feedback; the journal article is precast in a rigid form and feedback is slow and cumbersome. The talk can often only summarize the author's current work and prompt the listener to want to hear more; a good journal article should offer enough detail and "close" some aspects of the topic so that the reader feels that at least one question has been answered.

*The differences between these two media are basically due to a simple physiologic fact: the eye is quicker than the ear. We read twice as fast as we talk. But not only can the reader go forward in time faster, he or she can also go back in time—that is, review—*ad libitum*. As a result, the reader is in a position to be more critical; the itinerant eye, directly scanning the text, demands a continuity of thought and a level of supporting detail that the accepting ear, oriented 90° away from the source of its information, does not.*

How, then, can a spoken scientific presentation be converted into a good scientific paper? With great difficulty. Details have to be added, statements supported with references, subheadings inserted, digressions pared, and figures and tables condensed, and the line of thought or "critical argument" has to be clarified. . .

(Morgan[2])

that tables and figures, with their titles, captions, and footnotes, should be self-explanatory without reference to the body of the text.

An author should be familiar with the average level of knowledge of his or her audience and express titles, terminology, and abbreviations accordingly. Moreover, these tables and graphic elements must comply with the instructions to authors. Summarizing the results and preparing corresponding tables and figures, usually the first steps in compiling a scientific paper, allow the authors to determine whether more experiments should be performed or more data collected.

During this first round of table and figure creation, authors can and are encouraged to let their imaginations run free; to sketch several versions of tables and figures. The goal is to incorporate the maximum amount of information into the minimum amount of space, without overloading, ambiguity, or misleading layout. Occasionally tables or figures may be divided into panels to promote understanding. One such approach is to place a composite figure of two or more frames above each other, with the same x-axis in all frames. Figures presented in papers are often ill-suited to lectures (PowerPoint presentations) or posters, and tables in a paper are always ill-suited to other uses. Therefore it may be prudent for an author to prepare different versions of tables and figures for lectures, posters, and the pending paper.

As tables and figures should be self-explanatory, an author must often strike a compromise between brevity and comprehensiveness in captions and footnotes. For example, the details of methodology should be given in the Material and Methods section, but occasionally authors must include some details on the methods in the footnotes and captions of tables and figures so that they can stand independently of the text. Abbreviations should be avoided as much as possible, but if included, should be written out in full upon first use, or defined in a dedicated section in the paper, according to the instructions for authors. In addition, abbreviations must be redefined in each table and figure where they are used; again, so that tables and figures can stand independently of the text. Moreover, one must define the location (center of the distribution), the uncertainty measure, and the number of replicate analyses and replicate experiments. For example, "Means with their 95% confidence intervals are shown; $n = 7-8$ replicate determinations in each case. A total of three independent experiments with similar results were performed." If normalized data are presented, the range of raw data corresponding to 100% should be stated. When such information is given in the footnote to the first table or the first figure caption, one often can save space in the rest of the figures and tables by writing "See legend to Figure 1 for further information."

When the drafts of the tables and figures are complete, it is important to verify that they are presented in a uniform manner. Do the column headings show the same sort of data; are they the same in different tables? Likewise, have axes of the same variable been labeled the same way in different figures? One should go through all the tables and figures with a fine tooth comb, with a sharp eye on the tables for superfluous information that should be removed. Are there any words, parts of pictures, or nonessential data that can be deleted, or moved to figure or table captions or footnotes? And yet again, has a capable, critical, nonspecialist colleague gone through the tables and figures?

Finally, authors should check the tables and figures for internal consistency, as well as for consistency between these elements and the statements in the body of the text. Readers will have a poor impression of the work if in the Results section a contention is made that is not obvious in the tables and figures, or if the text and the tables or figures are contradictory.

Finally, the tables and figures must be numbered in their order of appearance in the paper. The author may mark the desired location of data presentations ("Figure 1 approx. here").

13.3.1 Tables

Below is an example of a typical table, but in all cases table format should comply with that stated in the instructions to authors.

13.3.2 Title

	Column heading[a]		
Row heading[b]	**Column 1**	**Column 2**	**Column 3[c]**
Row 1	Data field		
Row 2			
Row 3			

Footnote a
Footnote b
Footnote c

The table title should convey its main message, i.e., what it shows. Avoid vague, mundane phrases, such as "The effect of...." One should use the footnotes to convey more information on table content and interpretation.

The same sort of data should preferably be presented in columns, not in rows. And the columns should be arranged to facilitate the comparison of data sets, e.g., adjacent columns in the table. It is often convenient to start with the control data or normal values in the upper left-hand corner of the table. An author should keep in mind that the reader will absorb the information in the table from left to right, and from top to bottom.

BOX 13.5 Good Tables

- Write clear column headings.
- Create tables in Word using the Table tool—or possibly in Excel. Never create a table for a paper in the same manner as for a presentation, such as in PowerPoint. Tables for slides should be simple, without too many footnotes.
- Whenever possible, present the same sort of data in columns, not rows.
- Avoid conveying a false impression of precision, for example by listing more than two significant digits.
- Do not include sequences of numbers that can easily be calculated from other numbers in the table.
- Proofread the table against the text of the paper.
- Titles and footnotes should be succinct. Describe what is shown in the title. It is often wise to describe the purpose of the table. State the uncertainty measure or spread measure, such as SEM, 95% confidence interval, SD. State the *P*-values and the range of the control values that have been set to 100% in normalization.

If the exposure and outcome variables are both categorical, tables are presented as number of cases for each specified category of exposure and outcome. In that case the table may be presented with percentages in each cell. These percentages are calculated from the total number of exposures for each specified exposure category. Table 11.2 is an example of such a table.

Data that easily can be calculated from the raw data presented or monotonical data, such as multiple occurrences of 0, 100, + or −, may be moved to table footnotes or to the Results section (see Box 13.6). Be meticulous and uniform with the number of significant figures (see Box 13.6) and with the use of units.

Authors must choose the correct number of digits in the values they present to avoid creating a false impression of high precision. As a general rule, the number of significant digits is equal to the number known with some degree of confidence, so the choice should be guided by the uncertainty measure, be it SEM or 95% confidence interval. If the authors choose two significant digits for the confidence interval, they should not write *1.1224 (1.0015–1.2468)* $* 10^9$ *cells/L*, as this interval range has four significant digits (245.3 million/L). Present the same data to the correct number of significant digits *1.12 (1.00, 1.25)* $* 10^9$ *cells/L (median, 95%*

BOX 13.6 A Poor Table

Table 13.1 Cellularity and differential counts of peritoneal exudate cells, accumulated 1 h after injection of the chemoattractant FMLP

Dose of chemoattractant (μg)	Treatment time (h)	Total cellularity (× 10^6)	Differential counts (%)			
			Macrophages	Lymphocytes	Neutrophils	Eosinophils
Saline	1	5.8 ± 3.2	76.6 ± 8.6	16.5 ± 8.6	6.1 ± 4.7	0.7 ± 1.3
FMLP 25	1	14.0 ± 5.6	63.4 ± 8.6	26.8 ± 8.8	9.1 ± 5.2	0.7 ± 0.8
FMLP 50	1	21.1 ± 7	63.9 ± 11.7	23.0 ± 9.5	12.1 ± 7	1 ± 0.8
FMLP 100	1	25.5 ± 5.4	64.5 ± 10.1	16.8 ± 7.7	17.6 ± 8.8	1.2 ± 1.3

Column 1 (reading column) should be changed: the word "Saline" exchanged with 0 (zero), "FMLP" be deleted, and the column should be headed "Dose of FMLP" (preferably given in μmol or nmol, since the MW of FMLP, which is a microbial peptide, is known). Column 2 should be deleted (and the time, 1 h, given in a footnote). The variability (SEM is given) of replicate analyses (the number of which should be given in the legend or in the field) indicates that the data should be presented without decimals (as has in fact been done inconsistently—for part of three entries), according to the "significant digits" convention. The data furthermore demonstrate the inappropriateness of parametric methods (i.e., calculation of SEM) applied to data that are not normally distributed: Taken at face value, cell counts can apparently have negative values here!

confidence interval, N = 20), or alternatively *1.12 (1.00–1.25) nL*$^{-1}$. A correct example (undesignated) of a high-precision measurement might be *1.1224 (1.1215–1.1268)*, as the width of the confidence interval (0.0053) now has only two significant digits, whereas the mean—correctly—has five significant digits. Some authors go further in contending that whenever data are listed for comparison, as in a table, they should have no more than two significant digits (Ehrenberg[3]). This means, for instance, that statements of percentage greater than or equal to 10% never have decimals.

Good style, as well as most journal style guides, dictates aligning decimal points under each other in a column; the same applies to the plus/minus sign (±) in front of SD or SEM. Inside borders of tables should not be used (although used by Elsevier), but adhere to the instructions to authors. Occasionally, white space may be used to separate data presented, but otherwise the space between columns of data should be as small as practical, without disrupting overall appearance or legibility (Box 13.6).

SI units should be used as consistently as possible, and powers of ten should, whenever practical, be replaced with SI unit abbreviations (micro, nano, kilo, mega, etc.) (Box 13.7).

13.3.3 Figures

Russian writer Ivan S. Turgenev (1818–1883) is reported to have said: "A picture shows me at a glance what it takes dozens of pages of a book

BOX 13.7 Système International d'Unités (SI Units): Some Examples

Names and symbols for basic SI units

Physical quantity	Name of SI unit	Symbol for SI unit[a]
length	meter	m
mass	kilogram	kg
time	second	s
electric current	ampere	A
thermodynamic temperature	kelvin	K
luminous intensity	candela	cd
amount of substance	mole	mol

Special names and symbols for some derived SI units

Physical quantity	Name of SI unit	Symbol for SI unit	Definition of SI unit
energy	joule	J	$kg\ m^2\ s^{-2}$
force	newton	N	$kg\ m\ s^{-2} = Jm^{-1}$
power	watt	W	$kg\ m^2\ s^{-3} = Js^{-1}$
pressure	pascal	Pa	$kg\ m^{-1}\ s^{-2} = Nm^{-2}$
electric charge	coulomb	C	A s
electric potential difference	volt	V	$kg\ m^2\ s^{-3}\ A^{-1} = JA^{-1}\ s^{-1}$
electric resistance	ohm	Q	$kg\ m^2\ s^{-3}\ A^{-2} = VA^{-1}$
frequency	hertz	Hz	s^{-1}
concentration	mole per liter		$mol\ L^{-1}$ ($mol\ dm^{-3}$)
(radio)activity[b]	becquerel	Bq	s^{-1}
radiation dose	gray	Gy	Jkg^{-1} (1 Gy = 100 rad)

Prefixes for SI units

The following prefixes may be used to indicate decimal fractions or multiples of the basic or derived SI units.

Fraction	Prefix	Symbol	Multiple	Prefix	Symbol
10^{-1}	deci	d	10	deca	da
10^{-2}	centi	c	10^2	hecto	h
10^{-3}	milli	m	10^3	kilo	k
10^{-6}	micro	μ	10^6	mega	M
10^{-9}	nano	n	10^9	giga	G
10^{-12}	pico	p	10^{12}	tera	T
10^{-15}	femto	f			
10^{-18}	atto	a			

[a]Symbols for units do not take a plural form and should not be followed by a period; e.g., 5 cm, but not 5 cms or 5 cm. (except at the end of a sentence). Note that expressions like mg/kg/d and mmol/mL/s are not correctly written; the designations should be mg/kg*d (or $mg\ kg^{-1}\ d^{-1}$) and mmol/mL*s (or $mmol\ mL^{-1}\ s^{-1}$).

[b]1 μCi = 37 kBq.

BOX 13.8 Good Figures

- Function: Simple; illustrate a single point
- Accuracy: No disagreement with the text, neither of data nor of terminology
- Composition: Use the space available
- Contrast: Highlight differences between the test and the control, if it illustrates the point
- Legibility: No overloading; just a few curves or bars in each. Make large figures and consider reduction (this may be less important with Internet publication), but think of esthetics and harmony!
- Choice of figures or tables: Illustrate the relationship between two variables, such as the changes in concentration (*y*) with time (*x*)
- Uniformity both in creating figures and in their text: Mark coordinate axes with units and magnitudes; axes should be broken at discontinuities (even from the origin) or for changed scales. Use familiar abbreviations. Add arrows to the essential features in complex photos. Figure captions in the figure: preferably in slides; are often not allowed in set type.
- Figure caption: Be succinct. Describe what is shown, preferably in the first sentence of the caption, which may be in italics, like a subtitle. See also the last item in the bullet list of Box 13.5.

to expound." However, this is something that authors should refrain from trying, though some authors do, with different presentation graphics applications. It is important to remember that a figure should be quickly understood. Nonetheless, it is easy to agree with Tufte[4] that "Graphical excellence is that which gives to the viewer the greatest number of ideas in the shortest time with the least ink in the smallest space."

A checklist for good figures (graphs) is presented in Box 13.8. First, the author should choose the data to be presented, and then choose the most suitable type of figure. Curves (Box 13.9) are good for dynamic comparisons and for illustrating courses, such as weight (y) as a function of concentration (x) or time (x). A bar (horizontal) or column (vertical) chart (Box 13.10) is well suited to the presentation of discontinuous variables or ratios. Magnitudes often become clearer in figures than in tables. Scatter plots (Box 13.11), with or without regression lines, are often the best way to illustrate relationships between x and y variables. In any case, if the data are amenable, authors should make a scatter plot for their own use, as it may reveal interesting trends in the data, such as deviating subgroupings, they might otherwise overlook.

BOX 13.9 Line Graph

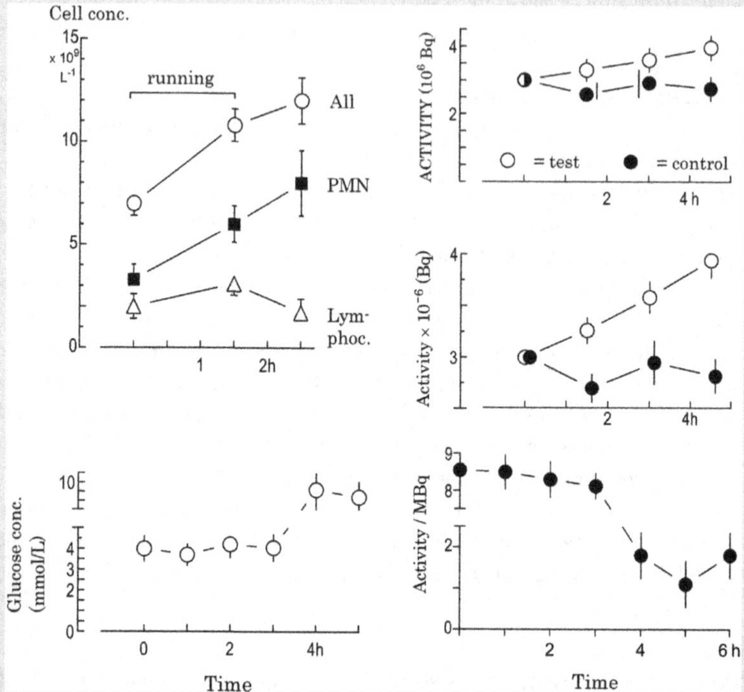

Figure 13.1 *Five panels illustrating various ways of constructing line graphs, some of them not recommendable.* Note (i) the various ways of labeling the axes, (ii) the misleading impressions that can be created by manipulating the scales (panel lower left and the two upper right), (iii) the acceptability of axis displacements, (iv) the marking of discontinuities, (v) the explanation of the symbols in the figure space, (vi) the displacement of variability bars (upper right) to avoid overlap, (vii) the utility of combining panels (two panels to the left), (viii) the usage of the SI system and avoidance of powers of 10, (ix) the better readability of words written in lowercase letters than in equal-height uppercase letters, and (x) the dispensability of the horizontal finials of the variability bars (upper left versus the others). The figure was designed with Graph Pad Prism®.

The advantage of photographs used to be that they could confirm the authenticity of a phenomenon, but it has become too easy to manipulate digital photos to give false impressions. A *Nature* editorial[5] called this "Beautification" and went on to explain that "the data are legitimately

BOX 13.10 Column Chart

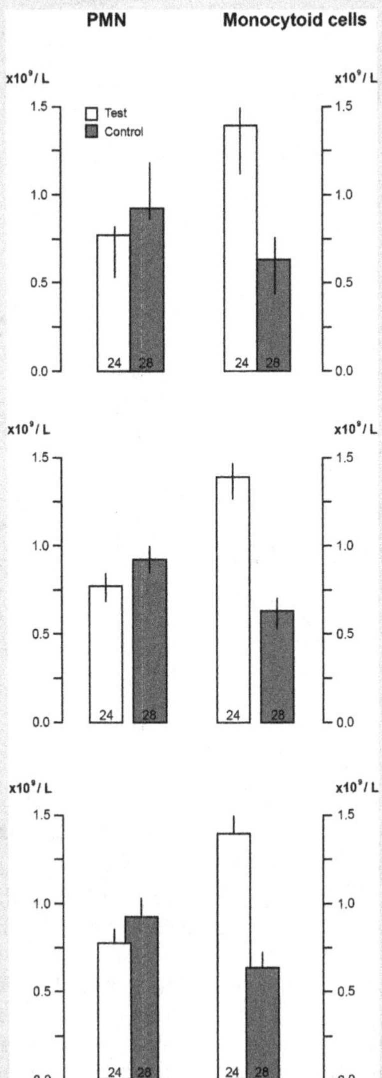

Figure 13.2 *Upper panel*: A conventional column chart. One must be sure that the left- and right-hand scales unequivocally refer to the left- and right-hand columns, respectively. Median values are shown with their 95% confidence intervals, estimated with a nonparametric method and therefore asymmetric around the medians, and number of replicate values. Alternatively, the *y*-axis designations may run vertically upwards, along the axes.
Middle panel: Variant design, with space between the columns (not having the same width as the columns!). SE(M)s are here shown as measures of uncertainty of mean values.
Lower panel: A third variant, with slightly overlapping columns and only 1 SE(M) drawn for each column.
The columns should be colored, hatched, or shaded so that the difference between tests and controls is conspicuous even in a black-and-white copy.

BOX 13.11 Scatter Plot

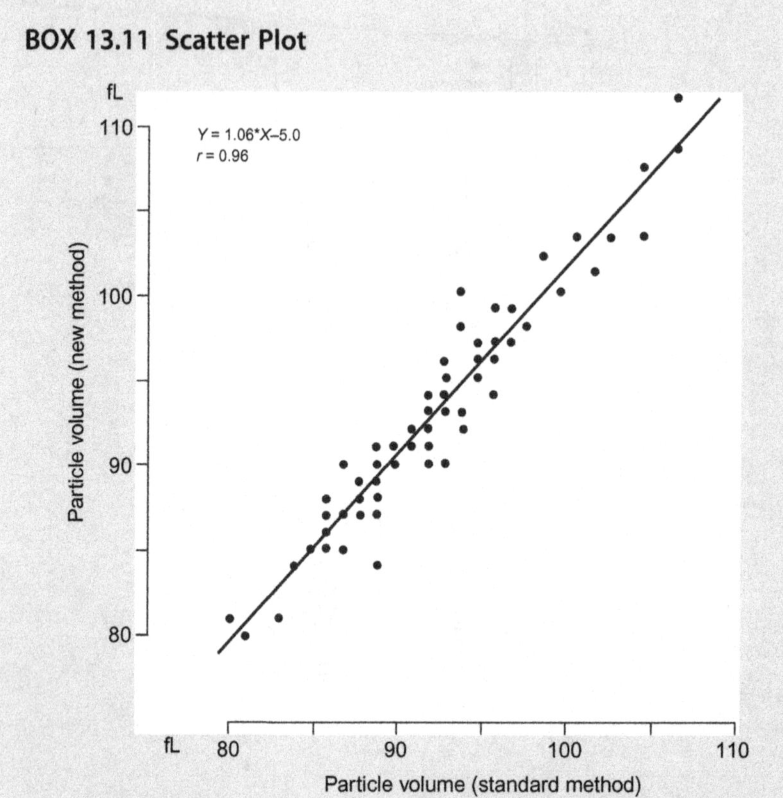

Figure 13.3 The regression line (here very close to the line of identity, $Y = X$), its formula, and the correlation coefficient (r) are indicated. Y = particle volume recorded with new method; X = particle volume recorded with standard method. Sometimes the line should be omitted and the formula and r-value given in the text or the figure legend. The P-value for testing the hypothesis that the correlation is different from zero should also be shown; alternatively, information on the 95% confidence intervals of the coefficients may also be presented (slope or declination: 1.00–1.13; intercept: −11 to +1). fL, femtoliter.

acquired but then processed to yield an idealized image. . . .a form of misrepresentation. . . .the *Nature* family of journals has developed a concise guide to approve image handling." One may find wit in the historical remark of a professor: "Video, sed non credo." (I see it, but I do not believe it), but nowadays the truth in this wry remark is revealed. Line

BOX 13.12 Questions Posed by Journal Editors

Is the article considered to be of sufficient scientific value to warrant publication?

Is the investigation well planned?

Is the scope of the paper suitable for the journal?

Is the paper clearly written, the language acceptable?

Is the experimental material adequate and controls sufficient?

Are the methods appropriate and satisfactorily described?

Is the statistical treatment adequate?

Are there any unresolved ethical issues raised by this paper or the work it reports?

If human or animal studies are included, is there an acceptable statement regarding informed consent or permission to perform the experiments?

Are tables and figures adequate and necessary? What can be deleted?

Is the discussion relevant—or too long?

Are the references to the literature satisfactory?

Are the conclusions justified?

What are the author contributions?

Is there any data deposition statement?

Have the authors completed the disclosure of potential conflicts of interest form?

In its instructions to authors, the journal *Nature* gives a checklist to ensure good reporting standards and to improve the reproducibility of published results (http://www.nature.com/authors/policies/checklist.pdf), as well as a subsection on "Reporting Life Sciences Research" that summarizes several elements of methodology that are frequently poorly reported (http://www.nature.com/authors/policies/reporting.pdf).

drawings are usually clearer, and convey points better. Pictures supported by sketches are often a good combination. Photographs of gels, gel prints, etc. must occasionally be retouched or otherwise processed; if so, the processing used should be stated in the Material and Methods section, or given as supplementary information. The authors should be sure to adequately scale and explain added graphic items, such as designations, letters, and arrows. Microscope photographs should have a scale in the original submitted, so the indication of size is preserved, regardless of whether the photo is enlarged or reduced by the journal or by a reader who zooms in on the article on the Internet (Box 13.12).

It is important not to overload figures! Authors should have no more than three curves in a graph, or four if they are far from each other and do not cross each other several times. The curves may be identified by different symbols (circles, triangles, squares—open or solid) or by different line styles (solid, dashed, etc.), but not both! The writer should not show dissimilar parameters (such as body temperature and sedimentation rate) on the same *y*-axis, but rather use a left axis and a right axis, or draw several panels under each other in the same graph. Authors should consider simplifying axis designations, with supplementary explanations in the caption. For instance, every other number could be deleted from the axes. Designations appearing in the figure space can promote understanding and are excellent in slides, but are not always allowed in print. The same holds for a short title in the figure space above the figure.

But one should resist extravagance: Can two figures be combined into one? Can unessential parts of a photo be cropped out? Might the axis scales be made more economical? The axes need not extend beyond the maxima of the data plotted. Make numerals, letters, and symbols large enough that they are easily deciphered when read in print. Changes in scale should be indicated by discontinuity marks on curves and axes. But it is best if the curves are plotted to avoid breaks that may easily connote disorder (example in Box 13.9). It should be easy to see if an axis does not start at zero (or 1.0 for a logarithmic scale). Axes may be displaced to avoid data points falling on them. The *x*-axis may be deleted in bar charts. The spaces between bars should be less than their widths. Here as elsewhere, the author should examine how figures are drawn in a journal renowned in the discipline.

13.4 THE FINISH

13.4.1 Acknowledgments

The Acknowledgment section is a short section in which one should extend thanks for financial support, for example: "Support to HBB from the Norwegian Cancer Society and the Research Council of Norway is gratefully acknowledged." Occasionally it is appropriate to underscore the contributions of assistants: "We also thank NN and MM for excellent technical assistance."

13.4.2 Quality Control

The structure outlines in Box 13.3 should be used, and if not, the author should be able to explain why this structure was not followed. Most important, all data, figures, tables, text, and references cited should be checked. Reference control is simple and straightforward using EndNote, ProCite®, PubMed®, or similar database-based applications. Likewise, spell checkers and grammar checkers are useful aids, but they cannot replace conscientious review. The error in "the colleges raised there voices" will not be spotted by a spell checker. Authors should have a goal every time they read through their papers. Can long sentences be divided into two or three sentences? Can superfluous words, expressions, or sentences be weeded out (see Box 13.1)? Are there any sentences with too many prepositions, nouns that can be replaced by verbs, passive voice that can be replaced by active? Are all verbs in the correct tense? Has the author really tried to improve the writing style, such as by avoiding redundant or repetitive use of his favorite verb? A thesaurus is a great help here, for example, the verb "record" can be replaced by monitor, measure, analyze, assess, examine, determine, study, investigate, scrutinize, evaluate, explore, or probe. Are all the references cited in the text present in the reference list, and vice versa? Authors should carry out a final check of the instructions to authors; be their own personal reviewer through multiple readings, and *require* that coauthors also thoroughly read the entire paper. Once this has been accomplished thoughts can turn to submitting the paper.

13.4.3 Submission Cover Letter

A typical submission letter might read:

> *Dear editor NN (whenever possible get the name of the editor from the journal's website),*
>
> *Please find enclosed our manuscript entitled "...", which we wish to submit for your consideration as an Original Article in the* International Journal of XXX. *The investigation concerns "... ", which we feel falls within the scope of your journal. All data presented in the article are original data of the authors. The data have not been published previously. The manuscript has been reviewed and approved by all coauthors. This manuscript has not been and will not be submitted simultaneously to another journal, in whole or in part, in any language, and reports previously unpublished work. All named authors have agreed to submit the paper to the* International Journal of XXX *in its present form, and all those named as authors have made a sufficient contribution to the work to be considered as such. No author has any conflict of interest to declare. We hope that you will find this manuscript suitable for publication in the* International Journal of XXX, *and look forward to receiving your decision.*
>
> *Yours sincerely,*

When creating the cover letters authors should also consult the instructions to authors, to determine what information the journal requires.

Online submission is common now, and in this case a pdf version of the manuscript is submitted. There may be forms to complete about the contribution of the various authors and about any vested interests. The author should also determine whether he can keep the copyright, so that the option of open access publishing is not lost.

After submission and peer review, many manuscripts are returned to authors with requests for revisions. Indeed, manuscripts are seldom accepted without changes. Authors should revise according to those aspects of the critique with which they agree, as well as those they feel do not diminish the value of the manuscript. Thereafter, the editor's instructions should be followed. This may involve answering each referee on a separate sheet, point by point, with references to the locations (page or line numbers) where changes have been made in the text. If a referee has done a conscientious job, or has provided valuable suggestions for improvement, the author should acknowledge his/her contribution. Authors must be polite in these replies, even if the referee, in the light of anonymity or due to unfamiliarity, has not grasped the important points. When this happens the author should tactfully and delicately explain why the suggested changes have not been made.

13.4.4 Authorship

Authorship is a sensitive subject, and the situation is often asymmetrical. The scholarship holder or research fellow thinks that he or she has done the work, which may have included many late nights away from family. The adviser, who has built up the laboratory, worked on procedures and quality control, and acquired funds for the research—and moreover most likely hatched the idea for the project and supervised the data collection—is aware of his own indispensable role. Clearly, both should be co-authors, the research fellow first, as he or she drafted the manuscript. The situation worsens when the head of the institute, or a colleague who merely performed a particular analysis or delivered a particular reagent, also insists on being listed as a co-author. This is often the situation, and happens despite rules to the contrary published by the International Committee of Medical Journal Editors (ICMJE). (In 1978, a group of editors of general medical journals met informally in Vancouver, British Columbia, Canada, to establish guidelines for the format of manuscripts submitted to their journals. The group became known as the Vancouver Group. It subsequently expanded and became the ICMJE, which has a website where the rules are revised when appropriate.)

Authorship should be based on the following four criteria:

1. Substantial contributions to the conception or design of the work; or the acquisition, analysis, or interpretation of data for the work; AND
2. Drafting the work or revising it critically for important intellectual content; AND
3. Final approval of the version to be published; AND
4. Agreement to be accountable for all aspects of the work in ensuring that questions related to the accuracy or integrity of any part of the work are appropriately investigated and resolved (http://www.icmje.org/recommendations/, retrieved December 9, 2014).

> *The authorship criteria are not intended for use as a means to disqualify colleagues from authorship who otherwise meet authorship criteria by denying them the opportunity to meet criterion number 2 or 3. Therefore all individuals who meet the first criterion should have the opportunity to participate in the review, drafting, and final approval of the manuscript.*
>
> ***(ICMJE)***

13.5 POSTERS

Most meetings and symposia have poster sessions. A researcher might feel thwarted at not being selected to give a speech based on his or her results, but chosen instead to present at a poster session. However, the poster session does not constitute a subculture, and justifiably not. If an author manages to interest a session audience in his or her subject and results, the informal discussion of the poster can be more thorough and rewarding than it would be in a nervous session following an oral presentation. Moreover, he or she can gain friends and colleagues for life.

Regrettably, poster sessions are often poorly organized; one may have to cut down on a lunch break or meet late in the evening, when one is weary. One's work might then be undervalued. A poster session might be expedient for the researcher who must present something to have travel expenses covered, but this is not the best of motivations. Poster sessions can be superb, when the audience has the opportunity to study the posters undisturbed by simultaneous sessions, and with the posters of a particular sector grouped together.

BOX 13.13 The Content of Posters

When studying a poster the reader can linger over it as long as he or she wishes, and in this sense a poster presentation is more like a printed page than a slide. In most cases, however, the reader will be confronted with a large number of posters, and he will be expected to study them while standing up. Information must therefore be presented clearly and concisely so that the essentials of the message are easily grasped. Studies in museums have shown that visitors tire of reading long caption panels very quickly, and while they may begin by systematically reading each one, they soon start to wander at random. After this point they are likely to spend more time on the most visually "attractive" displays. Posters should stimulate interest rather than present complex details.

In effect, a poster is an advertisement for the author's particular ideas or techniques. He stands in front of his display with something to sell. Good posters use the best techniques of salesmanship. The content should be succinct and to the point, with short pithy subheadings, and the design should be attractive in terms of color, lettering and layout.

(Reynolds and Simmonds[6])

When preparing a poster one should take time to plan and sketch alternative layouts on letter-size or A4 sheets of paper. Details on poster dimensions and orientation (landscape or portrait) are furnished by the conference organizers. The poster should be designed like a newspaper page, with the title, problem or purpose description, and final conclusion in large letters. The most common mistake is to overload the poster with information (Box 13.13).

In posters authors can be daring; they need not refrain from choosing shorter, catchier, and more challenging titles than would generally be used in a paper. Figures from a paper should go through the same simplification process employed when preparing a PowerPoint presentation, and tables should also be far simpler than those of a paper. The message must be discernible by a weary watcher, without strain. Therefore one can delete measures of spread and confidence intervals and just state significant differences, for example, with star symbols. One may also curtail the description of the methods or provide more detailed information in smaller type than used elsewhere on the poster, and use handouts to provide supplementary details for those interested. Sometimes a small version of the poster can serve as a handout. Anyone who wants to know the SD of measurements can ask, and thus spark a discussion.

The Physiological Society in London once published guidelines for producing posters, some of which are thought-provoking or challenging.

The space available is limited, so the poster should have just one main point. All information should be readable at a distance of 2 m, which means that the amount of information is also limited. One should have no more than six illustrations that together take up no more than 50% of the poster, and the text on the poster should be kept to a minimum. The message of the poster should be obvious without supplementary verbal explanations. Acronyms, abbreviations, and slang should be avoided. The order of reading the material presented should be obvious.

Many scientists prefer a three-column poster in the customary portrait orientation. The letters of the title should be at least 4 cm high, subheadings at least 1.5 cm high, and text in the first and last sections—introduction (problem description) and conclusion—at least 1 cm high. Colors are commonplace, but ponder them well, as there should be an educational reason for their use. For instance, comprehension is eased if all the curves or bars representing test group data are the same color in all illustrations, while control group data curves and bars have another. Colors may also be used to highlight key points: the problem the author started with and its solution (conclusion) at the end. One should let the figure and table captions convey the results that tell the story.

Computer applications, such as PowerPoint and the Goliat® poster printer, are routinely used to produce posters. The trick is to resist the temptation to overload the poster with beautiful, colorful items and too much text. As Goethe[7] advised: "In der Beschränkung zeigt sich erst der Meister" (Restraint is the hallmark of the master). The Adobe InDesign application can also be used for poster production; guidelines are available on the Internet.

Even though the poster must be inconveniently rolled up for the trip to the conference, be sure to take it as carry-on baggage.

A good mnemonic rule is POSTER: P for "prepared and planned," O for "one main theme," S for "simple pictures," T for "tables minimal," E for "explains itself" and R for "readable at 2 meters" (Brown[8]).

A sample poster sketch is shown in Box 13.14.

13.6 SOME FINAL POINTS AND CAVEATS

13.6.1 Reviews and Perspective Articles

These papers vary a great deal, from pedagogical overviews for the general scientist via analytical exposures of a theme, to detailed reviews of a specialized field. In review and perspective papers more than all others it

BOX 13.14 A Recommended Poster Design

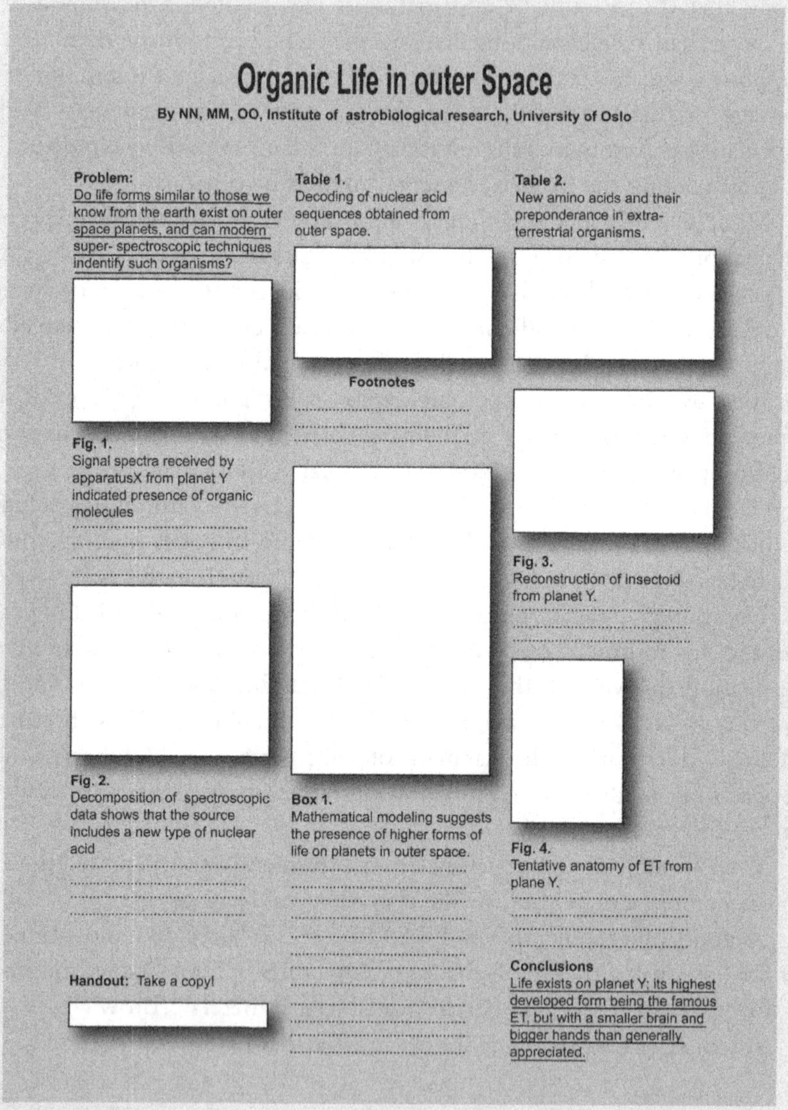

Organic Life in outer Space

By NN, MM, OO, Institute of astrobiological research, University of Oslo

Problem:
Do life forms similar to those we know from the earth exist on outer space planets, and can modern super- spectroscopic techniques indentify such organisms?

Fig. 1.
Signal spectra received by apparatusX from planet Y indicated presence of organic molecules

Fig. 2.
Decomposition of spectroscopic data shows that the source includes a new type of nuclear acid

Handout: Take a copy!

Table 1.
Decoding of nuclear acid sequences obtained from outer space.

Footnotes

Box 1.
Mathematical modeling suggests the presence of higher forms of life on planets in outer space.

Table 2.
New amino acids and their preponderance in extra-terrestrial organisms.

Fig. 3.
Reconstruction of insectoid from planet Y.

Fig. 4.
Tentative anatomy of ET from plane Y.

Conclusions
Life exists on planet Y; its highest developed form being the famous ET, but with a smaller brain and bigger hands than generally appreciated.

Figure 13.4 A Poster.

is important to consider the interests and needs of the target audience or readers, and adapt accordingly. Equally important is to study how these papers are done in the journal to which the author will submit—read the instructions to authors carefully. The general layout may be similar to that of short, scientific, original publications, which is detailed above, but there are certain differences. The text, after the introduction, is usually a "portioned story," with labeled topics and subtopics, and the Material and Methods and Results sections are replaced by paragraphs presenting new or different information.

The substitute for the Material and Methods section in reviews is a subsection that details the knowledge base of the review. Information on the literature databases used and the keywords used for literature search may be required (see Chapter 5), as well as the framing of the search (date performed, languages, type of studies, number of hits, selection criteria for the references chosen, etc.). The writer may even have to grade the quality of the evidence presented. A commonly used method is the Grading of Recommendations Assessment, Development and Evaluation (GRADE, http://ims.cochrane.org/revman/gradepro, retrieved December 5, 2013), which describes how evidence can be graded down or up. Downgrading is done when there is risk of bias, bad consistency between various results, a lack of direct relationships that the review finds in the database, or low data precision. Evidence is strengthened by strong associations in observational studies, clear dose-response effects, and certainty that all confounding factors would have reduced the estimated effect.

Results are most often given in general terms, without data presentation, and discussed as soon as they have been summed up.

The review may end with a section called Outlook or Conclusion (or both). And it may in fact have a layout that is similar to the one described in this chapter.

13.6.2 Accountability and Adaptation of the Message

Scientists are financed by either taxpayer or nongovernmental organizations, and they will increasingly be required to be accountable for their research and share their results with the lay public. They must carefully evaluate their target audience (Figure 13.5) to make sure their message is understandable; at least avoid jargon and acronyms. "What do or will they want to know?" is the all-important question.

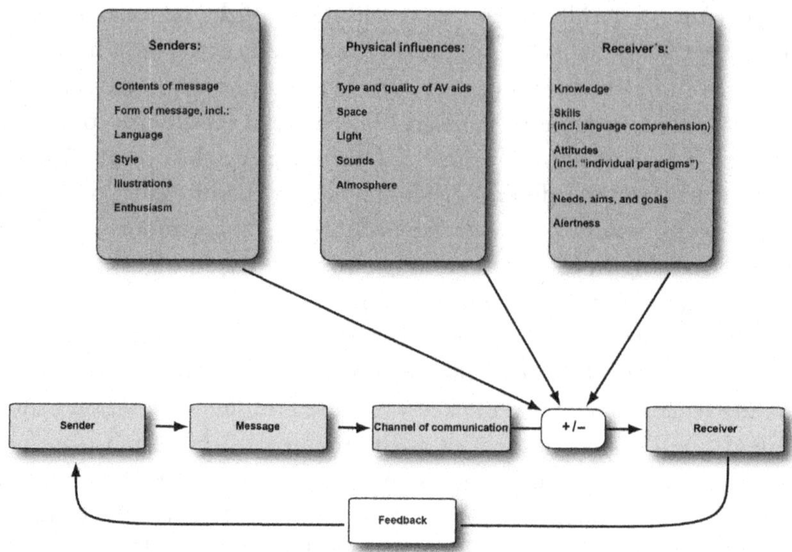

Figure 13.5 The communication process. Identity between original meaning of the sender and perceived meaning of the receiver is the ideal. The feedback may be formal (oral or written) or informal (spontaneous oral or behavioral responses); it may concern the communication process, the result (the learning), or both. AV, audiovisual.

13.6.3 First, Scientific Publishing!

Never approach a lay audience or a journalist with sensational discoveries before they have been published in a scientific journal! If an author breaks this rule, journal editors may refuse to consider the scientific version of the finds. There is a "gray zone" here, when an abstract is presented at a congress. What if a journalist attends the congress, grasps the potentially "revolutionary" meaning, adds parts of what the author has told the audience in his talk and turns it into a newspaper article? How can authors prevent the publication of (parts of) their results, which may exclude publication in a scientific journal? This may represent an unsolvable dilemma. Care must similarly be taken when publishing on the Internet (web page, Facebook, etc.).

13.6.4 New Formats

New formats for scientific articles increase the possibilities and choices for publication. Papers published on the Internet can be longer and more colorful. The same is true for supplementary material of print journals, which is often only published on the Internet. More graphs can be collected within the same figure, since details can be zoomed in and examined on the Internet.

13.6.5 Open Access

About half of all papers can now be accessed for free in some form, mostly in open access journals. Increasingly, funders require that scientific results from the projects they finance be published in an open access journal, made freely available on an official database, or freed from the publisher within 6 months to 1 year. To comply with such requirements, authors may have to secure the copyright to an article when submitting to a traditional journal. If an author submits to an open access journal, care should be taken to avoid the multitude of "pirate journals" that have recently arisen. These journals do not have proper peer review procedures, and their goals are purely financial. Open access publishing will—if institutional or other funds for it do not exist—have to be paid for by the author, so it is useful to include these expenses in applications to funding bodies.

13.6.6 Clinical Trial

In order to conduct a clinical trial, a researcher must publish the project description in a designed, official register such as www.clinicaltrials.gov before the trial begins. Otherwise he or she will not be allowed to publish results in traditional journals. And at least in the United States, clinical results must also be deposited within 1 year of completion.

13.6.7 New Types of Scientific Publications

New possibilities are continually arising. The journal *Nature*[9] states that "everyone wants better ways to make research data available and to give more credit to researchers who create and share data"; its owner has therefore launched *Scientific Data*, an open access, online-only journal (http://nature.com/scientificdata) for descriptions of data sets that are detailed enough to make data reusable.

QUESTIONS TO DISCUSS

1. Can you give some examples of verbose formulations and how they could be simplified?
2. Why and when should you do a power analysis or a Bonferroni multiple comparison correction?

3. When is it appropriate to give SD, and when SE(M)?
4. When is it best to present data as a table rather than a graph?
5. What determines the number of significant digits?
6. What do you think about the "Vancouver rules" for authorship?
7. What are the differences between manuscripts intended for scientific journals and for public outreach?
8. Name some important differences between a scientific article and a poster presentation.
9. What are the pros and cons of open access publishing?
10. What conditions would you require in order to submit your data, in a generally usable form, on a public database?

ACKNOWLEDGMENTS

I thank Kristin Larsen Sand for making the panel graphs of Figure 13.1 and Carina V. S. Knudsen for adapting and photographing all the figures, both of the University of Oslo.

REFERENCES

1. Strunk Jr. W, White EB. *The elements of style*. 4th ed. New York, NY: Longman; 2000.
2. Morgan PP. Can a speech make a good journal article? *Can Med Assoc J* 1983;**129**:317.
3. Ehrenberg K. Rudiments of numeracy. *J R Statist Soc A* 1977;**140**(Part 3):277–97.
4. Tufte ER. *The visual display of quantitative information*. Cheshire, CN: Graphic Press; 1983.
5. Editorial. Not picture-perfect. *Nature* 2006;**439**:891–2.
6. Reynolds L, Simmonds D. *Presentation of data in science. Publication, slides, posters, overhead projections, tape-slides, television. Principles and practices for authors and teachers*. The Hague: Nijhoff; 1982.
7. Goethe JW. *Das Sonnet*, 1802: written for the opening of the new theatre at Lauchstaedt on 26 June 1802, second part, line 13.
8. Brown BS. Communicate your science! ... Producing punchy posters! *Trends Cell Biol* 1996;**6**:37–9.
9. Anon. Launch of an online data journal. *Nature* 2013;**502**:142.

FURTHER READING

Briscoe MH. *A researcher's guide to scientific and medical illustrations*. New York, NY: Springer-Verlag; 1990.

Ebel HF, Bliefert C, Russey WE. *The art of scientific writing. From student reports to professional publications in chemistry and related fields*. Weinheim: VCH; 1987.

Gowers E. *The complete plain words*. Middlesex: Penguin Books; 1971.

Gustavii B. *How to write and illustrate a scientific paper*. Lund: Studentlitteratur; 2000.

Simmonds D, Reynolds L. *Data presentation and visual literacy in medicine and science*. Oxford: Butterworth-Heinemann; 1994.

Style Manual Committee/Council of Biology Editors. *Scientific style and format. The CBE manual for authors, editors, and publishers*. 6th ed. New York, NY: Cambridge University Press; 1994.

CHAPTER 14

Successful Lecturing

Heidi Kiil Blomhoff
Department of Biochemistry, Institute of Basic Medical Sciences, University of Oslo, Blindern, Oslo, Norway

14.1 INTRODUCTION

Today we are surrounded by computers, mass media, and audiovisual (AV) aids, but the spoken word remains the cornerstone of communication. There are daily reminders aplenty of this fact, as the success or failure of politicians, lawyers, and other public figures depends on what they say and how they say it. Verbal skills are no less important for us scientists in presenting our research results and hypotheses for more or less interested audiences. Indeed, the root of the word "lecture" is the Latin *lectura*, meaning "to read." Unquestionably, verbal communication skills remain vital.

Though lectures are often criticized as a teaching method, they are still used in most universities for the large majority of courses. Lecturing is considered to encourage mainly one-way communication, placing the students in a passive role, and as such it often stands in contrast to active learning. However, lectures can be successfully used to rouse interest in a subject, and are well suited for structuring and clarifying difficult subjects for an audience—be it students or other listeners. What remains clear is that successful lecturing requires effective writing and speaking skills.

Certainly, we know from experience that even the most highly motivated listener can have difficulties grasping everything that is presented in a lecture. We also know that audience attentiveness varies during the course of a lecture, as illustrated in the attentiveness curve for a typical 45 min lecture (Figure 14.1).

At first glance, the high-to-low swing of the attentiveness curve may discourage a lecturer, but it is actually a valuable aid when preparing a lecture. The highs of the attentiveness curve show that lecturers can convey their message most easily in the first few minutes and toward the end of a lecture. The course of the curve implies that there is always room for improvement in lecturing.

The goal of this short chapter on lecture methods is to give you tips and advice on being a successful lecturer, so that you can actually enjoy

Research in Medical and Biological Sciences
DOI: http://dx.doi.org/10.1016/B978-0-12-799943-2.00014-8

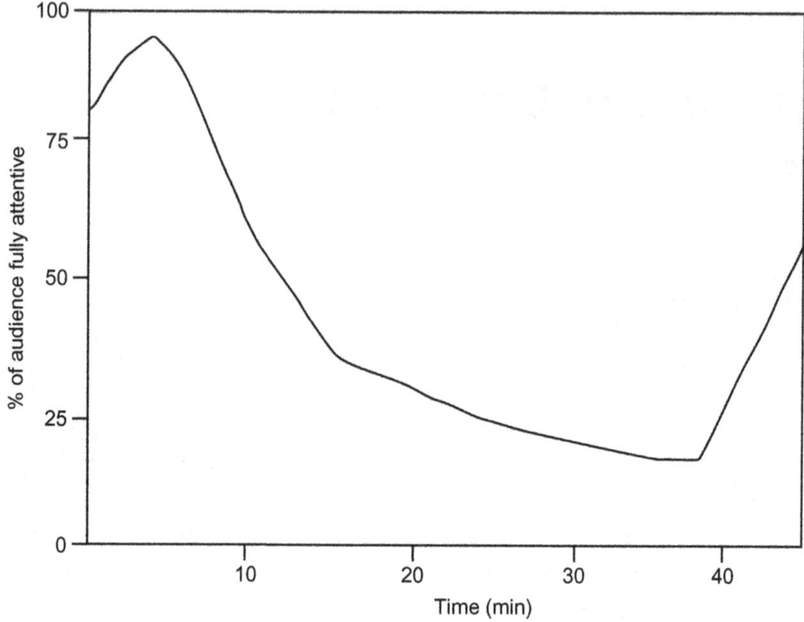

Figure 14.1 Audience attentiveness during a typical 45 min lecture.

lecturing, and most importantly so that an audience will enjoy listening to you. Success in lecturing requires preparation, preparation, and more preparation. You must prepare the lecture, and you must prepare and practice your lecture delivery.

14.2 PREPARING THE LECTURE

Before you start your preparation, be aware that preparation comprises two essential aspects. First, you must start well in advance, and second, you must gauge your audience (Box 14.1).

14.2.1 Take Time to Prepare

A good lecture is the result of a maturation process in which both the material to be presented and the manner of presentation improve with the time taken to prepare. Devote a few weeks to preparation, not continuously—none of us have time for that—but begin well in advance, so the material can mature in your subconscious. Unless you are a divinely inspired lecturer, a lecture prepared the evening before it is presented is seldom successful. The earlier you prepare the lecture, the more assured

BOX 14.1 Preparing the Lecture

- Take time
- Gauge your audience
  - type of lecture
  - audience level of knowledge

you will be about the message it conveys, the more time you will have to practice, and the more confident you will be of your delivery. Remember that the shorter the lecture, the more cautious your preparation should be. Indeed, even the most skilled of lecturers might do well to heed this remark by Winston Churchill: "If you want me to speak for two minutes, it will take me three weeks of preparation. If you want me to speak for thirty minutes, it will take me a week to prepare. If you want me to speak for an hour, I'm ready now."

14.2.2 Gauge Your Audience

Many lecturers follow the pretence that audiences are impressed by a wealth of details supported by intricate hypotheses. In fact the opposite is true; listeners are happiest when the material is so simple and clearly delivered that they can follow all parts of a lecture. Accordingly, as lecturers, it is vital that you gauge your audience. You must consider both the *type of lecture* to be given and the *level of knowledge* of the audience.

Whether the type of lecture is a university course lecture, a talk at a review seminar, or a presentation of your own work at an international meeting, the type of lecture dictates the way you assemble it. A university course lecture or a talk at a review seminar entails little reference to your own data, while a presentation at a meeting usually involves just your results, set in the perspective of a particular research sector.

In assessing the level of knowledge of your audience, you should determine whether they are experts in your field, biomedical researchers from a broad range of fields, clinicians, or students. The level of knowledge of your audience is decisive in determining the level of detail to use, the amount of background information to give, and the amount of data to present. The broader the audience, the more background information must be included and the more results must be explained along the way. Obviously, a lecture for experts in your field may include far more results and offer many more hypotheses than would a lecture for listeners with little background in your research field. REMEMBER: In lecturing, it is almost impossible to underestimate your audience; as a general rule, lecturers overestimate.

14.3 LECTURE CONTENT AND FORM

The rest of this chapter focuses on a scientific lecture presentation of research results. We lecturers in biomedical and clinical research are fortunate in that we can usually build our lectures around illustrations. In a lecture, illustrations may be used in the introduction, in the presentation of background material, results, and importantly, in the final summary of results. Even though illustrations ease the structure of the scientific lecture, there are still some general ground rules to follow.

14.3.1 Amount of Material

Assessing the right amount of material for a lecture is one of the greater challenges. Lectures are usually built around illustrations in the form of slides as part of a computer presentation. As a general rule, no more than 25–30 of these slides should be included in a 45 min lecture. The exact number varies, of course, depending on the complexity of the material presented. You can get a clue as to the right amount of material to present by reading the lecture aloud to yourself, speaking at a rate of no more than 80–100 words per minute. This will help you determine the right amount of material to include. That said, it is wise to leave yourself some leeway by preparing material that will require 10% less time to present than you are allocated. Indeed, unless you are not too nervous, live delivery always takes longer than reading aloud to yourself (Box 14.2).

14.3.2 Lectures Versus Journal Articles

Scientific lectures differ from scientific journal articles. Compared to a journal article, a lecture should place greater emphasis on introductory

BOX 14.2 Match the Amount of Material to the Time Allocated

- Maximum 25–30 slides per 45 min lecture
- 80–100 spoken words per minute
- Aim for 10% less time than allocated

BOX 14.3 Lectures Versus Journal Articles

Compared to an Article, a Lecture has:

- more introductory and background information
- a shorter description of methods
- a shorter discussion

10 min Lecture:

- 3 min for introduction or background information
- 5–6 min for results
- 1–2 min for summary and conclusions

BOX 14.4 Lecture Structure

- Start with the basics
- Build from the simple to the more complex
- Logical sequences of short points
- Build the lecture around illustrations
- Make summaries and conclusions along the way
- Repeat essential points

and background information, and less emphasis on describing methods (unless the lecture is on the development of a methodology) and on discussion of data collected. A common approach for a 10 min lecture is summarized in Box 14.3.

14.3.3 Lecture Structure

Start with the basics and go from the simple to the more complex. The obvious goal is for as many listeners as possible to grasp the lecture from the very start, and that is easier to achieve if you start with the basics. Overestimating the audience from the start may irritate and discourage them so much that their attention is lost for the rest of the lecture (remember the attentiveness curve in Figure 14.1) (Box 14.4).

A logical structure promotes understanding; do not jump back and forth between topics, but try to have a sequence of short points. The easiest way to do this is to build the lecture around the slides in your computer presentation. With the attentiveness curve of Figure 14.1 in mind, it is wise to include summaries and conclusions along the way. This strategy allows listeners who have lost the thread of the lecture to get back on track, and makes it easier for all listeners to remember your main points. It is also wise to repeat essential points several times during the course of a longer lecture.

14.3.4 Introduction

The introduction is without question the most important part of the lecture, so work hard on it. A good introduction rouses interest, and, as mentioned above, most easily conveys your message. Audience attentiveness will be high in the first few minutes, so if you first rouse attention you will have won half the battle.

A chairperson may introduce you and your topic, but if not, you may do that yourself at the start of your lecture. Here it is best to be succinct, before you launch into the relevant background details of your topic. The background details should enable the listeners to understand the data you will present, and should link directly to your aims or research questions. Take your time when delivering the background details. The shorter the lecture, the greater the portion devoted to background details should be. The aims should be presented as precisely and briefly as possible, point by point, using a bulleted list. If you can point to a broader implication or significance, do so (Box 14.5).

14.3.5 Illustrations

As mentioned above, it is beneficial to build lectures around illustrations, but one rule applies to all illustrations: Keep it short and simple (KISS)! (Figure 14.2) (Box 14.6).

Having few points per slide allows the audience to better follow what you are presenting. As a rule, it is best to present a message in a simple figure rather than in a bulleted list (which is the alternative). Most people can more easily grasp a graphic figure than a long text, as can be visualized by comparing the impact of Figure 14.3 with that of the text in Box 14.7.

BOX 14.5 Introduction

- Vital! Rouse audience interest.
- Start by introducing yourself and your topic.
- Take your time presenting the background details.
- Present the aims precisely, briefly, using a point-by-point list.

Figure 14.2 Illustrations: remember KISS—keep it short and simple.

BOX 14.6 Illustrations

KISS—Keep it Short and Simple

- Few points per slide
- Figures are preferable to text
- Figures are preferable to tables
- Colors and symbols are fine, but avoid excesses. Think of legibility!
- Use large fonts and clear typefaces, such as Arial or Comic sans, with uppercase and lowercase letters; do not use all capitals ("ALL CAPITALS").
- Place slides and computer presentations in landscape orientation.

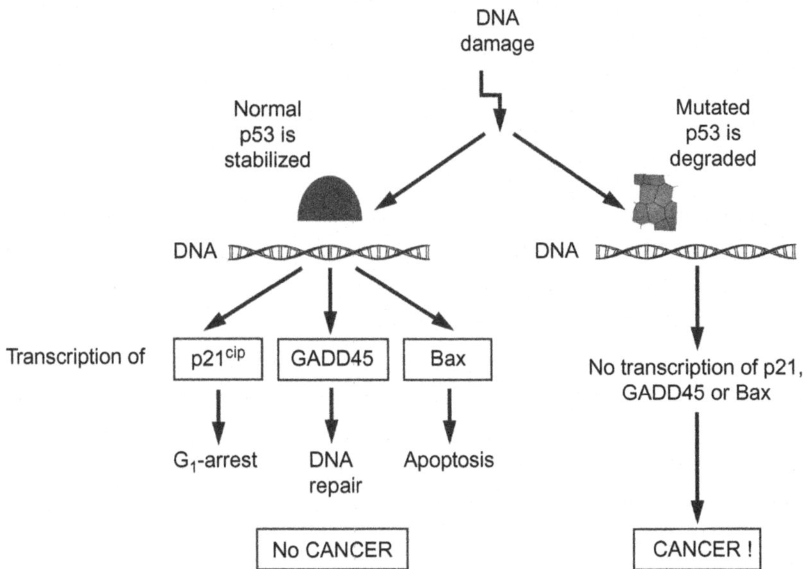

Figure 14.3 Figures are easier for an audience to grasp than text. Here: p53—the guardian of the genome.

BOX 14.7 Figure or Text?

p53—Genome Guardian

- p53—a transcription factor induced upon DNA damage, through protein stabilization.
- p53 induces transcription of p21^{cip} (CDK-inhibitor). Cells arrest in G_1.
- p53 induces transcription of GADD45 (DNA repair protein).
- p53 induces transcription of BAX (apoptosis-inducing protein). If the DNA damage is not repaired, the cell dies by apoptosis. Cancer development is hindered.

Upon Mutation of p53:

- p53 not stabilized upon DNA damage.
- p53 does not transcribe p21^{cip}, GADD45, or BAX.
- The cell replicates the damaged DNA, and the damaged cell does not die of apoptosis. Cancer may develop!

Furthermore, the take-home message in a figure is generally easier for an audience to comprehend and remember than is a table (compare Table 14.1 with Figure 14.4).

Font sizes and typefaces also are vital. Choose large fonts, do not use all capitals ("ALL CAPITALS"), and choose clear typefaces, such as Arial or Comic Sans, that are readily legible from a distance. Orientation should be natural; slides in a computer presentation should be prepared in landscape orientation.

14.3.6 Concluding a Lecture

As illustrated in the attentiveness curve of Figure 14.1, some listeners whose attentiveness may have waned on the way can be regained in the final minutes of a lecture. Since many listeners suddenly realize towards the end of the lecture that they have missed key parts of it (...sound familiar?), you now have a golden opportunity to reclaim this part of the audience. This is most easily accomplished by summarizing the key results you have presented, as well as stating their relevance to the initial aims or research questions. Thereby the wheel comes full circle, and even those who were attentive only at the start and at the end of the lecture will have received your message.

Table 14.1 Effect of forskolin on DNA-synthesis in B-cells

Stimulation of cells	DNA-synthesis (^{3}H-thymidine incorporation)		
	24	**48**	**72 (h)**
Medium	150	440	1560
SAC	1400	24,050	46,800
SAC/forskolin 10 μM	1050	17,360	30,654
SAC/forskolin 50 μM	850	12,406	16,388
SAC/forskolin 100 μM	650	8040	10,446

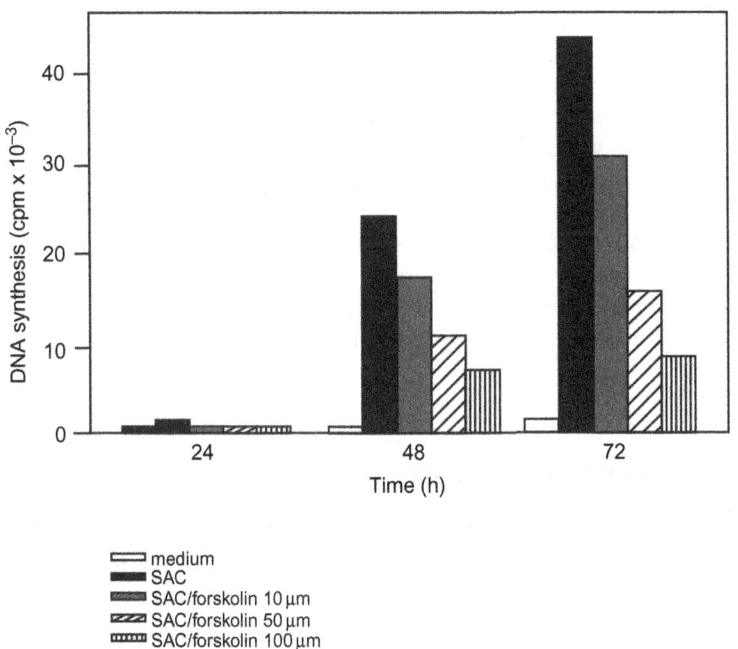

Figure 14.4 Figures are easier for an audience to grasp than tables. Here: effect of forskolin on DNA synthesis.

The summary and presentation of conclusions must be short and precise, and preferably presented in a point-by-point manner; simple cartoons also work well here. Your message can be underscored if you conclude the lecture by outlining the future prospects for your research, perhaps emphasizing their transfer value to other research areas (generalization). Do not forget to list your coauthors! (Box 14.8).

BOX 14.8 Concluding a Lecture

- Vital! Conclusions are remembered.
- Tie the conclusion to the introduction—the wheel goes full circle.
- Summary and conclusions presented as
  - short sentences (point-by-point)
  - simple figures
- Further prospects?
- List of coauthors

BOX 14.9 A Lecture Manuscript Is Reassuring!

- When compiling a lecture manuscript:
  - start with the "whole text"
  - highlight or underline the principal points
  - write a short version
  - write a catchword manuscript
- For a good catchword manuscript:
  - write only on one side of a sheet of paper
  - use large fonts—preferably marked in various colors

14.4 LECTURE MANUSCRIPT

Now you have finished structuring the lecture, but before you start working on delivery, you should decide what sort of lecture manuscript you will use. My advice to beginners is to have a lecture manuscript at hand during the lecture, but not to use it. Knowing that the lecture manuscript is readily accessible will provide reassurance (Box 14.9).

One of the best approaches to writing a lecture manuscript is to start with the "whole text," that is, the entirety of what you plan to say. Then highlight or underline the principal points, before you go on to a short version. Finally, compile a catchword manuscript (example in Box 14.10). Of course, you are free to choose the type of backup manuscript you prefer, but I recommend that you use a catchword manuscript if you need an aid in lecturing. Write the catchword manuscript only on one side of a sheet of paper. One handy trick is to use smaller, stiffer filing cards; another is to write with large fonts and use colors to highlight different sections.

BOX 14.10

Lecture manuscript—whole text	Catchword manuscript
First, **thanks** for inviting me to present my results at this meeting.	– Thanks for the invitation
In **our laboratory** we have worked for years with **regulating proliferation and cell death (apoptosis) of lymphoid cells**, and we have focused on studying **how elevated levels of cAMP influence these processes in normal versus malignant lymphocytes**.	– Our lab: Proliferation and apoptosis in lymphocytes – Special interest: Role of cAMP
cAMP is an important physiological regulator of normal lymphocytes. Induction of cAMP such as by prostaglandins, is known to inhibit proliferation of lymphocytes, and it has been shown that stimulation of normal lymphocytes causes secondary increase in cAMP level.	– cAMP: physiologic regulator in lymphocytes, prostaglandines inhibit proliferation
cAMP acts as a **secondary messenger** and activates **cAMP-dependent protein kinase (PKA)**, which is known to inhibit cellular activation of lymphocytes by activating tyrosine kinase ***csk***, which in turn inhibits the central tyrosine kinase ***lck***.	– cAMP: Secondary messenger, PKA – Csk activated, *lck* inhibited
Normal lymphocytes isolated from peripheral blood are in a **rest phase, called G_0**, and for the cells to divide, they must be **activated so they enter the G_1 phase, and subsequently the S, G_2 and M phases**.	– Normal lymphocytes in G_0 – Activation into G_1, S, G_2/M
The principal control point of this process is the **restriction (R) point in G_1**, in which cells elect to continue in the cell cycle or remain in G_1. For cells to pass the R point, they must receive signals from **mitogens, for example as in the form of growth factors**.	– R-point in G_1—mitogens/ growth factors
The R point itself is known to involve **phosphorylation of a tumor-suppression protein called pRB**. Phosphorylation of pRB causes **release of**	– R-point in G_1—phosphorylation of pRB – pRB phosphorylated-free E2F-S-phase genes – Phosphorylation of pRB—closely regulated

(*Continued*)

BOX 14.10 cont'd

a transcription factor (E2F), which is necessary for a series of S-phase-specific genes to be transcribed.

Phosphorylation of pRB is a **closely regulated process** that involves activation of the **cell-cycle machinery**. pRB is sequentially phosphorylated by various cyclin-dependent kinases **(CDKs)**, and these CDKs are in turn activated by specific cyclins, and inhibited by specific inhibitors called **CDK inhibitors (CKIs)**.

- Cell-cycle machinery:
  - pRB phosphorylated by CDKs
  - CDK activated by cyclins
  - CDKs inhibited by CKIs

- We have previously shown: cAMP causes G_1 arrest

We have previously shown that cAMP via PKA causes cells to **arrest in the G_1 phase** of the cell cycle, and the aims of the work I will present today are the following:

- How can increased levels of cAMP cause G_1 arrest in normal lymphocytes?
- How does cAMP influence the cell cycle machinery that regulates the R point in G_1?

Current aims:

- How can increased levels of cAMP cause G_1 arrest in normal lymphocytes?
- How does cAMP regulate the cell = cycle machinery at the R point in G_1?

The following components of cell-cycle machinery will be investigated:
pRB phosphorylation
levels of various cyclins in G_1
levels of various CKIs in G_1
activity of various CDKs in G_1
free versus bound E2F

Phosphorylation of pRB in G_1:

- Level of G_1 cyclins
- Level of CKIs in G_1
- Activity of CDKs in G_1
- Free versus bound E2F

14.5 DELIVERING A LECTURE

14.5.1 Preparation

As mentioned above, it is wise to start preparing well in advance. When you have finished preparing a lecture, it is time to practice its delivery. The more time you devote to practice, the more confident you will be as a lecturer. Again, thorough preparation is essential. Just think of how much actors and actresses rehearse their roles! (Box 14.11).

BOX 14.11 Delivering a Lecture

Prepare Well:

- Mentally—learn to master nervousness
- Rehearse—take your time, get feedback
- Be familiar with the lecture room and AV aids

BOX 14.12 Nervousness

- Everyone is nervous, more or less
- A higher adrenaline level gives you an extra, useful kick
- Learn to master nervousness
  - by being well prepared
  - be positive—boast to yourself!
  - keep a lecture manuscript at hand (but use it only in an emergency)

14.5.2 Mental Preparation to Counter Nervousness

Most people are nervous before they perform, be they actors, musicians, or lecturers; it is a natural, human reaction. However, it is vital that you do not allow nervousness to overpower you, rather that you learn to master it. Nervousness is marked by accelerated heart rate, butterflies in the stomach, shaky legs, and a dry mouth. These symptoms are due to an increased adrenaline level, and, in fact, to a moderate degree they are a healthy reaction for a lecturer. The adrenaline gives you the kick you need to go just a bit beyond your limits. But you need to learn to master your natural nervousness, by using a few simple tricks (Box 14.12).

The most important and easiest way to combat nervousness is to know that you are well prepared. If you have devoted enough time to preparation, your lecture will have matured in your subconscious, and you will have sufficient time to rehearse your delivery. Simple steps such as these increase confidence and combat nervousness. Boast to yourself. Tell yourself that you are the one in the lecture hall that knows your topic best and that you have been asked to lecture because of your professional expertise. Think positive thoughts, like what a joy it is to communicate and discuss your interesting results with others (Figure 14.5).

Having a lecture manuscript at hand also diminishes nervousness. If the "worst" happens and you lose the thread of your presentation, just leisurely pick up your lecture manuscript. Then you can continue

Figure 14.5 To reduce nervousness, boast to yourself: "I am the expert."

lecturing, aided by the manuscript, or just glance at it and then continue lecturing as if nothing happened. That certainty should make you confident. As long as you have a lecture manuscript at hand, you cannot go wrong!

14.5.3 Rehearse

The value of rehearsing the delivery of a lecture cannot be underestimated. Through rehearsing you gain confidence and combat nervousness. Rehearse aloud, so you are accustomed to hearing your own voice. Use students or colleagues as an audience and accept all the feedback they offer. Time yourself to be sure that you have correctly gauged the time required to cover your material. One good trick is to mark your lecture manuscript at the halfway point and check your timing against that. This will teach you to adjust your delivery speed as necessary during the lecture (Figure 14.6).

Figure 14.6 Time your lecture.

BOX 14.13 Familiarize Yourself with the Lecture Room

- Order the AV aids you need
- Check that they work
- Learn how to operate the lighting and the sound system
- Laser pointer—a fine aid, but learn to use it correctly

14.5.4 Familiarize Yourself with the Lecture Room

Many lecturers, even those who are more experienced, overlook the importance of being familiar with the lecture locale. Undoubtedly, you have seen lecturers delay seminars or lecture sessions because they are unfamiliar with the AV aids, such as microphones, remote controls, or laser pointers. The result can be embarrassing pauses, and you often see that the lecturer is unnerved. So always take the time to check such matters in advance. This makes you appear more experienced and less nervous, and means you will not cause unnecessary program delays. This is especially relevant to computer presentations, as the corresponding technologies still create problems for lecturers and meeting organizers alike. A small side note: If you use a laser pointer, turn it off when you are not using it to point at a board or screen. A laser beam can injure the audience's eyes if you let it wander over them as you lecture (Box 14.13).

Figure 14.7 Remember your appearance when you lecture.

BOX 14.14 Nonverbal Communication: the 55-38-7% Rule

- 55%—your stage presence, eye contact, movement, appearance, clothes, etc.
- 38%—your voice (pitch, intonation, regularity)
- 7%—what you say, your message

14.5.5 Nonverbal Communication

Before you go forth to lecture, be aware that the overall impression you give has a significant effect on how well the audience receives your lecture. It has been said that an audience gains its first impression of a lecturer within 90 s, and regrettably, the greater part of that impression depends not on what you have said, but rather on nonverbal communication. Hence the so-called 55-38-7% rule: 55% of the impression you give depends on your stage presence, your eye contact, your movements, your appearance, your clothes, etc.; 38% comes from your voice, that is, your pitch, intonation, and regularity; and a mere 7% comes from what you actually say! Perhaps this is discouraging, but it is all the more reason to carefully choose your style (Figure 14.7) (Box 14.14).

14.5.6 The Delivery Itself

Now that you have prepared your lecture in all the ways outlined in this chapter, and you know the lecture room and its AV aids, you are ready to deliver the lecture. Again, there are small tricks that can improve a lecture, both for you and for your audience. For example, a relaxed posture (within reasonable limits) helps you seem more at ease. And if you seem at ease, the audience may follow your lead, and the atmosphere will be friendly. Take a few deep breaths before you start your delivery in a relaxed, informal tone. This helps you to relax, combats nervousness, and helps you gain control of your voice (Box 14.15).

It is best to wait for the audience to quiet down before you start. Distracted listeners give you, the lecturer, an uninspiring start, and, as mentioned above, it is best if everyone is attentive from the beginning. Consciously make eye contact with the audience. If the light is not blinding, let your gaze wander over the audience, so you will have looked at everyone during the course of your lecture. However, be sure not to choose a "victim," a single person to whom you seem to lecture (Box 14.16).

Refer to a lecture manuscript if you wish, but do not read your lecture from it. The easiest way to avoid this is to prepare a catchword manuscript. Rapid speech is hard to understand, so talk at a rate of no

BOX 14.15 Delivering a Lecture, Part I

- Assume a relaxed posture.
- Take a deep breath before you begin, but do not sigh!
- Wait for the audience to quiet down before you start.
- Start with a relaxed, informal tone.
- Make eye contact with the listeners, also those in the back rows!
- Let your glance wander over the audience, but do not focus on a single "victim."

BOX 14.16 Delivering a Lecture, Part II

- If you have a lecture manuscript, do not read from it directly.
- Maximum of 80–100 words per minute
- Speak loud and clear, vary intonation.
- Take brief breaks and underscore essential points.
- Do not talk with your back to the audience.
- Point at the board or screen only when necessary.
- *Smile—show enthusiasm*!
- Be certain but not arrogant.

more than 100 words per minute. As you undoubtedly know from experience, it is easier to follow a lecture when the lecturer varies intonation, takes brief breaks, underscores key points, and overall tries to draw the audience along throughout the lecture. An audience will be more motivated if the lecturer smiles and seems enthusiastic. Remember that the lecturer is the salesperson of the lecture's message! Again, there is a fine line. An audience will be favorably impressed by a lecturer who is certain, but unfavorably impressed if the certainty becomes arrogance. Also note that it is OK to make mistakes, just fix them. If you drop your lecture manuscript or the laser pointer, or go backward instead of forward in your computer presentation, try to overlook it. Do not apologize every time or make funny noises, simply pick up what has dropped and go on with the presentations as if nothing had happened (Figure 14.8).

Allowing sufficient time for questions is also worthwhile. Answering questions well is an art in and of itself, but that art may be problematical at international meetings where it may be difficult to understand the questioners. Politeness is the solution; there is nothing wrong with asking that a question be repeated. Another check is for you, the lecturer, to repeat the question in your own words, as you have understood it. After repeating the question you can answer it, keeping in mind that the question and your answer should be addressed to the entire audience, not just to the listener who asked. Avoid arrogance in your answers. An answer such as "I believe I covered that earlier in the lecture" makes the

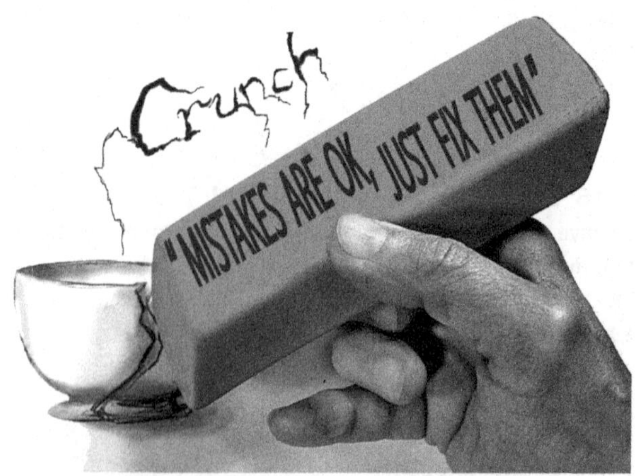

Figure 14.8 Mistakes are OK, just fix them.

questioner feel stupid and exposed to contempt, when in fact you may be responsible for the person not having grasped what you said. Instead, you might say: "Sorry that I didn't make that sufficiently clear, but the. . .." In this way, you lower the threshold for others to pose questions.

On the whole, the tips to keep in mind when you are lecturing also apply to any communication between people. Try to be as positive, polite, and humble as you can! But this is where the comparison with ordinary conversation ends, because in the environment of a lecture, you are trying to "sell" a message. You must, of course, be completely honest and informative in your account of results and hypotheses, but it is also essential that you try to appear certain in your presentation. It is always easier to convince an audience of your message if it is delivered well and with confidence. These skills must be learned, and the only way to learn is to prepare well when asked to lecture, and then practice, practice, practice . . . Good luck!

ACKNOWLEDGMENTS

Graphic illustrations were prepared with the expert assistance of Carina V. S. Knudsen, Institute of Basic Medical Sciences, University of Oslo.

QUESTIONS TO DISCUSS

1. At what time during a 45 min lecture is the audience most attentive?
2. Discuss with your fellow students how your own attentiveness is most effectively roused throughout a 45 min lecture.
3. What does KISS mean in relation to lecture preparation?
4. Prepare two to three slides that present selected parts of your own research project for your fellow students.
5. What kind of a lecture manuscript would you choose for yourself (if any)?
6. Prepare a lecture manuscript for presentation of your research project, and discuss its correct use with your fellow students.
7. What can you do to master nervousness when giving a lecture?
8. Practice—using the slides you already prepared for question 4—how to use a laser pointer while still making eye contact with your audience.
9. Make note of the speed (words per minute) at which you deliver a 15 min presentation of your own research project, and discuss with

your fellow students how well this coincides with the advisable speed for delivering a lecture.

10. What would you do if you "lost the thread" while delivering a lecture?

FURTHER READING

Bakshian Jr. A. *American speaker. Your guide to successful speaking*. Washington, DC: Georgetown Publishing House; 1998.

Benestad HB. Communicative skills—a matter of negligence? *Exp Hematol* 2000;**28**:1–2.

Kenny P. *A handbook of public speaking for scientists and engineers*. Philadelphia, PA: Institute of Physics Publishing; 1998.

Pihl A, Bruland ØS. Oral presentation in science and medicine—an art in decline? *Anticancer Res* 2000;**20**:2795–800.

CHAPTER 15

Guide to Grant Applications

Bjorn Reino Olsen[1], **Petter Laake**[2] **and Ole Petter Ottersen**[3]
[1]Harvard School of Dental Medicine, Harvard Medical School, Boston, MA, USA
[2]Oslo Centre for Biostatistics and Epidemiology, Department of Biostatistics, Institute of Basic Medical Sciences, University of Oslo, Blindern, Oslo, Norway
[3]Department of Anatomy, Institute of Basic Medical Sciences, University of Oslo, Blindern, Oslo, Norway

15.1 INTRODUCTION

The need to seek financial support for research projects is a harsh reality that makes itself known early in the career of a biomedical scientist. In many countries, PhD students are expected to apply for financial support, and in countries with a societal commitment to biomedical research the writing of competitive fellowship applications to national and international grant agencies and foundations is now commonplace at the postdoctoral level.

Even beyond the postdoctoral level, continued success in research for both junior and senior researchers requires a constant, intense effort to acquire financial support, regardless of whether one works at a university, a research institute, or a hospital. Writing successful grant applications is as essential to scientific success as formulating good research questions, testing them by experimentation, and publishing the results. But for many scientists writing and preparing grant applications is like a dark cloud over a bright landscape: necessary, but stressful and unpleasant.

One reason for the stress associated with grant writing is that grant applications are designed to support work that is planned but not yet done. Indeed, truly groundbreaking, innovative research cannot be planned; one can only optimize the general conditions in which it can take place. In some ways it is a form of self-deception to believe that the process of scientific discovery (at least in basic biomedical research) can be planned to the extent required by grant agencies, many of which mandate that applicants define and predict a timeline for discoveries (sometimes called "milestones" or "deliverables"). Therefore, writing grant applications that conform to the required formats frequently involves making promises that researchers know cannot be kept. In order to perform innovative basic biomedical research, this is an unfortunate necessary evil, which is at odds with the open and honest mindset of the most talented researchers, and thus generates stress.

Research in Medical and Biological Sciences
DOI: http://dx.doi.org/10.1016/B978-0-12-799943-2.00015-X

The review process can be another stressful aspect of grant writing. Indeed, during the review process scientific peer groups or administrators evaluate the personal ideas, speculations, hypotheses, and intuitive concepts the researcher puts forth about the unknown and how it should be explored. In essence the researcher must expose his or her scientific "soul" to convince reviewers that the ideas proposed are sufficiently important and innovative to warrant financial investment. In many cases, the rejection of a funding proposal is tantamount to a rejection of these personal ideas. Such rejection can shatter a researcher's confidence much more forcefully than the rejection of a manuscript submitted to a professional journal, as such rejection often amounts to disagreement between the author(s) and reviewer(s)/editor(s) about experimental evidence and how it is interpreted. Moreover, differences of opinion over a manuscript can sometimes be overcome by improving the evidence and rewriting the paper, or by submitting the manuscript to another journal. This is not the case for a grant application, though they can sometimes be revised and resubmitted. Revision of a grant application may increase the likelihood of receiving funding, but it in no way guarantees it.

Finally, most grant applications include indirect costs (overhead), i.e., the institutional costs of maintaining the infrastructure (space, electricity, water, gas, etc.) that make research possible. This, coupled with a continuous stream of calls for proposals issued by grant agencies in specific disease areas or on fashionable technologies, can lead some departmental and institutional administrators to place pressure on researchers to obtain grant funding for reasons that are more financial than scientific. Administrators often urge scientists to follow "the money trail" instead of their "idea trail," thereby shifting their focus from doing innovative science to obtaining large grants. Increasingly, institutions are even linking the research space allocated to scientists to the amount of overhead funding that they generate through grants.

That said, it is possible to structure grant writing in such a way that it becomes part of the creative aspect of scientific discovery. This may be difficult for some large grants involving multiple researchers, but for smaller, single-investigator-initiated grants, a few selected strategies can help make grant writing an intellectually rewarding exercise.

15.2 GETTING STARTED

PhD students and postdoctoral fellows once struggled to write grant applications, with little or no help, unless their advisor(s) or the Principal

Investigator were sufficiently experienced to offer guidance at all stages of the process. In many countries this process is now easier. University-based workshops and courses on grant writing are offered, and although such formal training may never fully supplant personal mentoring by a dedicated senior researcher, it certainly can help to demystify the process. Major research centers such as Harvard Medical School in Boston, MA, and Karolinska Institutet in Stockholm, Sweden, recognize the importance of early introduction to the grant writing process and have long organized one-semester courses in this subject for PhD students. These courses include grant writing and the critiquing of grant applications. At Karolinska Institutet the courses apparently end with the submission of a grant application to a Swedish grant agency.[1]

Assistance is also available from the grant agencies themselves; major grant agencies offer specific instructions both in print and online. For example, Britain's Wellcome Trust (http://www.wellcome.ac.uk) has a short article for first-time applicants available online: "Grantsmanship: signposts for competitive grant applications." The article comprises a short list of questions that researchers should ask themselves before submitting their grant application (Box 15.1). The US National Institutes of Health (NIH) offer a variety of informational services through their Office of Extramural Research (http://grants1.nih.gov/grants/oer.htm), including seminars on the drafting of grant applications given by NIH experts at major research centers in the United States. The National Science Foundation (NSF) in the United States organizes NSF Regional Grants Conferences (see http://www.nsf.gov/events/ for more information). The National Institute of Allergy and Infectious Diseases at the NIH has compiled an instructive, annotated example of a well-written R01 grant application and a comprehensive monograph, "All About Grants Tutorials" (available at http://www.niaid.nih.gov/ncn/grants/default.htm).

BOX 15.1 Questions to Ask Before Submitting a Grant Application

- Is the work novel, exciting, and necessary?
- Have the experiments proposed already been done by others?
- Is the requested funding for various aspects of the study justified?
- Have the instructions been read carefully and are the forms filled out correctly?
- Has the accuracy of spelling and grammar been checked?

BOX 15.2 Useful Publications on Grant Writing

- *How to Succeed in Academics.*
  Linda L. McCabe and Edward R.B. McCabe. Academic Press, 2000, Chapters 4–6.
- An evidence-based guide to writing grant proposals for clinical research.
  Sharon K. Inouye and David A. Fiellin. *Ann Intern Med* 2005;142:274–282.
- Getting funded. Career development awards for aspiring clinical investigators.
  Thomas M. Gill, et al. *J Gen Intern Med* 2004;19:472–478.
- How to get a grant funded.
  David Goldblatt. *BMJ* 1998;317:1647–1648.
- Applying for grant funds: there's help around the corner.
  Liane Reif-Lehrer. *Trends Cell Biol* 2000;10:500–504.
- Writing successful grant applications for preclinical studies.
  David Kessel. *Chest* 2006;130:296–298.
- How to increase your funding chances: common pitfalls in medical grant applications.
  Gideon Koren. *Can J Clin Pharmacol* 2005;12:e182–e185.

Moreover, several books and journal articles cover the general principles of grant writing and discuss features that may be specific to grants within different fields of biological and biomedical sciences (Box 15.2).

15.3 THE POSTDOCTORAL FELLOW AND THE JUNIOR SCIENTIST

With all of the electronic and print resources available, there is no shortage of advice on how to write a grant application. However, in the early stages of one's research career, a successful grant application is not only one that reviewers rank highly for its exciting and significant research plan, but also one that receives high scores for the training experience and scientific standing of the advisor or mentor, and for the quality of the institution. Postdoctoral fellowship applicants who propose research projects that do not depart significantly from their previous research and do not take place in a new environment (another institution, state, or country) are viewed unfavorably. For example, postdoctoral fellowship applicants who propose a continuation of previous PhD research, with no change in direction, usually receive lower rankings, both for NIH fellowships and fellowships from international organizations such as the Human Frontier Science Program.

Consequently, postdoctoral fellowship applicants should devote their time and energy to deciding the direction they would like their scientific career to take after graduate school. This is a critical decision, not only because of what it means for the postdoctoral fellowship application, but also what it means for the researcher, as postdoctoral research usually develops into a lifelong research commitment. As with all crucial decisions, it is difficult to define a set of advisory rules that applies to all cases. However, a decision that is made based on what one finds fascinating is unlikely to take the research in the "wrong" direction. Once this decision starts to take shape, it is time to consider to which grant agency to submit the postdoctoral fellowship application. Indeed, gaining financial support from a prestigious, highly competitive grant agency can help open the doors to institutions with outstanding mentors, laboratories, and research environments, which can stimulate significant accomplishments.

Having a recognized mentor in a strong institution with a significant commitment to research is crucial if one is to be successful in obtaining fellowships and grants that will support one's transition to independence. One example of such grants is the new Pathway to Independence Award from the NIH (K99/R00), because the experience and commitment of the mentor, the mentoring plan, and the institutional commitment are the factors that carry the most weight in the review process. This award provides financial support for up to 5 years. It is intended for junior scientists with no more than 4 years of postdoctoral research experience at the time of either the initial submission or resubmission of a revised application, who require at least 12 (up to 24) months of additional mentored research training and career development before launching their independent research project. The grant covers funding for up to 3 years. For junior scientists (at the level of Assistant Professor), the NSF offers a similar award (known as the CAREER award).

For the junior scientist, deciding where to submit that first grant application is also important. Grant agencies have distinct missions, and they are likely to reject proposals that are inconsistent with them. For example, in the United States the NIH supports both medical research and basic science projects. The NSF, on the other hand, rejects proposals that deal with the cause of disease, diagnostic and therapeutic medical procedures, drugs, or animal disease models. In contrast, the CAREER award supports activities that integrate research and education; in fact, plans for international cooperative research and education activities are encouraged. In many other countries, national research support for biomedical and more fundamental biological research is similarly divided between different grant agencies. Some good examples are the Medical Research Council (MRC) and the Biotechnology and Biomedical Research Council (BBSRC) in the United Kingdom.

To ensure that a grant application matches the mission of a potential grant agency, the applicant needs to seek out the themes these agencies are willing to consider. Submitting a good proposal to the wrong agency in the hope that it will succeed is arguably a waste of time. Fortunately, there are extensive web-based resources that provide details on grant agencies worldwide, as well as of the types of grant initiatives and opportunities available (Box 15.3). Whenever the information provided by these or other websites is insufficient to make a decision, a telephone call to the grant agency's office may be helpful.

Some grant agencies, such as the European Union (EU), provide lists of projects that have obtained funding in previous calls for proposals, which may also be of assistance when choosing where to submit. For large organizations such as the NIH, where the available funds are divided among numerous institutes, their respective programs often overlap. In this case, a telephone call to the program directors at the institutes in question may be helpful when deciding which institute is thematically the most appropriate target for the application.

In addition to a grant agencies mission, it is also important to note that some calls for proposals have specific requirements. For example, in past calls for proposals from the EU it was expected that the research project would address a number of issues that extended beyond the realm of science, into the economic and political spheres. In recent EU Framework Programmes (FPs), applications had to describe the "added value" of the proposed research, including the social and economic impact the results would have on the European community. Information on the most recent EU FP can be found in Section 15.7.

The effect of targeting a proposal to a specific institute at the NIH should also be considered, as it may have a significant effect on the likelihood of receiving funding. This is because NIH institutes have different budgets and make different decisions about how they allocate their resources between investigator-initiated grants and large multiproject Program and Center grants. Consequently, the funding rates for junior investigators vary from institute to institute. Whenever congressional funding for NIH is ample, this difference is less important. But when funds are meager, the dissimilarity in funding rates between institutes can translate into the inability to fund a particular grant application.

BOX 15.3 Useful Resources for Writers of Grant Applications

Resource	Website	Description
Computer Retrieval of Information on Scientific Projects (CRISP) Database	http://crisp.cit.nih.gov	Contains descriptions of all projects funded by the NIH
Community of Science (COS)	http://www.cos.com (main COS website); http://fundingopps2.cos.com/	Database of funding opportunities worldwide; available only to researchers at COS member institutions. Weekly Funding Alerts are sent by e-mail to keep researchers abreast of new opportunities in their disciplines
COS	http://www.cos.com/services/	Database of researcher profiles from over 190 leading universities Available only to researchers at COS member institutions
Deutsche Forschungsgemeinschaft (DFG)	http://www.dfg.de/en/research_funding/index.html	Provides a description of the various types of funding available through the DFG
EMBL Heidelberg	http://www.embl-heidelberg.de/training/index.html	Describes advanced training programs at EMBL and information about international fellowship programs
EMBO	http://www.embo.org/fellowships/index.html	Describes Long-term and Short-term EMBO Fellowships, as well as application forms and guidelines
EU	http://ec.europa.eu/research/index.cfm	Provides links to the research programs, multinational networks and business oriented research strategies organized and funded through the EU
Funders Online	http://www.fundersonline.org/ http://www.fundersonline.org/grantseekers/	Allows searches of Europe's OnLine philanthropic community (foundations and corporate funders)
Grantsnet	http://www.grantsnet.org (e-mail: grantsnet@aaas.org)	Website for young biomedical scientists (undergraduate to just beyond postdoctoral training). Part of *Science*'s "Next Wave" A large database of fellowships, links to websites of funding organizations, and information about—and tips from—previous funding recipients

Grantsnet "Global Links"		For scientists working in the United Kingdom, connects to funding databases around the world. "Next Wave" has websites for the United Kingdom, Canada, and Germany
Human Frontier Science Program	http://www.hfsp.org	Contains description of programs for postdoctoral fellows and young researchers
Institut national de la santé et de la recherché médicale (Inserm)	http://english.inserm.fr/	Provides links to the programs supported by Inserm
International Grants Finder	http://www.nature.com/	Database, maintained by *Nature*, for locating grants available in scientific fields worldwide; updated annually
National Natural Science Foundation of China	http://www.nsfc.gov.cn	Promotes and finances basic and applied research in China
Science's "Next Wave"	http://sciencecareers.sciencemag.org/career_development	Provides career information and reviews of useful resources for young scientists
The European Science Foundation	http://www.esf.org/	Has funding programs in many scientific fields, including medical, life, and environmental sciences
BBSRC	http://www.bbsrc.ac.uk/	The United Kingdom's leading funding agency for academic research in the nonmedical life sciences
MRC	http://www.mrc.ac.uk/	Provides links to the wide range of medical research opportunities that are funded by the MRC in the United Kingdom
The Wellcome Trust	http://www.wellcome.ac.uk/funding	Maintains a database of organizations in the United Kingdom supporting biomedical research
NIH website	http://www.nih.gov/	Provides links to all the information that is available through the individual NIH institutes

Modified from Reif-Lehrer.[2]

Note that most research institutions offer advice on how to identify appropriate grant agencies. Major universities have significant expertise in house, whereas smaller institutions often draw on the expertise of professional consultants with insight into the operation of specific grant agencies.

15.4 WHAT GOES INTO A SUCCESSFUL GRANT APPLICATION?

Once a grant agency has been identified and one has started writing the grant application, a recurring question is what can be done to maximize one's chance of success. With success rates ranging between 10% and 15% for grants to federal agencies such as the NIH and NSF, this question looms large in the minds of junior researchers trying to secure their first research grant. Another way to put the question is to ask what mistakes most commonly contribute to rankings that fall in the nonfundable category. The National Institute of Allergy and Infectious Diseases at the NIH has put together a list that highlights the most common negative phrases used by reviewers to describe unsuccessful grant applications. A summary of these can be found in Box 15.4; for the full list, visit the website of the National Institute of Allergy and Infectious Diseases.

The first comment in Box 15.4, "Problem not important," indicates that an applicant has utterly failed to convince the reviewer that the problem presented in the grant application is important in the context of contemporary understanding. Obviously, it is unconvincing for an applicant to simply state that the problem is significant. A critical analysis of current

BOX 15.4 Reasons Commonly Cited by Reviewers for an Application's Failure to Be Funded

- Problem not important; study not likely to produce useful information
- Study based on shaky hypothesis/data; alternative hypotheses not considered
- Unsuitable methods; controls not included
- Problem more complex than the researcher appears to realize
- Too little detail in research plan; no recognition of potential pitfalls
- Overambitious research plan
- Direction or sense of priority not clearly defined; lack of focus
- Lack of new ideas; fishing expedition; method in search of a problem
- Rationale for experiments not provided
- Not enough preliminary data; insufficient consideration of statistical needs

evidence and an identification of gaps in contemporary understanding must be presented in such a manner that the problem to be studied logically emerges as paramount, and to make clear that addressing and solving it using the experimental methods described in the application will help move the field forward. As discussed in Chapter 6 of this book, a research problem should not only be interesting; it must be important, such that the answer truly matters. The problem should permit a clear description of overall goals and specific aims, contain testable hypotheses, and allow the selection of efficient and relevant methods for exploring these hypotheses.

The effort expended in writing a succinct, compelling story about the context and rationale of the research planned becomes part of the intellectual exercise necessary for initiating worthwhile research. It should be viewed as the challenging first phase of the research process, not just as a necessary step to obtain funding. In so doing, the researcher is more able to convey her or his excitement to the reviewer. Indeed, the grant writing period is an exciting and ideal time for brainstorming. However, it should also focus on and include leeway for speculation and intellectual leaps, as well as take several steps back. It is the time to take in the big picture of the intellectual landscape, but also to zoom in on crucial details. The applicant who manages to view the grant writing process in this light will be stimulated by it and may actually enjoy it. Researchers who are preparing to write their first grant application may find it useful to ask senior colleagues to serve as a "grant committee," with whom they can discuss ideas for the application, and who can assist in defining the most compelling specific aims before the writing starts. In this way, the content and organization of the application can receive critical considerations at an early stage, making it easier to begin writing the first draft of the application.

Applications are assessed by reviewers who often have limited time available, as they must read several grant applications in a brief period. Therefore, a grant application should be well organized and well written, so the reader can grasp the flow of ideas and understand the experiments and potential outcomes. Simple diagrams summarizing the work flow of the project may prove helpful. The writing should be edited and formatted to enable the reader to easily follow the main points and thereby rapidly evaluate the significance and novelty of the proposed work without being mired in excessive detail. Simple, preliminary experiments should be identified and described along with preliminary results, to help instill confidence in the soundness of the proposed ideas and eliminate unnecessary speculation. Minor experimental details should not be mentioned, but all essential details should be included. The experimental strategies and the rationale for selecting some methods and not others must be explained.

In order to preclude many of the common mistakes listed in Box 15.4, a wise strategy is to ask friends and colleagues to read and critique a draft of the grant application. It may also be beneficial to seek the advice of a professional consultant, but this is only effective if sufficient time is allowed for revising and rewriting before the submission deadline. It may also be helpful to put the document aside for a few days and then try to read it again from the perspective of a potential reviewer. By anticipating and addressing questions that reviewers may ask (based on the comments in Box 15.4), one may be able to deflect them when the application is reviewed. For example, if there are potential weaknesses in the preferred experimental plan, it is better to acknowledge these in the proposal and offer alternative strategies than to have a reviewer point them out later. In fact, reviewers will often list discussions of potential weaknesses and alternative plans among the strengths of a grant application. Also, acknowledging that the results of the planned experiments may prove the hypotheses wrong is likely to be viewed favorably if the applicant can convince the reviewer that the result is important either way.

The title and abstract are often left to the last minute and accordingly receive the least attention. This is a mistake. A concise title and a clear abstract enable the reader to readily understand the project proposed and serve to attract attention. This is particularly important when reviewers meet as a group to discuss and rank the priority of a large number of proposals. Indeed, group members who do not serve as in-depth reviewers of an application may nonetheless be asked to rank its overall impact or priority. When the reviewers assigned to a particular application summarize their evaluations, the rest of the group may quickly scan the abstract, diagrams, and figures in the experimental plan to assess the application. To ensure that the abstract reflects all aspects of the proposal, it is useful to think of it as a means of providing answers to six germane questions: (1) What is proposed? (2) Why is it important? (3) What has been done already by the applicant and by others? (4) How will the study be done? (5) What will be learned? (6) What are the qualifications that provide the applicant with a competitive edge relative to other research groups addressing the same research questions?

The procedures for submitting grant applications, the relevant deadlines, and the ways in which applications are reviewed and funding decisions are made differ across countries and between grant agencies. Covering all the relevant rules and policies is beyond the scope of this chapter. As an illustrative example, we will describe the scenario of an investigator-initiated R01 NIH grant in the United States. We believe that the R01 grant mechanism is germane for two reasons. First, the sequence of steps, from submission to funding, is typical of the sequence of similar grants in most countries. Hence, with some modifications, the process for an R01 grant is applicable to most situations. Second, the peer-review process used to evaluate R01 grants has long served American biomedical research well. The process is not perfect, but it is useful and being imitated elsewhere as countries in other parts of the world strengthen their mechanisms for funding research.

15.5 INVESTIGATOR-INITIATED R01 NIH GRANTS

The R01 grant is the original, historically oldest, grant mechanism of the NIH. Today, the R01 grant is one of several types of grants available in the field of health-related research and training, yet it still accounts for a substantial portion of federal funding for biomedical research. As for all research project grants, R01 grants are awarded to sponsoring institutions on behalf of, and to facilitate research projects conceived by, a Principal Investigator. The awards are usually for periods of 3–5 years. When accepting an award, the sponsoring institution assumes the financial responsibility for the grant and for providing the laboratory and other facilities the research project requires.

R01 grant applications are submitted to the Center for Scientific Review, the NIH entity responsible for assigning applications to the institute with the mission that best fits the proposed research topic, and to a "Study Section" or scientific review group (SRG). Each application is given an ID number, and the applicant is notified that the application has been assigned to a specific SRG for the evaluation of scientific merit, and an institute or center for funding consideration. The applicant is provided with the name and contact information of a Scientific Review Administrator, to whom he or she can direct questions about SRG assignments or other issues that may arise before the SRG review takes place. For questions that arise after the review, the applicant is directed to program staff of the assigned institute or center.

Getting the application assigned to the right SRG and institute is important, and applicants have some control over this. They can request assignments at the time of submission, and they can request a reassignment should they feel that the assignments provided are not in their best interests. SRG rosters are available on the Internet; an applicant who feels that a given SRG does not have the required expertise to review the application, or that a member of the SRG has a competitive conflict of interest, can request reassignment to another SRG.

15.5.1 The Review Process

All successful grant applications have research plans that reviewers find exciting, significant, and well documented, be they applications for PhD or postdoctoral fellowships, mentored awards for transition to independence, or independent research awards. However, the extent to which the scientific evaluation contributes to the overall rating of the grant application varies according to the nature of the application. For research awards, evaluation of the scientific plan is of primary concern, although other aspects are also considered when assessing the overall merit of the grant application. The review criteria for R01 grants change depending on whether the application was submitted in response to a special initiative, such as a call for proposals or a program announcement in a specific area of biomedical research, or whether an investigator initiated it. The standard NIH evaluation criteria for an investigator-initiated R01 grant are summarized in Box 15.5. Reviewers are requested to give each of these five criteria a numerical score using whole numbers on a scale from

BOX 15.5 Review Criteria for Investigator-Initiated NIH R01 Grants

- *Significance*—Does the study address an important problem or critical barrier to progress in the field? How will scientific knowledge, technical capability, and/or clinical practice be improved, if the aims of the project are achieved?
- *Investigators*—Are the investigators appropriately trained and well suited to carry out the work?
- *Innovation*—Is the project original and innovative?
- *Approach*—Are the conceptual or clinical framework, design, methods, and analyses adequately developed, well integrated, well reasoned, and appropriate to the aims of the project?
- *Environment*—Does the scientific environment contribute to the probability of success?

1 to 9: (1) Exceptional, (2) Outstanding, (3) Excellent, (4) Very good, (5) Good, (6) Satisfactory, (7) Fair, (8) Marginal, and (9) Poor. In addition, reviewers provide an impact score that reflects their assessment of the likelihood that the project will exert a sustained, powerful influence on the field involved. When assigning an impact score, reviewers take the aforementioned scoring criteria into consideration, but there is no direct numerical relationship between these criteria and the impact score. Moreover, an application does not need to score high in all five scoring criteria to be given a high impact score. However, the most reliable route to a high (fundable) score is to take all the criteria into account when writing an investigator-initiated R01 grant. Obviously, an applicant should identify and address any threshold scores before submitting any proposal.

SRGs, composed of scientists with relevant expertise, are responsible for evaluating R01 applications for scientific and technical merit, and they meet (usually at NIH) to discuss the applications assigned to them. Note that this initial review is a peer review, with scientists evaluating the proposals of their colleagues. Health policy and funding issues are not considered at this stage; instead, the goal is to identify proposals that are scientifically the most promising and exciting. The following procedure helps the SRG accomplish this as rigorously as possible with the steadily increasing number of grant applications. Before the peer review meeting, certain reviewers are assigned applications on which they are to provide critiques, preliminary scores for each of the five scoring criteria (see Box 15.5), and an impact score. These preliminary scores are replaced by final scores following the discussion of each application during the meeting. All members of the SRG (except those who may have a conflict of interest; working at the same institution as the applicant, having mentored or collaborated with the applicant, etc.) are required to provide an impact score for each application discussed. After the meeting, individual impact scores are averaged and multiplied by 10, so that applicants receive a score between 10 and 90 in the review reports (Summary Statements). A score of 10–30 is considered high, a score of 40–60 is medium, and a score of 70–90 is a low score. In addition to the impact score, the Summary Statement summarizes the group discussion and the critiques and scores of the assigned reviewers.

Some applications may receive scores that are not numerical. For example, some SRGs elect not to discuss applications that are unanimously judged to be less competitive. In this case they may give the applications scores for the five criteria, but no impact score. Instead, the applicant is informed that the application was not discussed. For applications that lack substantial merit or present serious ethical problems (related to the use of human subjects, experimental animals, biohazards, or select agents), SRGs may also elect to prevent them from proceeding to the second level of review, eliminating them from consideration for funding. This option, known as "not recommended for further consideration," requires a majority vote by the SRG. Finally, applications may be deferred if important information is lacking or allegations of scientific misconduct are associated with the proposed project.

Within the institutes, a second-level review takes place at a meeting of the institute-specific National Advisory Council, composed of scientists and laypersons with special interests in health-related issues relevant to the mission of the institute. Council members review the appropriateness of the SRG Summary Statements and weigh in on issues that may affect the funding of applications, such as mission relevance, program goals, and availability of funds. This system of separating the assessment of scientific merit from public health-related political consideration has proven effective in maintaining high scientific standards for NIH-supported research.

If an application is not successful on the first try (the most common outcome), applicants can revise the application and try again. Over 50% of NIH applicants will ultimately receive funding after revisions and resubmissions.

15.6 MULTIPROJECT GRANTS

The days when researchers working alone in their laboratory made startling discoveries are, for the most part, over. The complexities of modern biomedical research require teamwork, with disparate team members contributing their different technical and theoretical skills to a research project. For example, in the case of research projects supported by R01 grants, the Principal Investigator may direct a team of PhD students and postdoctoral fellows, collaborate with several coresearchers, and have several consultants who provide expert advice and specialized reagents. For these larger team efforts there are other grant mechanisms, including grants to support Program Projects, Specialized Centers or Centers of Excellence, and multiproject Networks. Program Projects, which are supported by some NIH institutes, comprise several (more than three) Principal Investigator-directed projects focused on different, but complementary, objectives within a larger common goal, usually combined with specialized "Cores" that provide technical, administrative, and molecular/cellular/clinical resources for the program. An advantage of a Program Project is that the built-in mechanism for funding of Cores provides collaborating research groups with resources to enhance their collaboration and embark on research projects that are more complex than those that can be carried out by individual R01-funded institutions.

Center grants are used in many countries as mechanisms for stimulating collaborations among a large number of scientific teams and clinicians to enhance research activities that can result not only in improved understanding, but also in the detection, treatment, and prevention of human diseases. The Center concept may work particularly well when the collaborating teams of scientists and clinicians have worked together before. Indeed, researchers who work well together without a grant framework can respond easily to this grant requirement. In this case a Center grant provides a funding base for a natural extension and expansion of collaborative studies that have already been started. Whether Center grants in specific areas of biomedical research truly stimulate excellence based on entirely new collaborations is less clear. This is because Principal Investigators frequently tailor applications to compete for the available funding, but will continue to do more or less what they were doing before a grant is awarded. There is also a financial factor to consider, which often is counterproductive to starting new collaborative projects. Indeed, to be competitive for large Center grants, applications must define ambitious goals and objectives. These goals are often overly ambitious, given the available funding, but this becomes obvious only after the grant has been awarded and the funds are divided among all participating groups, as only then are they found to be insufficient to accomplish the planned work. The risk of ending up in such a situation can be curtailed if the number of groups included in a given proposal is tailored to the anticipated size of the grant. As a rule of thumb, for each participating group, the budget should accommodate a minimum of one full-time researcher plus reasonable running costs.

For large networks, there may be similar problems associated with imposed collaborations. It is unfortunate when collaboration fails to promote innovative science. For this reason, network grant mechanisms probably work best when they aim at clearly defined goals, such as development of new technologies or clinical research objectives. In these cases, it is essential to assemble teams that can cover many different technical approaches, and it is often necessary to generate and study large repositories of patient data and tissue samples from large populations.

Accordingly, the NIH has a Division of Program Coordination, Planning and Strategic Initiatives within the Office of the Director that is charged with identifying scientific opportunities, public health challenges, and resource development. This division, incorporating all the functions of the former Office of Portfolio Analysis and Strategic Initiatives, serves as an "incubator" for trans-NIH initiatives and support of priority projects. It thereby meets the needs for stimulation of collaborative research along such lines and helps to identify and support future research needs (Box 15.6).

The NIH Office of Strategic Coordination runs a large number of Common Fund programs, ranging from programs comprising New Tools and Methods, Databases and Libraries, High Throughput Analyses, Translational Research, Population Science and Training, to High Risk Research. The High Risk Research programs are designed to capture those highly innovative investigator-initiated ideas that have potential for high impact, but are judged to be too risky and too far off the beaten path to be funded through the R01 grant mechanism. Several awards of this type are available, such as the NIH Director's Pioneer Award, the New Innovator Award, and the Transformative Research Award. In addition, the NIH Director's Early Independence Award provides a mechanism for exceptional early career scientists to establish an independent research program without traditional postdoctoral training. To be eligible for this last

BOX 15.6 The Division of Program Coordination, Planning, and Strategic Initiatives

1. Office of AIDS Research—setting priorities in global fight against AIDS
2. Office of Behavioral and Social Sciences Research—furthering research in behavioral and social sciences
3. Office of Disease Prevention—fostering prevention and health promotion research
4. Office of Research Infrastructure Programs—supporting infrastructure programs and science education at NIH
5. Office of Research on Women's Health—supporting research on diseases affecting women
6. Office of Strategic Coordination—responsible for managing the NIH Common Fund (previously the NIH Roadmap) and identifying NIH-wide scientific opportunities

award, candidates must be within a year of their terminal degree or residency, and must be chosen by their institution through an internal selection process. Applications for a Pioneer Award (PioneerAwards@mail.nih.gov) are brief in that they consist primarily of a one-page description of the researcher's most significant research accomplishment and an essay of up to five pages about the proposed research strategy. Awardees are required to commit the major portion (at least 51%) of their research effort to activities supported by the Pioneer Award. Consequently, the award is intended to be a supplement to, not a substitute for, an R01 grant. It is designed to support creative individuals rather than a project, based on the assumption that if highly creative scientists are given some unrestricted money, significant findings may result. The New Innovator Award is intended to stimulate highly innovative research and support promising new researchers. The emphasis in the application for this award is therefore on innovation and creativity. Preliminary data are not required, and no detailed annual budget is requested (NewinnovatorAwards@mail.nih.gov). The primary emphasis of the Transformative Research Award is on bold, paradigm-shifting but untested ideas. Applications can be from individuals or teams of researchers, preliminary data are not expected, and the research strategies must focus on significance and innovation (TransformativeAwards@mail.nih.gov).

15.7 HORIZON 2020

As described above, multiple types of research grant opportunities are available to researchers in the United States. A similar rich offering of research support is now available in the EU. The new FP for research and innovation, Horizon 2020, started in 2014 and will run until 2020 (http://ec.europa.eu/programmes/horizon2020). It succeeds FP 7 and covers a number of different funding instruments, but its structure differs considerably from that of previous FPs. It is divided into three "pillars": "Excellent Science," "Funding Innovation," and "Societal Challenges." Within these pillars there are funding opportunities for individual researchers as well as groups of researchers. The choice of pillar and underlying program depends on the size and nature of the project and whether the research is basic or applied. Horizon 2020 is open to all actors involved in research and innovation, including those in academia, industry, and other stakeholders.

15.7.1 Pillar 1—Excellent Science

There are four different schemes in the Excellent Science pillar, largely for bottom-up funding for individual researchers or teams:

- The European Research Council (ERC) provides funding for excellent researchers at different stages of their career. This important instrument is discussed in a separate section below.
- Future Emerging Technologies (FET) is a funding mechanism for collaborative, "high-risk" research under three different streams, FET Open, FET Proactive, and FET Flagships.
- Marie Skłodowska-Curie Actions fund mobility, training, and career development in academia, industry, and other nonacademic sectors through individual mobility grants and projects.
- European Research Infrastructures includes funding for e-infrastructures and access to infrastructures for researchers.

15.7.2 Pillar 2—Funding Innovation

This pillar is focused on industry and innovation and offers ample opportunities for applied research. Funding is available for the so-called Key Enabling Technologies, including areas such as information and communication technologies, nanotechnologies, advanced materials, and biotechnology.

15.7.3 Pillar 3—Societal Challenges

This pillar reflects major concerns shared by citizens in Europe and elsewhere. The relevant calls for proposals follow a top-down approach with seven defined, challenge-based topics:

1. Health, demographic change, and well-being
2. Food security, sustainable agriculture, marine and maritime research, and the bioeconomy
3. Secure, clean, and efficient energy
4. Smart, green, and integrated transport
5. Climate action, resources, and raw materials
6. Inclusive, innovative, and reflective societies
7. Secure societies

Given that the overall aim of this pillar is to tackle Societal Challenges, most projects will require a broader approach in terms of disciplines and might require the inclusion of several different stakeholders.

15.7.4 How to Find Funding Opportunities in Horizon 2020

Funding opportunities available through Horizon 2020 are published on a central European Commission website—the Research and Innovation Participant Portal (http://ec.europa.eu/research/participants/portal/desktop/en/home.html). Together with the Online Manual (http://ec.europa.eu/research/participants/portal/desktop/en/funding/guide.html), these resources give access to practical and administrative details about funding opportunities, proposal submission, and project management, and contain links to official Horizon 2020 guidance and documentation. The Participant Portal is also the official site for all Horizon 2020 calls for proposals. There are several ways of searching for a call for proposals or funding topic, including a keyword search.

15.7.5 The Review Process

All grant applications submitted to the EU are evaluated by experts based on three criteria: "excellence," "impact," and "quality and efficiency of the implementation." The different aspects of these criteria are as set out in Box 15.7. Horizon 2020 differs from previous FPs in that it puts a strong emphasis on the criterion of impact.

Unless otherwise specified in the call for proposals, each criterion is given a score between 0 and 5. The threshold for individual criteria is 3. The overall threshold for moving forward in the process, applying to the sum of the three individual scores, is 10.

BOX 15.7 Evaluation Criteria

Excellence:	Clarity and pertinence of the objectives
	Credibility of the proposed approach
	Soundness of the concepts, including trans-disciplinary considerations, where relevant
	Extent to which the proposed work is ambitious, has innovation potential, and is beyond the state of the art
Impact:	The expected impacts listed in the Work Program under the relevant topic
Implementation:	Coherence and effectiveness of the work plan, including appropriateness of the allocation of tasks and resources
	Complementarity of the participants within the consortium
	Appropriateness of the management structures and procedures, including risk and innovation management

BOX 15.8 Scoring and Thresholds (Half Marks Are Possible)

Scoring:

0:	The proposal fails to address the criterion or cannot be assessed due to missing or incomplete information
1:	Poor: the criterion is inadequately addressed or there are serious inherent weaknesses
2:	Fair: the proposal broadly addresses the criterion but there are significant weaknesses
3:	Good: the proposal addresses the criterion well but with a number of shortcomings
4:	Very good: the proposal addresses the criterion very well but with a small number of shortcomings
5:	Excellent: the proposal successfully addresses all relevant aspects of the criterion; any shortcomings are minor

Thresholds:

Full proposal:	3 for each individual criterion, overall 10
Two-stage proposal:	Only excellence and impact are evaluated, with a threshold of 4 for each, overall 8.5

Under a two-stage submission procedure, the first stage considers only the criteria "excellence" and "impact." In this case the threshold for both individual criteria is 4. The scoring and thresholds are summed up in Box 15.8.

15.7.6 The European Research Council

The ERC is in Pillar 1 of the Horizon 2020 structure and aims to support excellence in frontier research through bottom-up, individual-based, pan-European competition. Quality is the only criterion: the applicant is free to choose the topic and team. Calls for proposals in the ERC are open to researchers of any nationality and age, with the prerequisite that at least 50% of the work (in terms of man-hours) be carried out in Europe.

Currently there are four ERC funding schemes. Starting grants provide up to a total of 2 million euro for 5 years for successful applicants that are early in their careers (2–7 years after PhD). Consolidator grants target researchers 7–12 years after they receive their PhD and offer up to 2.75 million euro for 5 years. Advanced grants are open to all, but they require a track record of significant research achievements in the last 10 years. The advanced grants award a maximum 3.5 million euro for 5 years. The fourth funding scheme is Proof-of-Concept, intended to bridge the gap between research and earliest stage of marketable innovation (up to 150,000 per year). Only ERC grant holders are eligible. The total ERC budget under Horizon 2020 is 13.1 billion euro.

The ERC funding schemes give a high degree of flexibility to the awardee. As stated above, the applicant is free to choose the topic and team. In addition, the successful awardee may move with the grant anywhere in Europe. This gives leverage to the grant holder when he or she negotiates the working conditions with the host institution.

Because quality is the sole criterion for success in ERC grants, the number of ERC grants awarded has become a yardstick for institutional performance. Researchers will therefore typically experience considerable backing from their own institutions in the preparation phase of these applications—as well as in the administration of their grant, should their application succeed. Most host institutions arrange mock interviews to prepare grant applicants for their presentations in Brussels. Scientists who consider applying for an ERC grant should acquaint themselves with the various types of support their institutions offer and should make full use of this support so as to enhance their chance of success.

Success rates for starting grants range between 9% and 16%, except for 2007, the first year they were offered, when the success rate was very low due to a massive number of applications. Advanced grants have success rates between 12% and 16%, with an average of 13.9%. For Proof-of-Concept applications the success rate is higher, with an average of 36.9% for the years 2011–2013. Potential applicants should also be aware that some countries provide national funding for the best runners-up, i.e., projects that score high on quality but that end up just below the threshold for ERC funding.

In the course of a few years, the ERC has established itself as an important funding opportunity, complementary to the other parts of the FP. Successful ERC grantees typically find that their grants help them to gain recognition and attract additional funding.

15.7.7 Why Apply for EU Grants?

In Europe, there is a bewildering array of funding initiatives and opportunities for researchers. Typically, national funding is available through one or several research councils or agencies, and through public institutions such as hospitals and health authorities. Depending on the country, there may also be a number of private foundations to which researchers may apply to support biomedical projects. So why apply for EU grants?

The answer is obvious if national funding—public and private—does not provide the financial means to carry out the project in question, in which case EU funding is a good alternative. But there is ample reason to apply for EU grants even if national funding is sufficient to cover the basic costs of the project.

First and foremost, EU grants typically provide access to complementary expertise and technologies that allow the successful grantee to set more ambitious goals for his or her research. This leads to more opportunities for multidisciplinary research and for interaction with relevant industrial partners. Second, some EU funding schemes—the ERC in particular—are willing to fund high-risk projects, provided that the risk is balanced by a potential high gain. This opens the door for highly original projects that fail to obtain support from national grant agencies. Third, EU grants facilitate recruitment and exchange of students and experienced scientists. Fourth, successful applicants will observe that their EU grants make their research more visible on the international scene and help attract additional funding from national and international funding bodies. Indeed, in the case of EU grants, success breeds success.

So the answer to why one should apply for EU grants is quite clear: EU grants give a return that by far exceeds the financial gain. They bolster the research environment as well as the host institution, and they boost internationalization. EU grants should also be considered as a potential funding source for scientists early in their careers. In fact, getting to know the EU funding system should be an important goal in any PhD program. Mentors should help ensure that their PhD students and postdoctoral fellows are aware of the funding opportunities in EU and that they develop a research profile that is conducive to success in the EU system. The requirements for independence, which are set out in the guidelines for ERC starting grants, are of particular importance in this regard.

Every so often one encounters the argument that applying for and managing an EU project is too much of a burden and that the administrative efforts involved are not commensurate with the size of the grant. One should then remember that an EU grant provides benefits far beyond the actual economic return, as discussed above. Also, year after year host institutions are becoming more efficient in managing EU projects and offering administrative support to grantees. The message is clear: Do not shy away from the EU funding system. It is there to stay, and it will grow in importance in the years to come.

15.8 INTERNATIONAL RESEARCH COLLABORATIONS

Science knows no borders, and international cooperation between scientists is not only common but also responsible for some of the most significant advances in biomedical research. Unfortunately, obtaining grant funding for international collaborations is not always straightforward, partly because most grant agencies primarily promote national research programs. Another reason is that rules that regulate research involving human subjects and experimental animals are not always the same in different countries, and this may generate problems when material collected in one country is used for research by a collaborator in another country. However, these are all problems that can be solved; it may simply require more effort than in cases where collaborators are within the same institution or country.

A Principal Investigator does not need to be a citizen or permanent resident of the United States to apply for an NIH R01 grant. Hence, researchers in other countries can obtain independent funding from the NIH for projects that may complement or match a project of an American collaborator. It is also possible for a Principal Investigator in the United States to include a foreign collaborator and her or his project as a subcontract on an NIH grant. Thus, in principle, NIH funding knows no borders. However, in the case of a foreign grant or foreign component of a United States grant, one has to convince the reviewers that the foreign research is unique and does not duplicate research that can be done as easily, or better, by researchers in the United States.

The NIH also has a Center that was specifically established (in 1968) to address global health problems by supporting innovative and collaborative research and training programs. This is the John E. Fogarty International Center for Advanced Study in the Health Sciences (FIC) (http://www/fic/nih/gov/), which currently supports research and training programs in over 100 countries, involving some 5000 scientists in the United States and other countries. For example, the FIC Global Health Research Initiative Program for New Foreign Investigators provides partial salary and research support in behavioral and social science or basic science for NIH-trained researchers who are returning to their home countries; and the FIC International Research Collaboration Award supports collaborations between NIH-supported American scientists and researchers in developing countries. Like the NIH, the NSF also has an office, the Office of International Science and Engineering

(OISE), for facilitating research overseas. The OISE supports international training programs for scientists of all levels and funds postdoctoral fellowships for training in international research.

In China, the National Natural Science Foundation (NSFC) manages the National Natural Science Fund, aimed at promoting and financing basic and applied research. A medical department of the NSFC that was launched in 2010 is comparable to the NIH. Applications submitted to NSFC go through a peer-review process that is similar to the evaluation of grant applications in many other countries. The approval rates for applications are about 20%; special funding mechanisms exist for young researchers in the early stages of their careers. Support for collaborative national and international research projects is available through the Collaborative Research Fund. In cooperation with the Research Grants Council of Hong Kong, the NSFC developed the Joint Research Scheme 2014, aimed at promoting collaboration between researchers and research teams in Hong Kong and the Mainland. Grants are available for 4-year projects in six focus areas: information technology, life sciences, new materials science, marine and environmental science, medicine, and management science. Also new for 2014 are funding opportunities for joint projects between French and Hong Kong scientists, and between German and Hong Kong scientists. In addition, the 2014 Collaborative Research Fund offers an increased budget for support of both Equipment Grants and Group Research Grants.

In the United Kingdom, the BBSRC has established an International Relations Unit, which promotes contacts with international scientists, provides advice on funding opportunities for collaborative projects, supports international visits, helps BBSRC-funded institutes and universities identify international sources of scientific expertise, and contributes to international science policy. Several types of grants are available for international travel and collaborations for researchers who are either supported by BBSRC or affiliated with BBSRC-supported institutions. These include awards to travel abroad (International Scientific Interchange Scheme); International Fellowships for "high profile" researchers to visit the United Kingdom for periods of up to 1 year, grants to support international workshops (particularly those that establish further links with the United States, Canada, EU member states, Japan, China, and India); and Partnering Awards for BBSRC-supported scientists to establish collaborations with researchers in Japan and China. The BBSRC website (see Box 15.3) also contains links to other international grant agencies and grant mechanisms, such as the EU FP, the Human Frontier Science Program, the European Science Foundation, COST, the British Council, Royal Society, and the Alexander von Humboldt Foundation.

For African scientists, the website of South Africa's National Research Foundation (NRF) contains useful information about opportunities for international collaboration and the availability of grants and fellowships administered by the NRF's International Science Liaison office (http://www.nrf.ac.za/division/irc/about). For researchers in Australia, the Australian Research Council (ARC, http://www.arc.gov.au) provides support for international collaboration and network building by funding Research Awards, International Fellowships, and Internationally Coordinated Initiatives. For international collaborations in clinical trials, funding is available in cases where the researchers involved are each eligible for funding by the MRC (United Kingdom), the Veterans Administration (United States), or the Canadian Institutes for Health Research.

Researchers in the Nordic countries should note that in 2005 a new funding agency, NordForsk (www.norden.org/forskning/sk/nordforsk.asp), was established to promote seamless research across the countries, particularly in scientific disciplines in which they already have leading positions. So far, NordForsk has rather limited resources at its disposal. Thus, the funding from this agency should be regarded as seed money or add-on funding for establishing or bolstering collaboration between Nordic research groups. NordForsk has supported several Nordic Centers of Excellence through its Molecular Medicine program.

Finally, several charities devote significant funds to international biomedical research. For example, the Wellcome Trust (United Kingdom) supports international biomedical research in developing and restructuring countries, provides funding for tropical and clinical research in developing countries, and supports international networks and partnerships. This support is available through Fellowships, Project and Program Grants, and other targeted mechanisms. In the United States, the Bill & Melinda Gates Foundation (http://www.gatesfoundation.org) is emerging as one of the world's most powerful charities, with a major focus on funding global health programs. The Foundation supports research and clinical initiatives in several priority disease areas, including child health, HIV/AIDS, malaria, tuberculosis, and vaccine-preventable diseases. It also will consider proposals to support research that is likely to achieve fundamental breakthroughs in three areas: (1) science and technologies to make advances against diseases in developing countries,

BOX 15.9 Final Words of Advice

- Study the call for proposals. Then, study it again.
- Make sure you know what is already done in the field.
- Make your application easy to read.
- Make your application intellectually accessible to a diverse group of readers.
- Never take a rejection personally.
- Resubmit whenever possible.

(2) technologies that can serve as platforms for accurate and affordable diagnostic tools, and (3) application of advances in genetics and molecular biology to global health problems.

15.9 SUMMARY AND PERSPECTIVE

Grants and other mechanisms of financial support are available to students, postdoctoral fellows, and researchers in most countries for training and research in the biomedical sciences. Knowing as much as possible about where to apply for grants and how to write a good grant application are critical to success in the funding arena. However, in a climate of increasing competition for grant funding, there are two simple rules to keep in mind: First, to be awarded a grant, you must apply. Second, if at first you don't succeed, try, try again!

Keep in mind the final words of advice in Box 15.9!

ACKNOWLEDGEMENTS

We are grateful to Senior Consultants Mette Topnes and Lars Øen, Office for International Relations and Research Support, University of Oslo, for helpful advice.

QUESTIONS TO DISCUSS

1. Based on your own research, start preparing an application for funding your present or future research project by writing a few sentences to define: (a) the overall goal of the project, (b) reasons why you think the project needs doing, (c) specific aims of the study, (d) major methods to be used, and (e) what advance in understanding or practice will be gained if the research is successful.

2. Most grant applications must contain an abstract. What are the important elements to include in this part of the application?
3. You have agreed to review the draft of a grant application written by another postdoctoral fellow in your department. In the background section of the application he explains that the aims of the study are designed to prove that the hypothesis he recently proposed for a severe genetic disorder is, in fact, correct. Therefore, he anticipates that the outcome of the study will lead to major changes in the way this disorder is treated. What are your major comments on this draft?
4. You recently read an article published in *Nature* that you found so exciting that you immediately went into your research mentor's office, suggesting that you drop some of the planned experiments in your dissertation project and instead test some of the exciting consequences of the data described in the *Nature* paper. What do you believe are the reasons she demanded that you instead return to your bench and complete the planned experiments as soon as possible?
5. In sequencing DNA obtained from members of a family with a severe bone disease, you are delighted to discover heterozygous mutations in a gene for an extracellular matrix protein. As part of the plan to characterize the mechanism by which the mutation causes the disease, you write a grant application for funding that will allow you to make a mouse knock-out model for this disease. Present some reasons this application would not be funded.

REFERENCES

1. Kreeger K. A winning proposal. *Nature* 2003;**426**:102–3.
2. Reif-Lehrer R. Applying for grant funds: there's help around the corner. *Trends Cell Biol* 2000;**10**:500–4.

INDEX

Note: Page numbers followed by "*b*," "*f*," and "*t*" refer to boxes, figures, and tables, respectively.

A

B

C

D

J

K

L

M

P

R

S

T

U

V

W

Z

Printed in the United States
By Bookmasters